FOOD MICROBIOLOGY:
PUBLIC HEALTH AND SPOILAGE ASPECTS

other AVI books

FOOD MICROBIOLOGY:
PUBLIC HEALTH AND SPOILAGE ASPECTS

Edited by   **MARIO P. DEFIGUEIREDO,**
*Vice President of Research
and Development
National Portion Control
Chicago, Illinois*

and   **DON F. SPLITTSTOESSER,**
*Professor, Cornell University
New York State Agricultural
Experiment Station
Geneva, New York*

THE AVI PUBLISHING COMPANY, INC.
WESTPORT, CONNECTICUT

Library of Congress Catalog Card Number: 76-27467
ISBN-0-87055-209-0

Printed in the United States of America

Contributors

FRANK L. BRYAN, Ph.D., Chief, Food Borne Disease Activity Center for Disease Control, U.S. Department of Health, Education, and Welfare. Atlanta, Georgia.

BLANCA G. CALANOG, Ph.D., Literature Scientist, National Dairy Council, Rosemont, Illinois.

LEE N. CHRISTIANSEN, Ph.D., Swift and Company, Oakbrook, Illinois.

DEAN O. CLIVER, Ph.D., University of Wisconsin, Department of Bacteriology and Food Research Institute, Madison, Wisconsin.

MARIO P. DEFIGUEIREDO, Ph.D., Vice President of Research and Development, National Portion Control, Chicago, Illinois.

CHARLES L. DUNCAN, Ph.D., University of Wisconsin, Department of Bacteriology and Food Research Institute, Madison, Wisconsin.

JAMES M. JAY, Ph.D., Wayne State University, Department of Biology, Detroit, Michigan.

ALLEN A. KRAFT, Ph.D., Iowa State University, Department of Food Technology, Food Research Laboratory, Ames, Iowa.

J. LISTON, Ph.D., University of Washington, Institute for Food Science and Technology, College of Fisheries, Seattle, Washington.

ELMER H. MARTH, Ph.D., University of Wisconsin, Department of Food Science, Madison, Wisconsin.

H.B. NAYLOR, Ph.D., Cornell University, Laboratory of Microbiology, Ithaca, New York.

HENRY J. PEPPLER, Ph.D., Universal Foods Corporation, Milwaukee, Wisconsin.

D.B. PREST, Ph.D., Kouka College, Department of Biology, Kouka Park, New York.

FLOYD R. SMITH, Ph.D., Retired, Pet Foods Corporation, Greenville, Illinois.

D.F. SPLITTSTOESSER, Ph.D., Professor, Cornell University, New York State Agricultural Experiment Station, Geneva, New York.

J.R. STAMER, Ph.D., Cornell University, New York State Agricultural Experiment Station, Department of Food and Technology, Geneva, New York.

R. BRUCE TOMPKIN, Ph.D., Swift and Company, Oakbrook, Illinois.

JOHN A. TROLLER, Ph.D., The Proctor and Gamble Company, Winston Hill Technical Center, Cincinnati, Ohio.

HOMER W. WALKER, Ph.D., Iowa State University, Department of Food Technology, Food Research Laboratory, Ames, Iowa.

Preface

The objective of this book is to provide information about the microorganisms that are important in foods either because they cause illness or because their activities may result in a loss in food quality. Since the organisms *per se* are the main concern, the arrangement of the book differs from that of many food microbiology texts in that the chapters are organized around individual species or groups of microorganisms, rather than according to subjects such as preservation methods or type of food.

Our intended audience is the individual in the food industry who safeguards company products against public health hazards and spoilage, as well as the student who is training for this responsibility. The contributors were encouraged, therefore, to concentrate on subject areas that would be especially helpful to people in these positions. Thus, topics that receive special emphasis are the organisms' habitats, the bases for their recognition, and the factors that affect their growth and death in foods.

The microorganisms that have been singled out are those believed to be the most troublesome; the book's size limitations prevented a treatise on every species that has been incriminated in a foodborne illness or a spoilage outbreak. For example, two disease-causing bacteria that might have been given greater attention are *Bacillus cereus* and enteropathogenic *Escherichia coli*. Special sections were not devoted to these organisms since information about them is presented in other chapters.

A number of chapters were completed before the 8th Edition of Bergey's "Manual of Determinative Bacteriology" became available. As a result, the classification and nomenclature used by some authors may not agree completely with that found in this edition of Bergey's Manual. We do not believe this to be a serious deficiency, however, since, in general, the epithets and taxonomy that have been adopted are the ones used in much of the food literature.

MARIO P. DEFIGUEIREDO
DON F. SPLITTSTOESSER

Contents

CHAPTER 1

Floyd R. Smith | # Statistics in Microbiological Control

Stearman (1955) commented on the use of statistics in micro-biology. His thoughts are of considerable interest and certain statements are particularly pointed in explaining communication problems between the two fields. He states: "Statistical methods are being used to an increasing extent in the field of microbiology. Whether we like it or not, once a science advances beyond the descriptive stage its problems become statistical even though we don't use formal statistical techniques." In addition he notes: "The training of the statistician is a long and involved program just as is the training of a microbiologist." Thus, the skilled microbiologist cannot be expected to be equally competent in the field of statistics. Although some familiarity with statistical terminology and concepts underlying the methods may be valuable there are inherent dangers. Stearman points this out when he states: "Statistical procedures like dynamite are beneficial if used properly but may be dangerous when used improperly . . . The statistical results obtained from the body of data are no better than the technique involved in the experiment. Further, statistical methods are not a substitute for sound pro-fessional judgment in interpretation of the results of an experiment."

The above comments say that statistics will be more and more involved in the field of microbiology and, as a result, cooperation between the two disciplines is essential. Speaking for microbiology, some appreciation of the terminology, and confidence and risk factors in statistical applications, seems essential. However, design of the mathematical treatment of data should be the responsibility of the statistician. In interpreting the data the microbiologist must understand the confidence and risk factors involved. This informa-tion can be provided by the statistician without resorting to confusing mathematical terminology.

STATISTICAL TERMS

Population.—A group or set having common detectable character-istics. Thus, if one were studying the microbiological content of cheese, the microflora during manufacture and after aging would be two different populations. Intermixing of data from each would lead to faulty analyses with incorrect conclusions.

Sample.—A definite number of units collected from the popula-tion and considered representative of the total population.

Random Sample.—One designed so there is an equal chance of any item in the population being chosen.

Frequency Distribution.—Data arranged in classes with the tabulation of the frequency of occurrence in each class.

Normal Distribution.—A distribution with a single peak number of occurrences (average) and an equal number of observations equidistant from this average. The data indicating a normal distribution results in a bell-shaped curve when plotted as a frequency distribution.

Average.—A measure of the central tendency. The sum of all observed values divided by the number of values.

Range.—The difference between the greatest and least value observed in the sample.

Standard Deviation.—The root mean square of the deviations from the average (mean).

Confidence Limits.—A definition of the amount a value may deviate from the true value. (Usually 0.95 confidence limits are set but greater or lesser values may be used).

Variables and Attributes.—Variables are the record made when actual quality characteristics are measured. A record by attributes, on the other hand, shows only the number of articles that conform or fail to conform to specified requirements (Grant 1964).

VALUE OF UNDERSTANDING TERMS

The difference between variables and attributes is fundamental to both experimental planning and data analysis. An example of an attribute plan would be the examination of canned food for the presence of *Clostridium botulinum*; here the finding of a single viable organism would result in the lot being rejected. The standard plate count is an example of a variable plan since we now are dealing with numbers. The normal distribution is not applicable to attributes but is possible with variable data.

In defining population, reference was made to *common detectable characteristics*. Recognition of difference in a lot, due to certain production or handling problems, can result in population changes. Thus, in food production, changes may be noted at starting or shut-down periods which vary from those of continuous operation. These factors must be considered in drawing conclusions regarding the product lot. This will be discussed later.

The frequency distribution is valuable where variables are involved. Advantages to preparing a frequency distribution curve are: (1) The data is arranged in an orderly manner which makes it more readily understandable. (2) Both the average and range can be roughly estimated. (3) The possible existence of a normal distribution can be noted.

The normal distribution is one of the most important distributions in statistics and many statistical procedures are based on it. The false assumption that one is dealing with a normal distribution, however, can lead to wrong conclusions. Dixon and Massey (1957) describe methods which may be used to determine whether the assumption of a normal distribution is correct; they also give an understandable explanation of several statistical procedures and applications.

The average and range are used extensively with variables and need no elaboration.

The standard deviation is extremely useful with a normal distribution. By knowing the standard deviation and the average, the frequency distribution can be constructed. One may also determine the probable number of units falling above or below a given value by use of the normal table (Dixon and Massey 1957).

The appreciation of confidence limits or the confidence interval is extremely important in considering controls or standards. If a limiting count for a product is proposed, one must recognize that, due to variations in samples and analytical aberrations, a lower or higher value or set of values would be expected on a new set of samples and analyses. If such standards, based on numerical data, are developed by statistical procedures, the statistician should be able to indicate the control count necessary to give a defined confidence of meeting the standard. This information permits a decision whether the process permits control at a value giving the required confidence of product acceptance.

SAMPLING

Thatcher and Clark (1968) discuss the purpose and choice of a sampling plan and point out some of the problems with foods. With respect to difficulties, they point out that the analytical laboratory can seldom cope with the number of samples needed to demonstrate the absence of pathogens in a lot of commercial size at a probability level that most scientists would consider desirable. They state that practical considerations often require that a lower degree of probability will have to be accepted. "The need for compromise, however, does not justify unnecessary lack of observance of sampling principles, a fault which is common."

All plans, whether based on variables or attributes, depend on random sampling. Although devices have been recommended for assurance of randomness, many may not be applicable to industrial operations which involve several hours of production and thousands of units. In this situation one may install a device in the line to drop out product at certain time intervals. This does not destroy randomness and has the advantage of setting up sublots covering the over-all production. Testing may involve the total sublot or one may randomly select from this sublot. The sample size (n) depends on the units tested and not on the number removed from the lot.

The size of the inoculum utilized must also be considered. If the test dosage were 1 ml removed from the finished product, one merely translates back to the quantity under test. If, however, the packaged product were incubated under conditions assuring growth of the organisms, one may state the number of sterile units. Prevention of growth would generally be considered in a variables plan, while incubation with the development of numbers would be utilized in an attribute plan involving sterility or lack of sterility. In the first instance, numbers per ml are generated, while in the latter the numbers of unsterile product are recorded. Interpretation of such data involves applicable statistical analyses which can only be touched on in this chapter.

Values of sampling size depending on lot size are given in the Military Standards Sampling Plans (MIL — STD-414 1957; MIL — STD-105D, 1963). These values are based to a certain degree on the empirical relationship between lot size and sample size, and recognition is given to the difficulty of securing small random samples from a large lot. It must be emphasized that the use of percentages or the square root of the lot are not based on statistical recommendations.

Frequently in production control one may find it necessary to deviate from the above-mentioned formalized plans. In this case one may base the sample size (number of units analyzed) upon a curve of acceptance sampling known as the *Operations Characteristic (OC) Curve* (Fig. 1.1). In developing the OC data, one sets a probability of accepting satisfactory material, while recognizing there is some risk of accepting substandard items. The probability of accepting substandard product can be developed. Methods of calculation, interpretation, and use are detailed by Grant (1964). The methods in general use in utilizing the OC method do not involve over-all lot size but define risk involved and conditions of rejection with a given sample size. Such calculations and methods are more widely used for attributes. The OC data and curve however, are applicable and utilized with the variables plans.

Variables Plan

The variables plan has not been widely used in microbiological control. It is essential that the data follow the normal distribution if one becomes involved with quantitative determinations and wishes to utilize averages and standard deviations to define risk factors. Assumption of normal distribution involves an unknown risk, since, as Thatcher and Clark (1968) note: "International diversity among food producers makes it unlikely that any common distribution can be assumed . . . "

If one wishes to use the variables plan as outlined in MIL—STD-414 (1957), it is important to note the limitations discussed in the introduction. The following discussion is of particular importance: "The variables sampling plans apply to a single quality characteristic which can be measured on a continuous scale, and for which quality is expressed in terms of per cent defective. The theory underlying the development of the variables sampling plans, including the operating characteristic curves, assumes that measurements of the quality characteristic are independent, identically distributed, normal random variables." Unless one can be assured that these conditions are met, use of this valuable procedure can lead to erroneous conclusions. Statisticians can determine whether such conditions are met, and if not, may advise on transformation of data permitting a modification of the variables plan. Note that such manipulations do not necessarily permit the use of Military Sampling Plan MIL—STD-414.

In discussing possible standards for frozen meat pies, a variables plan was suggested (AFDOUS 1966). The statisticians commented on their statistical methods: "Certain statistical techniques were applied to obtain appropriate estimates of variability in total aerobic plate counts of samples of *n* units. Data on this characteristic were transformed to logarithms to achieve homogenicity of variance and to satisfy the assumption of a normal probability distribution needed in carrying out analyses of variance."

The methods utilized involved tables not available in standard texts and mathematical manipulations which would be confusing to many microbiologists. However, the advantage of a reasonable sample size of ten with reasonable risk makes this development attractive. A more complete discussion of the method and its application could encourage acceptance and use.

Use of variables plans without statistical help or skill is not advisable. In addition, acceptance of the above procedure prior to more consideration of possible limitations and more thorough explanation than that given in the report does not appear likely.

Attributes Sampling Plan

The attribute plan most widely used distinguishes between satisfactory and unsatisfactory product. It often is referred to as a *go-no-go* plan. One application would be to determine product sterility or spoilage. In such usage the method of examination may involve destructive or nondestructive methods. If, on spoilage, a product exhibits obvious gas formation with container distension or an obvious change in physical state, such as coagulation, a defective unit may be selected without opening the container. In the evaporated milk industry, coagulation and/or gas formation has long been recognized as indication of product contamination and spoilage; here detection is possible without destruction of the container. If, however, one wishes to determine sterility, a destructive testing would be required since the integrity of the container must be destroyed.

A thorough, understandable, discussion on attribute sampling is given in MIL—STD-105D (1961). Familiarity with this reference is strongly recommended, since information is presented on sample size, acceptance value, confidence limit, and operation characteristic curves.

In developing an attribute plan certain decisions must be made. First one must select an *acceptable quality level* (AQL). Lots at this defect level may be accepted 95 times out of 100 and rejected 5 out of 100 times. Thus, at the AQL value selected one expects a 0.95 confidence of acceptance and a 0.05 confidence of rejection. In other words, the producer may risk rejection of lots meeting the AQL 5 times in 100. This 0.05 value is called the *producer's risk*. These risks for the producer have been generally accepted and any of several sample sizes may satisfy this requirement.

The next decision involves the risks in accepting lots exceeding the AQL values (*consumer's risk*). Since the risk of acceptance is inversely proportional to defective level present, a series of values result. The plotting of these values gives the OC curve. In control work one may select a single defect level which should result in a high degree of rejection. Generally, this has been referred to as the *Lot Tolerance Percent Defective* (LTPD). Assuming that one desires to reject 90 times out of 100 if a lot has a certain defect level, this then becomes the LTPD.

Criticism has been leveled at the idea that lots with any defect may be accepted. This has led to acrimonious discussions on zero tolerance. Actually in mass production one can only hope to approach, but not reach a zero tolerance. Such a goal would necessitate 100% inspection by some infallible method. It is, of course, possible to set a defect level of zero defects in a given number or amount of sample. However, this does not assure absence of defects in the total production. All one can say is that a control has been developed to accept a given defect risk, and reject if defects reach a defined level. Thus, in all sampling methods, one accepts a risk of error but maintains a calculated risk.

Concerning sample size, as the number of samples tested increases, the OC curve steepens and the LTPD approaches the AQL.

One may develop plans to suit any sample size desired and risk considered feasible. The calculations are not complex if one accepts the mathematical derivation and calculations required to develop the necessary tables. These tables involve calculations derived from the equation utilized for the Poisson distribution. This approximation is accepted by statisticians and is widely used in industrial control. In the author's opinion a number of probability studies involved in microbiological control or general process control involving attributes may be determined by use of the Poisson data.

Information required for the application of the Poisson distribution are a Poisson table for *C values* (risk factors) as given in Table 1.1, data showing the effect of sample size on the producer's risk (Table 1.2), and an OC curve (Fig. 1.1).

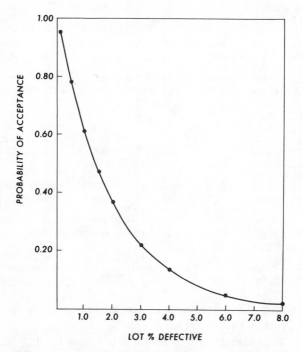

FIG. 1.1. NUMBER OF SAMPLES ANALYZED (n) = 50; ACCEPTANCE QUALITY LEVEL (AQL) = 0.10%; ACCEPTANCE VALUE (C) = 0.

TABLE 1.1
SUMMATION OF POISSON TERMS
(1000 x probability of C or less events)

	C					C			
np	0	1	2	3	np	0	1	2	3
0.02	980	1000	—	—	3.0	050	199	423	647
0.04	961	999	1000	—	3.2	041	171	380	603
0.06	942	998	1000	—	3.4	033	147	340	558
0.08	923	997	1000	—	3.6	027	126	303	515
0.10	905	995	1000	—	3.8	022	107	269	473
0.15	861	990	999	1000	4.0	018	092	238	433
0.20	819	982	999	1000	4.2	015	078	210	395
0.25	779	974	998	1000	4.4	012	066	185	359
0.30	741	963	996	1000	4.6	010	056	163	326
0.35	705	951	994	1000	4.8	008	048	143	294
0.40	670	938	992	999	5.0	007	040	125	265
0.45	638	925	989	999	5.2	006	034	109	238
0.50	607	910	986	998	5.4	005	029	095	213
0.55	577	894	982	998	5.6	004	024	082	191
0.60	549	878	977	997	5.8	003	021	072	170
0.65	522	861	972	996	6.0	002	017	062	151
0.70	497	844	966	994	6.2	002	015	054	134
0.75	472	827	959	993	6.4	002	012	046	119
0.80	449	809	953	991	6.6	001	010	040	105
0.85	427	791	945	989	6.8	001	009	034	093
0.90	407	772	937	987	7.0	001	007	030	082
0.95	387	754	929	984	7.2	001	006	025	072
1.0	368	736	920	981	7.4	001	005	022	063
1.1	333	699	900	974	7.6	001	004	019	055
1.2	301	663	879	966	7.8	000	004	016	048
1.3	273	627	857	957	8.0	000	003	014	042
1.4	247	592	833	946	8.5	000	002	009	030
1.5	223	558	809	934	9.0	000	001	006	021
1.6	202	525	783	921	9.5	000	001	004	015
1.7	183	493	757	907	10.0	000	000	003	010
1.8	165	463	731	891	10.5	000	000	002	007
1.9	150	434	704	875	11.0	000	000	001	005
2.0	135	406	677	857	11.5	000	000	001	003
2.2	111	355	623	819	12.0	000	000	001	002
2.4	091	308	570	779	12.5	000	000	000	002
2.6	074	267	518	736	13.0	000	000	000	001
2.8	061	231	469	692	13.5	000	000	000	001
					14.0	000	000	000	000

Source: Grant 1964.

The following illustrate some of the calculations (more complete information can be found in Grant's text):

Assuming that one desires to set an acceptable quality level (AQL) for product sterility of 0.10% and will risk rejection of product at this level 5 times in 100, what sample size or sizes are recommended? In addition, what is the risk of accepting lots exceeding this value at a 10% level?

TABLE 1.2
EFFECT OF N AND C ON PRODUCER'S RISK

AQL 0.10%

n = 50	c = 0	n = 316	c = 1	n = 800	c = 2
p	Accept	p	Accept	p	Accept
0.001	0.95	0.001	0.95	0.001	0.95
0.007	0.705	0.0034+	0.699	0.00237+	0.704
0.016	0.449	0.0057−	0.463	0.00375	0.423
0.032	0.202	0.0095−	0.199	0.00525	0.210
0.048	0.091	0.012+	0.092	0.00675	0.095

Source: Grant 1964.
Used with permission of McGraw-Hill Book Co.

In calculating this problem one sets out certain values. First the AQL is 0.95 at a defect level of 0.10% (p=0.001). Sample size (n) must be calculated. In addition one must decide whether a plan with zero defects or one with a certain number of defects is preferable.

In Table 1.1 locate under the C = O column the value 950 which is a 0.95 confidence of acceptance (interpolation between 961 and 942 is required). Thus, the np value (left hand column) is approximately 0.05. Since p has been set at 0.10% (0.001), n times 0.001 equals 0.05 or n equals 50. Thus the answer to the first question is acceptance with a sample size of 50 if no rejects are discovered.

The second question refers to defective content accepted at a 0.10 confidence level. Under the C = O column, locate 010 which is the desired confidence level. Again interpolation is required; between 111 and 091, the np value of 2.3 would be a close approximation. Since n has already been set at 50, we have 50p = 2.3 or p=4.6%. Thus one may accept product containing 4.6% defectives 10 times in 100 lots.

If the risk of accepting 1 in 10 times (10 in 100) at a 4.6% level is too great, it would be necessary to recalculate for larger sample sizes. If the AQL value must be retained, larger samples with larger C values will permit preservation of the AQL values but will lessen the defectives accepted at the 0.10 confidence level.

Table 1.2 gives risk values for different n and c values. As n and c increase note the change in the LTPD defect values. When the sample is increased to 316 and c = 1, this defect level becomes 1.2% with acceptance approximately 9 times in 100. If n = 800 and c = 2, the defect level drops to approximately 0.7% (0.675) at a risk of approximately 9% acceptance.

Figure 1.1 is an OC curve for an AQL of 0.10% with c = O. Other OC curves may be drawn from the values in Table 1.2.

Heterogeneous Foods

Thatcher and Clark (1968) point out that: "The degree of homogeneity of the food being tested greatly influences the choice of sampling plan, more samples being needed if the food is not homogeneous." This is particularly important with such products as frozen meat pies, since an adequate sample of the contents is difficult to obtain. Here one deals with a mixture generally involving crust, gravy, meat and vegetables and thus one expects greater risk of sampling errors. In preparation for analysis, adequate mixing to get a representative quantity of each ingredient often is not obtained. One would expect greater variance than with a more homogenous product, such as milk.

In the case of ingredients received in large containers, one should recognize that one expects lack of uniform distribution throughout the lot or even throughout a single container. Several samples from a single container cannot be relied upon to determine the condition of the lot. Sampling problems are particularly difficult with frozen product. Generally samples, while still in the frozen state, are taken around the periphery and central area. The use of a sterile agar for sample collection has been a common practice. Cost of sampling becomes a factor in deciding on the number of samples.

STATISTICAL CONCEPTS

Thatcher (1971) discusses the development of sampling plans by the International Committee on Microbiological Specifications for Foods. He states: "Use of a common sampling plan is as important as use of a standard method of analysis. For each sampling plan the probabilities of acceptance should be known. In the opinion of the committee, the probabilities of accepting a substandard lot and of rejecting a lot that actually meets the acceptance criteria should be clearly specified and appreciated both by control agencies and the food companies."

One cannot take exception to the above statement. However many of the methods utilized in the data analysis may not be comprehended by leading microbiologists. In addition, over-enthusiasm by statisticians in controlling errors in acceptance may result in prohibitive analytical cost. It is important in all phases of control that those responsible be able to determine: (1) The number of samples involved with the associated analytical cost, (2) The process quality level required to prevent costly lot condemnations, (3) Possible use of nondestructive testing in general control, (4) That control mechanisms are designed to protect the consumer but must also recognize serious loss to the producer unless realistic values are utilized, (5) That procedures and analytical methods are understandable and workable. In addition, microbiologists must recognize that statistical approaches cannot be ignored.

BIBLIOGRAPHY

AFDOUS. 1966. Microbiological examination of precooked frozen foods. Bull. Assoc. Food Drug Officials U.S., Suppl. Issue.

DIXON, W. J., and MASSEY, F. J. 1957. Introduction to Statistical Analysis. McGraw-Hill, New York.

GRANT, E. L. 1964. Statistical Quality Control, 3rd Edition. McGraw-Hill, New York.

HOEL, P. G. 1960. Elementary Statistics. John Wiley & Sons, New York.

MIL—STD-414. 1957. Sampling procedures and tables for inspection by variables for percent defective. U.S. Govt. Printing Office, Washington, D.C.

MIL—STD-105D. 1963. Sampling procedures and tables for inspection by attributes. U.S. Govt. Printing Office, Washington, D.C.

STEARMAN, R. L. 1955. Statistical concepts in microbiology. Bacteriol. Rev. *19*, 160–215.

THATCHER, F. S. 1971. The international committee on microbiological specifications for foods: Its purposes and accomplishments. J. Assoc. Offic. Anal. Chem. *54*, 836–841.

THATCHER, F. S., and Clark, D. S. 1968. Microorganisms in Foods. Univ. of Toronto Press, Toronto, Canada.

CHAPTER 2

Frank L. Bryan | # Staphylococcus Aureus

Certain strains of *Staphylococcus aureus* (alias *Micrococcus pyogenes aureus, M. aureus, S. pyogenes, S. pyogenes aureus, S. albus, M. pyogenes albus, M. albus, S. pyogenes albus, M. pyogenes citreus, S. pyogenes citreus, S. citreus, M. citreus*) produce one or more exotoxins, called enterotoxins A, B, C, D, or E, which act on the human intestinal lining and give rise to one of the more common types of foodborne disease—staphylococcal intoxication (or more correctly referred to as staphyloenterotoxicosis). To prevent occurrence of staphylococcal intoxication, the nature of *S. aureus* and its enterotoxins, the methods of isolating and identifying the organism and detecting its toxin, the ecological relationships of the organism and man, and the nature and epidemiology of the disease must be understood. With this understanding, effective preventive and control methods can be devised. This chapter reviews these features.

NATURE OF THE ORGANISM

History

Staphylococcal infections are diseases of the present and of antiquity. The Bible contains citations of boils; ancient Egyptian, Greek, and Roman medical writings describe carbuncles and wound infections; Egyptian mummies have shown evidence of osteomyelitis and dental abscesses (Hare 1967). But knowledge of the cause of these diseases awaited the development of the science of bacteriology.

Staphylococci had been isolated for more than a decade before Pasteur studied the cause of pyogenic lesions; he, however, proved that staphylococci were pathogenic. Using staphylococci that he isolated from pus that was associated with boils and osteomyelitis, he continued to substantiate the germ theory by growing these bacteria in broth and by intradermal inoculation of rabbits. The rabbits later developed abscesses at the site of inoculation, and staphylococci were recovered from the abscesses (Pasteur 1880). At about the same time, Ogston (1880, 1881) isolated staphylococci from 69 acute abscesses, but not from nonsuppurating tissues. He injected pus that contained staphylococci into animals and produced septicemia or local suppuration.

Staphylococcal intoxication is probably also of great antiquity, but records of this disease are relatively recent. A report of a typical outbreak was recorded in Paris as early as 1830 (Ollivier 1830).

12

Members of a family became ill 3 hours after eating a ham pie that they purchased from a pastry shop. Their illness was characterized by a generalized uneasiness, followed by cold sweats, shivering, violent pain in the stomach, frequent vomiting, burning thirst, extreme tenderness of the belly, and profuse diarrhea (typical symptoms of staphylococcal intoxication). Cases were also reported by other customers of the shop. The shop was inspected and found to be clean, and the pie was examined and found to be free of common metals. Although the cause was not determined, as was usual in such investigations at this time, the outbreak was analogous to certain others, reported in Germany, that had followed the ingestion of sausage, cheese, or ham. Other early episodes of food poisoning in which symptoms and incubation periods suggested staphylococcal intoxication are reviewed by Dack (1956A).

Denys (1894) was the first to incriminate staphylococci as a cause of food poisoning. During an investigation of an outbreak involving at least eight cases of gastroenteritis, he found that the victims had eaten meat from a cow that had died from puerperal fever. Enormous numbers of staphylococci were isolated from the meat and from the spleen of a fatal case. Denys, therefore, attributed the harmful qualities of the meat to *S. aureus.*

In the U.S., Owen (1907) incriminated staphylococci as a cause of an outbreak of food poisoning involving 19 people from several families. Typical symptoms of staphylococcal intoxication developed within a few hours after the victims ate dried beef. The beef had been purchased by members of the various families from the same butcher. A pure culture of a white pigmented *Staphylococcus* was isolated from both aerobic and anaerobic cultures of the meat. Beef-tea emulsions of the meat were intraperitoneally injected into rats. The rats died, and staphylococci were recovered from their blood. Illness was not produced when small animals were fed the contaminated meat. Efforts to detect a filterable toxin failed.

Barber (1914) reported on cases of gastroenteritis that stemmed from a single farm over a period of four years. Attacks of gastroenteritis occurred 1¾ to 2½ hours after cream from one particular cow had stayed at room temperature and was then ingested. Barber drank some cream from the incriminated cow and developed similar symptoms. Large numbers of white staphylococcal colonies were isolated from the cow's milk. Pure cultures of these organisms were inoculated into preserved milk which was incubated for 8½ hours and then ingested by Barber. After 1¾ hours, he developed a gastroenteritis similar to that which resulted from ingestion of milk from the incriminated cow. When raw milk from the infected cow was no longer ingested, the illnesses ceased. As a result of this investigation (although filtrates were not tested), Barber concluded, " . . . that the illness was due to a poison formed by the white staphylococcus in milk."

Although Barber suggested that staphylococcal food poisoning was caused by a toxin, this fact was not proved until 1929 when Dack *et al.* (1930) discovered a filterable enterotoxin as the cause of the illness. After a sponge cake was incriminated as the vehicle in an outbreak, bacteriological analysis revealed that the predominant organism in a sample of the cake was a yellow, hemolytic staphylococcus. Isolates were grown in broth, and filtrates were fed to human volunteers and injected intravenously into a rabbit. All volunteers developed gastrointestinal distress within 3 hours; the person receiving the largest quantity, 25 ml, became severely ill. The rabbit developed a profuse watery diarrhea, lost weight, and died within 12 hours. Other rabbits inoculated with heated filtrates of the staphylococcal culture either developed diarrhea or died.

More complete historical surveys of staphylococci and their enterotoxins can be found in the books by Dack (1956A) and Elek (1959) and the article by Dolman (1956).

Nomenclature and Classification

Because of their characteristic microscopic shape and clustering, Ogston (1882) named the "grouped micrococci"—"staphylococci" from the Greek word *staphyle*, meaning a bunch of grapes. He also suggested the existence of different species of micrococci by stating that micrococci were similar in appearance and growth but different in the effects they produced.

Rosenbach (1884) classified staphylococci on the basis of the colony pigmentation they produce, because, at that time, it was believed that a yellow pigment was an indication of pathogenicity. He called organisms that produced golden yellow pigmented colonies *Staphylococcus aureus* and those that produced white colonies *Staphylococcus albus*. A year later, Passet (1885) added a further species, *Staphylococcus pyogenes citreus*, which produced a lemon-yellow colored colony. Pigmentation as a basis of classification, however, proved to be unsatisfactory and was later replaced by a more useful criterion for classification—the ability of *S. aureus* to produce coagulase—based on observations by Loeb (1903) and Much (1908).

The eighth edition of Bergey's "Manual of Determinative Bacteriology" (Buchanan and Gibbons 1974) divides the family *Micrococcaceae* into three genera: *Micrococcus* Cohn 1872, *Staphylococcus* Rosenbach 1884, and *Planococcus* Migula 1894. The genus

Staphylococcus consists of aerobic, facultatively anaerobic cocci that form irregular masses; species ferment glucose anaerobically to produce acid. *Micrococcus* consists of strictly aerobic cocci that form irregular masses and sometimes tetrads; their action on glucose is only oxidative. *Planococcus* consists of motile cocci which form tetrads; their action on glucose is only oxidative. Colonies show a yellow-brown pigment.

The genus *Staphylococcus* is divided into three species: *Staphylococcus aureus* Rosenbach 1884, which ferments mannitol and is coagulase-positive; *Staphylococcus epidermidis* (Winslow and Winslow 1908) Evans 1916, which does not ferment mannitol and is coagulase-negative; and *Staphylococcus saprophyticus* (Fairbrother) *emend. mut. char.* Shaw, Stitt and Cowan 1951, which resembles *Micrococcus* in many ways, but because of its DNA and cell wall composition and ability to grow slowly anaerobically is now classified as *Staphylococcus.* Organisms formerly named *S. albus* and *S. pyogenes citreus* are classified as *S. aureus.*

Based upon physiological and biochemical tests, Baird-Parker (1963) recognized six subgroups of *Staphylococcus.* Subgroup I, which is analogous to *S. aureus*, was defined as coagulase-positive, phosphatase-positive organisms which produce acid from mannitol both aerobically and anaerobically. Subgroups II through VI are analogous to *S. epidermidis* (Baird-Parker 1965). Smith and Farkas-Himsley (1969) studied 46 characters of *Staphylococcus* and concluded that *Staphylococcus* cannot be divided satisfactorily into demarcated species, but that strains within the genus form a continuous spectrum from the two extremes (*S. aureus* and *S. epidermidis*), with new subtypes evolving as characters are lost. They agreed that coagulase-positive strains could be classified as *S. aureus*, but they did not feel that strains could be classified with accuracy as *S. epidermidis* on the basis of the absence of a single character, coagulase.

Characteristics

Staphylococci are Gram-positive, nonmotile, nonsporeforming, spherical cells ($<1\mu$ in diameter) occurring singly, in pairs, and divide in more than one plane to form irregular clusters resembling bunches of grapes. Clusters occur more frequently in colonies on solid media than in liquid media.

Cultural and Biochemical.—Growth on agar appears as smooth, round, opaque, glistening, convex, circular colonies with entire edge and a white, yellow, gold, or orange color. Under conditions inhibitory to normal growth, rough or dwarf colony variants may be produced. The name *aureus*, a Latin word meaning golden, was chosen for the species named by Rosenbach (1884) because of the orange chromogenesis seen in many colonies. Strains that on original isolation form colonies with typical orange chromogenesis readily develop variants that develop dirty-white colored colonies. Strains that are white on original isolation, however, do not develop orange variants. By prolonging incubation, uncolored colonies may show pigments and lightly colored colonies become darker. Growth in broth media is turbid, but subsequently becomes clear and forms a surface ring and fine sediment. In broth containing pigmented strains, the ring and sediment are yellow. When gelatin is stabbed, saccate liquefaction, white to yellowish pellicle, and white or yellow to orange sediment develops. Morphological, cultural, and biochemical characteristics of *S. aureus* are listed in Table 2.1.

Strains of *S. epidermidis* or subgroups of *Staphylococcus* other than Subgroup I vary in most characters. They are characterized, however, by an inability to grow anaerobically in a standardized complex medium containing glucose, to produce coagulase, or to ferment mannitol; they require biotin for growth and uracil for anaerobic growth; they reduce nitrate but not beyond nitrite (Jones and Niven 1964, Baird-Parker 1965).

Toxins and Enzymes of Staphylococci.—*Staphylococcus aureus* produces a number of enzymes and toxins. These substances enable staphylococci to survive; to split proteins, carbohydrate, and fats to obtain essential nutrients; to resist drugs; and to function as a pathogen. Some of the more important substances are briefly reviewed.

Coagulase.—Coagulase or staphylocoagulase is an extracellular enzyme that accelerates the clotting of plasma. Citrated, oxalated, or heparinized human, pig, and rabbit plasmas are most susceptible to coagulase, but horse, dog, goat, calf, donkey, and goose plasmas are also clotted by this enzyme. The coagulating principle, coagulase-thrombin, is the result of the interaction of coagulase-reacting factor in plasma.

Coagulase is closely related to pathogenicity and is used to differentiate *S. aureus* from other staphylococci. In infections, coagulase may have a role in either the initiation of a lesion or the protection of a lesion already formed. It aids staphylococci by assisting in forming a mechanical barrier to passage of foreign particles; thus, it impedes the elimination of staphylococci from the lesion and resists the bactericidal action of defibrinated blood. Detailed reviews of coagulase are in the literature (Elek 1959; Tager and Drummond 1965).

TABLE 2.1
CHARACTERISTICS OF *STAPHYLOCOCCUS AUREUS*[1]
OR *STAPHYLOCOCCUS* SUBGROUP I[2]

Characteristic	Reaction[3]
Gram reaction:	
Positive	+
Negative	−
Size:	
$<1\mu$	+
$1-1.5\mu$	−
$>1.5\mu$	−
Colony appearance:	
Smooth, convex, entire edge	+
Nodular irregular edge	−
Granular surface, entire edge	−
White pigment	−
Yellow pigment	−
Orange (gold)	+(±)
Appearances in broth:	
Fine Deposit	+
Granular deposit	−
Mucoid deposit	−
Metabolic products:	
Coagulase	+
Phosphatase	+
Catalase	+
Acetoin	+
Oxidase	−
Growth:	
Presence of no added NaCl	+
Presence of 10% NaCl	+
Presence of 15% NaCl	+
At 10°C	±
At 45°C	+
Media containing ammonium phosphate as sole nitrogen source	−
Media containing glucose as sole carbon source	−
Anaerobic fermentation (acid produced):	
Glucose	+
Mannitol	+
Oxidative action on sugars (acid produced):	
Glucose	+
Mannitol	+
Lactose	+
Sucrose	+
Maltose	+
Mannose	+
Glycerol	+
Galactose	±(+)
Salicin	−(±)
Arabinose	−
Celobiose	−
Dulcitol	−
Dextrin	−

TABLE 2.1 *(Continued)*

Characteristic	Reaction[3]
Inositol	−
Raffinose	−
Rhamnose	−
Xylos	−
Nitrogen cycle:	
Ammonia from peptone	+
Ammonia from Arginine	+
Nitrates reduced to nitrites	−
Nitrates reduced beyond nitrites	+
Hydrolysis:	
Sodium hippurate	+
Esculin	−
Lard	+
Butter	+
Tween 20	+
Tween 40	+
Tween 60	+
Tween 80	±
Starch	−
Gelatin	+
Urea	+
Litmus milk:	
Acid	+
Acid and coagulation	±
Digestion	−
Litmus dicolorized	±
Methylene blue reduced	+
Clearing of casein	+
Clearing of egg yolk (lipase)	+
Methyl red (pH 4.3–4.6, average 4.4)	+
Voges-proskauer	+
Indole	−
$H_2 S$ produced	±

[1] From Buchanan and Gibbons (1974)
[2] From Shaw *et al.* (1951) and Baird-Parker (1963, 1965)
[3] + = greater than 80% positive; − = less than 20% positive; ± between 20 and 80% positive; ( ) different reaction from strains studied by different investigators

Bound Coagulase.—Bound coagulase, or clumping factor, is responsible for the thick clumps of fibrin which occur when *S. aureus* cells are mixed with plasma on a slide. The coagulase-reacting factor is not necessary for clumping as it is for clotting when "free" coagulase reacts with plasma in tubes. Bound coagulase is attached to the cell and not liberated during growth. There is neither evidence that bound coagulase bears any relation to free coagulase nor that it is a precursor of free coagulase (Tager and Drummond 1965).

Hemolysins.—Hemolysins are extracellular substances that liberate hemoglobin from red blood cells. Strains of *Staphylococcus aureus* produce one or more of at least three types of hemolysins (Elek 1959; Arbuthnott 1970; Wiseman 1970).

α-Hemolysin.—Alpha-hemolysin, or alpha-toxin, hemolyzes rabbit, sheep, cow, and goat, but not human, red blood cells. On rabbit blood agar, α-hemolysin produces a clear zone immediately surrounding a colony. Most strains of coagulase-positive staphylococci from human sources produce α-toxin (Elek and Levy 1950; Jay 1966).

Alpha-toxin, also called lethal toxin, is the most potently toxic product produced by staphylococci. After alpha-toxin is intravenously injected into a rabbit or other susceptible animal, the animal becomes unsteady, paralysis of hind legs develops, respiration becomes irregular and gasping occurs, movements become uncoordinated, pupils become dilated, violent convulsions occur, and death follows within a few minutes (occasionally a few hours). On autopsy, serosanguineous exudates are found in the body cavities; hyperemia and hemorrhage with intravascular hemolysis are noted. Tissue destruction ranges from swelling to intense necrosis.

When α-toxin contacts the skin of man or animals, necrosis occurs. Thus, the term dermonecrotoxin has been used. The dermonecrotic and lethal action of α-toxin appears to be caused by constriction of the small veins followed by capillary stasis and ischemic necrosis.

Alpha toxin has leukocidin activity. It destroys platelets, lyses spheroplasts and L-forms of bacteria, and damages a number of cells including cells in tissue culture. It also causes contraction and subsequent paralysis of several smooth muscle prepartitions and may affect skeletal muscles (Arbuthnott 1970).

β-Hemolysin.—Beta-hemolysin acts on sheep and oxen, but not on rabbit or human, red blood cells. On sheep blood agar, beta-hemolysin produces a wide zone of discoloration around colonies at 37° C incubation. If this incubation is followed by overnight incubation at 10 to 20° C, the lysis becomes complete. This reaction is known as "hot-cold" hemolysis. Beta-hemolysin is much less toxic than alpha-hemolysin. Strains of staphylococci that produce β-hemolysin are frequently isolated from animals (Wiseman 1970).

σ-Hemolysin.—Delta-hemolysin produces narrow, well-defined, clear zones of laking around colonies on human, rabbit, monkey, sheep, horse, rat, mouse, and guinea pig blood agar. When this hemolysin is injected into rabbits, nephrotic changes occur and the animal dies. Most human strains of coagulase-positive staphylococci produce σ-toxin; coagulase-negative strains do not (Elek and Levy 1950).

Leukocidins.—Leukocidins are extracellular substances that are toxic to polymorphonuclear leukocytes of rabbit or man. They kill the cells with or without lysis. Three distinct types of leukocidins are produced by one or another strain of *S. aureus.* Two of these are identical to α-hemolysin and σ-hemolysin. The third, which consists of two proteins (F and S components), is not associated with hemolytic activity. This third type has a cytotoxic effect on, but does not lyse, leukocytes (Woodin 1965; Woodin 1970).

Lipases.—Lipase catalyzes the hydrolysis of ester linkages between fatty acids and glycerol of triglycerides and phospholipids. Staphylococcal lipases, however, can attack more complex lipids such as Tween polyoxyethylene esters and egg yolk (Gillespie and Alder 1952).

The action of a staphylococcal lipase, tributyrinase (known as egg yolk factor), can be observed on egg yolk agar as a clearing reaction, a zone of precipitate and opacity around a colony, and a liberation of fatty material (Shah and Wilson 1963; Lundbeck and Tirunarayanan 1966). Most coagulase-positive staphylococci are lipolytic; a much lower percentage of coagulase-negative staphylococci produce lipase (Gillespie and Alder 1952; Elek 1959; Jay 1966). Lipase also plays a role in pigment formation. Lipolytic action is caused by the combined action of a lipase, an esterase, and possibly a phosphatidase (Stewart 1965).

Hyaluronidase.—Hyaluronidase catalyzes hydrolysis of hyaluronic acid, the cement-like substance of tissues. It is referred to as the spreading factor. Coagulase-negative staphylococci do not produce hyaluronidase, but 87% of coagulase-positive staphylococci produce this enzyme (Schwabacher *et al.* 1945).

Fibrinolysin.—Fibrinolysin, or staphylokinase, catalyzes digestion of fibrin, the insoluble protein formed from fibrinogen by the action of thrombin which is responsible for clotting of blood. Coagulase-positive staphylococci may or may not have fibrinolytic activity; coagulase-negative staphylococci, however, seldom exhibit fibrinolysin (Elek 1959; Jacobs *et al.* 1963).

Deoxyribonuclease.—Deoxyribonuclease catalyzes the depolymerization of deoxyribonucleic acid. A high correlation exists between deoxyribonuclease activity and coagulase activity in staphylococci of human origin (Brandish and Willis 1970; Stickler and Freestone 1971). Jacobs *et al.* (1963) and Lachica *et al.* (1969) reported that only a few coagulase-negative strains show a weakly positive deoxyribonuclease reaction, but Blair *et al.* (1967) and Stickler and Freestone (1971) observed that up to 21% of coagulase-negative strains produce deoxyribonuclease upon prolonged incubation. This enzyme has a high thermal stability (Chesbro and Auborn 1967). It has a D value of 16.6 minutes at 266°F (130°C), 34 minutes at 248°F (120°C), and 180 minutes at 212°F (100°C) (Erickson and Deibel 1973).

Penicillinase.—Penicillinase acts by opening up the β-lactan ring of penicillin and hence destroys its antibiotic activity (Elek 1959). Many strains of *S. aureus* have become resistant to penicillin by this mechanism.

Phosphatase.—Phosphatase hydrolyzes monophosphoric esters with liberation of inorganic phosphate. Coagulase-positive staphylococci produce acid phosphatase, but so do some strains of coagulase-negative staphylococci (Blumenthal and Pan 1963; Baird-Parker 1963).

Lysozyme.—Lysozyme is an enzyme which has lytic or antibacterial activity against some organisms. Jay (1966) found that 95% of coagulase-positive staphylococci produce lysozyme, and that only 8% of coagulase-negative staphylococci produce this enzyme.

Catalase.—Catalase, an enzyme found in almost all bacteria except anaerobic bacteria, catalyzes the breakdown of hydrogen peroxide into water and oxygen. *Staphylococcus aureus* is strongly catalase positive.

Proteases.—Gelatinase causes gelatin to liquefy. This enzyme is produced by 97% of coagulase-positive staphylococci from human lesions, by 98% of these organisms from animal lesions, and by 60% of coagulase-negative staphylococci (Elek 1959). Other proteases are also produced by staphylococci.

Enterotoxins.—Staphylococcal enterotoxins are simple proteins, composed of 18 amino acids and primary amide, which upon ingestion by people or susceptible animals or injection into susceptible animals cause vomiting and frequently diarrhea within a few hours. The complete amino acid sequence of enterotoxin B has been determined by Huang and Bergdoll (1970A, 1970B, 1970C). The physiochemical nature, amino acid composition, and terminal amino acids of enterotoxins are reviewed by Bergdoll (1967, 1970).

Enterotoxins are exotoxins which are produced within the cell and diffuse into a food or culture media; they can be separated from the cells by filtration. Friedman (1968) suggested that the cell wall is the site of enterotoxin synthesis; but Forsgren *et al.* (1972) found a thousandfold lower enterotoxin B content in digested cell walls than in extracellular fluids, indicating that enterotoxin is a true exotoxin. Markus and Silverman (1969) suggested that enterotoxin is produced from a precursor pool within the cell, and they observed that growth per se was not necessary for its formation.

Enterotoxigenic staphylococci almost always produce coagulase, but not all coagulase-positive staphylococci synthesize enterotoxin. On rare occasions, however, strains of coagulase-negative staphylococci have been found that synthesize enterotoxin (Thatcher and

Simon 1956; Omori and Kato 1959; Bergdoll *et al.* 1967; Breckinridge and Bergdoll 1971).

Several different types of enterotoxins are known; others are yet to be investigated. After the discovery that staphylococci produce more than one enterotoxin, the first two known enterotoxins were called E (because of its history of implication in enteritis following antibiotic therapy) and F (because of its implication in foodborne outbreaks). When more enterotoxins were discovered, it became apparent that some other system would have to be devised for their nomenclature. Thus, to provide greater flexibility, to facilitate naming enterotoxins which may be identified in the future, and to conform with rules of bacteriological nomenclature, a designation with sequential letters of the alphabet was decided upon by a committee of the American Society for Microbiology (Casman *et al.* 1963A). To date, the following enterotoxins have been identified: A (formerly called F) (Casman 1960), B (formerly called E) (Bergdoll *et al.* 1959), C (Bergdoll *et al.* 1965), D (Casman *et al.* 1967), E (Bergdoll *et al.* 1971). Two slightly different enterotoxins C have been identified; they appear to be identical in most respects, including the fact that they react to the same antibody, but their movements in an electrical field are quite different (Avena and Bergdoll 1967; Borja and Bergdoll 1967).

Enterotoxins are antigenic. Antitoxins prepared from purified enterotoxins by injecting rabbits or other animals are used to identify enterotoxins in foods. (A discussion of these procedures is given in that part of the chapter designated as "Laboratory Methodology.")

Enterotoxins are rather stable. They have been shown to be resistant to 0.3% formalin for up to 48 hours (Minett 1938); to proteolytic enzymes such as trypsin, chymotrypsin, rennin, papain, and pepsin (Schantz *et al.* 1965); to pH values from 3 to 10 (Dack 1956A); to storage for as long as 67 days in a refrigerator (Jordan *et al.* 1931); to storage in the freeze-dried state for months (Schantz *et al.* 1965); and to 3 minutes contact with a 915 ppm chlorine solution (Jordan *et al.* 1931). A dose of 5 Mrad was required to reduce an enterotoxin B concentration of 31 μg/ml in veronal buffer to less than 0.7 μg/ml, but 20 Mrad was needed for this same reduction in milk (Read and Bradshaw 1967).

Crude enterotoxins are relatively heat resistant. They are gradually destroyed by boiling but persist after 30 minutes (Jordan *et al.* 1931; Dack 1956A). Ordinary cooking, spray drying, and pasteurization will not inactivate enterotoxin but canning might (Denny *et al.* 1966; Read and Bradshaw 1966B). Denny *et al.* (1966) found that concentrated crude enterotoxin B had a Z value of 48°F. When kittens were tested, the toxin was destroyed at 250°F (121.1°C) in 11 minutes; and when monkeys were tested, the toxin was destroyed at 250°F (121.1°C) in 8 minutes. Hilker *et al.* (1968) reported that 21μg/ml of enterotoxin A in veronal buffer (pH 7.2) was reduced to less than 1 μg/ml when heated at 212°F (100°C) for 130 minutes. A Z value of 27.8°C was obtained. In this same buffer with 30 μg/ml of crude enterotoxin B, Read and Bradshaw (1966A) reported D values of 64.5 minutes at 210.2°F (99°C) and 11.4 minutes at 250°F (121°C) for enterotoxin B.

Studies of the thermostability of purified enterotoxins have given varying results. Purified enterotoxin A is relatively heat labile. A decrease of 50% in the reaction of enterotoxin A with its specific antibody resulted when this toxin was heated in a 0.05 M solution of sodium phosphate (pH 6.85) for 20 minutes at 140°F (60°C). Heating a similar solution at 158°F (70°C) for 3 minutes resulted in a 60% decrease in reaction. No antigen-antibody reaction was obtained after heating this enterotoxin at 176°F (80°C) for 3 minutes or at 212°F (100°C) for 1 minute (Chu *et al.* 1966). The Z value for enterotixin A in beef bouillon was approximately 50°F (27.8°C) with three different concentrations of toxin. Less heat was required for inactivation in pH 7.2 phosphate buffer than in beef bouillon. Inactivation was dependent on concentration (Denny *et al.* 1971). The biological activity of enterotoxin B was retained after heating in a 0.05 M phosphate solution (pH 7.3) at 140°F (60°C) for as long as 16 hours. At 212°F (100°C) for 5 minutes, less than 50% of the biological activity was destroyed, but the toxin coagulated at this time-temperature value (Schantz *et al.* 1965). Read and Bradshaw (1966A) reported that 30 μg/ml of enterotoxin B in veronal buffer was reduced to less than 0.7 μg/ml in 103 minutes at 204.8°F (96°C) and 16.4 minutes at 250°F (121°C). At 204.8°F (96°C) the D value was 61.9 minutes and at 250°F (121°C) was 9.9 minutes. The Z value was 32.4°C. These same authors (1966B) also reported that 30μg/ml of enterotoxin B in milk was reduced to 90% of its activity (D value) at 210°F (98.9°C) in 68.5 minutes and at 250°F (121.1°C) in 9.4 minutes. The Z value was 46.6°F. When solutions of enterotoxin C were heated at 140°F (60°C) for one half hour, there was no loss of protein. When the heating was continued for one hour, the solution became turbid (Borja and Bergdoll 1967). The reaction of enterotoxin C with antienterotoxin was reduced to about 20% of normal when the protein was heated at 212°F (100°C) for 1 minute (Avena and Bergdoll 1967). These data show that purified enterotoxins are more sensitive to heat than are crude enterotoxins or that enterotoxins are protected from effects of heat when mixed in food substrates.

LABORATORY METHODOLOGY

Enumeration

Staphylococci are enumerated in foods to determine the food's possible role as a vehicle in the foodborne disease outbreak, to demonstrate the possible risk that is associated with a food if it is eaten or used as an ingredient in another food, to ascertain postprocessing contamination, and to evaluate previous storage conditions. With the exception of foods responsible for outbreaks, staphylococci usually make up only a small part of the bacterial flora in foods. Because of this, selective media have been developed that allow staphylococcal growth but inhibit the growth of other bacteria. The inhibitory substances, however, also interfere with optimum growth of staphylococci. In the selective media that are commonly used, other bacteria are inhibited by 7.5% sodium chloride, lithium chloride, potassium tellurite, polymyxin, mercuric chloride, sorbic acid, or sodium azide. Most media also contain ingredients that help to differentiate staphylococci from other bacteria. In the more recently developed media, differentiation is based on the organism's ability to utilize mannitol, to reduce potassium tellurite, or to produce lipase, phosphatase, deoxyribonuclease, or coagulase.

The first media developed for selective enumeration of staphylococci were staphylococcus medium 110, Chapman Stone medium, and mannitol salt agar (Chapman 1945, 1946, 1948). Selectivity and identification of staphylococci in these media are based on the ability of staphylococci to grow in 7.5% NaCl, to utilize mannitol, and to form pigmented colonies. Gelatin liquefication can also be detected on staphylococcus medium 110 and Chapman Stone medium. Several investigators cited limitations of these media (Raj and Liston 1961; Baird-Parker 1962A; DeWaart et al. 1968; Crisley et al. 1965; Gilbert et al. 1969).

The next major advancement in selective-differential media was made by Ludlam (1949) who developed tellurite lithium chloride agar. Staphylococci form large black colonies on this agar; thus, subjective judgement of pigmentation was eliminated. Several investigators, however, reported that this medium was inhibitory to S. aureus (Chapman 1949; McDivitt and Husseman 1954; Innes 1960; Baird-Parker 1962A). Modifications in this medium were made to remove the inhibitory effects by adding glycine (Zebovitz et al. 1955; Vogel and Johnson 1960; Moore and Nelson 1962). Proteus vulgaris, unfortunately, closely resembles staphylococci on these media (Baird-Parker 1962B; Crisley et al. 1965).

Carter (1960) and Herman and Morelli (1960) improved staphylococcus medium 110 and Innes (1960), Baird-Parker (1962A, 1962B), and Crisley et al. (1964) improved tellurite-containing agars by the addition of egg yolk and other ingredients. In these improved media, characteristic zones of precipitation form around staphylococcal colonies.

Fibrinogen has been added to or on media so that coagulase-positive staphylococci could be detected during primary isolation (Duthie and Lorenz 1952; Klemperer and Haughton 1957; Deneke and Blobel 1962; Blair et al. 1967). Orth and Anderson (1970A, 1970B) developed two media, polymyxin coagulase mannitol agar and polymyxin coagulase deoxyribonuclease agar, which purportedly give direct counts for coagulase-positive staphylococci and inhibit Gram-negative organisms and S. epidermidis. False positive reactions by lipolytic coagulase-negative staphylococci were reduced by gel filtration of plasma to remove the coagulase-reacting factor. This filtration process, however, makes the preparation of the media too difficult for routine use.

Other media that have been developed for selective isolation of staphylococci and used in laboratories in various countries are milk salt agar (Meshalova and Mikhailova 1964), egg yolk azide agar (Hopton 1961; Lundbeck and Tirunarayanan 1966), and phenolphthalein phosphate agar with polymyxin (Hobbs et al. 1968). Staphylococci are also sometimes enumerated by a most probable number procedure with three or five tubes of enrichment broth, followed by plating. Procedures and typical reactions of S. aureus on solid media that have been mentioned are listed in Table 2.2.

Several investigators have observed that heat-shocked cells are inhibited by selective media, such as those containing 7.5% NaCl (Busta and Jezeski 1963; May and Kelly 1965; Stiles and Witter 1965; Iandolo and Ordal 1966). Sublethally-heated organisms need special nutrients and a period of adjustment; thus, they have a longer lag than unheated organisms (Jackson and Woodbine 1963). For maximum recovery of heated cells, blood and pyruvate (Baird-Parker and Davenport 1965), glucose and galactose (Stiles and Witter 1965), yeast extract (Allwood and Russell 1966), and amino acids (Iandolo and Ordal 1966) have been recommended. For optimum recovery, a temperature of 32°C and a pH of 6 was suggested by Allwood and Russell (1966).

Comparative evaluations of differential media have been conducted by a few investigators with varying results. Some factors that affect the results are: the type of food analyzed, level of contamination with staphylococci, and type and level of contamination with competitive flora. The data are often not statistically analyzed,

TABLE 2.2

PROCEDURES AND CHARACTERISTIC REACTIONS OF MEDIA USED FOR THE ISOLATION OF *S. AUREUS*

Media	Technique	Incubation Period and Temperature	Colony Appearance	Reaction Around Colony	Reference
Baird-Parker agar (Egg tellurite glycine pyruvate agar)	Spread	24 and 48 hrs at 37°C	Black and shiny with narrow white margins	Clear zone around colony (24 hr), zone of precipitation after 48 hrs, clear zone remains	Baird-Parker (1962A, 1962B)
Blood agar or azide blood agar	Streak	24 hrs at 37°C	White, yellow or orange	Wide or narrow clear zone or incomplete zone of hemolysis around colonies	Rammell and Howick (1967)
Chapman Stone medium agar	Streak	48 hrs at 30°C	White, yellow or orange	Clear zone around areas of picked colonies when plates flooded with saturated solution of ammonium sulfate—change in color when added bromcresol purple	Chapman (1948)
DNase agar	Streak	24 hrs at 37°C	White, yellow or orange	Clear zone around streak after adding 1N HCl to agar surface	Jeffries *et al.* (1957) Lachica and Diebel (1969)
DNase agar with methyl green	Streak or spot	Overnight at 37°C	White, yellow or orange	Clear zone on green background	Smith *et al.* (1969)
Egg yolk azide agar	Spread	24 and 48 hrs at 35 to 37°C	White, yellow or orange	Clear, oily areas around colonies with or without white granular precipitation	Lundbeck and Tirunarayanan (1966) Hopton (1961)

Medium	Pour, spread or streak	Depends on agar used	Depends on agar used	Halo around colony	Reference
Fibrinogen medium	Pour, spread or streak	Depends on agar used	Depends on agar used	Halo around colony	Klemperer and Haughton (1957) Deneke and Blobel (1962)
Mannitol salt agar	Streak or spread	36 hrs at 37°C	White, yellow or orange	Yellow zones around colony	Chapman (1945)
Milk salt agar	Spread	24 and 48 hrs at 37°C	White, yellow or orange		Meshalova and Mikhailova (1964)
Phenolphthalein phosphate agar with polymyxin	Streak or spread	24 and 48 hrs at 37°C	Pink on exposure to ammonia		Hobbs et al. (1968) Barber and Kuper (1957)
Polymyxin coagulase deoxyribonuclease agar	Spread	15 and 48 hrs at 37°C	White, yellow or orange	Fibrin halos around colonies—zone of fading green round colonies, green background after 1 min application of methyl green and 1 hr room temperature incubation	Orth and Anderson (1970A)
Polymyxin coagulase mannitol agar	Spread	15 and 48 hrs at 37°C	White, yellow or orange	Fibrin (opaque) zone around colonies—opaque, gray zone or halo radiating outward from colony—yellow area around colony	Orth and Anderson (1970B)
Polymyxin staphylococcus medium	Streak	24 hrs at 37°C	White, yellow or orange		Finegold and Sweeney (1961)
Salt mannitol plasma agar	Streak	24 to 48 hrs at 35°C	White, yellow or orange	Clear areas or blue zone around colony—white precipitate around colony—medium turned yellow (acid type)	Blair et al. (1967)

TABLE 2.1 (Continued)

Media	Technique	Incubation Period and Temperature	Colony Appearance	Reaction Around Colony	Reference
Staphylococcus medium 110 agar	Spread or streak	48 hrs at 30°C 43 hrs at 37°C	White, yellow or orange	Clear zone around areas of picked colonies when plates flooded with saturated solution of ammonium sulfate—change in color when add bromcresol purple	Chapman (1946)
Staphylococcus medium 110 egg yolk agar	Spread	24 and 48 hrs at 35 to 37°C	White, yellow or orange	Zone of precipitation around colony	Hermann and Morelli (1960) Carter (1960)
Tellurite Egg yolk agar	Spread	30 to 48 hrs at 37°	Dark gray	Opaque zones around colony	Innes (1960) Alder et al. (1962)
Tellurite glycine agar	Spread	24 hrs at 37°C	Black		Zebovitz et al. (1955) Moore and Nelson (1962)
Tellurite polymyxin egg yolk agar	Spread	24 and 48 hrs at 35 to 37°C	Jet black or dark gray 1.0–1.5 mm in size after 24 hrs	Discrete zone of precipitation around and beneath colony—clear zone or halo around colony, often with zone of precipitation or precipitation beneath colony	Crisley et al. (1964, 1965)
Vogel-Johnson agar (Tellurite glycine red agar)	Streak	48 hrs at 35 to 37°C	Black	Yellow zone around colony	Vogel and Johnson (1960) Baer et al. (1966)

and if they were, statistically significant differences between media would not be observed very frequently. Innes (1960) reported that tellurite egg yolk agar was superior to both tellurite glycine and staphylococcus medium 110 agars. A higher percentage of meat samples was positive for *S. aureus* when staphylococcus medium 110 egg yolk agar and mannitol sorbic acid broth were used (Jay 1961, 1963). Decreasing numbers of positive samples were yielded by mannitol salt, tellurite glycine, polymyxin, and Vogel-Johnson agars.

Statistical analysis of the efficiency of recovering staphylococci from pure cultures revealed efficiencies in the following descending order: tellurite polymyxin egg yolk agar, Baird-Parker agar, tellurite glycine agar, tellurite egg yolk agar, staphylococcus medium 110 (Crisley *et al.* 1965). The investigators also observed the following descending order of inhibition of coagulase-negative cocci: Baird-Parker agar, tellurite egg yolk agar, staphylococcus medium 110, tellurite glycine agar, and tellurite polymyxin egg yolk agar. Recovery on tellurite polymyxin egg yolk agar was influenced less than the other media by the type of food examined, but each medium was good for specific items of food. Efficiency of recovery of staphylococci on tellurite glycine, staphylococcus medium 110, and tellurite polymyxin egg yolk agars was not dependent upon the level of contamination, but recovery on tellurite polymyxin egg yolk agar decreased with increased numbers of contaminants.

Baer *et al.* (1966) detected *S. aureus* in a variety of foods with greater frequency when using Vogel-Johnson agar than when using staphylococcus medium 110, staphylococcus medium 110 egg yolk, tellurite glycine, polymyxin B, and mannitol egg agars. After evaluating mannitol salt agar and Baird-Parker agar, DeWaart *et al.* (1968) recommended Baird-Parker agar for general use for enumerating *S. aureus* in foods. Sessoms and Mercuri (1969) did not observe any statistically significant difference between staphylococcus medium 110, two commercial formulae of tellurite polymyxin egg yolk, tellurite glycine, and Vogel-Johnson agars for the recovery of *S. aureus* in the presence of the other species of bacteria.

Chou and Marth (1969) observed that trypticase soy broth containing 8 or 10% NaCl, mannitol salt agar, and staphylococcus medium 110 agar yielded more *S. aureus* from feed grade frozen meat and liver than did tellurite polymyxin egg yolk, tellurite glycine, or Vogel-Johnson agars. Mannitol salt, tellurite glycine, and Vogel-Johnson agars uniformly recovered coagulase-positive staphylococci from enrichment broths from the highest dilutions. Erratic results were obtained with staphylococcus medium 110 and tellurite polymyxin egg yolk agars.

Gilbert *et al.* (1969) evaluated the media recommended by the International Association of Microbiological Societies' Committee on Microbiological Specifications for Food (Thatcher and Clark 1968). They concluded that for routine analysis—if a cheap, simple to prepare, and stable medium is required—phenolphthalein diphosphate agar with polymyxin and milk salt agar were quite satisfactory, but for research and development studies Baird-Parker medium and tellurite polymyxin egg yolk agar were probably the best.

Tardio and Baer (1971) found that Baird-Parker agar detected 25% more positive samples than were detected with Vogel-Johnson agar. Baer *et al.* (1971) did not observe a statistically significant difference in the detection of *S. aureus* from four types of foods when Vogel-Johnson agar, Baird-Parker agar, tellurite polymyxin egg yolk agar, and staphylococcus medium 110 egg yolk agar were compared as plating media in an enrichment isolation procedure.

The Food Protection Committee (1971) of the National Research Council developed recommendations for reference methods for the microbiological examination of foods, but they felt that there was insufficient basis for recommending a reference method for *S. aureus* at that time. The results of comparative studies of media and methods support this suggestion, but there is a distinct edge in favor of Baird-Parker agar. Because of this, and because of the finding that Vogel-Johnson agar was more inhibitory than Baird-Parker agar, there was a recommendation to change the recommended AOAC method for the isolation and enumeration of *S. aureus* by replacing Vogel-Johnson agar with Baird-Parker agar (Baer 1971).

Enrichment

An enrichment procedure is more efficient than a direct plating procedure for isolating *S. aureus* from foods containing only a few of these organisms. Gilden *et al.* (1966) and Tardio and Baer (1971), for instance, detected *S. aureus* from foods more often by enrichment methods than by direct plating methods. Direct plating methods are just not appropriate for samples containing fewer than 50 to 100 *S. aureus* per gm; sometimes, depending on procedure, these methods are not appropriate unless the sample contains 3,000 organisms per gm or more. Enrichment procedures are also useful when a food or specimen contains large numbers of other bacteria or when *S. aureus* cells are injured by some processing procedure such as freezing, drying, or mild heat treatment.

Brain heart broth with 7.5% NaCl (Wilson *et al.* 1959), 6.5% salt (beef extract-peptone) broth, 10% salt broth, glucose broth (Meshalova and Mikhailova 1964), cooked meat broth with 10% NaCl (Baer *et al.* 1966), mannitol sorbic acid broth (Raj 1966), anaerobic tellurite mannitol glycine broth (Giolitti and Cantoni 1966), and trypticase soy broth containing 8% or 10% salt (Baer *et al.* 1966; Chou and Marth 1969) have been devised as enrichment media for staphylococci. Baer *et al.* (1971) detected *S. aureus* more frequently with trypticase soy broth with 10% NaCl than with the tellurite broth of Giolitti and Cantoni.

When enrichment methods are employed, a swab, 1 gm or ml of a food, or a 1 ml of a homogenate of a 1:10 dilution of a food is put into 10 ml of the enrichment broth. If larger quantities of food are put into larger volumes of media, the likelihood of positive findings will be increased. If three tubes each of three consecutive ten-fold dilutions are used, a most probable number can be calculated. After initial enrichment, broth cultures are plated on a selective-differential agar and typical staphylococcal colonies are picked and tested for coagulase production and other characteristics.

Confirmation

After staphylococci are isolated on solid media, they must be confirmed as *S. aureus*. The ability of *S. aureus* to grow anaerobically, to form catalase, to liquify gelatin, and to utilize mannitol are useful tests, but testing for the presence of coagulase is the most common means of identifying this organism. Methods of coagulase testing are given in the manuals edited by Sharf (1966) and Thatcher and Clark (1968) or in the article by Blair (1970).

Microscopic Examination

Direct microscopic counts indicate the number of Gram-positive micrococci that are or have been in a food. Because this method detects both living and dead micrococci, it is necessary to analyze the processing history of the food which is being tested. In foods that have undergone heat processing or prolonged storage or that have been held under situations in which other microorganisms may have overgrown the micrococci, microscopic counts may exceed plating counts. Because *S. aureus* cannot be differentiated from other micrococci by direct microscopic method, fluorescent antibody techniques have been used to specifically detect and count *S. aureus* (Carter 1959; Smith *et al.* 1962). Microscopic procedures are given in "Standard Methods for the Examination of Dairy Products" (Hausler 1972).

Tests that Reflect Staphylococcal Growth

Chesbro and Auborn (1967) reported that measurement of heat-stabile nuclease, which is produced by *S. aureus*, is a sensitive means for detecting evidence of *S. aureus* growth in foods. The test uses common reagents and a spectrophotometer and only takes 3 hours but is laborious because it involves prior purification of the enzyme. The investigators reported that whenever 0.34 or more units of nuclease were detected, it was possible to isolate *S. aureus*, unless destroyed by heat treatment, and to detect enterotoxin. Quantities of nuclease as low as 0.005 μg/ml were detected by a metachromatic agar-diffusion microslide technique in 3 hours (Lachica *et al.* 1972). Because of the heat stability of nuclease, evidence of growth can be detected in a cooked product in which viable cells are no longer present.

Phage Typing

The primary value of phage typing is to trace sources of staphylococcal contamination. It also has value in suggesting association between a food vehicle and the victims of an outbreak. Because of the ubiquity of *S. aureus*, the finding of coagulase-positive staphylococci in a food and in the nares, in a lesion, or on the skin of a food handler does not fix either the food as a vehicle or the food handler as the source, even if these organisms are found in vomitus or stools of patients. Phage patterns of isolates from a carrier, a food, and patients must be similar to prove transmission. Phage typing is a useful method of labeling strains so that their spread may be followed over a relatively short period of time.

The International Association of Microbiological Societies', Subcommittee on Typing of Staphylococcus (Nomenclature Committee), proposed a set of 22 basic phages for routine typing of human *S. aureus* strains (Blair and Williams 1961; Blair and Parker 1967; Parker and Rountree 1971). These are:

Group I: 29, 52, 52A, 79, 80
Group II: 3A, 3C, 55, 71
Group III: 6, 42E, 47, 53, 54, 75, 77, 83A, 84, 85
Group IV: 42D
Not Alloted: 81, 187

Additional phages, however, are used in certain typing centers (Smith 1970).

It is now recommended that typing be done with the basic set of phages at routine test dilutions (RTD) and cultures not lysed by any of the basic-set phages at RTD should be typed with the same phages (except 83A, 84, and 85) at RTD × 100 (Parker and Rountree 1971). This is a change from the previous practice of retesting all untypable cultures at RTD × 1000 (Blair and Williams 1961). Routine test dilution is defined as the highest dilution that just fails to give confluent (complete) lysis. Standardized methods for propagating phages, determining RTD, and typing are given by Blair and Williams (1961) and Smith (1970).

A working group of this subcommittee also established a basic set of 16 phages for typing bovine *S. aureus* strains (Parker and Rountree 1971; Davidson 1971). This set includes 9 phages of the basic set used for typing *S. aureus* from human sources and 7 other phages. These are:

Group I: 29, 52A
Group II: 3A, 883
Group III: 6, 42E, 53, 75, 84
Group IV: 42D, 102, 107, 1363/14
Miscellaneous: 78, S1, S6

It is recommended that phages should be used at routine test dilutions.

Most typable strains of *S. aureus* are lysed by more than one phage, resulting in a phage pattern. A phage pattern that is reported as 3A/3C/55 was lysed strongly by these three phages; one reported as 6/47/53/54/75+ was lysed strongly by these five phages and weakly by others. Isolates from different sources often show slightly different patterns. If the isolates from patients and from sources are epidemiologically linked, they will be lysed in common by several, but not necessarily all of the same phages. In a food poisoning outbreak usually only minor differences are noted in phage patterns that are epidemiologically associated. Interpretations of phage patterns are reviewed by Anderson and Williams (1956).

Microbial Sensitivity Tests

A large proportion of strains of *S. aureus* are resistant to one or more antibiotic. Therefore, comparisons of antibiograms of cultures taken from patients with antibiograms of cultures obtained from foods and food workers may suggest associations. These comparisons, however, are not as definitive in epidemiologically identifying strains as is phage typing. Microbial sensitivity tests are of primary value as a guide for treatment. Methods for microbial sensitivity tests are given by Bodily and Updyke (1970).

Detection of Enterotoxin

The only ways that a food can be proven to be a vehicle of staphylococcal enterotoxin are to feed it to human volunteers or animals, to inject animals with boiled or otherwise treated filtrates from cultures isolated from foods, or to detect enterotoxin from food extracts or from filtrates from cultures isolated from foods by serological reactions. Procedures used in human volunteer studies have been reviewed by Dolman (1934) and Dack (1956A). The animal assay methods most frequently used for detection of enterotoxin are intraperitoneal or intravenous injections of cats and kittens (Dolman *et al.* 1936; and Hammon 1941) and feeding young rhesus monkeys (Surgalla *et al.* 1953; Dack 1956A; Bergdoll 1970). Because of many obvious problems associated with human volunteer tests, feeding monkeys is the most reliable and practical bioassay method. In this test, six monkeys are each given 50 ml of a sample of enterotoxin by catheter, and they are observed for 5 hours. Vomiting by at least two monkeys is considered a positive reaction. Bioassay methods for enterotoxin are difficult to perform, of variable reliability, and expensive. They lack either sensitivity or specificity. Because of these reasons, they have been replaced with specific serological procedures.

Several in vitro methods have been devised for the serological detection of staphylococcal enterotoxins. These include gel diffusion, capillary tube, quantitative precipitin, hemagglutination, and fluorescent antibody procedures. The gel diffusion tests are performed in several ways: single diffusion tube test (Oudin 1952; Surgalla *et al.* 1952, 1954; Bergdoll *et al.* 1959; Silverman 1963; Hall *et al.* 1963, 1965; Weirether *et al.* 1966), double diffusion tube test (Oakley and Fulthrope 1953; Bergdol *et al.* 1959; Hall *et al.* 1965; Read *et al.* 1965A, 1965B) and micro-slide or Ouchterlony plate double diffusion tests (Ouchterlony 1949, 1953; Wadsworth 1957; Crowle 1958; Casman and Bennett 1965; Casman 1967; and Casman *et al.* 1969).

In the single diffusion tube test, antitoxin is mixed with agar in the bottom half of a tube and unconcentrated or concentrated liquid food extract containing enterotoxin is layered over the solidified antitoxin-containing agar. Upon incubation, the enterotoxin migrates down into the agar and a band of precipitate forms. The rate of migration in both tests is directly related to concentration of enterotoxin and is a function of distance over a given time as determined by a standard curve which is derived by plotting known enterotoxin concentrations against their migration times. This procedure detects as little as 1 μg of enterotoxin per ml and requires incubation from 1 to 7 days.

In the double diffusion tube test, antitoxin is mixed with agar in the bottom half of a tube and an unconcentrated or concentrated liquid food extract is mixed with agar in the top half of the tube. The two layers are separated by a layer of clear buffered agar.

Enterotoxin diffuses downward through the layer of buffered agar and antitoxin diffuses upward through the same layer of agar. A band of precipitate forms where the two substances meet. A standard curve is also used to estimate concentration of enterotoxin present in the food. This procedure detects as little as 0.05 μg of enterotoxin per ml and requires incubation of at least 1 week. Both tube methods require rather large quantities of antisera (0.2 ml per analysis).

In the micro-slide gel double diffusion test, antitoxin is placed in the central well of a thin layer of agar on a microscope slide, and known (reference) enterotoxin solutions and food extracts are put adjacently in peripheral wells. Lines of precipitate form between the central and any peripheral well where an antigen-antibody reaction occurs. These lines are easily identified by their coalescence with lines formed by the reference toxins and their antienterotoxins. This feature of coalescence of lines of identity, or specificity, makes the procedure superior to other techniques because unknowns can be directly compared with controls. By procedures outlined by Casman et al. (1969) approximately 0.01 μg of enterotoxin per ml can be detected. The slides require room temperature incubation for 1 to 3 days. This test is relatively simple and economical in the use of reagents; only 0.02 to 0.025 ml of appropriately diluted antiserum and reference enterotoxin are required. It is the procedure used by most laboratories that routinely test foods for the presence of enterotoxin. This procedure is also used to confirm the results of other serological procedures. The Ouchterlony plate method is performed similarly, but it requires more antitoxin.

Enterotoxin must be extracted from insoluble food constituents before gel diffusion tests can be performed. Extracts containing enterotoxin are usually separated from soluble extractives and then concentrated. Methods that have been used are froth flotation with rhodamine B-labeled antibodies followed by gel filtration through Sephadex (Hopper 1963), a series of acid precipitations, centrifugation, filtration, and chloroform extractions (Read et al. 1965A, 1965B); phosphate precipitation, centrifugation and chromatography with ion exchange resin Amberlite CG50 (Hall et al. 1965); and centrifugation, chloroform extraction, chromatography with carboxymethyl cellulose (Casman and Bennett 1965; Casman 1967; and Casman et al. 1969). The last three methods concentrate the extracts with polyvinylpyrolidone or polyethylene glycol 20,000. In the procedure developed by Casman and coworkers, which is the more widely used method, 100 gm of food in 500 ml of 0.2 molar NaCl are concentrated between 0.1 and 0.5 ml. Extraction and concentration require 3 to 4 days.

Capillary tube tests as used for the Lancefield grouping of streptococci have been modified to detect staphylococcal enterotoxins (Fung and Wagner 1971; Gandhi and Richardson 1971). Microamounts of antitoxins are introduced into closed 1-mm diameter capillary tubes and microamounts of purified and concentrated extracts of food or staphylococcal cultures are layered onto the surface of the antitoxins. As little as 1 μg of staphylococcal enterotoxin has been detected in less than 1 hour by this procedure.

The quantitative precipitin test gave results for enterotoxin B that were comparable to gel diffusion test (Silverman 1963). In the precipitin test, antigen and antitoxin are mixed and the solution incubated for 4 hours at 37°C and for 4 days at 4°C. Enterotoxin content is calculated from a standard curve prepared with purified toxin and antitoxin.

Another in vitro method, passive hemagglutination, known also as hemagglutination inhibition, has given rapid, sensitive, reproducible results with the use of small quantities of reagents (Robinson and Thatcher 1965; Brown and Brown 1965; Morse and Mah 1967; and Johnson et al. 1967). After the test is set up, hemagglutination patterns can be read in 3 hours. With experience in performing the test, results are easy to read. Latex particles, instead of erythrocytes, coated with specific antitoxin have been used in similar titrations. As little as 0.0002 μg of enterotoxin B has been detected by these procedures in 1 day (Salomon and Tew 1968).

By attaching antitoxin globulin directly to erythrocytes, a direct measure of toxin by hemagglutination, rather than the two-step procedure of passive hemagglutination, can be made (Silverman et al. 1968). This method is known as reversed passive hemagglutination. It can detect quantities of enterotoxin as small as 0.0007 μg or 0.0015 μg/ml. The investigators reported that neither elimination of interfering proteins from food extracts nor concentration of the sample is required, as in gel diffusion tests. Limitations of this procedure are that antienterotoxic sera which are sufficiently free of nonspecific antibodies are not yet available for all enterotoxins and that the test does not compare an unknown with a known enterotoxin (Casman et al. 1969).

Detection of enterotoxin in vitro by immunofluorescence has been tried (Friedman and White 1965; Genigeorgis and Sadler 1966A, 1966B; Stark and Middaugh 1969). Only 4 to 5 hours were required to detect enterotoxin in food smears or extracts, and less than 1 μg/ml can be detected without any special extraction procedures (Genigeorgis and Sadler 1966B). Forsgren et al. (1972), however, observed that S. aureus strains that did and did not produce enterotoxin B showed cell fluorescence, probably as a result of a protein A-immunoglobulin G interaction. Soluble enterotoxin B, however, was detected by immunofluorescence.

A radiobioassay method (Hodoval *et al.* 1966) and a radio-immunoassay method (Johnson *et al.* 1971; Collins *et al.* 1972) have also been developed to detect enterotoxin. The tests are very sensitive and quantitative. The methods, however, are not yet practical for routine laboratory use.

Enterotoxin detection has practical application in the investigations of outbreaks (Casman and Bennett 1965; Wolf. *et al.* 1970; Weimann *et al.* 1971), in evaluating the hazards of specific foods (Casman 1965; Donnelly *et al.* 1967), and in testing incriminated lots of processed food to see if part of the production lot can be released (Zehren and Zehren 1968A). As an example of this last application, 4.07 million pounds (2,112 lots) of cheese were detained in storage as a result of an outbreak. Each lot was tested, and 59 lots were found contaminated with enterotoxin A. The nontoxic cheese was released and consumed without reported illness.

Interpretation of Bacteriological Findings

Before a food can be responsible for a staphylococcal intoxication, an enterotoxigenic strain of *S. aureus* must grow in the food for sufficient time to produce an accumulation of enterotoxin. This toxin is not produced in detectable amounts until staphylococci are in the middle or late exponential or the stationary growth phase. At this time, the food is more or less uniformly contaminated with millions of coagulase-positive staphylococci per gm. If the food has been heated after the formation of enterotoxin, then either a low number of or no staphylococci will be found. Significant numbers of staphylococci may not be found if they become overgrown by other bacteria such as *Proteus* (Dolman 1943). Yet, a food may be toxic if such overgrowth occurred after enterotoxin was formed. Also, staphylococci decline as a result of exhaustion of essential nutrients, a change in the pH of a food, and other adverse conditions. Thus, the staphylococci found during laboratory isolation may be survivors of larger numbers initially present. Staining of a loopful of food homogenate and its microscopic examination or detection of nuclease may give an indication of the large numbers of staphylococci that were once present in a food. Microscopic examination, however, will not differentiate between coagulase-positive staphylococci and coagulase-negative staphylococci or between staphylococci and micrococci.

The finding of large numbers of staphylococci in a food casts suspicion on handling techniques, plant sanitation, and storage practices that have been associated with the food. If a half million or more staphylococci per gm are isolated from a food during an outbreak investigation, there is presumptive evidence that the food served was the vehicle. The isolates must be confirmed as coagulase-positive staphylococci, but, even then, not all coagulase-positive staphylococci are enterotoxigenic. The recovery of S. aureus with the same phage pattern from both an epidemiologically incriminated food and the vomitus or stools of victims is strong circumstantial evidence that the food served was the vehicle. Enterotoxin, however, must be found in a food to prove that the food was the vehicle.

It should be expected to find small numbers of staphylococci in raw foods and in foods that have been handled after heat processing. Any standard or guideline that specifies the number of staphylococci that are allowed in a food should be interpreted on the history of the food and on the number of samples that were taken from the production lot. The numbers of coagulase-positive staphylococci present in a food at any specific time is not always a valid index as to the risk involved when the food is eaten or used as an ingredient in another food, because enterotoxins accumulate during periods of staphylococcal growth and remain in the food after staphylococci are destroyed.

ECOLOGY

Ecology is the study of the interrelationships of organisms with their environment and with other organisms. An organism's environment must be considered as the sites occupied by the organism in a reservoir or host, and as the vehicles, such as food or fomites, in or on which organisms survive or multiply. Optimum combinations of environmental factors lead to rapid growth of staphylococci and toxin formation; other combinations of the factors that are less than optimum lead to slower growth or no growth. When environmental factors become more adverse to these organisms, death occurs. A review of some of the more important sources of S. aureus and the factors that affect its growth and survival follow.

Reservoirs and Sources

The incidence of S. aureus will vary with the method of sampling and the medium used, as well as with differences in sources, reservoirs, vehicles, or fomites. As the area or number of small areas swabbed is increased, or the length of time air samples are collected is increased, the number of positive samples will increase. Repeated swabbings of the same area will also increase the percentage of positive findings. Swabs planted in enrichment broths will yield a higher number of positives than swabs streaked on solid agar. These factors must be considered when interpreting results of a particular study or when comparing two or more studies. Because of variation in methods, a range of results or a few representative studies are cited to illustrate the importance of a specific source.

Human.—From the epidemiologic standpoint of human disease transmission, man is the most important reservoir of *S. aureus*.

Nose.—In man, the principal source of *S. aureus* is the nose (Williams 1963). Colonization in the nose begins the first few days of life and within two weeks nearly all infants become nasal carriers. This rate drops during the first 2 years of life but by the time a child reaches 4 to 6 years of age, the level of infection approaches the adult rate (Cunliffe 1949).

Numerous studies, too many to review in this article, have been made of the incidence of *S. aureus* in the anterior nares of general and specific population groups, so only surveys of several investigations and studies of food handlers will be mentioned. In a survey of over 80 investigations, the annual nasal carriage rate in general population groups ranged from 21.5 to 49.2% with a mean of 34%; in hospital groups the nasal carriage rate ranged from 32 to 59.2%, with a mean of 47.5% (Munch-Petersen 1961). In a study of adults selected randomly from applicants for work at food service establishemnts, *S. aureus* nasal carrier rates of 36.7% for white females, 38.4% for white males, 19.5% for Negro females, and 9.7% for Negro males were observed. The difference between males and females, age groups of 18 to 29, 30 to 49, and 50 to 69, and seasons of the year was not significant. The difference between rates for white and Negro persons was significant (Millian *et al.* 1960).

Untermann (1972) isolated coagulase-positive staphylococci from 99 (36.9%) of 268 food handlers. Enterotoxin was produced in 18.6% of the isolates of *S. aureus* that were selected for testing. Other studies of the incidence of nasopharyngeal carriers of staphylococci in persons in the hotel industry and food handling trades showed carrier rates of 19.5 and 60.5% (Denes and Rampazzo 1967; Ricciardi 1967).

Jones and Bennett (1965) found *S. aureus* in the anterior nares of 27% of 48 dairymen on 34 farms. Seven of the infected persons harbored the same phage types of staphylococci as the types that were isolated from milk from the respective farms. Smith and Crabb (1960) isolated *S. aureus* from 35 to 57% of nose and skin swabs of attendants of pigs and chickens. Tests disclosed that 32% of workers at slaughterhouses and meat, poultry, and fish processing plants had *S. aureus* in their nasal passages (Ravenholt *et al.* 1961).

The average carrier rate for *S. aureus* is in the range of 20 to 60%; this, however, does not give a complete picture of the magnitude of the number of nasal carriers. From 80 to 90% of people examined over a period of several weeks were nasal carriers on one or more occasions (Williams 1946; Roundtree and Barbour 1951). Staphylococci are often found in the nasal accessory sinuses and are frequently associated with sinusitis and postnasal drips associated with colds (Dack 1956A).

Three kinds of nasal carriers of *S. aureus* are known: persistent carriers who harbor the same phage type upon 90% or more of examinations over a period of months or years; intermittent carriers who harbor a phage type for a few weeks, then become free of it, and then once again become carriers, but usually of a different phage type; and occasional carriers who harbor staphylococci less than 10% of the time—but the staphylococci that they harbor from time to time are different phage types (Williams 1946; Gould and McKillop 1954; Roodyn 1960; McNamara *et al.* 1966; Williams 1967). Some individuals appear to be resistant to *S. aureus* carriage (Brodie *et al.* 1956, Hutchinson *et al.* 1957). Nasal carriers usually harbor only one phage type at any given occasion (Roundtree and Barbour 1951). Persons who harbor either coagulase-positive staphylococci or coagulase-negative staphylococci on the first examination tend to carry the same type on subsequent occasions (Lepper *et al.* 1955; Silberg *et al.* 1967). These observations suggest some predisposing factors in the hosts.

Throat.—Staphylococci are not found in the pharynx as commonly as in the nasopharynx. Surveys have shown that the throat is positive for *S. aureus* in from 1 to 64% of persons examined (Commission on Acute Respiratory Diseases and Plummer 1949; Martin and Whitehead 1949; Hare and Thomas 1956; Loh and Abiog 1957; Williams 1963; Denes and Rampazzo 1967; Noble *et al.* 1967; Jayakar and Bhaskaran 1968). Low percentages, however, are most frequently reported, and many investigators feel that the throat is an insignificant source of these organisms (Niven and Evans 1955; Hare and Thomas 1956). On the other hand, pathogenic conditions of the throat are sometimes associated with *S. aureus* (Campbell 1948).

Skin and Hands.—Surveys indicate that *S. aureus* is a frequent contaminant of human skin (Martin 1942; Williams 1946; Martin and Whitehead 1949; Hare and Thomas 1956; and Munch-Petersen 1963). A single swab taken from a small area of normal skin of adults has a 5 to 10% chance of yielding *S. aureus*, but this is a great underestimate of the frequency of carriage. Examination of a number of different areas of skin on one day or the same area on different days increases the proportion of skin carriers to between 50 and 80% (Williams 1965). The strains isolated from the skin are frequently the same phage types that are isolated from the nose (Williams 1946; Hare and Thomas 1956). Horwood and Minch (1951) recovered beta-hemolytic staphylococci from 29 of 30 samples of water in which food handlers had washed their hands, and Kallander (1953) isolated *S. aureus* from 30% of the hands of food handlers.

Skin bacteria may be either resident or transient flora (Price 1938). The resident flora are rather firmly attached to the skin and can multiply there. Transients are acquired from touching the nose or contacting other contaminated areas, and they are abundant on exposed skin and under nails. Resident flora are comparatively stable and may persist on the skin for weeks (Williams 1946); transients are free on the surface or loosely attached along with dirt or fat, hence they can be removed by washing. Resident flora may be located superficially or deeply. The deeply located flora do not appear in appreciable numbers until after 15 minutes of washing (Price 1951).

Staphylococci are readily dispersed in the atmosphere on small fragments of skin by rubbing hands together and by arm movements, walking, exercise, washing, scrubbing, and showering (Hare and Thomas 1956; Bethune *et al.* 1965; Noble and Davies 1965).

Perineum and Axilla.—Several investigators have reported that the perineal region frequently harbors *S. aureus* (Hare and Ridley 1958; Ridley 1959; Kay 1962; Selwyn *et al.* 1967). Staphylococci persist in this region for months, and they are dispersed in air when persons exercise (Ridley 1959). Staphylococci thrive and multiply in body areas rich in apocrine sweat glands such as the axilla and perineum (Strauss and Kligman 1956; Lotter *et al.* 1968).

Infections.—Staphylococcal infections are some of the most common bacterial diseases affecting man. General practitioners have reported that 1.5 to 9% of their patients had one or more staphylococcal lesions each year (Gould and Cruikshank 1957; Kay 1962; Nahmias *et al.* 1962). Most of these lesions were superficial, and the commonest sites were head, neck, hands, and forearms. Often infections are so trivial that people never seek medical treatment; hence a large proportion of the incidence of these infections goes unrecorded. When skin lesions on meat processing plant workers were cultured, 58% yielded *S. aureus* (Ravenholt *et al.* 1961).

Staphylococci can be transferred to an abrasion or a wound via blood from a remote lesion or the nasopharynx, by direct implantation from contaminated hands, droplets, fomites, or airborne droplet nuclei; or from contaminated skin at the site of an incision (Calia et al. 1969). Because of the presence of staphylococci on the skin, burns are readily invaded by these bacteria (Lowbury 1960). Staphylococcal lesions are a primary result of high nasal carriage rates. In most infections the same phage type of staphylococci was found in the lesion and the individual's nose (Nahmias et al. 1962; Blowers and Hodgkin 1967). Fresh wounds and healing cuts, as well as septic lesions (such as boils, pimples, impetigo, and septic wounds, cuts, and rashes) containing pus, frequently harbor S. aureus (Williams and Miles 1945; Blowers and Hodgkin 1967).

Hair.—The hair of 10 to 40% of people studied yielded S. aureus (Hare and Thomas 1956; Summers et al. 1965; and Noble 1966). A higher percentage of carriers was observed in hospitalized patients and in patients with skin diseases. Summers et al. (1965) observed that hair yielded staphylococci more often than did the nose. Phage types that were recovered from the hair were usually different from those isolated from the nose.

Feces.—Staphylococcus aureus are frequently found in the feces of infants (Duncan and Walker 1942; Martyn 1949; Buttiaux and Pierret 1949). The number of carriers was apparently reduced when they got less milk in their diets. Studies have also shown that from 5 to 40% of adults shed these bacteria in their feces (Chapman 1944; Buttiaux and Pierret 1949; Fairbrother and Southall 1950; Brodie et al. 1956; Matthias et al. 1957; Williams 1963; Agafenova and Tkrachenko 1964; and Murakami and Asakawa 1966).

Animal.—Because of the frequency of S. aureus in man and the epidemiologic association of strains of human origin with human disease, the animal reservoir is often neglected. Animals, however, are also important reservoirs of S. aureus, and they or their carcasses may serve as sources of staphylococci to man and his foods.

Bovines.—Bovines, compared with man and most other animals, are unique in the sites at which S. aureus is primarily carried. The teats and udder are the most important sources of this organism. Price et al. (1954) isolated S. aureus from the udders of 46% of over 1000 cows. Staphylococci multiply both on the udder surface and inside the udder. In a 6-year study of a dairy herd, in which hundreds of swab samples were taken from various sites on cows. more isolations and higher numbers of coagulase-positive staphylococci were obtained from teats and the udder than from other sites (Davidson 1961). Teats and udders were positive 38% and 36.2% of the time, respectively. Over 20% of the swabs from the sacral region, belly, and chest were positive; and over 10% of swabs from the poll, vagina, and caudal folds were also positive. Staphylococci were able to multiply in the perineum and possibly the vagina. Only 4.8% of the swabs from the nose yielded *S. aureus*. In a survey of two herds of cows, one extensively affected with staphylococcal mastitis, *S. aureus* was not isolated from cows' noses (Roundtree *et al.* 1956).

Staphylococcal mastitis is common in cows. Either acute or chronic mastitis can be produced when *S. aureus* is inoculated into teats (Slanetz and Bartley 1953). In several surveys of cows with mastitis, 10 to 64% of clinical cases were caused by staphylococci (Davidson 1961). Terplan and Zaadhof (1969) studied five herds and found *S. aureus* in 41% of 442 cows. This organism was also recovered from 16% of 1,763 milk samples from individual quarters of cows' udders; isolates from these samples produced enterotoxin A, B, C, or D.

From the infected quarters or teat sores, staphylococci are disseminated to hands of workers, cloths used to wash udders, teat cups in milking machines, and to the udders of other cows. Neave *et al.* (1962) observed that 50% of milkers' hands were contaminated with staphylococci before milking, but 100% became contaminated during milking. Cloths used to wash udders became contaminated with *S. aureus*, and a high percentage of these cloths retained staphylococci even after they remained in a disinfectant solution for 3 minutes (Newbould 1968). Teat cups become contaminated from contact with both skin and milk (Loken and Hoyt 1962). Newbould (1968) reported that an average number of 52,000 staphylococci were recovered from teat cup liners after they were removed from healthy cows and over 100,000 were recovered from those liners removed from cows with sores or chapped skin. A low vacuum reserve (less than 8 cu ft per minute at 15″ Hg—particularly when less than 1 cu ft per minute) was significantly associated with a high incidence of mastitis (Nyhan and Cowhig 1967).

Strains of *S. aureus* that have been isolated from cattle frequently produce beta toxin. They are also frequently lysed by phage 42D (Macdonald 1946; Loken and Hoyt 1962), but phages of group III also frequently lyse these isolates (Price *et al.* 1954; Frost 1967).

Poultry.—Staphylococcus aureus is commonly associated with infections in domestic poultry and, to a lesser extent, in wild birds. For instance, the skin and nasal sinuses of 49% of 276 chickens and turkeys from 162 farms were infected with *S. aureus*; these same tissues were infected in 14% of 122 wild birds (Harry 1967A). Phage types of *S. aureus* that were found on skin and in upper respiratory tract of poultry were similar to those from their internal lesions (Harry 1967B). Smith and Crabb (1960) isolated *S. aureus* from 95% of the noses and skin of chickens that were fed feed containing antibiotics, and from 58% of chickens that were fed feed without antibiotics. It was also isolated from 71% of buccal cavities of turkeys (Smart *et al.* 1968).

This organism is commonly found in bruised tissues of poultry (McCarthy *et al.* 1963). Healthy tissues possess a clearing mechanism that rids them of staphylococci in a short period of time, but bruised tissues stimulate and support the growth of staphylococci and allow them to persist for a long time (Hamdy and Barton 1965). *Staphylococcus aureus* was isolated from poultry feed, feathers, skin and guts of birds; fecal droppings; air sacs; and from bruised tissue (Hamdy *et al.* 1965).

Arthritis (Jungherr and Plastridge 1941), abscesses (Ravenholt *et al.* 1961), synovitis (Miner *et al.* 1968), and systemic infections in adult birds (Sahu and Munro 1969) and yolk sac infections in embryos and chicks (Harry 1967B) are associated with *S. aureus*. Birds with such infections may introduce *S. aureus* into the food chain.

Swine.—Smith and Crabb (1960) isolated *S. aureus* from 71.8% of swabs from the nose and skin of pigs fed feed containing tetracycline and from 56.8% of swabs from these sites on pigs not fed such feed. Phage types from attendants usually were identical with those isolated from the animals. Apparently most staphylococci on the skin of pigs are transient; only 8% of washed pig skin revealed *S. aureus* (Baird-Parker 1962C). These organisms are also a common cause of abscesses in swine (Ravenholt *et al.* 1961; Courter and Galton 1962). Staphylococci were isolated from the air of pigs' weaning environment and from pig feed (Hill and Kenworthy 1970). Wilssens and Vande Casteele (1967) isolated *S. aureus* from 19% of samples of feces from piglets, but not from feces from adults.

Pets.—Healthy dogs are frequent, but intermittent, nasal carriers of *S. aureus* (Mann 1959; Rajulu *et al.* 1960; Morrison *et al.* 1961; Blouse *et al.* 1964; Hajek and Marsalek 1969). Investigations have shown that from 42 to 80% of dogs are such carriers. Throat and skin swabs also frequently yield *S. aureus* (Rajulu *et al.* 1960; Morrison *et al.* 1961). Cats also harbor *S. aureus* in their noses and on their skin (Mann 1959; Morrison *et al.* 1961).

Vectors.—*Staphylococcus aureus* have been isolated from flies (Hewitt 1914; Moorehead and Weiser 1946; West 1951; Quevedo and Carranza 1966; Greenberg 1971). After houseflies were fed sugar solutions containing *S. aureus*, these bacteria survived in the digestive tract for up to 8 days and were found in the fly excreta and vomitus.

Footprints and proboscis marks also yielded these bacteria for 3 days after the flies were fed the contaminated sugar solution (Moorehead and Weiser 1946).

In a comprehensive review of the medical importance of cockroaches, Roth and Willis (1957) cited several investigations in which *S. aureus* was isolated from feces, intestinal contents, antennae, legs, or hemolymph.

Environmental.—Staphylococci reach the environment from human or animal sources, and fomites may transmit staphylococci to people or to foods. Fomites only transmit infections to people when they are heavily contaminated, and, in general, they can be considered as of relatively minor importance (Gonzaga *et al.* 1964). When *S. aureus* is subjected to natural drying, a significant number die, but some remain viable for long periods (Maltman *et al.* 1960). Although they may survive, they suffer sublethal damage, cumulative with time, and have an overall decrease in infective potential (Hinton *et al.* 1960).

Clothing of nasal carriers frequently harbors *S. aureus* (Hare and Thomas 1956). Air becomes contaminated with *S. aureus* more readily from dust liberated from clothing than from droplet nuclei created by sneezing (Duguid and Wallace 1948). Air contamination with dust-borne bacteria from clothing was reduced by only about half when a sterile, loose cotton gown was worn over clothes. Air is readily contaminated when a person dresses or undresses, engages in vigorous activity, or brushes clothes. Hare and Ridley (1958) observed that two-thirds of nasal carriers had sufficient *S. aureus* on their skin and the fronts of their clothes to render them as staphylococcal donors. They found work clothes contaminated 4.5 to 10.8% of the time and dress clothes contaminated 0.6 to 3.2% of the time.

Food preparation equipment may transfer *S. aureus* from contaminated food to previously uncontaminated foods. A slab of roast pork, for instance, was contaminated with *S. aureus* and then it and other meat was sliced by machine. This organism was isolated from the sliced meat up to the 41st slice (Gilbert 1969). *Staphylococcus aureus* was also isolated from cloths rubbed over the slicer blade. Contaminated cleaning cloths could readily spread the organism to other equipment utensils.

Major Transmission Cycle of *S. aureus*

Staphylococcus aureus passes from person to person. Staphylococci multiply readily in nasal saline secretions. These secretions flow toward the nasal orifice and eventually reach the skin of the upper lip in a liquid condition, or they remain in the nose as dry crust. From the lip and nose, staphylococci are transported by hands to other parts of the body, to clothing, or to objects nearby. Besides touching the nose, hands become contaminated by touching handkerchiefs or paper tissues that have absorbed nasal secretions, by touching other body parts (such as the perineum) or infected lesions, and, to a lesser extent, by handling animals or foods of animal origin. Hands can contaminate all that they touch—one's own face, body, clothes, or those of others, and food. Once on the skin, staphylococci may only be of a transient nature or they may become part of the resident skin flora and multiply there, or they can enter wounds or minute abrasions and cause pyogenic infections. Body movements and shaking clothing liberates skin or dust particles that carry the organisms into the atmosphere. Droplets from the nose may also contaminate clothing, and droplet nuclei expelled during coughing and sneezing carry staphylococci into the atmosphere, but these methods of egress are much less important than hand contamination (Hare and Thomas 1956). (Aerial dissemination of staphylococci is reviewed by Williams 1966.) Shaking dressings and bedding, sweeping, and vigorous activity resuspend in the air staphylococci that have settled on floors or other surfaces. Aerial transmission by air currents or picking with fingers leads to contamination of the anterior nares with small numbers of staphylococci. Those which are not mechanically removed and do not fall prey to desiccation or local antibacterial substances are probably able to multiply only if the site of implantation is not already occupied or encroached upon by some other organism (O'Grady and Wittstadt 1963). Thus, the cycle as illustrated in Figure 2.1 is completed. Compared to most other

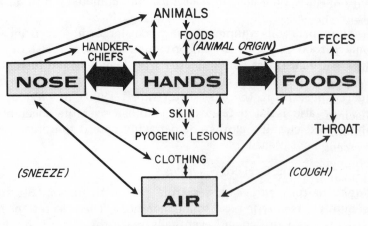

FIG. 2.1 TRANSMISSION OF *S. AUREUS*

vegetative bacteria, *S. aureus* remains viable with little or no moisture and this helps maintain this cycle (Roundtree 1963).

Factors that Affect Growth and Survival

Competition and Amensalism.—Many kinds of organisms reach a piece of food, but only a few survive and even fewer thrive. Each organism must obtain from its surroundings needed space, nutrients, water, and bound or free oxygen. These essentials for growth are not in unlimited supply. Competition—the struggle or interaction between organisms, which occurs during their growth, for the essentials that by their limited supply check growth—has a profound impact on selection, succession, and type of organisms in a contaminated food. When, because of its faster growth rate under the prevailing conditions (type of food, water activity, pH, temperature), one organism has an advantage, it will predominate and inhibit other organisms.

Some organisms, as they grow, synthesize and excrete inorganic or organic compounds which are injurious to the development of others. The substances produced by one organism may diminish the growth rate of another by making the medium unfavorable or may be toxic for this second organism. Various bacteria have demonstrated an inhibitory effect on *S. aureus* when these bacteria are grown together in culture media or food substrates. Those organisms in which an inhibitory effect on *S. aureus* has been established are listed in Table 2.3.

Type of medium, temperature, and ratio of inhibitor to staphylococci were important factors in determining the amount of inhibition that occurred in the studies cited in Table 2.3. The pH and water activity of the medium also directly affected the growth of individual species. The closer the environment was to the optimum for an individual organism, the better the organism grew (Troller and Frazier 1963A; Peterson *et al.* 1964A, 1964B). Inhibition was also greater the farther the temperature was from the optimum for *S. aureus*; for instance, greater inhibition occurred at 5°C to 25°C than at 30°C and 37°C (Peterson *et al.* 1962A; Troller and Frazier 1963A; DiGiacinto and Frazier 1966; Kao and Frazier 1966).

Inhibition of *S. aureus* is caused by organisms, such as *Serratia marcescens, Pseudomonas,* and *Streptococcus diacetilactis,* that outcompete them for essential nutrients (Troller and Frazier 1963B; Iandolo *et al.* 1965). Lactic acid bacteria inhibit staphylococci by the acid they produce (Kao and Frazier 1966) or by the production of hydrogen peroxide (Dahiya and Speck 1968). Other organisms, such as *Proteus vulgaris, Escherichia coli, Enterobacter (Aerobacter)*

TABLE 2.3
BACTERIA THAT HAVE AN INHIBITORY EFFECT ON *S. AUREUS*

Inhibitory Organisms	Substrate	Temperature	References
Proteus vulgaris	Weiner sausages	22°C	Dolman (1943)
Normal flora	Ground pork	18.3°C	Miller (1955)
Lactic acid and fecal streptococcal bacteria	Milk	25°C, 30°C, 37°C	Jones *et al.* (1957)
Escherichia coli *Streptococcus faecium* *Streptococcus faecalis* *S. faecalis* var. *liquefaciens* nisin-producing *Streptococcus* Meat lactobacilli	Meat infusion broth	15°C, 30°C, 44°C	Oberhofer and Frazier (1961)
Normal bacteria population	Frozen pot pies	35°C	Dack and Lippitz (1962)
Saprophytic and psychrophilic bacteria	Frozen chicken pot pies, frozen macaroni, frozen cheese dinners	defrosted at 5°C, 20°C, 37°C	Peterson *et al.* (1962A)
Saprophytic and psychrophilic bacteria	Culture media	0°C, 5°C, 10°C, 20°C, 30°C, 37°C	Peterson *et al.* (1962B, 1964A, 1964B, 1964C)
Saprophytic bacteria	Crab meat	12°C, 22°C	Stabyj *et al.* (1965)
Aerobacter aerogenes *Serratia marcescens* *Bacillus cereus* *Achromobacter* sp. *P. vulgaris* *E. coli* *Pseudomonas* sp.	Nutrient broth	15°C, 20°C, 25°C, 30°C, 37°C	Troller and Frazier (1963A, 1963B)

Organism	Substrate	Temperature	Reference
Lactic acid bacteria *Streptococcus* *Leuconostoc* *Lactobacillus* *E. coli* *Citrobacter*	APT broth	25°C	Graves and Frazier (1963)
Streptococcus diacetilactis Lactic acid bacteria	10% nonfat skim milk trypticase soy broth	30°C	Iandolo *et al.* (1965)
S. diacetilactis	Milk and cream fillings	25°C, 30°C	Radich *et al.* (1967)
Pseudomonas aeruginosa *P. fluorescens* *Pseudomonas* sp. *Micrococcus flavus* *M. freudenreichii* *P. vulgaris* *P. mirabilis* *Salmonella enteritidis* *Serratia pymuthicum* *S. faecalis* *S. faecium* *A. aerogenes* *E. coli* *S. marcescens*	Beef and pork	25°C, 35°C	McCoy and Faber (1966)
Meat lactobacilli *Leuconostoc dextranicum* *S. faecalis* var. *liquefaciens*	Trypticase soy broth	10°C, 15°C, 20°C, 25°C, 30°C, 37°C	Kao and Frazier (1966)
Pseudomonas	Trypticase soy broth	15°C, 18°C, 22°C, 25°C, 30°C	Seminiano and Frazier (1966)
Proteus Coliform	Trypticase soy broth	15°C, 22°C, 30°C, 42°C	DiGiacinto and Frazier (1966)
Lactobacillus lactis *Lactobacillus bulgaricus*	Trypticase soy broth	32°C	Dahiya and Speck (1968)

TABLE 2.3 (Continued)

Inhibitory Organisms	Substrate	Temperature	References
Pediococcus cerevisiae Streptococcus lactis	All purpose medium with tween broth	25°C, 30°C	Haines and Harmon (1973A)
Streptococci P. cervisiae	APT broth	30°C, 41°C	Haines and Harmon (1973B)
P. aeruginosa	H and K medium	37°C	Collins-Thompson et al. (1973)

aerogenes, Bacillus cereus, Achromobacter, inhibit staphylococci by antibiotic substances (Troller and Frazier 1963B).

Some organisms have a stimulatory effect on *S. aureus*. Graves and Frazier (1963) found that 56.6% of cultures isolated from foods were inhibitory; 43.4% were stimulatory. The following genera or species have been mentioned as stimulatory to the growth of *S. aureus*: *Streptococcus, Lactobacillus, Candida*, some strains of *E. coli* (Graves and Frazier 1963), and *B. cereus* (McCoy and Faber 1966); however, certain strains of most of these organisms have also been mentioned as inhibitory (Table 2.3). Kao and Frazier (1966) observed that some lactic acid bacteria were consistently stimulatory to *S. aureus* at the early stages of incubation, some stimulated staphylococcal growth only at higher temperatures, some inhibited staphylococci only at lower temperatures, some inhibited staphylococci at all temperatures that were tested, and some killed staphylococci during the later stages of incubation by means of the acid they produced. *Bacillus megaterium* and *Brevibacterium linens* were inhibited by staphylococci.

In general, suppression of staphylococcal growth also suppresses enterotoxin production. Enterotoxin was detected in milk samples containing 10^3 to 10^4 total aerobic organisms per gm at temperatures from 68° to 95°F (20°–35°C) when 10 to 10^6 *S. aureus* per ml were added, but enterotoxin could only be detected in samples of milk containing 10^6 total aerobic organisms at 95°F (35°C) when the same amount of *S. aureus* were present (Donnelly *et al.* 1968). This study shows the inhibition of both staphylococcal growth and enterotoxin production in the presence of natural competitive flora. McCoy and Faber (1966) observed inhibition of enterotoxin A production by *S. marcescens* and *E. coli*; Haines and Harmon (1973B) observed inhibition of enterotoxin production by streptococci and *Pediococcus cervisiae*; and Collins-Thompson (1973) observed inhibition of enterotoxin production by *Pseudomonas aeruginosa*.

In raw foods containing a mixed bacterial flora, staphylococci are poor competitors and are inhibited by metabolites that are produced when competitive bacteria grow. In low acid, protein-containing foods that have been cooked or that have above-normal levels of salt or carbohydrates, however, the situation may be reversed.

Initial Numbers.—The larger the initial contamination by *S. aureus* in a food the better its chances of surviving and multiplying to a point that it becomes the predominating organism and of producing enterotoxin. As the ratio of inhibitor to *S. aureus* was increased from 1:10 to 1:1 to 10:1 and to 100:1, the amount of inhibition was markedly increased. Even when *S. aureus* outnumbered the inhibitor by 10:1 as compared to growth of a pure culture, inhibition of at least one log cycle occurred (Troller and Frazier 1963A; DiGiacinto and Frazier 1966; Kao and Frazier 1966; and Seminiano and Frazier 1966). Enterotoxin was detected in about half the time when milk containing 10^3 to 10^4 bacteria per ml was inoculated with 10^6 *S. aureus* cells per ml as when this product was inoculated with 10^4 *S. aureus* cells per ml (Donnelly *et al.* 1968).

Phase of Growth.—If bacteria are taken from a piece of meat in which they are in the lag phase of the growth curve and inoculated onto another freshly cut surface of meat, the lag phase is long. On the other hand, if bacteria are taken from the logarithmic phase and inoculated on a meat surface, there is no apparent lag (Jensen 1945). The lag phase is greatly affected by environmental conditions. Under optimum conditions, the lag phase for *S. aureus* may be as short as 1 hour (Iandolo *et al.* 1964) or under adverse conditions, as long as several days (Angelotti *et al.* 1961A).

Before enterotoxin is produced, staphylococci must pass through the lag phase and their growth must become exponential. McLean *et al.* (1968), Markus and Silverman (1968), and Morse *et al.* (1969) observed that the maximum production of enterotoxin B occurred at the beginning of the stationary growth phase; Markus and Silverman (1969) observed that 95% of enterotoxin B appeared during the latter part of the exponential growth (logarithmic) phase. Enterotoxin A was produced during exponential growth, as well as during the stationary phase, and was directly related to cell numbers (Markus and Silverman 1970). In studies by Donnelly *et al.* (1968), 5 × 10^7 *S. aureus* cells/ml were reached before enterotoxin was detected.

Nutrition.—Nutritional requirements of *S. aureus* are somewhat complex. Minimum requirements are inorganic salts, thiamine, nicotinic acid, and up to 11 amino acids (glycine, valine, leucine, threonine, phenylalanine, tyrosine, cysteine, methionine, proline, arginine, and histidine) (Mah *et al.* 1967). In a medium containing glucose, ammonium sulfate, inorganic salts, thiamine and nicotinic acid, only arginine and cystine were required; without the ammonium ion, glycine was also required; and in a glucose-free medium up to eight amino acids were required (Peters 1966). The mineral requirements are for calcium, magnesium, and potassium (Shooter and Wyatt 1955, 1956). Biotin is needed when glutamic acid, but not glucose, is used as a carbon source (Gretler *et al.* 1954; Mah *et al.* 1967).

Enterotoxin is readily produced when enterotoxigenic strains of *S. aureus* are grown in meat infusion broths, such as brain heart infusion (Casman and Bennett 1963), or in protein or casein hydrolysates supplemented with nicotinic acid and thiamin (Segalove 1947; Kato *et al.* 1966; Reiser and Weiss 1969).

Temperature.—Optimal temperature for growth of *S. aureus* and the production of enterotoxin is around 98.6° F (37°C). At this temperature, Iandolo *et al.* (1964) observed a lag of 1 hour and a 19-minute generation time in a medium containing 0.5% NaCl. Higher yields of staphylococci were also obtained at 98.6° F (37°C) incubation, as compared to 59.9° F (15.5°C), 80.6° F (27°C), and 113° F (45°C).

The temperature extremes that permit growth vary with the food substrate and environmental factors. Angelotti *et al.* (1961A) found that an inoculum of *S. aureus* (10^7 per gm) would grow in custard at temperatures as low as 46° F (7.8°C) and as high as 114° F (45.6°C) and in chicken-a-la-king at temperatures as low as 44° F (6.7°C) and as high as 112° F (44.4°C). In ham salad, growth did not occur below 50° F (10°C) or above 112° F (44.4°C). At the extreme cold end of the temperature range that permitted growth of *S. aureus*, a lag of 4 to 5 days occurred before growth commenced. Growth at 50° F (10°C) but not at 41° F (5°C) was reported in chicken gravy (Gunderson 1960). Williams-Smith (1957) reported growth of *S. aureus* in raw milk at 61.9° F (16.5°C) after a 24-hour lag; at 86° F (30°C) to 109.4° F (43°C) growth occurred after an 8-hour lag, but no growth was reported at 114.8° F (46°C). Clark and Nelson (1961) observed an increase of *S. aureus* up to a thousandfold when raw milk was stored at 50° F (10°C) for 7 days. An actively growing staphylococcal culture failed to grow or produce enterotoxin when it was inoculated into meat and held at 43° F (6.1°C) for as long as 2 weeks (Gross and Vinton 1947). In a study of competitive growth, Peterson *et al.* (1962A, 1962B, 1964A, 1964B, 1964C) did not observe growth of *S. aureus* at 50° F (10°C) in thawed frozen foods, culture media, or media that was adjusted to various pH values, various salt concentrations, or different types and concentrations of sugars, starch, whole egg, and corn oil.

Enterotoxin production occurs at a similar but somewhat more restricted optimum and range. Segalove and Dack (1941) found that enterotoxin was produced in veal infusion agar in 12 hours at 98.6° F (37°C) and in 3 days at 64.4° F (18°C) but not at 48.2° F (9°C) in 7 days or at 59° F (15°C) in 3 days. Enterotoxin was not produced at 39.2° F (4°C) or at 44.1° F (6.7°C). Enterotoxin A was detected in 6 to 9 hours at 95°F (35°C) incubation, 9 to 12 hours at 86°F (30°C), 18 hours 77°F (25°C), and in 36 hours at 68°F (20°C) when raw and pasteurized milks containing 10^3 to 10^4 bacteria per ml were inoculated with 10^6 S. aureus per ml (Donnelly et al. 1968). The optimal temperature for production of enterotoxin B and C was found to be 104°F (40°C) by Vanderbosch et al. (1973) and 98.6° (37°C) by Dietrich et al. (1972). Genigeorgis et al. (1969) observed enterotoxin B production in hams that were held at 50°F (10°C) for 2 weeks longer. Enterotoxin production was greater at 86°F (30°C) than at 68°F (20°C) and greater at this temperature than at 50°F (10°C). Tatini (1973) observed that enterotoxins A, B, C, and D occurred in cooked ground beef, ham, and bologna when incubated aerobically at 50°F (10°C). The upper limit for enterotoxin production is between 7.2 to 7.8°F (45-46°C) (Tatini et al. 1971; Scheusner et al. 1973).

Water Activity.—Bacteria, as all forms of life, require water for growth. Water in a food or medium that is available to microorganisms for growth is known as water activity (a_w). The water activity value is the point where the relative humidity of an atmosphere around a food neither gains nor loses water. Thus, the two humidities are in equilibrium, and the water vapor pressure of the food is the same as the atmosphere. The water vapor pressure of a food expressed as a ratio of the vapor pressure of pure water at the same temperature is numerically equal to the water activity. When the a_w of the solution is altered, the relative concentrations of water and of solutes also change. The concept of water activity is reviewed in greater detail by Scott (1957).

Staphylococcus aureus can grow at lower water activities or in higher solute concentrations than most bacteria. Aerobic growth has been reported from a_w 0.999 to 0.86 (Scott 1953). Anaerobic growth produces lower yields of cells and probably does not occur below a water activity value of 0.90. Scott (1953) reported that a reduction of the a_w below the optimum of 0.99 caused a progressive decrease in both the rate of growth and the maximum yield of cells. In brain heart infusion medium, the lag phase for staphylococci increases from 1 hour at 0.995 a_w to 3 to 6 hours at 0.96 a_w, and to 3 days at 0.90 a_w. In casamino acid yeast-extract casitone medium, the lag periods were much shorter; for instance, less than 24 hours at 0.90 a_w (Scott 1953).

Iandolo et al. (1964) found that at or near optimum pH and temperature a lag of only 1 hour occurred in a medium with only 0.5% NaCl; a lag of 3 to 4 hours was observed in a medium containing 4.6% NaCl; and a 5 to 6 hour lag occurred in a medium containing 8% NaCl. Genigeorgis and Sadler (1966C) also reported rapid growth in broth without salt and slow growth in media at or above 6% NaCl. Parfentjev and Catelli (1964) reported that *S. aureus* grew in tryptose phosphate broth that was saturated with NaCl. Distilled water is detrimental to *S. aureus* (Lechowich *et al.* 1956; Parfentjev and Catelli 1964).

The a_w growth range stated by Scott (1953), Hucker and Haynes (1937), Segalove and Dack (1951), and Genigeorgis and Sadler (1966C) agree. Nunheimer and Fabian (1940) found that in 50 to 60% sucrose and in 35 to 45% glucose *S. aureus* growth was inhibited. Disagreements on the upper salt concentration limit that permits growth can be explained as a result of the interaction of different environmental factors (such as nutrients, pH, and temperature) and of differences in strains tested. Labuza *et al.* (1972) reported growth in baby food pork when the a_w was depressed to 0.84 to 0.75 by glycerol. They concluded that a_w, total water content, and chemical structure of the food ingredients must be considered when evaluating bacterial growth in foods.

Staphylococci survive normal curing operations for ham until the hams are smoked (Lechowich *et al.* 1956; Buttiaux and Moriamez 1958; Silliker *et al.* 1962; Patterson 1963). Staphylococci survived in bacon curing brine for 80 days at 33.8°F (1°C), 41°F (5°C), and 50°F (10°C) (Eddy and Ingram 1962). These organisms also survived but did not multiply in salted herring for 4 months at 50 to 57°F (10 to 14°C) (Hempel and Maleszewski 1968).

Increased concentrations of salt, even relatively low concentrations—as in cured meats—reduce enterotoxin production more than they reduce cell growth (Genigeorgis and Sadler 1966C; McLean *et al.* 1968; Hojvat and Jackson 1969; Genigeorgis and Prucha 1971; Troller 1971). Enterotoxin B production was inhibited in broths containing 4% and 8% NaCl at six temperatures from 39.2 to 95°F (4 to 35°C) and in broth containing 12% NaCl at all temperatures tested (Hojvat and Jackson 1969). Enterotoxin B production was inhibited in two media at a_w levels below 0.97 (Troller 1971). The minimum a_w value permitting enterotoxin A production was 0.90 (Troller 1972).

Oxidation-Reduction.—Although *S. aureus* is facultative, the rate and quantity of growth and enterotoxin produced under aerobic conditions are much higher than that produced anaerobically (Genigeorgis *et al.* 1969, 1971; Barber and Deibel 1972). In aerated brain heart infusion broth, 65 µg/ml enterotoxin B was produced in less than 24 hours at 98.6°F (37°C). In static cultures at the same temperature, 10 days were required to produce half this amount

(McLean *et al.* 1968). Vacuum packaging markedly inhibited the growth and total numbers of *S. aureus* on sliced hams (Christiansen and Foster 1965). Under the same incubation conditions, less enterotoxin was produced in vacuum packaged products than when the same products were stored aerobically (Thatcher *et al.* 1962). But in foods, such as bacon, vacuum packaging had a greater inhibiting effect on spoilage organisms than on staphylococci and allowed enterotoxin production to occur without spoilage.

pH.—Optimum pH for growth of *S. aureus* is around 7 (Iandolo *et al.* 1964). The pH extremes that permit growth will vary with several environmental factors: type of food, type of acid, competition, oxygen-reduction potential, and water activity. Growth in cheese-cake fillings was gradually inhibited by addition of acetic acid to pH 5.2, and completely stopped at pH 4.67 (Cathcart *et al.* 1947). Growth at pH 6, but not at pH 5.2, was observed while staphylococci were grown in competition with saprophilic bacteria (Peterson *et al.* 1964A). Growth of *S. aureus* occurred at pH 5 when all other conditions were optimum, but there was a lag of 6 hours and a generation time of 30 minutes (Iandolo *et al.* 1964). Lechowich *et al.* (1956) showed that anaerobiosis affects the minimum pH for *S. aureus* to initiate growth; under aerobic conditions, growth diminished at pH 5.6 and ceased at pH 5; under anaerobic conditions, growth ceased at pH 5.6. Growth has been observed at pH 8 but not at pH 9 (Peterson *et al.* 1964A; and Iandolo *et al.* 1964).

In addition to pH, organic acids have bacteriostatic action on *S. aureus* (Ryberg and Cathcart 1942; Cathcart *et al.* 1947). Growth of *S. aureus* was effectively inhibited by lemon, orange, pineapple, apricot, strawberry, peach, and raspberry fruit fillings (Ryberg and Cathcart 1942; Cathcart *et al.* 1947). Ingredients such as milk, however, may have a buffering action on these natural acids.

As the pH decreases from the lowest value that permits growth or increases from the highest value that permits growth, bacteriostatic effects occur; these are followed by bactericidal effects (Hewitt 1957). In laboratory-made mayonnaise at pH 5, *S. aureus* survived for 144 hours; but in commercial mayonnaise with a pH 3.8, it survived for only 96 hours. In laboratory-made salad dressing at pH 4.4, *S. aureus* survived for 120 to 168 hours; but in commercial salad dressing with a pH 3.2, it survived for only 30 hours (Wethington and Fabian 1950). In salad dressing containing dried egg, *S. aureus* survived for over 24 hours at pH 3.95 to 4.56, but it died off between 4 and 24 hours at pH 3.4 (Kintner and Mangel 1953A). In meat-curing pickle (60 salometer brine) with sucrose, $NaNO_2$ and

$NaNO_3$, *S. aureus* died within 72 hours, but it was protected when meat juices were added to the pickle medium (Lechowich *et al.* 1956).

The pH range permitting enterotoxin production is narrower than that for *S. aureus*. In experiments on the effect of pH on the production of enterotoxin A, a maximum yield of toxin was produced in brain heart infusion agar at an initial pH of 5.3 to 5.5 (Casman and Bennett 1963) and in casein hydrolysates at pH 6, 6.5, and 7 (Kato *et al.* 1966). Markus and Silverman (1969) reported that the optimal pH for enterotoxin B production in a nitrogen-free medium was 8 to 8.5, and in a medium containing nitrogen was 7 to 7.5. Optimal production of enterotoxin A was at a pH more likely to be associated with foods, 6.5 to 7.0 (Markus and Silverman 1970). Both the growth of *S. aureus* and the production of enterotoxin B is better when the pH is near optimum and when salt concentration is low. For instance, good growth occurred at pH 6.9 with 16% NaCl, but no staphylococci survived 10 days at this salt concentration when the pH was 5.1 (Genigeorgis and Sadler 1966C). This pH (5.1) was also bactericidal at 12% NaCl. Enterotoxin B was produced in broth at pH 6.9 containing up to 10% NaCl and in the same broth at pH 5.1 containing up to 4% NaCl. High concentrations of NaCl and extreme pH values delay or prevent growth and diminish the total numbers of cells and the amount of enterotoxin produced. Furthermore, small concentrations of NaCl can inhibit staphylococcal growth at pH values remote from the optimum (Genigeorgis *et al.* 1971A).

Humidity.—Relative humidity affects the survival of *S. aureus*. On silk sutures at temperatures of 98.6°F (37°C), 86°F (30°C), and 68°F (20°C), McDade and Hall (1963A) found that survival of *S. aureus* was greatest at relative humidities of 95 to 98%. Multiplication occurred at these relative humidities, probably due to a carryover of media. Other than at these high relative humidity values, the death rate increased with increased relative humidity. On glass, ceramic tile, asphalt tile, rubber tile, polished stainless steel, and silk suture surfaces that are commonly found in hospitals or in food processing and service establishments, survival was best at 77°F (25°C) with relative humidities of 11% and 33% (McDade and Hall 1963B). With an inoculum of 10^2 cells, die-off was progressive and accelerated at 77°F (25°C) with relative humidities of 53% or 85%. There were frequently no organisms recovered after 4 to 7 days at 77°F (25°C) with relative humidities that were greater than 50%. On fabrics, *S. aureus* persisted longer at relative humidity of 35% than at 78% (Wilkoff *et al.* 1969). The ability to survive on surfaces gives *S.*

aureus a selective advantage over other organisms. Thus, contaminated surfaces can constitute a source of these organisms for several days after initial contamination.

Freezing.—Freezing and frozen storage can be detrimental to vegetative bacteria. The rapidity of freezing, temperatures of freezing and frozen storage, length of frozen storage, and type of food substrate influence survival of bacteria. Compared with other non-spore-forming organisms, *S. aureus* is relatively resistant to freezing (Haines 1938; Ahn *et al.* 1964). Experiments have shown that this organism survives freezing, frozen storage, and thawing quite well in chicken-a-la-king, ham-a-la-king, creamed tuna, and creamed salmon at $-0.4°$ F $(-18°$ C) (Phillips and Proctor 1947), in beef at $15.8°$ F $(-9°$ C) and $0°$ F $(-17.8°$ C) (Hartsell 1951), in ground pork at 0 to $-8°$ F $(-17.8$ to $22.2°$ C) (Miller 1955), in corn syrup, rice flour and egg white at $12.2°$ F $(-11°$ C), $-5.8°$ F $(-21°$ C), $-22°$ F $(-30°$ C) (Woodburn and Strong 1960). Higher rates of destruction of *S. aureus* occurred during frozen storage of strawberries at $-4°$ F $(-18°$ C) (McCleskey and Christopher 1941), peas at $15.8°$ F $(-9°$ C) and at $0°$ F $(-17.8°$ C) (Hartsell 1951), phosphate buffer at $12.2°$ F $(-11°$ C), $-5.8°$ F $(-21°$ C), $-22°$ F $(-30°$ C) (Woodburn and Strong 1960), and turkey at $-20.2°$ F $(-29°$ C) (Kraft *et al.* 1963).

Numbers of staphylococci generally decrease at a fairly rapid rate during freezing and during the early frozen storage period. Declines in populations after a month of frozen storage are slow, but progressive. The colder the freezing and storage temperatures the better survival is, generally. Haines (1938) observed greater mortality of *S. aureus* when substrates were stored at $30.2°$ F $(-1°$ C) or at $28.4°$ F $(-2°$ C) than when stored at $14°$ F $(-10°$ C) or at $4°$ F $(-20°$ C); Woodburn and Strong (1960) observed greater mortality at $12.2°$ F $(-11°$ C) and at $5°$ F $(-15°$ C) than at $-22°$ F $(-30°$ C). Nutritional requirements are also altered after freezing and prolonged frozen storage (Hartsell 1951).

Heat.—In relation to other vegetative bacteria, *S. aureus* has higher-than-average heat resistance. Its greater heat resistance might be attributed to its tendency to form clumps of numerous individual cells. Theoretically, during exposure to lethal temperatures, the survivor curve for such clumps of organisms should follow a pattern of an initial lag and become exponential only when essentially all clumps have been reduced to only one viable cell per clump (El-Bisi and Ordal 1956). Some studies, however, have shown curves with a different appearance. Walker and Harmon (1966), for instance, observed exponential death of *S. aureus* in milk, whey, and phosphate buffer until about 99.99 to 99.999% of the cells were destroyed; then, a decline in the rate of destruction occurred.

Various investigators have reported conflicting results on the heat resistance of *S. aureus*. Resistance will vary with the strain (Angelloti *et al.* 1960; Thomas *et al.* 1966), ingredients (Kadan *et al.* 1963), type of heat (dry or wet) Iwaswa and Ishihara 1967, pH, age of culture (Gross and Vinton 1947; Walker and Harmon 1966), temperature (Gross and Vinton 1947; Bhatt and Bennett 1964), and number of cells initially heated (Gross and Vinton 1947). Investigations indicate that survival is favored by more viscous substrates, sugar concentrates exceeding 14%, serum solids above 9%, dry heat, old cultures, low temperatures, and initially high numbers of cells. The recovery medium also influences the results of heat resistance studies. Because sublethally-heated organisms are weakened, they have a longer lag and are inhibited by selective media (Jackson and Woodbine 1963; Busta and Jezeski 1963; Iandolo and Ordal 1966).

Variation in heat resistance is shown by several studies. Over 3 million *S. aureus* cells survived 13.8 minutes but were destroyed at 18.8 minutes at 149°F (65°C) in broth at pH 7 (Beamer and Tanner 1939). One million of these organisms were destroyed in buffer at pH 7 in 2 minutes at 150°F (65.6°C) (Stritar 1941). Bhatt and Bennett (1964) found that 1.5% of 171 strains of *S. aureus* survived 30 minutes, but not 45 minutes, at 143°F (61.7°C) in milk; 0.38% survived 15 seconds, but not 35 seconds, at 161°F (71.7°C). Other studies have shown that *S. aureus* did not survive in milk heated for 10 minutes at 143°F (61.7°C) (Heinemann 1957; Williams-Smith 1957). Recent investigations have recorded more valuable data (Z, D, and F values) that can be used in calculating or interpreting thermal processes. Highlights of these studies are listed in Table 2.4.

Chemicals.—Chemicals affect *S. aureus* in several ways.

Antibiotics.—After the introduction of penicillin into medical practice, reports of penicillin-resistant *S. aureus* appeared. Similar reports were made for other antibiotics as they were brought into use. Resistance to different antibiotics is caused by blockages of metabolic pathways which are genetically controlled. Recently there have been reports of trends toward increasing sensitivity to many antibiotics and a decrease in the incidence of multiple-resistant strains (Bulger and Sherris (1968).

Nitrites and Nitrates.—The antimicrobial effect of nitrite is due to the formation of undissociated nitrous acid; nitrates have no effect other than after they are reduced to nitrites. A concentration of 0.02% (200 ppm $NaNO_2$) retarded growth of *S. aureus*, but after a prolonged lag it was able to multiply and reach the same total count as controls (Tarr 1942). A concentration of 10% $NaNo_2$ reduced *S. aureus* counts by 4 logs after 30 minutes exposure and killed 10^9 cells within 6 hours (Tarr 1942). Growth of staphylococci was

TABLE 2.4
OBSERVED AND CALCULATED Z, D, AND F VALUES FOR S. AUREUS
IN VARIOUS LABORATORY MEDIA AND FOOD SUBSTRATES

Food	Strain	Number of Cells	Z^1	D^2	F^3	Reference
Beef Bouillon	196E	10^{7-8}	10	$D_{140} = 2.2–2.6$		Thomas *et al.* 1966
	MS149	10^{7-8}	10	$D_{140} = 2.2–2.6$		
Cheese Whey	161-C	10^4		$D_{140} = 1.3$		Walker and Harmon 1966
	S-1	10^4		$D_{131} = 1.8$		
Chicken a-la king	196E	10^7	10.5	$D_{140} = 5.37$	$F_{140} = 47$ $F_{145} = 15.5$ $F_{150} = 5.2$	Angelotti *et al.* 1960
	MS 149	10^7	9.3	$D_{140} = 5.17$	$F_{140} = 40$ $F_{145} = 11.5$ $F_{150} = 3.4$	
Custard	196E	10^7	10.5	$D_{140} = 7.82$	$F_{140} = 59$ $F_{145} = 19.5$ $F_{150} = 6.6$	
	MS149	10^7	9.95	$D_{140} = 7.68$	$F_{140} = 53$ $F_{145} = 16.5$ $F_{150} = 5.2$	
Green pea soup	196E	10^{7-8}	10	$D_{140} = 6.7–6.9$		Thomas *et al.* 1966
	MS149	10^{7-8}	10	$D_{140} = 7.9–10.4$		
Meat, sterilized		10^4	9		$F_{150} = 6.2–13$	Gross and Vinton 1947
Milk, skim	196E	10^{7-8}	10	$D_{140} = 3.1–3.4$		Thomas *et al.* 1966
	MS149	10^{7-8}	10	$D_{140} = 3.3$		
	161-C	10^{7-8}	10	$D_{140} = 1.3$		Walker and Harmon 1966
	S-1	10^{7-8}	10	$D_{140} = 1.6$		
			9.2	$D_{140} = 3.7–6.5$	$F_{140} = 24$ $F_{145} = 6.8$ $F_{150} = 1.9$	Heinemann 1957

Substrate	Strain	No. of cells	Z	D	Reference
Milk, whole	MF31		12.2–13.0	$D_{145} = 1.0-1.7$	Walker and Harmon 1966
	161-C	10^4		$D_{140} = 0.8$	
	S-1	10^4		$D_{131} = 1.3$	
	B-120	10^4		$D_{131} = 0.7$	
	S-18	10^4		$D_{131} = 0.5$	
		10^8	5.11	$D_{140} = 59$	Amin and Olson 1967
				$D_{171} = 0.25$	Evans et al. 1970
				$D_{180} = 0.7$	
				$D_{143} = 14.5$	Bhatt and Bennett 1964
				$D_{161} = 0.1$	
0.5% NaCl	196E	10^{7-8}	10	$D_{140} = 2.2-2.5$	Thomas et al. 1966
	MS149	10^{7-8}	10	$D_{140} = 2.0$	
Phosphate buffer	161C	10^4		$D_{140} = 0.6$	Walker and Harmon 1966
	S-1	10^4		$D_{131} = 1.1$	
	B-120	10^4		$D_{131} = 1.0$	
	S-18			$D_{140} = 0.4$	
	MF31		14.5–19.8	$D_{140} = 0.5-2.5$	Stiles and Witter 1965
				$D_{145} = 0.4-1.1$	
Turkey stuffing		10^7	12.3	$D_{140} = 15.4$	Webster and Esselen 1956
				$D_{150} = 2.45$	
				$D_{160} = 0.38$	
				$D_{165} = 0.15$	

[1] Z = Degrees Fahrenheit required for the thermal destruction curve to traverse one log cycle. Mathematically, equal to the reciprocal of the slope of the thermal destruction curve.

[2] D = The time required at any temperature to destroy 90% of the spores or vegetative cells of a given organism. Numerically, equal to the number of minutes required for the survivor curve to traverse one log cycle. Mathematically, equal to the reciprocal of the slope of the survivor curve.

[3] F = The equivalent in minutes at a particular temperature ($140°F = F_{140}$), of all heat considered with respect to its capacity to destroy spores or vegetative cells of a particular organism.

reported in broth containing up to 500 ppm nitrites and in broth containing 2500 ppm nitrate (Lechowich *et al*. 1956). Neither NaNO$_3$ in concentrations up to 1000 ppm or NaNO$_2$ at levels up to 200 ppm appears to have affected growth of staphylococci or enterotoxin A or B production under aerobic conditions (McLean *et al*. 1968; Markus and Silverman 1970). Under anaerobic conditions, lower than optimal pH causes greater inhibition of enterotoxin production (Genigeorgis *et al*. 1971B; Genigeorgis and Prucha 1971; Troller 1972; Tompkin *et al*. 1973).

Disinfectants.—Numerous disinfectants, such as chlorine and chlorine compounds (Dychdala 1968), iodine (Gershenfeld 1968), quarternary ammonium compounds (Lawrence 1968), alcohols (Morton 1968), and phenolic compounds (Prindle and Wright 1968), are effective *in vitro* against *S. aureus*. The effectiveness is sometimes indicated by the phenol coefficient. The phenol coefficient is designed to determine the highest dilution of a germicidal compound as compared to phenol which will kill *S. aureus*, or other test organisms, in 10 minutes but not in 5 minutes under specified conditions. The concentration of germicide that is usually used is 20 times the phenol coefficient, or equivalent to a 5% phenol solution, but a use dilution method should be used to confirm the results (Bass and Stuart 1968; Shaffer and Stuart 1968).

Mallmann and Schalm (1932) observed that *S. aureus* was killed faster with increasing concentrations of hypochlorite solutions (at constant pH of 9). At 2 ppm available chlorine, complete kill occurred in 5 minutes; at 1.2 ppm, complete kill occurred in 10 minutes; and at 0.3 ppm, complete kill did not occur in 30 minutes. A solution containing 200 ppm available chlorine was about seven times as rapid as one with 25 ppm available chlorine in killing 99% of *S. aureus* on metal trays with dried milk on the surface (Neave and Hoy 1947). Lasmanis and Spencer (1953), however, were not able to obtain complete kill of staphylococci when 3% skim milk was added to hypochlorite solutions; smaller amounts of milk exhibited progressively lesser effect on the bactericidal action. The addition of 0.2% whole milk to a hypochlorite solution containing 200 ppm available chlorine did not significantly reduce its efficiency within 30 minutes (Neave and Hoy 1947).

THE DISEASE

Staphylococcal Intoxication

The initial symptoms of staphylococcal intoxication are excessive salivation and nausea. These are followed by a sudden onset of vomiting which is the predominant and the most severe symptom.

Vomiting usually occurs 2 to 4 hours after a food that contains staphylococcal enterotoxin is ingested; however, it may occur in as short a time as 1 hour or as long as 8 hours. Several episodes of vomiting occur. In an early human volunteer study, for instance, McBurney (1933) recorded 3 to 14 spells of vomiting per person, and an average of 9. These occurred at 10- to 20-minute intervals and persisted for 1 to 4 hours. The initial vomitus consists chiefly of food, but that which is expelled during later episodes consists of mucus containing blood and bile (Dolman 1943). Vomiting is accompanied, and sometimes replaced, by retching. The site of the emetic action of enterotoxin is the abdominal viscera. From this site, a stimulus is passed along the vagus and sympathetic nerves to the vomiting center of the brain (Sugiyama et al. 1961; Sygiyama and Hayama 1965). This stimulus triggers the emetic response.

Abdominal cramps and diarrhea are often very marked. Diarrhea, consisting of mucoid-watery fecal excretions, will often occur at the same time as the vomiting or within a few minutes to an hour after the onset of vomiting; usually it follows the initial spell of vomiting by about 15 minutes. Two to thirteen and an average of five episodes of diarrhea occurred in the volunteers studied by McBurney (1933). These episodes persisted for 2 to 7 hours. The mechanisms of enterotoxin-induced diarrhea have been studied by Hayama and Sugiyama (1964) and by Shemano et al. (1967).

Other symptoms observed in patients during outbreaks, or in human volunteers, are sweating, cold and clammy skin, tetanic muscular contractions—especially in flexors of legs—collapse while attempting to walk, weakness, lassitude, depression, anorexia, shock, and dehydration (Dack et al. 1930; McBurney 1933; Dolman 1934, 1943; Denison 1936). Reports on temperature are contradictory, but temperature generally is described as normal or subnormal. In severe attacks, respiration is shallow, pulse rapid, and blood pressure has been reported to drop to 60 systolic and 40 diastolic (Denison 1936). An increase in white blood count also has been observed.

McBurney (1933) observed that volunteers who ingested 10 ml of a filtrate from a 48-hour broth culture did not experience relief until 3 to 10 hours had elapsed and did not completely recover for 5 to 48 hours. The volunteers commented that, if death was necessary for relief, they were ready to die. Complaints that followed recovery were general weakness, soreness of abdominal muscles, slight headache, and hunger. A classical acount of staphylococcal intoxication is given in the review on food poisoning by Dolman (1943). Research on the mode of action of enterotoxin is reviewed by Bergdoll (1970).

Mortality from staphylococcal intoxication is low, but there are a few fatal cases on record (Denys 1894; Blackman 1935; Fanning

1935; Kodama *et al.* 1940; Dorling 1942; Weed *et al.* 1943; Meyer 1953; Elek 1959; Kienitz 1964; Asante 1969; Currier *et al.* 1973). Most of the deaths have occurred in children or the aged. On the other end of the spectrum of intoxication, mild attacks of just nausea or belching without vomiting or diarrhea also occur.

Information on toxicity of enterotoxins is meager, but it is known that very small amounts of enterotoxin can cause illness. In human volunteer studies, Dack *et al.* (1931) found that as little as 0.5 ml of crude filtrate caused symptoms in one individual, but 23 ml failed to cause symptoms in another individual. A dose of 20 to 25 μgm of purified enterotoxin B produced typical symptoms in two adults (Raj and Bergdoll 1969). Even smaller amounts may have made the subjects ill, and other enterotoxins are more potent than enterotoxin B. From estimating the amount of enterotoxin that was in cheese that had been ingested by human volunteers, Bergdoll (1970) suggested that less than 1 μgm of enterotoxin could possibly cause illness in sensitive individuals. Foods from various outbreaks contained from 0.01 to 0.9 μgm/gm (Casman and Bennett 1965; Gilbert *et al.* 1973). If 100 gm of food were ingested, the dose would be 1 to 9 μgm.

Enterocolitis

Antibiotic-resistant strains of *S. aureus* that became the predominant organisms in the intestinal tract—as a result of antibiotic therapy which accompanied abdominal surgery—have produced mild to extremely severe gastrointestinal and systemic symptoms. Mild to severe diarrhea, excessive fatigue, exhaustion, fever, anorexia, nausea, vomiting, distention, and varying degrees of shock occur in such patients (Dearing and Heilman 1953; Dearing 1956; Hallander and Korlof 1967). These symptoms are attributed to *in vivo* enterotoxin production. A frequently fatal form of postoperative intestinal illness is known as pseudomembranous enterocolitis. Its course is characterized by fever (101 to 107°F; 38.3 to 41.7°C), frequent copious water discharges, dehydration, sudden profound shock, and death within a few days. Autopsy shows that the jejunum and ileum is lined by a continuous yellow-tan fibrinous pseudodiphtheritic cast which can be easily lifted intact from the bowel lumen. Cultures from the membrane grow *S. aureus* that are resistant to the antibiotics used in therapy (Brown *et al.* 1953; Speare 1954; Dack 1956B; Dearing *et al.* 1960). A scarlatiniform rash has also been described in patients with staphylococcal gastroenteritis (Nagaki *et al.* 1955).

Treatment

Treatment for persons with staphylococcal intoxication is symptomatic. Emetics and purgatives are not necessary because vomiting and diarrhea are so severe that most of the enterotoxin is eliminated. Replacement of salt loss and fluids by oral, when vomiting has ceased, or parenteral administration of saline solution or intravenous administration of hypertonic glucose may be required in cases of shock or dehydration. Antibiotics are not effective in treating the intoxication (Ager and Top 1968). Details on treating persons with food poisoning are given by Wenzel (1974).

EPIDEMIOLOGY

Descriptive Statistics

Incidence.—Staphylococcal intoxication has been an important foodborne disease in the United States for many years. It was once thought to be the most common cause of foodborne illness. In the 1940s and 1950s, it accounted for over 50% of the reported foodborne outbreaks of known etiology (Dauer 1952, 1953, 1958, 1961; Dauer and Sylvester 1954, 1955, 1956, 1957; Dauer and Davids 1959, 1960). In the years 1945–1947, for instance, Feig (1950) attributed 80% of the reported foodborne outbreaks to staphylococcal intoxication. In the 1960s, its incidence dropped to about 35% of the reported foodborne outbreaks of known etiology. So far in the 1970s, it has accounted for 25% of the confirmed foodborne outbreaks (U.S. Department of Health, Education, and Welfare 1967-1974). The decrease in percent of outbreaks is not entirely attributable to improved procedures in handling food although there were some notable strides in this area, such as increased use of refrigerated storage and display of cream-filled pastry. Interest in, and improved reporting of, *Salmonella* and *Clostridium perfringens* foodborne illnesses were undoubtedly responsible for at least some of the change. Case rates have shown little significant decrease. Table 2.5 lists reported cases, outbreaks, and rates of staphylococcal intoxication by year from 1938, the time this disease was first officially reported, through 1971. Outbreaks that occurred before 1938 were summarized by Stone (1943).

During the last decade, most outbreaks occurred in August and fewest in January and February, but there was no definite seasonal pattern. Outbreaks occur throughout the year.

Size of Outbreaks.—Outbreaks of staphylococal intoxication usually involve only a few persons (Table 2.6). During a 6-year

TABLE 2.5
REPORTS OF OUTBREAKS AND CASES OF
STAPHYLOCOCCAL INTOXICATION IN THE UNITED STATES
1938 TO 1973

Year	Reported Outbreaks Number	Percent[1]	Reported Cases Number	Percent[1]	Rate per 100,000
1938	11	14	393	18	0.3
1939	28	18	487	13	0.4
1940	–[2]	–	–	–	–
1941	74	31	1982	15	1.5
1942	61	24	2558	22	1.9
1943	86	29	4231	30	3.1
1944	100	20	5108	35	3.8
1945	95	34	5464	47	3.9
1946	103	34	2722	22	2.0
1947	109	19	3443	27	2.4
1948	96	29	2599	26	1.8
1949	121	33	2958	33	2.0
1950	111	32	3295	32	2.2
1951	63	24	–	–	–
1952	77	54	3798	55	2.4
1953	81	42	4045	41	2.5
1954	100	42	4868	42	3.0
1955	102	53	4130	43	2.5
1956	111	53	4313	39	2.7
1957	58	23	1660	15	1.0
1958	62	26	2291	23	1.3
1959	89	28	4138	39	2.3
1960	54	30	2088	28	1.2
1961	45	23	1496	19	0.8
1962	40	18	2028	19	1.1
1963	64	17	4028	25	2.1
1964	25	12	428	5	0.2
1965	17	18	523	7	0.3
1966	26	14	860	11	0.4
1967	32(55)[3]	12(20)	1339(1914)	6(9)	0.8(1.0)
1968	51(82)	20(32)	3994(4419)	26(29)	2.0(2.2)
1969	55(94)	15(25)	2809(3481)	10(12)	1.4(1.7)
1970	42(102)	11(28)	2881(4699)	12(20)	1.4(2.3)
1971	26(92)	8(29)	930(5115)	7(38)	0.5(2.5)
1972	34	11	1948	13	0.9
1973	20	7	1272	10	0.6

[1] Percent of all reported foodborne outbreaks and cases, both of known and unknown etiology.
[2] Data not available.
[3] Data in parenthesis are outbreaks, cases, percentages, and rates, of episodes that were not laboratory confirmed, but were clinically similar to staphylococcal intoxication as well as those that were confirmed—some of the reports in the earlier years also included unconfirmed outbreaks and cases.

period (1966-1971), over 60% of these outbreaks were reported to affect fewer than 10 persons each (U.S. Department of Health, Education, and Welfare 1967-1972). In some instances, however,

TABLE 2.6
SIZE OF STAPHYLOCOCCAL INTOXICATION OUTBREAKS REPORTED FROM 1966 THROUGH 1971

Year	Size of Outbreak							Unknown	Total
	1–3	4–10	11–30	31–100	101–300	301–1000	1000+		
1966	10	7	8	2	4	0	0	0	31
1967	19	17	7	5	5	1	0	1	55
1968	24	24	12	11	9	1	1	1	82
1969	21	35	13	14	10	1	0	0	94
1970	38	23	19	13	6	1	0	2	102
1971	25	37	6	16	5	0	0	3	92
Total	137	143	65	61	39	4	1	6	456
Percent	30.0	31.4	14.3	13.4	8.5	0.9	0.2	1.3	
Cumulative percent	30.0	61.4	75.7	89.1	97.6	98.5	98.7	100	

large numbers of people were stricken. In one city, 1,364 children became ill after eating lunch served at schools (Peavy *et al.* 1968). The meal was prepared in a central kitchen. In another outbreak, 406 children were affected a few hours after a school lunch (Blakey *et al.* 1968). In yet another outbreak, more than 1,000 persons became ill at a company picnic (Googins *et al.* 1961).

Foods Involved in Outbreaks.—The foods that are usually involved in staphylococcal intoxications are protein-containing products which are contaminated after cooking. In the U.S., ham has been responsible for most (21.5%) reported outbreaks during the years 1961 through 1973 (U.S. Department of Health, Education, and Welfare 1967-1974). During the same period, cream-filled pastry, chicken, turkey, and beef each accounted for over 8% of the outbreaks. Other foods that less frequently caused outbreaks were fish and shellfish salads, potato salad, macaroni salad, meat mixtures, processed meats, pork, milk, milk products, and egg products (Table 2.7). Of the staphylococcal intoxication outbreaks that Hodge (1960) surveyed, cooked high-protein foods were responsible for 99% and foods containing a mixture of two or more ingredients were responsible for 67% of the outbreaks.

In England and Wales, most reported outbreaks have been associated with processed meats, such as pressed jellied meat, meat pies, tongues, and hams, that were handled after cooking (Manufactured Meat Products Working Party 1950). Contamination frequently occurred during addition of gelatin to pies, during skinning of tongues, or during slicing of hams (Belam 1947; Hobbs and Thomas 1948; Hobbs and Freeman 1949).

Place of Contamination and Proliferation.—Most of the outbreaks that were reported from 1967 through 1973 were caused by meals served in food service establishments such as restaurants, schools, and medical institutions (U.S. Department of Health, Education, and Welfare 1968-1974). A large number of outbreaks also stemmed from meals prepared in homes (Table 2.8). A few outbreaks have been attributed to processed foods, such as cream-filled pastry, cheese, dried milk, ice cream, butter, canned peas, liver sausage, and Genoa salami (Bryan 1974A; Bryan 1974B).

Incidence, Growth, and Enterotoxin Production in Foods

Surveillance of the incidence of S. aureus in foods must be considered along with surveillance of human illness. A review of the incidence of *S. aureus* and of the potential for staphylococcal growth and of enterotoxin production in foods that have been implicated in outbreaks follows.

TABLE 2.7
FOODS INCRIMINATED IN STAPHYLOCOCCAL FOOD
POISONING OUTBREAKS, 1961 THROUGH 1973

Vehicle	Number of Outbreaks	
MEAT	251	
Ham		137
Baked			76
Unspecified		40
Sandwiches		8
Salad		4
Barbecued		3
Loaf		2
Mixed with vegetables		2
Ground		1
Casserole		1
Beef		60
Cooked (including sandwiches)		19
Ground meat with sauce[1]		14
Hamburger		12
Corned		11
Jerky		2
Meatloaf		1
Beef pie		1
Pork (uncured)		27
Cooked		13
Barbecued		8
Sausage		3
Salad		1
Pigs feet		1
Raw		1
Combination or Unidentified		27
Luncheon meat		10
Salami		6
Stews and hash		5
Gravy		2
Ham and corned beef sandwiches		2
Meat		1
Liver sausage		1
Chopped Liver		1
POULTRY	102	
Turkey		52
Roast (or not specified)		32
Dressing		6
Gravy or with gravy		5
Salad		6
With vegetables		2
Broth		1
Chicken		50
Salad or Salad Sandwiches		18
Cooked		12
Mixtures[2]		8
Barbecued		5
Broth or soup		3
With dressing		2
Gizzards		1
Pot pie		1

TABLE 2.7 (*continued*)

CUSTARDS AND CREAM-FILLED PASTRIES	55	
Cream-layered cakes		18
Cream pies		18
Eclairs, cream-filled puffs, or doughnuts		16
Combinations		2
Rice pudding		1
FISH AND SHELLFISH	34	
Salads[3]		11
Lobster or shrimp		6
Gefilte fish		2
Fish (unspecified)		3
Clams		2
Fishsticks		1
Fish cakes		1
Sardines		1
Oyster dressing		1
Crab		1
Stew (crayfish)		1
Canned tuna		1
Cocktail		1
Salmon		1
Squid		1
SALADS (other than meat, poultry, fish)	31	
Potato		22
Macaroni		7
Combination		2
EGGS AND EGG PRODUCTS[5]	17	
Egg salads and deviled eggs		13
Baked Alaska (whites)		1
Hollandaise		1
Eggs (unspecified)		1
French toast		1
MILK AND MILK PRODUCTS	14	
Cheese[4]		4
Ice cream		4
Butter		3
Milk or buttermilk		2
Milk formula		1
VEGETABLES	9	
Canned vegetables		1
Vegetable soup		1
Potato patties		1
Burrito		1
Creamed vegetable		1
Green beans		1
Corn		1
Coleslaw		1
Lima beans		1

TABLE 2.7 (*continued*)

CEREAL PRODUCTS	6	
Spanish rice		1
Noodle casserole		1
Macaroni and cheese		1
Dressing		1
Rice in seaweed		1
Mostaciolli		1
OTHER FOODS	13	
Cake (pineapple or unspecified[6])		9
Icing[6]		
Popsicles		1
Doughnuts[6]		1
Pie[6]		1
MIXTURES	46	
Chinese foods		7
Japanese food		1
Mexican food (unspecified)		1
Combinations[7]		37
UNSPECIFIED OR QUESTIONABLE	60	
TOTAL	638	

[1] Tortillos, enchiladas, spaghetti, stuffed pepper, lasagna, tacos
[2] a la king, with vegetables, spaghetti, dumplings, casseroles
[3] Tuna, shrimp, and unspecified
[4] Macaroni and cheese listed in cereal products category
[5] Note that many salads (tuna, potato, chicken, ham, etc.) and custards contain eggs but these are not included under this category
[6] May be cream-filled but report did not specify
[7] Two to three food items listed, including ham, potato salad, eggs, turkey, chicken, TV dinners, pizza, etc.

TABLE 2.8
PLACES WHERE FOOD WAS MISHANDLED IN STAPHYLOCOCCAL
INTOXICATION OUTBREAKS, 1966 THROUGH 1973

Place	Number of Outbreaks	Percent
Food service establishments	202	46.5
Homes	72	16.6
Food processing establishments	19	4.4
Unknown or unspecified	141	32.5
TOTAL	434	100.0

Raw Milk.—Raw milk from individual cows has been incriminated as a vehicle in numerous outbreaks of staphylococcal intoxication before the almost universal practice of pasteurization began. On rare occasions, milk contaminated after pasteurization has been implicated (Hackler 1939; Caudill and Meyer 1943). *Staphylococcus aureus* is frequently isolated from raw milk (Gibson and Abd-El-Malek 1957; Smith 1957; Clark and Nelson 1961; Galton *et al.* 1961; St. George *et al.* 1962; Sharpe *et al.* 1965; Terplan and Zaadhof 1969). It has also been isolated from pasteurized milk, cream, and buttermilk (Foltz *et al.* 1960; Sharpe *et al.* 1965).

For raw milk to cause outbreaks, cows must have staphylococcal udder infections, the cows must be milked under hygienic conditions, the contaminated milk must not be mixed with milk from other cows in a quantity sufficient to dilute the toxin below an intoxication threshold level or to add a significant number of competitive organisms, the contaminated milk must be kept at temperatures—77°F (25°C) to 112°F (44.4°C)—that permit rapid growth of *S. aureus*, and the milk must be ingested without heat processing (Jones *et al.* 1957; Smith 1957).

Spray-Dried Milk.—On at least two occasions, spray-dried milk has been a vehicle in outbreaks. In one episode, 8 outbreaks involving 1,190 cases resulted when a spray-dried milk product was used to make a cream product (Anderson and Stone 1955); in the other episode, 19 outbreaks involving 775 children in 16 schools stemmed from three lots of a spray-dried milk product (Armijo *et al.* 1957). In the investigation of the processing plant operation, Anderson and Stone (1955) found that milk from cold-storage tanks had a high bacterial count and 10,000 coagulase-positive staphylococci per ml; but the bacterial count decreased and all staphylococci were killed during heat treatment. One hundred million bacteria including 5 million *S. aureus* per gm were found in a feed tank for the atomizer of a spray-dryer. Enterotoxin was probably produced in this tank. Although the bacterial count decreased in the spray-dried product, enterotoxin was not destroyed and some staphylococci survived.

In a study of the influence of spray-drying manufacturing processes on *S. aureus*, Crossley and Campling (1957) observed that vacuum evaporation reduced numbers of these organisms, but survivors grew in concentrated milk. One strain multiplied at 109.4°F (43°C), but at temperatures above 120.2°F (49°C) staphylococci declined rapidly, and at 132.8°F (56°C) they were entirely eliminated. A small proportion of the original contamination survived spray drying. George and Olson (1960) observed that staphylococci grew rapidly at 90 to 113°F (32.2 to 45°C) in condensed skim milk. Growth was reported after 14 days at 45°F (7.2°C) but not after 52 days at 40°F (4.4°C).

Cheese.—Staphylococcal intoxication has resulted from the ingestion of cheese (MacDonald 1944; Hendricks et al. 1959; Allen and Stovall 1960; Hobbs 1964), and S. aureus has been isolated in or enterotoxin detected in many varieties of cheese (Thatcher and Simon 1955; Thatcher et al. 1959; Mickelsen et al. 1961; Donnelly et al. 1964; Rivas et al. 1965; Sharpe et al. 1965; Donnelly et al. 1967; Zehren and Zehren 1968A). Since S. aureus is present in raw milk, it is not surprising to find it in cheese made from raw milk. Tuckey et al. (1964) observed that Swiss, cheddar, Colby, Limberger, and cottage cheese supported growth of S. aureus. During cheese making most staphylococci become trapped in the curd. The number of staphylococci in pressed curd has a direct relationship to the number of staphylococci in milk at the start of cheese making (Thatcher and Ross 1960). These investigators cited four contributing factors to the development of large numbers of staphylococci in cheese: (1) substantial contamination with staphylococci originating from cows' udders; (2) inadequate cooling of raw milk; (3) presence of antibiotic residues; and (4) contamination with strains of staphylococci resistant to the antibiotics present in milk. Bacteriophages may suppress starter cultures, and a higher than normal pH may selectively permit multiplication of S. aureus.

Studies have shown that staphylococci increase during the making of cheese up to hooping, but they decline during aging (Takahashi and Johns 1959; Roughly and McLeod 1961; Walker et al. 1961; Sharpe et al. 1962; Zottola and Jezeski 1963; Tuckey et al. 1964). Enterotoxin developed only in cheese with less than normal (40%) acid (Zehren and Zehren 1968B). In a cheese of low acidity resulting from starter failure, no decrease in staphylococcal counts occurred over a curing period of 179 days (McLeod et al. 1962). Swiss-type cheese also showed increased staphylococcal counts during the first few weeks of storage (Tuckey et al. 1964). This was probably caused by delayed equilibrium of salt. The number of staphylococci in cheese is not an indication of presence or absence of enterotoxin, however, because enterotoxin remains after the staphylococci that produced it have died out. Prompt cooling of raw milk, and pasteurization of this milk before cheese making, and rapid growth of starter cultures are effective ways of preventing a staphylococcal problem from developing in cheese.

Mickelsen et al. (1963) found that 14% of 66 samples of cottage cheese were contaminated with S. aureus. Staphylococcal counts were low in cottage cheese made by the short-set method, but staphylococci multiplied in cottage cheese made by the long-set method. The short-set method consists of 86 to 88°F (30 to

31.1°C) incubation and 3-1/2% lactic starter culture and 4-1/2 hours are required before the pH reaches 4.7; the long-set method consists of 70 to 72°F (21.1 to 22.2°C) incubation and 1% lactic starter culture, and 11 hours are required before the pH reaches 4.7. During heat processing at 120 to 130°F (48.9 to 54.5°C) for 30 to 40 minutes at an acid pH, high levels of destruction of staphylococci occurred (Mickelsen et al. 1963; Tuckey et al. 1964). Low counts of S. aureus were obtained in cream cottage cheese after 8 to 10 days storage at 50°F (10°C). At the end of this period the pH had dropped to 4.5 (Lyons and Mallmann 1954; Tuckey et al. 1964). Survival of S. aureus in cottage cheese is directly related to pH.

Other Dairy Products.—Ice cream has been incriminated as a vehicle of staphylococcal intoxication (deWitt 1935; Geiger et al. 1935; New York State Department of Health 1940; Williams et al. 1946). Usually, in these outbreaks, ice cream mix was contaminated at the time of preparation, and staphylococci multiplied before or during the long period of cooling before the mix became frozen.

Butter has also been incriminated as a vehicle (Fanning 1935; Stone 1943; Wolf et al. 1970).

Custard and Cream-Filled Pastry.—Cream-filled pastries are notorious in their involvement in staphylococcal food poisoning (Coughlin and Johnson 1941; Tanner and Tanner 1953; Dack 1956A, Bryan 1974B). Synthetic cream fillings have been developed to alleviate this problem and to prolong product shelf life. Fillings made from only cooking fat, emulsifying agents, sugar, and salt (no protein) did not support the growth of S. aureus (Hobbs and Smith 1954). When water only was added to commercially available synthetic fillings, however, the fillings supported growth of S. aureus (Crisley et al. 1964; McKinley and Clarke 1964). At the interface of pie crust and cream fillings, S. aureus also grew profusely (Crisley et al. 1964). Silliker and McHugh (1967) observed that S. aureus could grow neither in the crumb portion of devil's food cake nor in butter-cream fillings containing as low as 1.8 parts sucrose to 1 part water. Water activity in both was low enough to inhibit growth of S. aureus (Scott 1953). But staphylococci grew in cream fillings containing as much as 2.7 parts sucrose, dextrose, or invert sugar to 1 part water when these fillings were put between layers of devil's food cake. This growth was attributed to migration of water from the cake to the interface of the cake and filling and to creation of localized areas of high water activity. After a cream pie is cut, there is marked surface dehydration. This dehydration retards bacterial growth, but the final a_w will determine whether or not staphylococci will grow (Preonas *et al.* 1969).

Ingredients affect the growth of *S. aureus* in cream-filled products. When nutrients, such as milk, whole eggs, or butter, were added to synthetic cream fillings, staphylococcal growth was enhanced (Hobbs and Smith 1954; Crisley *et al.* 1964). Staphylococcal growth was inhibited in cream fillings made with natural chocolate or cocoa (Cathcart and Merz 1942). This inhibition was attributed to nonfat components of chocolate and to pH.

Surkiewicz (1966) isolated up to 100 *S. aureus* per gm in 16 of 287 frozen cream-filled products produced in processing plants with marginal or poor sanitation, but he did not isolate this organism from 0.1 gm portions of 172 samples from processing plants that followed good sanitation practices. He observed evidence of staphylococcal growth in imitation cream pies that were held at room temperature for 24 and 48 hours. In general, it is difficult to predict whether or not particular cream fillings, icings, or toppings will support staphylococcal growth, because migration of water can result in creating areas of higher water activity than that of either of the components (Silliker 1969).

Fresh Meat.—*Staphylococcus aureus* has frequently been isolated from raw meat and from meat processing environments. Ludlam (1952), for instance, isolated *S. aureus* from 113 (63.5%) of 178 carcasses in a large slaughterhouse and from 8 of 11 carcasses in a small slaughterhouse. He also found this organism on 10 of 15 knives, 5 of 13 aprons, 3 of 6 hatchets, 2 of 7 floor swabs, and all of 3 samples of cloths and of 2 swabs of saws. Isolates from carcasses and implements were usually the same phage types. This finding emphasizes the opportunities for cross contamination within slaughterhouses. Vanderzant and Nickelson (1969) isolated coagulase-positive staphylococci from 35% of 33 beef, lamb, and pork carcasses. Jay (1962A) isolated *S. aureus* from 67 (38.7%) of 173 meat samples from 27 stores. An average of 7 different phage types per market were found; 60% of the phage types were group III. One phage pattern was isolated from 11 markets and from 7 kinds of meat (Jay 1962B). With the use of two different media, Sinell and Kusch (1969) isolated coagulase-positive staphylococci from 29 and 39% of scraped meat and from 39 and 46% of chopped meat. Messer *et al.* (1970) isolated 50 to 3,500 *S. aureus* from 35% of 30 market samples of hamburger. Surkiewicz *et al.* (1972) found 100 or less *S. aureus* per gm in 75% of fresh pork sausage samples. Similar findings were reported by Hill (1972), but he observed that cheeks and jowls, which are used in sausage, sometimes contained millions of staphylococci per gm.

Raw materials used in producing a fermented meat product contained staphylococci of animal origin as the dominant type of staphylococci. But, as the processing continued, staphylococci of animal origin decreased; staphylococci of human origin increased. At the end of the process, staphylococci of human origin dominated the microbial flora (Siems *et al.* 1971).

Dack (1956A) reported that *S. aureus* grew but did not produce toxin when heavily inoculated liver was held under typical conditions of a butcher shop. Most investigators, however, have reported that the natural flora of raw meat prevent the growth of *S. aureus* (Miller 1955; Casman *et al.* 1963B; McCoy and Faber 1966). Casman *et al.* (1963B) only obtained growth of *S. aureus* and enterotoxin production on aseptically cut portions of raw meat or on cooked meat that had the natural flora destroyed; they did not obtain growth on naturally contaminated surfaces.

Cured Meats.—Cured ham is the most common vehicle attributed to staphylococcal intoxication outbreaks that have been reported in the U.S. Typical outbreaks are cited by Slocum and Linden (1939), Wain and Blackstone (1956), and Googins *et al.* (1961).

Hams selectively favor the growth of *S. aureus* because of their relatively low water activity. Commercial hams varied in their percentage salt (w/v) from 2 to 5% (Genigeorgis *et al.* 1969). These investigators detected enterotoxin B in hams with a pH over 5.3 containing up to 9.2% NaCl and 0.54 ppm nitrous acid at temperatures of 86°F (30°C), 71.6°F (22°C), and 50°F (10°C). The hams were inoculated with 10^3 to 10^6 *S. aureus* and incubated aerobically and anaerobically. At 50°F (10°C) enterotoxin was not produced until after 2 weeks incubation, but usually it was not detected until after 8 weeks. Hams containing enterotoxin had more than 4×10^6 cells per gm and less than 10^5 other organisms per gm. Hams containing enterotoxin looked normal, even after 2 months incubation at 50°F (10°C). Casman *et al.* (1963B) demonstrated aerobic production of enterotoxin A in hams inoculated with 250 cells/cm^2 and incubated at 86°F (30°C) for 72 hours. Christiansen and Foster (1965) observed less growth of *S. aureus* on sliced ham under anaerobic conditions than under aerobic conditions. Manipulation of salt content of hams within acceptable taste levels will not prevent enterotoxin production.

An abscess on sliced bacon was responsible for a case of staphylococcal intoxication (Daffron *et al.* 1972). Twelve million coagulase-positive staphylococci per gm were isolated from the abscess. Cured bacon has yielded *S. aureus* (Eddy and Ingram 1962; Baird-Parker 1962C). Baird-Parker (1971) was unable to detect enterotoxin B in vacuum-packed bacon inoculated with 10^2 *S. aureus* cells per gm after the bacon was stored for up to 14 days at $77°F$ $(25°C)$. On the other hand, Thatcher *et al.* (1962) demonstrated enterotoxin in Canadian bacon (containing 0.3 to 3.2% NaCl and 50 to 80 ppm NO_2) that had been inoculated with 10^6 *S. aureus* per gm and stored at $37°C$ under vacuum, in air, exposed to N_2, and in 5% CO_2 and O_2.

Genoa salami has recently been incriminated as a vehicle in staphylococcal food poisoning outbreaks (Weimann *et al.* 1971; Eckhoff *et al.* 1971; Hopf *et al.* 1971). Genoa and an unknown type of sausage were found to have from 10^4 to 10^7 *S. aureus* per gm in the outer 6 mm of the sausage but less than 100 to 230 per gm in the core. Their growth was localized in the outermost areas of sausage where the oxygen tension was highest (Barber and Deibel, 1972).

Studies of the microbiology of the processing of Italian-style, dry salami paste indicated that 72.3% of 36 lots contained *S. aureus* at levels of 225 to 16,875 per gm. Salami manufactured from certified pork had significantly fewer staphylococci than salami manufactured from noncertified pork (Genigeorgis 1974). Follow-up studies at two plants showed that samples of raw salami were contaminated with *S. aureus* 70% and 62% of the time and that samples of the finished product were contaminated with this organism 53% and 64% of the time. When the fermentation was controlled by a starter culture, a smaller number of staphylococci was found in the finished product.

Although the manufacturing process of the incriminated product has not been described in detail, the use of head meat contaminated by nasal sinuses or staphylococcal abscesses, the use of frozen meat, high concentrations of salt, higher than normal pH, lack of using an acidulent, lack of using a starter culture, favorable temperatures, or presence of lactic acid bacteria that initially stimulate the growth of *S. aureus*, may give *S. aureus* a selective advantage over competitive organisms in the early stages of fermentation.

Poultry.—Chicken and turkey meats, particularly when made into salads, are frequently implicated in outbreaks of staphylococcal intoxication. As previously mentioned, staphylococci may be on fowl as they enter processing plants, but contamination spreads as the carcasses are handled during processing. DaSilva (1967) found *S. aureus* on 11 of 24 samples of turkeys after picking; 3 of 12 samples of turkeys after evisceration, washing, and spin chilling; and 10 of 20 samples after overnight chilling, just before packaging. Surkiewicz *et al.* (1969) isolated *S. aureus* in low numbers (26 per cm^2) from chicken carcasses at various stages of processing. They also found low numbers (less than 15 geometric mean per ml) of staphylococci in chill-tank waters. During a survey of market foods, Messer *et al.* (1970) isolated *S. aureus* from 80% of 30 chicken samples (2 to 620/ml of rinse). Salzer *et al.* (1967) found coagulase-positive staphylococci on 62 (17%) of 360 turkey livers before washing but on only 3 of 360 turkey livers after washing. Coagulase-positive staphylococci were isolated from 40 (1.6%) of 2,513 samples of chicken livers (Genigeorgis and Sadler 1966D). Half of those tested produced enterotoxin, and two-thirds of those tested were typable with human type phages; most were group III. All of nine samples of raw turkey rolls yielded coagulase-positive staphylococci, but these organisms were not recovered from any of nine samples of cooked rolls (Woodard 1968). In another study, *S. aureus* was found on 60% of raw turkey rolls, but not on cooked rolls (daSilva *et al.* 1967). Bryan and McKinley (1974) isolated *S. aureus* from 5 of 24 raw turkey carcasses but not from cooked turkeys.

Fish and Shellfish.—Fish and shellfish, usually in the form of salads are occasionally involved in outbreaks of staphylococcal intoxication (Bryan 1973A). Contamination of freshly caught fish is of far less importance than the contamination that occurs during handling and processing on board ship and on shore (Shewan 1962; Liston and Raj 1962). Appleman *et al.* (1964) isolated *S. aureus* from the skin of ungutted fish from vessels (2/12), skin of gutted fish (1/12), flesh of skinless fillets from shops (1/6), unfrozen fish cakes (4/30), frozen fish cakes (4/28), boneless kippers (4/8) and smoked yellow haddock (2/11). Surkiewicz *et al.* (1968) isolated coagulase-positive staphylococci from 51 of 276 samples of raw breaded fish and from 14 of 123 samples of fried breaded fish in processing plants reported as having poor sanitation and from 8 of 110 samples of the raw product and 1 of 64 samples of the cooked product prepared in plants reported as having good sanitation. Blocks of frozen fillets (4/34) also contained staphylococci, but these organisms were not found in dry batter or breading. On the other hand, Liston and Raj (1962) frequently isolated *S. aureus* from batter and breading but not from dry batter mix.

Silverman *et al.* (1961) found coagulase-positive staphylococci on 75% of 91 samples of frozen shrimp; and Virgilio *et al.* (1970) found them on 140 (35.7%) of 392 samples of precooked frozen shrimp from Chilean processors. Surkiewicz *et al.* (1967B) isolated coagulase-positive staphylococci from 317 of 564 samples of frozen breaded shrimp prepared in plants reported as having poor sanitation practices and from 128 of 297 samples of this product prepared in plants reported as having good sanitation practices. The staphylococcal counts were less than 1000 per gm in 95% of the samples. These organisms were also isolated from dry batter (3/35), from liquid batter (4/40) in plants that chilled the batter and discarded it at least once during the work period, and from liquid batter (10/39) in plants that did not chill or discard the batter. At inoculum levels of 10^3 and 10^6 per gm, aerobically incubated prawns contained enterotoxin after 7 days at 78.8°F (26°C) and at higher temperatures, but the prawns were considered spoiled at the time enterotoxin was detected (Baird-Parker 1971).

Canned Foods.—Canned foods have been associated with outbreaks of staphylococcal intoxication. Usually, contents of the cans were contaminated after opening, but sometimes the organisms entered through minute defects in seams when cans cooled or when they were handled while still wet after cooling such as during labeling (Hobbs 1968). Dack and his coworkers reported on growth and elaboration of enterotoxin by *S. aureus* in canned corn, oysters, roast beef, and potted meat (Davison and Dack 1952; Segalove *et al.* 1943; Surgalla and Dack 1945).

Salads and Sandwiches.—Both salads and sandwiches have been implicated in staphylococcal outbreaks. Christiansen and King (1971) isolated 1×10^2 to 2×10^5 staphylococci from 39% of 66 samples of salads and 1×10^2 to 3×10^6 of them from 60% of 62 samples of sandwiches from retail outlets.

Frozen Foods.—Although *S. aureus* can survive freezing and frozen storage better than most vegetative bacteria, frozen foods have rarely been incriminated in outbreaks. Abrahamson (1958) cited a few outbreaks of staphylococcal intoxication that were traced to frozen foods.

Shelton *et al.* (1960) found *S. aureus* in 11% of 1,631 samples of frozen precooked foods. Of foods (such as cakes and cream pies) that did not require any further cooking, 4% were positive for this organism; of foods (such as macaroni and cheese) that had ingredients added after heat processing and that needed warming, 12% were positive; of foods (such as crab cakes and Chinese foods cooked just before packaging and freezing) that needed warming before serving, 10% were positive; of foods (such as pizza and chicken pot pies) that required cooking before eating, 21% were positive. Surveys of operations and products prepared in frozen food processing plants showed the presence of coagulase-positive staphylococci in cream pies, breaded shrimp, potatoes, breaded fish, and cooked meat and gravy (Surkiewicz 1966; Surkiewicz *et al.* 1967A; Surkiewicz *et al.* 1967B; Surkiewicz *et al.* 1968; Surkiewicz *et al.* 1973). Gunderson and Rose (1948), Buchbinder *et al.* (1949), Canale-Parola and Ordal (1957), Huber *et al.* (1958), Gunderson (1963), and Gunderson and Peterson (1964) also isolated *S. aureus* from frozen foods.

Splittstoesser *et al.* (1965) isolated coagulase-positive staphylococci from 39% of 51 samples of frozen peas, 39% of 31 samples of frozen french-style green beans, 31% of 16 samples of frozen cut green beans, and 64% of 14 samples of frozen corn. All of these products contained an average (MPN) of less than 2 *S. aureus* per gm, except the peas; the peas averaged 7.3 *S. aureus* per gm. Line sampling revealed a higher percentage of positive samples from areas at which a high degree of human contact occurred (such as inspection belts, hoppers, filling machines, and final packaging). This finding suggests that workers' hands are a major source of *S. aureus* for processed products. The gravity separator used in processing peas was identified as a site at which staphylococci accumulated.

Foods in General.—When enterotoxigenic staphylococci grow in foods, they elaborate enterotoxin without producing abnormal odor or taste. Thus, enterotoxin-bearing foods will be accepted and eaten without suspicion.

Types of *S. aureus* Associated with Foodborne Outbreaks

Most of the strains of *S. aureus* that have been associated with foodborne outbreaks belong to phage group III (Allison 1949; Williams *et al.* 1953; Anderson and Williams 1956; Munch-Petersen 1963). Munch-Petersen (1963) reviewed 666 reports of staphylococcal food poisonings that occurred through the world, and he reported that phage group III staphylococci were associated with 77% of the outbreaks, group I with 1.5%, group II with 1.0%, group IV with 4.6%; only 4.4% of the organisms were untypable. Similar results were noted in a 9-year survey of outbreaks that occurred in Great Britain.

In comparison with other enterotoxins, enterotoxin A is more frequently detected in outbreak investigations. Casman *et al.*

(1967), for example, detected enterotoxin A from 49%, enterotoxin A and D from 25%, enterotoxin B from 3.8%, enterotoxin C from 2.5%, enterotoxin D from 7.6%, of 80 isolates from food poisoning outbreaks. Enterotoxins A and D were also more frequently recovered from strains that were isolated from nasal specimens.

Conditions Necessary for Outbreaks to Occur

Staphylococcal intoxication occurs only when several conditions exist:

1) *A source of an enterotoxigenic strain of* S. aureus *must be in a food production, processing, or preparation environment.* The source is usually the nares or hands, but sometimes skin infections, hair, or clothing, of food workers. Animals or foods of animal origin, such as raw milk, can also be sources.

2) *The organism must be transferred from the source to a food.* Transfer can be completely indirect—through air as in the case of cough or sneeze droplet nuclei or as in the case of organisms falling from clothing, hair, or skin onto foods. It can be relatively indirect, as from nose or lesion of a food worker—to his hands—to a food. It can be more direct, as in the case of an infected udder contaminating milk, or entirely direct, as in the case of arthritic infections of tissue in poultry. From a worker's nose—to his hands—to food is the most common mode of transmission. Whenever foods are cut, mixed, blended, diced, or ground, they are subjected to increased opportunities for contamination.

3) *The food must be contaminated with thousands of* S. aureus *per gm, or, more usually, the food must be heated before it becomes contaminated, or it must contain high levels of salt or sugar.* In exceptional situations, such as milk from a mastitic udder, a raw food can be contaminated with large enough numbers of staphylococci so that they outgrow small numbers of other organisms. Ordinarily, however, foods are cooked or heat processed, which destroys organisms that normally compete with and outgrow staphylococci, and then the foods are contaminated, or the foods must contain high levels of salts or sugars which inhibit the growth of competing organisms but allow staphylococci to grow.

4) *The organism must survive in the food; it must not be outgrown or inhibited by competing organisms or killed by heat, low pH, or other adverse conditions before it can produce enterotoxin.* Many organisms grow faster than S.

aureus, produce substances which are toxic to it, or change the medium so that it does not thrive. If the food, before toxin is elaborated, is subjected to temperatures of at least 165°F for short periods of time or to lower lethal temperatures for longer periods, no problem will exist unless the foods become recontaminated after the heat process. Cooking procedures, however, will not destroy any preformed staphylococcal enterotoxins.

5) *The food, after it becomes contaminated, must support the growth of* S. aureus. For food to support the growth of *S. aureus* and the production of enterotoxin, it must contain several amino acids, thiamine, and nicotinic acid. The food must not be too acid or too dry. Dry foods, however, again become susceptible to attack upon reconstitution, water migration, or condensation.

6) *The contaminated food must stay within the temperature range that is suitable for proliferation of* S. aureus *long enough for this organism to multiply and produce enterotoxin.* The mere presence of enterotoxigenic strains of *S. aureus* in food is not enough to cause illness; staphylococci must elaborate enterotoxin during multiplication in the food. The minimum time required is usually 5 hours or longer, depending on temperature, nature of the food, and competition. After a lag period, staphylococci can grow at temperatures between 44°F (6.7°C) and 114°F (45.6°C). They grow rapidly at temperatures near 100°F (37.8°C) and slowly at the low end of the growth range.

7) *A sufficient quantity of enterotoxin-bearing food must be ingested to exceed the enterotoxin susceptibility threshold of persons eating the food.*

Certain practices in food processing plants, food service establishments, and home kitchens allow these conditions to occur. These practices as they relate to outbreaks have been reviewed by Hodge (1960), Munch-Petersen (1963), Bryan (1972), and Bryan (1974A).

Hodge (1960) surveyed 95 outbreaks of staphylococcal food poisonings that were reported in 1955 and 1956. Visible infections on food handlers were reported in only 9 of these outbreaks, and insanitary practices were reported in only 13 outbreaks. Healthy people who conducted their activity in a routine manner in clean establishments were responsible for contamination of foods with *S. aureus* in 50 of the outbreaks. In only 5 outbreaks was the offending food cooked on the day it was eaten. In 79 of 83 outbreaks, the food was held at room temperature, or refrigerated in large masses, or held in warming ovens, between the time it was cooked and served. Inadequate refrigeration was reported in 89% of the outbreaks, and holding of foods in warming devices but at temperatures too low to prevent bacterial growth was reported in 16% of the outbreaks. In 1956, Dauer (1957) stated that lack of refrigeration and food handlers with infections were contributing factors to half of the reported outbreaks.

In a review of 110 outbreaks of staphylococcal intoxication, food handlers' nares were determined as the only source of enterotoxigenic *S. aureus* in 44 of these outbreaks and among several possible sources in 73 outbreaks; hands were determined as the only possible source in 27 outbreaks and among several possible sources in 47 outbreaks; and lesions were determined as the source in only 4 outbreaks (Munch-Petersen 1963).

In a search of journals and official records for outbreaks of staphylococcal intoxication from 1961 through 1970, 99 outbreaks with data about contributing factors were uncovered (Bryan 1972). The most significant contributing factors were failure to properly refrigerate foods, preparing foods a day or more in advance of serving, and an infected person who handles and thus contaminates food. These were reported in 74%, 51%, and 54% of the reviewed outbreaks, respectively. If more complete information had been obtained in the investigations or cited in the reports and if more intensive laboratory work had been conducted, there is little doubt that two factors—inadequate refrigeration and contamination by infected workers—would appear in nearly all these outbreaks. Two other factors that contributed significantly to outbreaks of staphylococcal intoxication were holding foods at warm (bacterial incubating) temperatures, which was reported 16% of the time, and inadequate cleaning of kitchen or processing equipment, which was reported 9% of the time. Inadequate cooking, inadequate reheating, and cross contamination were mentioned in 3%, 6%, and 3%, respectively, of these outbreaks. Another possible factor, seldom investigated, is that foods of animal origin may, when they arrive in a kitchen, harbor *S. aureus*. In this case, either inadequate cooking or cross contamination may explain the presence of staphylococci on the incriminated foods. Data for the years 1971 and 1972 confirm these factors (Bryan 1974A).

PREVENTION

Outbreaks of staphylococcal intoxication can be prevented by interfering with any of the conditions that are necessary for outbreaks to occur with special emphasis on eliminating those factors that experience has shown most frequently contribute to these outbreaks. (These conditions and factors are described in the section on Epidemiology.) As a result of epidemic investigations, specific recommendations have been made for the prevention of staphylococcal intoxication (Hobbs and Thomas 1948; Allison *et al.* 1949; Hobbs and Freeman 1949; Bryan *et al.* 1971; Bryan 1972; Bryan 1974B; Bryan and McKinley 1974). Preventive measures are: (1) prevent or limit contamination of foods with *S. aureus*, (2) destroy *S. aureus* cells in foods before they have a chance to elaborate enterotoxin, and (3) inhibit the growth of *S. aureus* and their subsequent toxin production in foods. Techniques of prevention and their limitations are discussed below.

Prevention and Limitation of Contamination

Preventing and limiting contamination includes attempts to keep the source and reservoir of enterotoxigenic strains of *S. aureus* away from the environment where food is produced, processed, or prepared and to prevent the dissemination of *S. aureus* from its source to foods. Inspecting food processing plants and food service establishments for conformity to sanitation standards, and testing foods to see if they conform to quality and bacteriological standards help to limit contamination. Constant supervision of operations is critical if contamination is to be minimized, and training is required to make supervisors and workers aware of hazardous situations and corrective procedures.

Isolating, Removing, or Destroying Organisms at Their Source.—Because man is a reservoir of *S. aureus* and principal sources are his nose, hands, clothing, and lesions, and because animals, also, are reservoirs, it is obvious that attempts to preclude *S. aureus* from environments where food is produced, processed, or served are impossible and that attempts to prevent contamination of foods are extremely difficult.

People who have open lesions, recurrent boils, acute sinus infections, or diarrhea should not handle cooked foods or the types of uncooked foods which readily support growth of *S. aureus* (U.S. Department of Health, Education, and Welfare 1962). Regulations that prohibit ill people from handling food are difficult to enforce and do not entirely solve the problem because healthy people are just as apt to carry *S. aureus* and contaminate food. Watertight bandaids and dressings can prevent dissemination of *S. aureus* from localized lesions (Hare and Cooke 1961), but they must be kept dry and replaced when they become soiled.

Chemotherapeutic agents used for treating infections should be chosen on the basis of an antibiogram of a culture isolated from the infected site. Chloramphenicol, erythromycin, kanamycin, vancomycin, oxacillin, methicillin, novobiocin, and neomycin are antibiotics for which *S. aureus* strains have shown minimal resistance (Schneierson 1965; Bulger and Sherris 1968). Other drugs that have been recommended for treating staphylococcal infections are cloxacillin, nafcillin, cephalothin, sodium fusidate, and linocomycin (Dineen 1970). In most instances of localized staphylococcal infections, however, there is no need to use antimicrobial drugs because simple incision and drainage will effect a cure. When such treatment is used for infection on food handlers, extreme care should be made to ensure that the person does not handle foods until the incision stops draining or that the incision site is covered with a watertight dressing. The site of the infection and the job situation will dictate which course of action should be pursued.

The nasal carrier state in a high percentage of hospital patients or staff has been temporarily eliminated by the use of antibiotic creams and sprays (Gould and Allan 1954; Gould 1955; Stratford *et al.* 1960; Varga and White 1961). Williams *et al.* (1967), however, observed that *S. aureus* was cleared from the nares of only 37 of 71 persistent carriers after local use of antibiotic creams for 1 week. Noone *et al.* (1970) reported that the carrier state was eliminated by 2% hexachlorophene cream when it was applied to the nares twice daily for 1 week. Coagulase-positive strains were frequently replaced by coagulase-negative strains. Lindborn and Laurell (1967) reported that antibiotic creams applied to nostrils two to three times daily neither eliminated a persistent carrier state nor protected persons from implantation. On the other hand, Denes and Rampazzo (1967) reported that 80% of food handlers were cleared as carriers by use of antibiotic nasal spray, nose drops, and tablets taken orally. Application of antibiotics to the nostrils of infected food handlers has occasionally been a part of screening programs for food handlers (McGirr 1969; Virgilio *et al.* 1970). There are several limitations, however, to the use of antibiotic creams for control of nasal carriers of *S. aureus* in the food industry. Antibiotic resistant strains may develop as a result of such treatment (Gould 1955). Frequent applications of such creams or sprays must be made for a few days to a week to rid most carriers of *S. aureus*. Once treatment has ceased, reacquisition of *S. aureus* begins within a short time. Martin and White (1968) observed that in people who had a relatively low number (10^5) of competitive organisms in their nares, *S. aureus* reappeared in over 50% of the nares of people within a week. In people with larger numbers of competitive flora, *S. aureus* reappeared in the nares of 50% of subjects who were treated with one antibiotic in a week and in the nares of 40% of subjects who were treated with another antibiotic in a month. It is very doubtful that ridding food processing and preparation environments of nasal carriers is either economically or theoretically feasible.

Attempts have also been made to rid hands of *S. aureus* with creams and soaps containing 2 to 3% hexachlorophene. Lowbury *et al.* (1964A) reported that repeated use of hexachlorophene soaps for handwashing was successful in ridding hands of staphylococci; but repeated application of hexachlorophene cream to dry skin caused no reduction in skin flora. The physical action of washing probably accounted for much of the difference. No difference in the numbers of bacteria on hands of nurses, however, was observed when hexachlorophene soaps were used intermittently with ordinary soaps (Weatherall and Winner 1963). Similar results would be expected if food handlers were required to use germicidal soaps at work. Gram-negative bacteria are less affected by hexachlorophene than are Gram-positive bacteria, such as *S. aureus*, and the continual use of hexachlorophene may promote colonization of skin by enteric bacteria or pseudomonads. (Bruun *et al.* 1968). Thus, one problem may be replaced by another. Also, the Food and Drug Administration and the National Academy of Sciences reported no evidence that hexachlorophene soaps were an effective aid to personal hygiene and stated that prolonged use of such soaps may be harmful (U.S. Department of Health, Education, and Welfare 1971).

Gloves are sometimes used to prevent hand contact with foods. Gloves are a barrier between contaminated hands or infected lesions and food, but they can become as contaminated with transient organisms as hands are if they are worn very long or while handling contaminated foods. Bacteria rapidly multiply on hands that are covered by gloves so that punctures and tears in gloves, which readily occur, result in leakage of large numbers of bacteria. Rubber gloves, in addition, are seldom washed as frequently as hands or as well as other pieces of equipment at the end of a day's operation and so bacteria will have harborage and may grow on glove surfaces. In poultry processing operations where most of the product contamination comes from incoming birds, Bryan *et al.* (1968) observed that rubber gloves were as readily contaminated with salmonellae as were hands. The same may be true in situations where staphylococci are present on raw foods or if workers touch gloves to their face or clothing. In a survey at the same type of processing plant, however, less staphylococci were found on gloves than on hands of workers (daSilva *et al.* 1967). Disposable plastic gloves may be valuable in limiting contamination of foods by contaminated hands. These gloves should be used mainly when cooked foods are handled. When used, they should be changed frequently—perhaps every 1/2 or 1 hour—during continued service, whenever unrelated tasks are performed, or when they become punctured or torn.

Washing can effectively remove transient staphylococci—those organisms that are acquired from fingering the nose, brushing the hair, and handling food—from the skin. Lowbury *et al.* (1964B) reported that 99% of transient bacteria were removed from hands with a 30-second wash. Price (1968), however, stated that it takes 5 to 10 minutes washing with soap and water to remove transient bacteria from hands. Ordinary washing will not remove many resident staphylococci (Lowbury *et al.* 1963). Rinsing without friction has practically no effect. Washing by rubbing hands together is less effective than scrubbing with a brush. Vigorous scrubbing with a stiff brush is the most effective method. Routine handwashing tends to bring staphylococci to the surface, and more of them may be found on hands after washing than before (Brodie 1965). Price (1938) estimated that a log rate of one half resident organisms are removed with each 6 minutes of surgical scrubbing. There is not any practical handwashing procedure that will ensure the removal of resident staphylococci.

Soaps, in dilutions comparable to their concentrations in lather, show little, if any, bactericidal action on *S. aureus*. At such concentrations, there was no evidence that two brands of household soap, four brands of toilet soap, green soap, or fawn germicidal soap killed *S. aureus* (Morton and Klauder 1944). The concentration of a soap solution, the degree of lather, and the time of exposure are pertinent factors in possible germicidal action of soaps. To obtain a good lather, washing must be prolonged. It is doubtful that even germicidal soaps contact the skin long enough to have much effect on bacteria. The action of soaps primarily depends on mechanical removal of bacteria from the washed skin by emulsification of lipids on the surface.

The use of disinfectants after handwashing, unfortunately, does not yield dramatic results in the control of *S. aureus* (Lowbury 1961; Brodie 1965). Lowbury *et al.* (1964B) reported that hypochlorite solutions had little or no effect on eliminating even transient staphylococci from hands. Quaternary ammonium compounds are effective against *S. aureus in vitro*, but on skin, under conditions of ordinary use, their antiseptic action is not great. Their loss of action is due, in part, to the neutralizing action of traces of soap regularly left on the skin after ordinary washing. Prolonged (1 to 3 minute) rinsing, after washing, of hands with 70% alcohol or with 70% alcohol containing 1% iodine is effective in degerming (Price 1968). After the numbers of bacteria on the skin are reduced by some type of disinfectant, resident bacteria are reestablished at a rate as represented by a sigmoid curve (Price 1938).

The so called "self-sterilizing power" attributed to human skin is caused by desquamation of outer layers of the skin, desiccation, pH of the skin, lipids, fatty acids, water-soluble substances on skin surfaces, and, possibly, some antagonistic effects of resident flora (Pillsbury and Rebell 1952; Muller 1967). Showering can effectively remove recently acquired bacterial contamination, but it causes some persons to shed more than normal resident staphylococci for some time after the shower (Cleton et al. 1968).

Hands frequently become soiled with discharges from nose and mouth, as well as in other ways, and hands frequently touch food. They, therefore, should be washed to remove as many transient staphylococci as possible after a person goes to the toilet, coughs or sneezes, fingers his nose or a sore, smokes, and handles raw poultry and meats. The frequency with which people touch their nose and mouth and possibly contaminate hands with staphylococci was reported by Hare and Thomas (1956) who observed nine students during a lecture that lasted 1 hour; the students touched their nose or mouth on 23, 19, 15, 15, 14, 11, 10, 7, or 6 separate occasions. Thus, even at best, handwashing or attempts at hand disinfection can readily become nullified by one swipe across the nose, brush of the hair, scratch of an itch, or reach into a pocket.

Prevent Dissemination of Staphylococci.—Contamination of foods can be minimized by strict practices of personal hygiene and by sanitary food preparation. Hands, even when recently washed, should be deemed contaminated and should not touch foods which are to be served without subsequent cooking. Utensils and automated equipment that prevent hand contact with foods should be used whenever possible. Clean work clothes or coats donned over street clothes will limit direct contamination of food from clothing, but they will not prevent dissemination of staphylococci into air.

To minimize cross contamination from raw or contaminated foods to cooked foods, utensils and equipment should be washed with hot water and detergent, rinsed, and disinfected at least at the end of each day's operation or after they are used for raw foods and before they are used for cooked foods. Gilbert and Maurer (1968) recommended that kitchen equipment be cleaned at least twice daily (midway through the operation and at closing). This practice should be practical in food processing and food service operations. In studies of cleaning slicing machines, carving knives, can openers (Gilbert and Maurer 1968), and cutting boards (Gilbert and Watson 1971), investigators concluded that bacteria are more effectively removed from equipment surfaces by detergent activity and the physical action of scrubbing than by disinfection. They suggested, nevertheless, that using a suitable disinfectant provides a second line of defense. Detergents and disinfectants should be applied with single-use disposable paper cloths instead of sponges or rags to avoid recontamination of equipment that is subsequently wiped (Gilbert 1969). Equipment sanitation is more effective if a cleaning schedule (which defines cleaning responsibility, frequency, time, and method) is followed.

Plant or kitchen layout and operational flow should be designed to segregate raw food operations from prepared food operations. Whenever possible, separate equipment should be used to process raw and cooked foods.

Control of Mastitis.—Mastitis in cows can be controlled by hygiene, sufficient reserve vacuum during milking, and drying-off therapy. Effective hygiene measures consist of hand disinfection between handling different cows, udder washing with a disinfectant, using paper towels or individual cloths for hand and udder disinfection, teat cup pasteurization with circulating 185°F (85°C) water or chemical disinfectant, and dipping cows' teats in a disinfectant immediately after milking (Newbould 1968).

Sanitation, Quality Control, and Bacteriological Standards and Guidelines.—In general, staphylococcal contamination should be less in food processing and food service establishments in which good sanitation practices are followed than in insanitary environments. Frozen food products, for instance, processed in plants which had good sanitation practices contained a lower percentage of samples with *S. aureus* than products from plants which had poor sanitation practices (Surkiewicz 1966; Surkiewicz *et al.* 1967A; Surkiewicz *et al.* 1967B; Surkiewicz *et al.* 1968). Poor practices, however, did not always coordinate with a high percentage of products positive for staphylococci. Good sanitation practices were described as having several of the following: handwashing basins available in processing area; use of hand sanitizing solutions; use of rubber gloves; equipment in good repair and cleaned, rinsed, and sanitized each day or before refilling; inplant chlorination; clean-up schedules and crews; training seminars; bacteriological quality control; and supervisory control over personal hygiene. Sanitation standards are specified for processing plants in "Good Manufacturing Practices" (U.S. Department of Health, Education, and Welfare 1968) and for food service establishments in the "Food Service Sanitation Manual" (U.S. Department of Health, Education, and Welfare 1962).

Many companies require that the ingredients they purchase meet bacteriological specifications. Small numbers of staphylococci must be expected to be on raw foods of animal origin and on foods that are handled by workers after heat processing. Raw materials, in-line product samples, scraps and tailings, swabs of equipment, swabs or rinses from workers' hands, and environmental samples, as well as finished product samples—coupled with sanitary surveys—are necessary for identifying sources of contamination and places where multiplication occurs. For such a determination, a flow-sheet type analysis should be made as typified in reports by Jensen (1945) and by Bryan and McKinley (1974). A practical quality control program for food processing plants is outlined by deFigueiredo (1969).

Some official agencies have advocated the use of bacteriological standards as administrative aids to enforcement. Some of these standards were reviewed by Elliott and Michener (1961) and by Thatcher (1963). Standards for *S. aureus* proposed for various foods have ranged from their complete absence (Abrahamson *et al.* 1959) to not over 1000 per gm of food (Abrahamson 1958; Thatcher 1963; Genigeorgis *et al.* 1971). The Commission on Microbiological Specifications for Foods has published recommended methods for isolating staphylococci (Thatcher and Clark 1968). A second book describes a sampling plan and microbiological specifications for many classes of foods (International Commission on Microbiological Specifications for Foods 1974). The specifications *set by the commission for staphylococci in foods are: Fresh and frozen fish (including fish frozen at sea, fish blocks, comminuted fish blocks, and scallops), fresh water fish, cold-smoked fish (including kippered herring) cooked prior to eating, frozen raw shrimp, prawns, and lobster tails, frozen raw breaded shrimp and prawns, n 5, c 3, m 10^3, M 2×10^3; breaded pre-cooked fish products (including fish sticks and fingers), fish portions, and fish cakes, n 5, c 2, m 10^3, M 2×10^3; cold-smoked fish eaten uncooked, frozen cooked shrimp, prawns, and lobster tails, cooked picked crabmeat n 5, c 1, m 10^3, M 2×10^3; special dietary dried foods of animal origin (for infants, aged, or relief), n 10, c 1, m 10, M 10^2; dried pasta with eggs, dried pudding with eggs n 5, c 1, m 10, M 10^3; dried potatoes, dried soups not to be cooked, dried meats including gelatin and fish protein concentrate, dried sea foods, n 5, c 1, m 10^2, M 10^4; frozen deserts n 5, c 2, m 10, M 10^3; dried milk n 5, c 1, m 10, M 10^2; dried milk for relief work n 10, c 1, m 10, M 10^2; ice cream n 5, c 1, m 10, M 10^2; cheese (hard and semi-hard types) n 5, c 1, m 10^3, M 10^4.

---

*The letter n refers to the number of samples which must be examined for each lot or production batch; c refers to number of defective samples which are acceptable or those within the m or M range; m refers to the number of organisms that separates good quality from marginally acceptable quality; M refers to the number that separates marginally acceptable quality from unacceptable quality.

Administrative microbiological guidelines concerning the number of coagulase-positive staphylococci and of other bacteria are used by the Food and Drug Administration. Interpretations are based on a minimum of 10 samples and should correlate with establishment inspection. Guidelines for deviled, cooked, frozen crabs, and cooked, frozen crab cakes, for example, are that coagulase-positive staphylococci should not exceed 3.6 per gm in 10% or more of ten samples from a plant having insanitary conditions (Angelotti 1971). But in a situation where there are 13 consecutive negative samples from a large lot, one can only say with 95% confidence that the lot is less than or equal to 20% contaminated (U.S. Department of Agriculture 1966). Thus, one item of food in five may be contaminated and yet not be detected with such a sampling approach.

Supervision and Training.—Day-by-day supervision of food processing and service operations is the key to the prevention of outbreaks of staphylococcal intoxication. Supervisors must be aware of specific factors related to their industry that most frequently contribute to foodborne disease outbreaks and they must know how to prevent these factors from occurring so they can direct and train their employees. Supervisors can acquire this information through training courses (Bryan 1969; U.S. Department of Health, Education, and Welfare 1969; Bryan 1973B; Bryan, 1973C; Goldenberg and Edmonds 1973), seminars, programmed training guides (Carter *et al.* 1964), texts (Longree 1972; Hobbs 1968), brochures or pamphlets (Zottola 1968), and fact sheets or newsletters. Most materials developed for training food service personnel deal with bacterial control in general, but they usually include techniques of personal hygiene that will limit dissemination of staphylococci. For the prevention of staphylococcal intoxication, however, food workers must not only know that staphylococcal contamination results primarily from infected, but usually healthy, workers touching foods, but that foods so contaminated must be handled in a way that will prevent staphylococci from multiplying and, thus, from elaborating enterotoxin.

Destruction of the Organism

Because *S. aureus* is a non-spore-forming organism, it is subject to destruction by routine cooking and heat processing, but, as previously mentioned, enterotoxins are not destroyed by such treatments. Therefore, prevention of staphylococcal intoxication requires killing staphylococci before they reach large numbers, as in the late stages of the logarithmic growth phase (the time when enterotoxin is first detected). Enterotoxin has been reported as being produced in as few as 4 or 5 hours (Barber 1914; Jensen 1945; Dack 1956A; Donnelly *et al.* 1968; Genigeorgis *et al.* 1969). Thus, heat treatment should be applied within 3 hours after potential contamination. If foods are held at temperatures at which *S. aureus* cannot grow, as in the case of chilled (40° F) raw milk, obviously, the time before beginning heat treatment can be safely extended.

It is generally accepted that if moist foods are heated to an internal temperature of 165° F (73.9° C) or higher, large numbers of vegetative organisms such as *S. aureus* will be either killed or greatly reduced in number. Heating food to a lower temperature and holding it at that temperature for a longer time may have the same lethal effect. In custard and chicken-a-la-king, for instance, more than 10 million cells of enterotoxigenic strains of *S. aureus* were reduced to nondetectable levels (7D reduction) at temperatures of 140° F (60° C) in 53 minutes and at 150° F (65.6° C) in 6 minutes (Angelotti *et al.* 1961A).

A number of investigators have studied the effect of heat on *S. aureus* in foods. Cathcart *et al.* (1942) observed that *S. aureus* did not survive the proper baking of custard, and that bringing custard to a boil after adding thickening ingredients killed *S. aureus*. Hussemann and Tanner (1947) reported that staphylococci in sponge cake batter were killed when the temperature of the cake reached 167° F (75.2° C). Temperatures of 185 to 199.4° F (84.9–93° C) rendered puddings, custards, and salad dressing containing eggs free of *S. aureus*, but it survived at the minimum temperature for eggs to coagulate (Kintner and Mangel 1953B).

Castellani *et al.* (1953) found that a temperature of 160° F (71.1° C) in the center of poultry stuffing was sufficient to kill *S. aureus*. They also observed a post-oven temperature rise of 5 to 10° F when the cooked turkey stuffing was left at room temperature for 20 minutes. They recommended that stuffed turkeys be cooked until the stuffing reached an internal temperature of 165° F (73.9° C). Rogers and Gunderson (1958) confirmed this recommendation. They observed that minimum temperatures of 160° F (71.1° C) must be reached in stuffings of large turkeys and 165° F (73.9° C) in stuffings of smaller turkeys to kill 10^7 *S. aureus* per gm.

Wiedeman *et al.* (1956) found that large numbers of inoculated *S. aureus* survived internal temperatures of around 145° F (62.8° C) in baked chicken casseroles. Wiedeman *et al.* (1957) found that 10^7 *S. aureus* cells per gm in meat loaf were reduced to 100 per gm when an internal temperature of 160° F (71.1° C) was reached. In a study of frozen meat pies, Kerleuk and Gunderson (1961) observed that large numbers of *S. aureus* cells survived freezing and then baking at 425°F (218°C) for 20 minutes; the internal temperature rose to 124°F (51.1°C) during baking. Few survived, however, when the baking time was lengthened to 30 and 40 minutes. Internal temperatures of 170°F (76.7°C) and 214°F (101.1°C), respectively, were reached. Wilkinson *et al.* (1965) found that *S. aureus* survived in eastern-type turkey rolls when the rolls reached internal temperatures of 140°F (60°C) but did not survive when temperatures of 150°F (65.6°C) were attained.

Other studies have been clouded by the possible presence of micrococci or enterococci that have a higher heat resistance than *S. aureus* and by the use of a laboratory method that did not differentiate these bacteria from *S. aureus*. For instance, Gross and Vinton (1947) observed that *S. aureus* in pasteurized and fresh meat samples, which may have contained heat resistant micrococci or enterococci, had a tenfold higher thermal resistance than *S. aureus* in inoculated sterilized meat. Deskins and Husseman (1954) observed that the counts of micrococci in ham loaves inoculated with *S. aureus* but containing other micrococci decreased by two logs when internal temperatures of 176°F (80°C) were reached. Large numbers of micrococci were still found in ham croquettes that were cooked to internal temperatures of only 139°F (59.4°C).

Besides observations from experimental data that large numbers of *S. aureus* are killed within a short period of time when foods are heated to internal temperatures of 160°F (71.1°C) or higher, calculations show that temperatures between 160°F (71.1°C) and 170°F (76.7°C) should kill large numbers of *S. aureus* in a minute or less. Table 2.4 shows D_{140} values ranging from 1.3 to 14.5 minutes for *S. aureus.* The D value, known as the decimal reduction time, is the time required at any temperature to destroy 90% of a given bacterial population. A 12D reduction of *Clostridium botulinum* is used in calculating canning operations, and 7D reduction of Salmonella is used in egg pasteurization. The destruction of 99.9999% of *S. aureus* is equivalent to 6D. Theoretically, if a million organisms were initially present, one organism would remain alive after the heat treatment. If a D value of 7.8 is chosen as an example, the time required for 12D destruction of 140°F (60°C) would be 93.6 minutes; for 9D, 70.2 minutes; for 7D, 54.6 minutes, and for 6D, 46.8 minutes. Experiments of Angelotti *et al.* (1961A) observed 7D destruction of *S. aureus* in custard and chicken-a-la-king in 53 to 59 minutes. These results are consistent with the calculated values.

D values decrease in a fairly regular way as the temperature is raised; thus, less time will be needed to kill a given number of organisms as the temperature is increased. The decimal reduction time (D) will decrease a tenth of its value when the temperature is raised by the amount equal to the Z value. The Z value for *S. aureus* is approximately $10°F$ (Table 2.4). Thus, for an equivalent of destruction at $150°F$ ($65.6°C$), 9.4 minutes would be required for 12D destruction; 7.0 minutes for 9D; 5.5 minutes for 7D; and 4.7 minutes for 6D. For an equivalent destruction at $160°F$ ($71.1°C$), 0.9 minutes would be required for 12D; 0.7 minutes for 9D; 0.6 minutes for 7D. For an equivalent destruction at $170°F$ ($76.7°C$), 0.09 minutes would be required for 12D and 0.06 minutes for 7D.

From this type of calculation, it can be seen that a 12D destruction of a product at the highest D value listed in Table 2.4 would occur at 1.74 minutes at $160°F$ ($71.1°C$) and 0.17 minutes at $170°F$ ($76.7°C$). A 12D value, however, is a much higher level of contamination than would be expected under any condition of natural contamination and growth. Thus, an internal temperature of $165°F$ ($73.9°C$) should be adequate to kill any expected levels of contamination of *S. aureus* cells in foods in less than a minute.

Cooking in microwave ovens has been shown to be as effective in killing *S. aureus* in custard, scrambled eggs, and ground beef patties as cooking in conventional ovens or broilers, but the time for destruction in microwave ovens only took one-half to one-ninth as long (Dessel *et al.* 1960). Woodburn *et al.* (1962) found that over a million *S. aureus* cells inoculated in 4-3/4 × 6-1/2 inch pouches of chicken, and chicken and broth, were killed in 120 seconds in microwave ovens. In the same amount of white sauce, these organisms were killed in 150 seconds. The internal temperatures were 195.8 to $206.6°F$. (91 to $97°C$). The same amount of destruction occurred when the pouches were immersed in boiling water for 10 minutes.

Terminal heat treatment, an old established public health principle, has been advocated as a means of preventing staphylococcal food poisoning outbreaks. At one time, reheating cream-filled pastry after filling was recommended (Stritar *et al.* 1936; Gilcreas and Colman 1941). It was suggested that cream puffs or eclairs be put into a pan, covered with an inverted pan, and heated in an oven at a temperature of $420°F$ ($215.6°C$) for 20 minutes. Although this procedure was effective in destroying staphylococci (and preventing staphylococcal intoxications if it was used before enterotoxin was produced), it lost favor as refrigeration became more readily available. Enterotoxins would not be destroyed by this procedure. Terminal heating procedures for other foods were recommended by Abrahamson and Clinton (1960). They found that contaminated meats (such as chicken, turkey, ham, roast beef, and corned beef) and seafoods (such as shrimp, crabmeat, and lobster meat) which were handled after cooking could be rendered free of S. *aureus* and indicator bacteria by boiling for 30 to 60 seconds in appropriate stock.

Some foods are ordinarily cooked to temperatures that will kill S. *aureus*, for example: breads, cakes, cookies, soups, gravies, stews, omelets, souffles, hard-cooked eggs, custards, bacon, pork, and chicken (Longree 1972). Staphylococci, however, will survive in other foods that are usually given lower temperature heat treatments (such as soft scrambled eggs, rare roasts, meat patties, meat loaves, oyster stew, and hollandaise sauce). Still other foods (such as leftovers, meat and fish patties and croquettes, and meringue pies) are frequently undercooked because of carelessness, lack of time, or lack of knowledge. Cooked proteinaceous foods which are depleted of microbial competitors readily support the growth of S. *aureus* if they are stored within the temperature range at which staphylococci grow.

Inhibition of Growth

Contemporary practices of personal hygiene and environmental sanitation do not ensure that the ubiquitous S. *aureus* will be kept out of foods. Therefore, foods must be treated as if they are contaminated. The mere presence of S. *aureus* in a food, however, will not cause illness. An enterotoxigenic strain of S. *aureus* must grow in a food long enough to allow production and accumulation of enterotoxin.

The most crucial procedure to prevent staphylococcal intoxication is to keep S. *aureus* from multiplying, and thus keep it from elaborating enterotoxin. This can only be accomplished by allowing foods to remain within the temperature range at which staphylococci can grow for such a short time that these organisms stay in the lag phase, by rapidly lowering temperatures of foods and storing them at temperatures at which staphylococci cannot grow, by holding hot foods above temperatures that allow staphylococci to grow, or by altering foods to make them unacceptable for supporting staphylococcal growth.

Time.—Potentially hazardous foods should not be held within the temperature range, 44 to 114°F (6.7 to 45.6°C), that permits the growth of S. *aureus* for more than 3 cumulative hours, including chilling time. This objective is difficult to obtain in many foods unless special procedures are taken to cool foods rapidly. Straka and

Combs (1952) showed the effect of the lag phase in controlling growth of *S. aureus*. When 211 × 200 size cans containing inoculated creamed chicken were held at room temperature for 2 hours before freezing, low bacterial counts were observed after the contents were thawed in less than 11 hours; but when these cans were held for 5 hours before freezing, and thawed in a similar manner, enormous bacterial counts were observed. During the first 2 hours, the organisms were in the lag phase, later they multiplied at an exponential rate. To prevent staphylococcal growth during food processing or preparation operations, it is essential that they be processed or prepared during the bacterial lag period (approximately 1 to 2 hours under ideal conditions for growth).

Cold Storage.—If potentially hazardous foods are stored for less than 3 days, they should be kept at temperatures of 45°F (7.2°C) or below (U.S. Department of Health, Education, and Welfare 1962). If they are stored longer, temperatures of 40°F (4.4°C) or less should be used.

Rapid Cooling.—Temperature alone considered for food storage can be misleading. The rate of cooling (and hence the interval during which a food remains at temperatures suitable for growth of staphylococci) is influenced by the nature of the food (composition, state, geometry, bulk, and viscosity), the cooling environment, and the characteristics of the food container. Solid foods cool by conduction which is slow cooling. Liquid foods cool by convection; such currents are sluggish in highly viscous foods. Sandwiches, potato salad, stuffing, bread pudding, custards, gravy, croquettes, meat loaves, casseroles, meat, and poultry—all of which have been involved in staphylococcal foodborne outbreaks—cool slowly.

Prompt cooling of processed or leftover foods is essential for preventing staphylococcal growth. Foods should never be cooled to room temperature before being refrigerated; they should be re-frigerated before their temperature drops below 140°F. Cooked foods should be cooled immediately after cooking, or they should be held at room temperature for not more than 30 minutes before being refrigerated (Black and Lewis 1948). The period of time it takes foods to cool to temperatures at which staphylococci cease multiplication is increased as the time foods are held at room temperature before they are refrigerated is increased (Longree and White 1955).

The size and shape of the container and its fill greatly influence cooling time. The farther heat must travel from the center of a food to the food surface or container wall where it is taken up by the cooling medium, the longer it takes the food to cool. The internal temperature of food cooled in shallow pans falls faster than that of foods cooled in large, deep containers. For instance, Lewis *et al.* (1953) observed that it took 9 hours for a 100-serving quantity of chicken salad to cool from 78° F (25.6° C) to 50° F (10° C) in a deep can in a walk-in refrigerator; it took 3 hours for a 24-serving quantity of chicken salad in a round bowl to cool to the same temperature. *Staphylococcus aureus* multiplied in the salad even though the deep can was refrigerated. No growth, however, was observed in the round bowl or in 50 servings of salad in a shallow pan. It took 24 hours for chicken salad in a 30-pound can to cool from 79° F (26.1° C) to 47° F (8.3° C), the temperature of the refrigerator unit; but it took only 3 hours for chicken salad piled to a height of 4 inches in a shallow pan to cool to the same temperature (Weiser *et al.* 1954). McDivitt and Hammer (1958) observed that *S. aureus* grew in custard that was stored in 100-serving quantities in 20-quart and 14-quart stock pots in a refrigerator (40° F; 4.4° C), but no growth occurred in custard that was stored at the same temperature in shallow pans.

Cooling rate is also affected by the material of which the container is made. Foods stored in containers which are made of good conductors of heat, such as stainless steel, cool faster than foods stored in crockery, glass, or plastic. Miller and Smull (1955) reported that 3 gallons of pie filling cooled faster in enamel pans than in earthenware crocks and that 1 gallon of potato salad cooled faster in stainless steel pans than in crocks.

Foods will cool faster in refrigerators than at room temperatures; they cool faster in freezers than in refrigerators and faster in ice or water baths than in refrigerators or freezers (Miller and Smull 1955). Foods of the same type and quantity will cool more rapidly in refrigerators that have the coldest temperature (Longree and White 1955). Movement of air around foods dissipates heat faster than still air. Thus, placing foods so that there is space around each item, the use of wire racks, and the use of fans in refrigerators enhances air circulation (Moragne *et al.* 1960; Newcomer *et al.* 1962). Walk-in refrigerators dissipate heat faster than do reach-in refrigerators. Stirring foods in vertical mixers increases cooling rates. Practical application of this technique is given for cornstarch pudding by McDivitt and Hammer (1958) and for turkey stock by Bryan and McKinley (1974). Other ways to cool foods rapidly in food service operations include cutting roasts into 1-1/4 pound portions before chilling (Bryan and Kilpatrick 1971), slicing cooked meats into cold pans—pans in contact with ice (Bryan and McKinley 1974), immersing pans of hot foods in pans or bags of ice (Miller and Smull 1955; Bryan and McKinley 1974), immersing pans of hot food into water baths (Miller and Smull 1955; Moragne *et al.* 1959; Bryan and

McKinley 1974), by scraper-lifter agitation (Moragne *et al.* 1963), cooling by cold tube agitator (Longree *et al.* 1960; Moragne *et al.* 1961), by agitation during refrigeration (Longree *et al.* 1963), and by putting foods into freezers (Hodge 1960; Bryan and McKinley 1974). Bryan and McKinley (1974), for instance, showed that 2-1/2 gallons of turkey stock (in a 5-gallon pot) at a temperature of 160°F (71.1°C) could be cooled to less than 65°F (18.3°C) in an ice bath or a water bath in an hour. When this volume of 160°F (71.1°C) stock was mixed in a vertical mixer for an hour, it cooled to less than 95°F (35.3°C). This is in contrast to cooling this volume of stock at room temperature or in a refrigerator; within an hour, only a 30°F (16.7°C) drop in temperature occurred. Staphylococcal growth can be essentially controlled if methods of rapid cooling are used and if these methods are followed by refrigerated storage.

Hot Holding.—If foods are held warm in steam tables, in hot air cabinets, or under infrared lamps, they should be kept at an internal temperature that prevents staphylococcal growth. A temperature of 140°F (60°C) is recommended (U.S. Department of Health, Education, and Welfare 1962). This temperature is close to the desired eating temperature for most foods (Blaker *et al.* 1961; Thompson and Johnson 1963) and, if given enough time (1 hour), will kill large numbers of staphylococci that may have survived cooking or that recontaminated the product after cooking (Angelotti *et al.* 1961B). To maintain internal temperatures of 140°F (60°C), foods must be hotter than this temperature when they are put in warming devices, and the water or air temperature must be higher than 140°F (60°C). It is unlikely that any arrangement of infrared lights can satisfactorily maintain temperatures of foods on open display. Bryan and Kilpatrick (1971) observed that a 200-watt bulb 8-1/2 inches from the surface of roast beef kept the surface at an average of only 115°F (46.1°C). Bulbs are seldom located this close to foods, and the bottom and shaded areas of foods receive little, if any, heat.

Temperature Evaluations.—Food temperatures can be satisfactorily evaluated only by inserting a bayonet-type thermometer into the center of the food, not by checking refrigerator, oven, or steam table temperatures. The recorded temperatures should be interpreted in reference to the preparation and storage history of the food.

Moisture Control.—Staphylococcal growth can be inhibited by drying a food until a low moisture level is attained. Although *S. aureus* cannot grow in foods that have a water activity (a_w) lower than 0.85 or in dried foods, such products may selectively promote its growth over other organisms before this a_w is attained. Areas of moisture migration or condensation accumulation and rehydrated foods readily support the growth of S. aureus.

Competition.—In fermentations, such as cheese making, it is essential that lactic acid starter culture multiply at a rapid rate during the early stages of processing to inhibit any staphylococci present in the product. Rapid growth of competitive organisms in any food inhibit staphylococci and cause the food to become organoliptically unsatisfactory before enterotoxin is formed.

pH Control.—In principle, the growth of S. aureus can be controlled by the use of acid ingredients. In practice, however, it is difficult to lower the pH of most foods to a value that will prevent growth of S. aureus without adversely affecting their taste. Ingredients that have been found to lower the pH and yet result in a palatable product for soups are tomatoes, green peppers, or okra (Longree et al. 1957); those for poultry stuffings are apricot, cranberry, orange juice, or raisins (Longree et al. 1958); those for sandwich fillings are mayonnaise, sweet pickles, lemon juice, pineapple juice, or cranberry juice (Longree et al. 1959B); those for potato and turkey salads are mayonnaise or pickles (Longree et al. 1959A); those for entrees are large amounts of catsup or tomato juice (Longree 1972); those for custards are fruit fillings (Ryberg and Cathcart 1942; Castellani et al. 1955); and that for cheese cake is lactic acid (Cathcart et al. 1947). Acetic acid has a definite inhibitory effect on S. aureus, and concentrations as low as 0.2% are lethal if given enough time (Hucker and Haynes 1937).

To inhibit staphylococcal growth, acid ingredients must be added in sufficient quantities and at the proper time. A pH of 4.5 in a product must be achieved to ensure the inhibition of S. aureus although there will be little growth at pH values below 5. This requires a relatively high proportion of acid ingredients which produce a fairly acid product and make many dishes unpalatable. The pH is lowered more successfully when ingredients are marinated in an acid substance such as vinegar, fruit juices, or salad dressing (Longree et al. 1959A, 1959B). In bulk items, protein ingredients must be finely divided and thoroughly mixed to distribute the acid ingredients over the individual pieces to allow penetration of acid. Mayonnaise or other acid-type spreads should be applied so as to contact the meat, cheese, or egg in sandwiches instead of being applied to lettuce or bread (McCroan et al. 1964). Applying acid-type spreads onto both sides of the meat enhances inhibitory effects. Acid ingredients should be added to a food as soon as possible after there is a likelihood of contamination, as when cutting, chopping, and mixing. Because of so many limiting factors in its use, acidification of a product should be used as an aid to supplement other measures (such as proper time-temperature control) to inhibit the growth of S. aureus.

Thus, to prevent staphylococcal intoxication efforts must be made to minimize contamination of foods, to inhibit the growth of staphylococci that do contaminate foods, and to kill staphylococci in foods by thorough cooking, or other effective means, before they have a chance to elaborate enterotoxin. Inhibition of growth is the most important and practical of these measures. Preventive measures must be applied at food processing plants, food service establishments, and homes.

BIBLIOGRAPHY

ABRAHAMSON, A. E. 1958. Frozen foods: Their importance on health. Assoc. Food Drug Officials U.S. Quart. Bull. 22, 7–16.
ABRAHAMSON, A. E., BUCHBINDER, L., GUENNEL, J., and HELLER, M. 1959. A study of frozen precooked foods: Their sanitary quality and microbiological standards for control. Assoc. Food Drug Officials U.S. Quart. Bull. 23, 63–72.
ABRAHAMSON, A. E., and CLINTON, A. F. 1960. The control of bacterial populations in foods. Assoc. Food Drug Officials U.S. Quart. Bull. 24, 31–38.
AGAFENOVA, L. I., and TKRACHENKO, A. M. 1964. Staphylococci of intestinal origin and their properties. Lab. Delo 2, 106–110. Cited by Excerpta Med. Sect. 4, 766, 1965. (Russian)
AGER, E. A., and TOP, F. 1968. Food poisoning. In Communicable and Infectious diseases, 6th Edition, F. Top (Editor). The C. V. Mosby Co., St. Louis.
AHN, T. H., NISHIHARA, H., CARPENTER, C. M., and TAPLIN, G. V. 1964. Viability and metabolism of Staphylococcus aureus after freezing, lyophilization, and irradiation. J. Bacteriol. 88, 545–552.
ALDER, V. G., GILLESPIE, W. A., and WALLER, L. J. 1962. The use of tellurite-egg yolk medium for the isolation and identification of staphylococci in hospitals. J. Appl. Bacteriol. 25, 436–440.
ALLEN, V. D., and STOVALL, W. D. 1960. Laboratory aspects of staphylococcal food poisoning from colby cheese. J. Milk Food Technol. 23, 271–274.
ALLISON, V. D., HOBBS, B. C., and MARTIN, P. H. 1949. A widespread outbreak of staphylococcal food poisoning. Monthly Bull. Min. Health Public Health Lab. Serv. (British) 8, 38–47.
ALLWOOD, M. C., and RUSSELL, A. D. 1966. Some factors influencing the revival of heat-damaged Staphylococcus aureus. Can. J. Microbiol. 12, 1295–1297.
AMIN, V. M., and OLSON, N. F. 1967. Factors affecting the resistance of Staphylococcus aureus to hydrogen peroxide treatments in milk. Appl. Microbiol. 15, 97–101.
ANDERSON, E. S., and WILLIAMS, R. E. O. 1956. Bacteriophage typing of enteric pathogens and staphylococci and its use in epidemiology. J. Clin. Pathol. 9, 94–127.
ANDERSON, P. H. R., and STONE, D. M. 1955. Staphylococcal food poisoning associated with spray-dried milk. J. Hyg. 53, 387–397.
ANGELOTTI, R. 1971. Catered convenience foods: Production and distribution problems and microbiological standards. J. Milk Food Technol. 34, 227–231.
ANGELOTTI, R., FOTER, M. J., and LEWIS, K. H. 1960. Time-temperature effects on salmonellae and staphylococci in foods. II. Behavior at warm

holding temperatures. Thermal-death-time studies. U.S. Dept. Health, Education, and Welfare, Public Health Serv., Cincinnati, Ohio.

ANGELOTTI, R., FOTER, M. J., and LEWIS, K. H. 1961A. Time-temperature effects on salmonellae and staphylococci in foods. I. Behavior in refrigerated foods; II. Behavior at warm holding temperatures. Am. J. Public Health *51*, 76–88.

ANGELOTTI, R., FOTER, M. J., and LEWIS, K. H. 1961B. Time-temperature effects on salmonellae and staphylococci in foods. III. Thermal death time studies. Appl. Microbiol. *9*, 308–315.

APPLEMAN, M. D., BAIN, N., and SHERMAN, J. M. 1964. A study of some organisms of public health significance from fish and fishery products. J. Appl. Bacteriol. *27*, 69–77.

ARBUTHNOTT, J. P. 1970. Staphylococcal α-Toxin. *In* Microbial Toxins, Vol. III, T. C. Montie, S. Kadis, and S. J. Ajl (Editors). Academic Press, New York.

ARMIJO, R., HENDERSON, D. A., TIMOTHEE, R., and ROBINSON, H. B. 1957. Food poisoning outbreaks associated with spray-dried milk: An epidemiologic study. Am. J. Public Health *47*, 1093–1100.

ASANTE, R. O. 1969. Staphylococcal food poisoning. Ghana Med. J. *8*, 51–52.

AVENA, R. M., and BERGDOLL, M. S. 1967. Purification and some physiochemical properties of enterotoxin C, *Staphylococcus aureus* strain 361. Biochem. *6*, 1474–1480.

BAER, E. F. 1971. Isolation and enumeration of *Staphylococcus aureus*: Review and recommendation for revision of AOAC method. J. Assoc. Offic. Anal. Chem. *53*, 732–735.

BAER, E. F., FRANKLIN, M. K., and GILDEN, M. M. 1966. Efficiency of several selective media for isolating coagulase-positive staphylococci from food products. J. Assoc. Offic. Anal. Chem. *49*, 267–269.

BAER, E. F., GILDEN, M. M., WIENKE, C. L., and MELLITZ, M. B. 1971. Comparative efficiency of two enrichment and four plating media for isolation of *Staphylococcus aureus*. J. Assoc. Offic. Anal. Chem. *54*, 736–738.

BAIRD-PARKER, A. C. 1962A. An improved diagnostic and selective medium for isolating coagulase-positive staphylococci. J. Appl. Bacteriol. *25*, 12–19.

BAIRD-PARKER, A. C. 1962B. The performance of an egg yolk-tellurite medium in practical use. J. Appl. Bacteriol. *25*, 441–444.

BAIRD-PARKER, A. C. 1962C. The occurrence and enumeration, according to a new classification, of micrococci and staphylococci in bacon and on human and pig skin. J. Appl. Bacteriol. *25*, 352–361.

BAIRD-PARKER, A. C., 1963. A classification of micrococci and staphylococci based on physiological and biochemical tests. J. Gen. Microbiol. *30*, 409–427.

BAIRD-PARKER, A. C. 1965. Staphylococci and their classification. Ann. N.Y. Acad. Sci. *128*, 4–25.

BAIRD-PARKER, A. C. 1971. Factors affecting the production of bacterial food poisoning toxins. J. Appl. Bacteriol. *34*, 181–197.

BAIRD-PARKER, A. C., and DAVENPORT, E. 1965. The effect of recovery medium on the isolation of *Staphylococcus aureus* after heat treatment and after the storage of frozen dried cells. J. Appl. Bacteriol. *28*, 390–402.

BARBER, L. E., and DEIBEL, R. H. 1972. Effect of pH and oxygen tension on staphylococcal growth and enterotoxin formation in fermented sausage. Appl. Microbiol. *24*, 891–898.

BARBER, M. A. 1914. Milk poisoning due to a type of *Staphylococcus albus* occurring in the udder of a healthy cow. Philippine J. Sci. *9B*, 515–519.

BARBER, M., and KUPER, S. W. A. 1951. Identification of *Staphylococcus pyogenes* by the phosphatase reaction. J. Pathol. Bacteriol. *63*, 65–68.

BASS, G. K., and STUART, L. S. 1968. Methods of testing disinfectants. *In* Disinfection, Sterilization, and Preservation, C. A. Lawrence, and S. S. Block (Editors). Lea and Febiger, Philadelphia.

BEAMER, P. R., and TANNER, F. W. 1939. Resistance of non-sporeforming bacteria to heat. Zentr. Bakteriol. Parasitenk. Abstr. II. *100*, 81-98. (German)

BELAM, F. A. 1947. Staphylococcal food poisoning by meat-pie jelly. Lancet *1*, 64-65.

BERGDOLL, M. S. 1967. The staphylococcal enterotoxins. InBiochemistry of Some Foodborne Microbial Toxins, R. I. Mateles, and G. N. Wogan (Editors). M.I.T. Press, Cambridge, Mass.

BERGDOLL, M. S. 1970. Enterotoxins. In Microbial Toxins, Vol. III, T. C. Montie, S. Kadis, and S. J. Ajl (Editors). Academic Press, New York.

BERGDOLL, M. S., BORJA, C. R., and AVENA, R. M. 1965. Identification of a new enterotoxin as enterotoxin C. J. Bacteriol. *90*, 1481-1485.

BERGDOLL, M. S., BORJA, C. R., ROBBINS, R. N., and WEISS, K. F. 1971. Identification of enterotoxin E. Infect. Immunity *4*, 593-595.

BERGDOLL, M. S., SURGALLA, M. J., and DACK, G. M. 1959. Staphylococcal enterotoxin: Identification of a specific precipitating antibody with enterotoxin neutralizing property. J. Immunol. *8*, 334-338.

BERGDOLL, M. S., WEISS, K. F., and MUSTER, M. S. 1967. The production of staphylococcal enterotoxin by a coagulase-negative microorganism. Bacteriol. Proc. p. 12, Am. Soc. Microbiology, Washington, D.C.

BETHUNE, D. W., BLOWERS, R., PARKER, M., and PASK, E. A. 1965. Dispersal of *Staphylococcus aureus* by patients and surgical staff. Lancet *1*, 480-483.

BHATT, U. A., and BENNETT, F. W. 1964. Thermal death time studies of staphylococci in milk. J. Dairy Sci. *47*, 666.

BLACK, L. C., and LEWIS, M. N. 1948. Effect on bacterial growth of various methods of cooling cooked foods. J. Am. Dietet. Assoc. *24*, 399-404.

BLACKMAN, S. S. 1935. Acute staphylococcal infection of jerjunum and ileum. Bull. Johns Hopkins Hosp. *57*, 289-295.

BLAIR, E. B., EMERSON, J. S., and TULL, B. S. 1967. A new medium, salt mannitol plasma agar, for the isolation of *Staphylococcus aureus*. Am. J. Clin. Pathol. *47*, 30-39.

BLAIR, J. E. 1970. Staphylococcal infections. In Diagnostic Procedures for Bacterial, Mycotic, and Parasitic Infections, 5th Edition, H. L. Bodily, E. L. Updyke, and J. O. Mason (Editors). Am. Public Health Assoc., New York.

BLAIR, J. E., et al. 1967. Report of the subcommittee on phage-typing of staphylococci of the international committee on nomenclature of bacteria, Moscow, July 1966. Intern. J. Syst. Bacteriol. *17*, 113-125.

BLAIR, J. E., and WILLIAMS, R. E. O. 1961. Phage typing of staphylococci. Bull. World Health Organ. *24*, 771-784.

BLAKER, G. G., NEWCOMER, J. L., and RAMSEY, E. 1961. Holding temperatures needed to serve hot foods hot. J. Am. Dietet. Assoc. *38*, 455-457.

BLAKEY, D. L., ANDREWS, R. H., and MOORE, S. L. 1968. Food Poisoning—Morton, Mississippi. Morbidity Mortality Weekly Rept. (U.S.) *17*, 348-349.

BLOUSE, L., HUSTED, P., McKEE, A., and GONZALES, J. 1964. Epizootiology of staphylococci in dogs. Am. J. Vet. Res. *25*, 1195-1199.

BLOWERS, R., and HODGKIN, K. 1967. Spread of hospital staphylococci in healthy families—a study from general practice. A report to the M.R.C. committee for research in general practice. Brit. Med. J. *4*, 642-644.

BLUMENTHAL, H. J., and PAN, Y. L. 1963. Acid phosphatase production by mercuric ion- or penicillin-resistant and -sensitive *Staphylococcus aureus*. Proc. Soc. Exptl. Biol. Med. *113*, 322-326.

BODILY, H. L., and UPDYKE, E. L. 1970. Microbial sensitivity tests. In Diagnostic Procedures for Bacterial, Mycotic, and Parasitic Infections, 5th Edition, H. L. Bodily, E. L. Updyke, and J. O. Mason (Editors). Am. Public Health Assoc., New York.

BORJA, C. R., and BERGDOLL, M. S. 1967. Purification and partial characterization of enterotoxin C produced by *Staphylococcus aureus* strain 137. Biochem. *6*, 1467-1473.

BRANDISH, J. M., and WILLIS, A. T. 1970. Observations on the coagulase and deoxyribonuclease tests for *Staphylococci.* J. Med. Lab. Technol. *27*, 355-354.

BRECKINRIDGE, J. C., and BERGDOLL, M. S. 1971. Outbreaks of food-borne gastroenteritis due to a coagulase-negative enterotoxin producing staphylococcus. New Engl. J. Med. *284*, 541-543.

BRODIE, J. 1965. Hand hygiene. Scot. Med. J. *10*, 115-125.

BRODIE, J., KERR, M. R., and SOMMERVILLE, T. 1956. The hospital staphylococcus: A comparison of nasal and fecal carrier states. Lancet *1*, 19-20.

BRODIE, J., SOMMERVILLE, T., and WILSON, S. G. 1956. Coagulase-positive staphylococci: A serial survey for nasal carriers during the first six months of nursing training. Brit. Med. J. *1*, 667-669.

BROWN, G. R., and BROWN, C. A. 1965. Sensitization of erythrocytes with staphylococcal enterotoxin by means of tolylene-2, 4-diisocyanate for the passive hemagglutination reaction. Bacteriol. Proc. p. 72, Am. Soc. Microbiology, Washington, D.C.

BROWN, W. J., WINSTON, R., and SOMMERS, S. C. 1953. Membranous staphylococcal enteritis after antibiotic therapy; report of two cases. Am. J. Digest. Diseases *20*, 73-75.

BRUUN, J. N., BOE, J., and SOLBERG, C. O. 1968. Disinfection of the hands of ward personnel: A comparison of six disinfectants. Acta Med. Scand. *184*, 417-423.

BRYAN, F. L. 1969. Use of visual aids in effective training of food service management in foodborne disease control. J. Milk Food Technol. *32*, 245-250.

BRYAN, F. L. 1972. Emerging foodborne diseases. Part II. Factors that contribute to outbreaks and their control. J. Milk Food Technol. *35*, 632-638.

BRYAN, F. L. 1973A. Activities of the Center for Disease Control in public health problems related to the consumption of fish and fishery products. *In* Microbial Safety of Fishery Products. Academic Press, New York.

BRYAN, F. L. 1973B. Training public health workers and food service managers. *In* The Microbiological Safety of Food, B. C. Hobbs, and J. H. B. Christian (Editors). Academic Press, New York.

BRYAN, F. L. 1973C. Control of foodborne diseases in the food service industry: Training kit. Center for Disease Control, Atlanta, Ga.

BRYAN, F. L. 1974A. Microbiological food hazards today—based on epidemiological information. Food Technol. *28*, 52-64, 84.

BRYAN, F. L. 1974B. Public health aspects of cream-filled pastries. N.Y. State Assoc. Milk Food Sanitarians Ann. Rept. *48*, 85-100.

BRYAN, F. L., AYRES, J. C., and KRAFT, A. A. 1968. Salmonellae associated with further-processed turkey products. Appl. Microbiol. *16*, 1-9.

BRYAN, F. L., and KILPATRICK, E. G. 1971. *Clostridium perfringens* related to roast beef cooking, storage, and contamination in a fast food service restaurant. Am. J. Public Health *61*, 1869-1885.

BRYAN, F. L., and McKINLEY, T. W. 1974. Prevention of foodborne illness by time-temperature control of thawing, cooking, chilling, and reheating turkeys in school lunch kitchens. J. Milk Food Technol. *37*, 420-429.

BRYAN, F. L., McKINLEY, T. W., and MIXON, B. 1971. Use of time-temperature evaluation in detecting the responsible vehicle and contributing factors of foodborne disease outbreaks. J. Milk Food Technol. 34:576-582.

BUCHANAN, R. E., and GIBBONS, N. E. 1974. Bergey's Manual of Determinative Bacteriology, 8th Edition. The Williams & Wilkins Co., Baltimore, Md.

BUCHBINDER, L., LAUGHLIN, V., WALTER, M., and DANGLER, G. 1949. A survey of frozen pre-cooked foods with special reference to chicken a la king. J. Milk Food Technol. *12*, 209-213.

BULGER, R. J., and SHERRIS, J. C. 1968. Decreased incidence of antibiotic resistance among *Staphylococcus aureus*. A study in a university hospital over a 9-year period. Ann. Intern. Med. *69*, 1099-1108.

BUSTA, F. F., and JEZESKI, J. J. 1963. Effect of sodium chloride concentration in an agar medium on growth of heat-shocked *Staphylococcus aureus*. Appl. Microbiol. *11*, 404-407.

BUTTIAUX, R., and MORIAMEZ, J. 1958. The behavior of indicator organisms of fecal contamination in cured meats. *In* Proc. Second Intern. Symp. Food Microbiol. H. M. Stationary Office, London (French). *Cited by* Eddy and Ingram, 1962.

BUTTIAUX, R., and PIERRET, J. 1949. Origin of pathogenic staphylococci in the feces of normal infants. Ann. Inst. Pasteur *76*, 480-484. (French)

CALIA, F. M. *et al.* 1969. Importance of the carrier state as a source of *Staphylococcus aureus* in wound sepsis. J. Hyg. *67*, 49-57.

CAMPBELL, A. C. P. 1948. The incidence of pathogenic staphylococci in the throat, with special reference to glandular fever. J. Pathol. Bacteriol. *60*, 157-169.

CANALE-PAROLA, E., and ORDAL, Z. J. 1957. A survey of the bacteriological quality of frozen poultry pies. Food Technol. *11*, 578-582.

CARTER, C. H. 1959. Staining of coagulase-positive staphylococci with fluorescent antisera. J. Bacteriol. 77, 670-671.

CARTER, C. H. 1960. Egg yolk agar for isolation of coagulase-positive staphylococci. J. Bacteriol. *79*, 753-754.

CARTER, E. J., MOORE, A. N., and GREGORY, C. L. 1964. Can teaching machines help in training employees? J. Am. Dietet. Assoc. *44*, 271-276.

CASMAN, E. P. 1960. Further serological studies of staphylococcal enterotoxin. J. Bacteriol. *79*, 849-856.

CASMAN, E. P. 1965. Staphylococcal enterotoxins. Ann. N.Y. Acad. Sci. *128*, 124-131.

CASMAN, E. P. 1967. Staphylococcal food poisoning. Health Lab. Sci. *4*, 199-206.

CASMAN, E. P., and BENNETT, R. W. 1963. Culture medium for the production of staphylococcal enterotoxin A. J. Bacteriol. *86*, 18-23.

CASMAN, E. P., and BENNETT, R. W. 1965. Detection of staphylococcal enterotoxin in food. Appl. Microbiol. *13*, 181-189.

CASMAN, E. P., BENNETT, R. W., DORSEY, A. E., and ISSA, J. A. 1967. Identification of a fourth staphylococcal enterotoxin, enterotoxin D. J. Bacteriol. *94*, 1875-1882.

CASMAN, E. P., BENNETT, R. W., DORSEY, A. E., and STONE, J. E. 1969. The micro-slide gel double diffusion test for the detection and assay of staphylococcal enterotoxins. Health Lab. Sci. *6*, 185-198.

CASMAN, E. P., BERGDOLL, M. S., and ROBINSON, J. 1963A. Designation of staphylococcal enterotoxins. J. Bacteriol. *85*, 715-716.

CASMAN, E. P., McCOY, D. W., and BRANDLEY, P. J. 1963B. Staphylococcal growth and enterotoxin production in meat. Appl. Microbiol. *11*, 498-500.

CASTELLANI, A. G., CLARKE, R. R., GIBSON, M. I., and MEISNER, D. F. 1953. Roasting time and temperature required to kill food poisoning microorganisms introduced experimentally into stuffing in turkeys. Food Res. *18*, 131-138.

CASTELLANI, A. G., MAKOWSKI, R., and BRADLEY, W. B. 1955. Inhibiting the growth of food poisoning bacteria in meringue-topped fruit cream pies. Bacteriol. Proc. p 20, Am. Soc. Microbiology, Washington, D.C.

CATHCART, W. H., GODKIN, W. J., and BARNETT, G. 1947. Growth of *Staphylococcus aureus* in various pastry fillings. Food Res. *12*, 142-150.

CATHCART, W. H., and MERZ, A. 1942. Staphylococci and salmonella control in foods. III. Effect of chocolate and cocoa fillings on inhibiting growth of staphylococci. Food Res. 7, 96-103.

CATHCART, W. H., MERZ, A., and RYBERG, R. E. 1942. Staphylococci and salmonella controls in foods. IV. Effect of cooking bakery custards. Food Res. 7, 100-103.

CAUDILL, F. W., and MEYER, M. A. 1943. An epidemic of food poisoning due to pasteurized milk. J. Milk Technol. 6, 73-76.

CHAPMAN, G. H. 1944. The isolation of pathogenic staphylococci from feces. J. Bacteriol. 47, 211-212.

CHAPMAN, G. H. 1945. The significance of sodium chloride in studies of staphylococci. J. Bacteriol. 50, 201-203.

CHAPMAN, G. H. 1946. A single culture medium for selective isolation of plasma coagulating staphylococci and for improved testing of chromogenesis, plasma coagulation, mannitol fermentation and the Stone reaction. J. Bacteriol. 51, 409-410.

CHAPMAN, G. H. 1948. An improved Stone medium for the isolation and testing of food-poisoning staphylococci. Food Res. 13, 100-105.

CHAPMAN, G. H. 1949. Comparison of Ludlam's medium with staphylococcus medium 110 for the isolation of staphylococci that clot blood. J. Bacteriol. 58, 823.

CHESBRO, W. R., and AUBORN, K. 1967. Enzymatic detection of the growth of Staphylococcus aureus in foods. Appl. Microbiol. 15, 1150-1159.

CHOU, C. C., and MARTH, E. H. 1969. A comparison of direct plating and enrichment methods for detection and enumeration of coagulase positive staphylococci in frozen foods of animal origin. J. Milk Food Technol. 32, 398-403.

CHRISTIANSEN, L. N., and FOSTER, E. M. 1965. Effect of vacuum packaging on growth of Clostridium botulinum and Staphylococcus aureus in cured meats. Appl. Microbiol. 13, 1023-1025.

CHRISTIANSEN, L. N., and KING, N. S. 1971. The microbial content of some salads and sandwiches at retail outlets. J. Milk Food Technol. 34, 289-293.

CHU, F. S., THADHANI, K., SCHANTZ, E. J., and BERGDOLL, M. S. 1966. Purification and characterization of staphylococcal enterotoxin A. Biochem. 5, 3281-3289.

CLARK, W. S., JR., and NELSON, F. E. 1961. Multiplication of coagulase-positive staphylococci in grade A raw milk supplies. J. Dairy Sci. 44, 232-236.

CLETON, F. J., VANDER MARK, Y. S., and VANTOORN, M. J. 1968. Effect of shower-bathing on dispersal of recently acquired transient skin flora. Lancet 1, 865.

COLLINS, W. S., METZGER, J. F., and JOHNSON, A. D. 1972. A rapid solid phase radioimmunoassay for staphylococcal B enterotoxin. J. Immunol. 108, 852-856.

COLLINS-THOMPSON, D. L., ARIS, B., and HURST, A. 1973. Growth and enterotoxin B synthesis by S. aureus S-6 in associative growth with Ps. aeruginosa. Can. J. Microbiol. 19, 1197-1201.

COMMISSION ON ACUTE RESPIRATORY DISEASE, and PLUMMER, N. 1949. A comparison of the bacterial flora of the pharynx and nasopharynx. Am. J. Hyg. 50, 331-336.

COUGHLIN, F. E., and JOHNSON, B. 1941. Gastroenteritis from cream-filled pastry. Am. J. Public Health 31, 245-250.

COURTER, R. D., and GALTON, M. M. 1962. Animal staphylococcal infections and their public health significance. Am. J. Public Health 52, 1818-1827.

CRISLEY, F. D., ANGELOTTI, R., and FOTER, M. J. 1964. Multiplication of Staphylococcus aureus in synthetic cream fillings and pies. Public Health Rept. (U.S.) 79, 369-376.

CRISLEY, F. D., PEELER, J. T., and ANGELOTTI, R. 1965. Comparative evaluation of five selective and differential media for the detection and enumeration of coagulase-positive staphylococci in foods. Appl. Microbiol. *13*, 140–156.

CROSSLEY, E. L., and CAMPLING, M. 1957. The influence of certain manufacturing processes on the *Staphylococcus aureus* content of spray-dried milk. J. Appl. Bacteriol. *20*, 65–70.

CROWLE, A. J. 1958. A simplified micro double-diffusion agar precipitation technique. J. Lab Clin. Med. *52*, 744–787.

CUNLIFFE, H. C. 1949. Incidence of *Staphylococcus aureus* in the anterior nares of healthy children. Lancet *2*, 411–414.

CURRIER, R. W. II, TAYLOR, A. Jr., WOLF, F. S., and WARR, M. 1973. Fatal staphylococcal food poisoning. Student Med. J. *66*, 703–705.

DACK, G. M. 1956A. Food Poisoning, 3rd Edition. Univ. of Chicago Press, Chicago.

DACK, G. M. 1956B. The role of enterotoxin of *Micrococcus pyogenes* var. *aureus* in the etiology of pseudomembranous enterocolitis. Am. J. Surg. *92*, 765–769.

DACK, G. M., CARY, W. E., WOOLPERT, O., and WIGGERS, H. 1930. An outbreak of poisoning proved to be due to a yellow hemolytic staphylococcus. J. Prevent. Med. *4*, 167–172.

DACK, G. M., JORDAN, E. O., and WOOLPERT, O. 1931. Attempts to immunize human volunteers with staphylococcus filtrates that are toxic to man when swallowed. J. Prevent. Med. *5*, 151–159.

DACK, G. M., and LIPPITZ, G. 1962. Fate of staphylococci and enteric microorganisms introduced into slurry of frozen pot pies. Appl. Microbiol. *10*, 472–479.

DAFFRON, D., LOVEJOY, G. S., and ARNOLD, W. 1972. Staphylococcal food poisoning—Tennessee. Morbidity Mortality Weekly Rept. (U.S.) *21*, No. 20, 169–170.

DAHIYA, R. S., and SPECK, M. L. 1968. Hydrogen peroxide formation by lactobacilli and its effect on *Staphylococcus aureus*. J. Dairy Sci. *51*, 1568–1572.

DaSILVA, G. A. N. 1967. Incidence and characteristics of *Staphylococcus aureus* in turkey products and processing plants. Ph.D. Thesis. Iowa State Univ., Ames.

DaSILVA, G. A. N., KRAFT, A. A., and AYRES, J. C. 1967. The occurrence of *Staphylococcus aureus* on cooked and uncooked turkey rolls. Bacteriol. Proc. p. 12, Am. Soc. Microbiology, Washington, D.C.

DAUER, C. C. 1952. Food and waterborne disease outbreaks. Public Health Rept. (U.S.) *67*, 1089–1095.

DAUER, C. C. 1953. 1952 Summary of foodborne, waterborne, and other disease outbreaks. Public Health Rept. (U.S.) *68*, 696–702.

DAUER, C. C. 1958. 1957 Summary of disease outbreaks. Public Health Rept. (U.S.) *73*, 681–686.

DAUER, C. C. 1961. 1960 Summary of disease outbreaks and a 10-year resume. Public Health Rept. (U.S.) *76*, 915–922.

DAUER, C. C., and DAVIDS, D. J. 1959. 1958 Summary of disease outbreaks. Public Health Rept. (U.S.) *74*, 715–720.

DAUER, C. C., and DAVIDS, D. J. 1960. 1959 Summary of disease outbreaks. Public Health Rept. (U.S.) *75*, 1025–1030.

DAUER, C. C., and SYLVESTER, G. 1954. 1953 Summary of disease outbreaks. Public Health Rept. (U.S.) *69*, 538–546.

DAUER, C. C., and SYLVESTER, G. 1955. 1954 Summary of disease outbreaks. Public Health Rept. (U.S.) *70*, 536–544.

DAUER, C. C., and SYLVESTER, G. 1956. 1955 Summary of disease outbreaks. Public Health Rept. (U.S.) *71*, 797–803.

DAUER, C. C., and SYLVESTER, G. 1957. Summary of disease outbreaks. Public Health Rept. (U.S.) *72*, 735–742.

DAVIDSON, I. 1961. Observations on the pathogenic staphylococci in a dairy

herd during a period of six years. Res. Vet. Sci. *2*, 22-40.
DAVIDSON, I. 1971. Working Group on phage-typing of bovine staphylococci. Intern. J. Syst. Bacteriol. *21*, 171.
DAVISON, E., and DACK, G. M. 1942. Production of staphylococcus enterotoxin in canned corn, salmon, and oysters. Food Res. *8*, 80-84.
DEARING, W. H. 1956. Micrococcic enteritis and pseudomembranous enterocolitis as complications of antibiotic therapy. Ann. N.Y. Acad. Sci. *65*, 235-242.
DEARING, W. H., BAGGENSTOSS, A. H., and WEED, L. A. 1960. Studies of the relationship of *Staphylococcus aureus* to pseudomembranous enteritis and to postantibiotic enteritis. Gastroenterol. *38*, 441-451.
DEARING, W. H., and HEILMAN, F. R. 1953. Micrococci (Staphylococcic) enteritis as a complication of antibiotic therapy; its response to erythromycin. Proc. Staff Meetings Mayo Clinic *28*, 121-134.
DeFIGUEIREDO, M. P. 1969. Staphylococcal control and the food processor. *In* Symposium on Staphylococci in Foods, D. L. Downing (Editor). N.Y. State Agr. Expt. Sta. Res. Circ. *23*.
DENEKE, A., and BLOBEL, H. 1962. Fibrinogen media for studies on staphylococci. J. Bacteriol. *83*, 533-537.
DENES, G., and RAMPAZZO, G. 1967. The incidence of nasopharyngeal carriers of staphylococci in persons in the hotel industry and food handling trades. Igiene Sanita Pubblica *11*, 637-647. *Cited by* Abs. Hyg. 4389, 1968. (Italian)
DENISON, G. A. 1936. Epidemiology and symptomatology of staphylococcus food poisoning. Am. J. Public Health *26*, 1168-1175.
DENNY, C. B., HUMBER, J. Y., and BOHRER, W. 1971. Effect of toxin concentration on the heat inactivation of staphylococcal enterotoxin A in beef bouillon and in phosphate buffer. Appl. Microbiol. *21*, 1064-1066.
DENNY, C. B., TAN, P. L., and BOHRER, C. W. 1966. Heat inactivation of staphylococcal enterotoxin A. J. Food Sci. *31*, 762-767.
DENYS, J. 1894. Presence of the pyogenic staphylococcus in a meat that has caused some cases of poisoning. Bull. Acad. Royale Med. Belgique *8*, 605-614. (French)
DESKINS, B. B., and HUSSEMAN, D. L. 1954. Effect of cooking on bacterial count of ground ham mixtures. J. Am. Dietet. Assoc. *30*, 1245-1249.
DESSEL, M. M., BOMERSOX, E. M., and JETER, W. S. 1960. Bacteria in electronically cooked foods. J. Am. Dietet. Assoc. *37*, 231-233.
DeWAART, J., MOSSEL, D. A. A., TEN BROEKE, R., and VAN DE MOOSDIJK, A. 1968. Enumeration of *Staphylococcus aureus* in foods with special reference to egg-yolk reaction and mannitol negative mutants. J. Appl. Bacteriol. *31*, 276-285.
DeWITT, R. F. 1935. A recent outbreak of food poisoning in Shoreham, Vermont. New Engl. J. Med. *213*, 1283-1284.
DIETRICH, G. G., WATSON, R. J., and SILVERMAN, G. J. 1972. Effect of shaking speed on the secretion of enterotoxin B by *Staphylococcus aureus*. Appl. Microbiol. *24*, 561-566.
DiGIACINTO, J. V., and FRAZIER, W. C. 1966. Effect of coliform and Proteus bacteria on growth of *Staphylococcus aureus*. Appl. Microbiol. *14*, 124-129.
DINEEN, P. 1970. Antibacterial drugs. *In* Drugs of Choice 1970-1971, W. Modell (Editor). The C. V. Mosby Co., St. Louis.
DOLMAN, C. E. 1934. Investigation of staphylococcus exotoxin by human volunteers with special reference to staphylococci food poisoning. J. Infect. Diseases *55*, 172-183.
DOLMAN, C. E. 1943. Bacterial food poisoning. Can. J. Public Health *34*, 97-111, 205-235.
DOLMAN, C. E. 1956. The staphylococcus: Seven decades of research (1885-1955). Can. J. Microbiol. *2*, 189-200.
DOLMAN, C. E., WILSON, R. J., and COCKCROFT, W. H. 1936. A new

method of detecting staphylococcus enterotoxin. Can. J. Public Health 27, 489-493.
DONNELLY, C. B., BLACK, L. A., and LEWIS, K. H. 1964. Occurrence of coagulase-positive staphylococci in cheddar cheese. Appl. Microbiol. 12, 311-315.
DONNELLY, C. B., LESLIE, J. E., and BLACK, L. A. 1968. Production of enterotoxin A in milk. Appl. Microbiol. 16, 917-924.
DONNELLY, C. B., LESLIE, J. E., BLACK, L. A., and LEWIS, K. H. 1967. Seriological identification of enterotoxigenic staphylococci from cheese. Appl. Microbiol. 15, 1382-1387.
DORLING, G. C. 1942. Staphylococcal food-poisoning due to contaminated soup. Lancet 1, 382.
DUGUID, J. P., and WALLACE, A. T. 1948. Air infection with dust liberated from clothing. Lancet 2, 845-849.
DUNCAN, J. T., and WALKER, J. 1942. Staphylococcus aureus in the milk of nursing mothers and the alimentary canal of their infants. J. Hyg. 42, 474-484.
DUTHIE, E. S., and LORENZ, L. L. 1952. Staphylococcal coagulase: Mode of action and antigenicity. J. Gen. Microbiol. 6, 95-107.
DYCHDALA, G. R. 1968. Chlorine and chlorine compounds. In Disinfection, Sterilization, and Preservation, C. A. Lawrence, and S. S. Block (Editors). Lea and Febiger, Philadelphia.
ECKHOFF, J. D. et al. 1971. Gastroenteritis associated with Genoa salami—United States. Morbidity Mortality Weekly Rept. (U.S.) 20, No. 29, 261, 266.
EDDY, B. P., and INGRAM, M. 1962. The occurrence and growth of staphylococci on packaged bacon with special reference to Staphylococcus aureus. J. Appl. Bacteriol. 25, 237-247.
EL-BISI, H. M., and ORDAL, Z. J. 1956. The effect of sporulation temperature on the thermal resistance of Bacillus coagulans var. thermoacidurans. J. Bacteriol. 71, 10-16.
ELEK, S. D. 1959. Staphylococcus pyogenes and its relation to disease. E. S. Livingstone, Ltd., Edinburg and London.
ELEK, S. D., and LEVY, E. 1950. Distribution of haemolysins in pathogenic and non-pathogenic staphylococci. J. Pathol. Bacteriol. 62, 541-544.
ELLIOTT, R. P., and MICHENER, H. D. 1961. Microbiological standards and handling codes for chilled and frozen foods: A review. Appl. Microbiol. 9, 452-468.
ERICKSON, A., and DEIBEL, R. H. 1973. Production and heat stability of staphylococcal nuclease. Appl. Microbiol. 25, 332-336.
EVANS, D. A., HANKINSON, D. J., and LITSKY, W. 1970. Heat resistance of certain pathogenic bacteria in milk using a commercial plate heat exchanger. J. Dairy Sci. 53, 1659-1665.
FAIRBROTHER, R. W., and SOUTHALL, J. E. 1950. The isolation of Staphylococcus pyogenes from feces. Monthly Bull. Min. Health Public Health Lab. Serv. (British) 9, 170-172.
FANNING, J. 1935. An unusual outbreak of gastro-enteritis. Brit. Med. J. 1, 583-584.
FEIG, M. 1950. Diarrhea, dysentery, food poisoning, and gastroenteritis. Am. J. Public Health 40, 1372-1394.
FINEGOLD, S. M., and SWEENEY, E. E. 1961. New selective and differential medium for coagulase-positive staphylococci allowing rapid growth and strain differentiation. J. Bacteriol. 81, 636-641.
FOLTZ, V. D., MICKELSEN, R., MARTIN, W. H., and HUNTER, C. A. 1960. The incidence of potentially pathogenic staphylococci in dairy products at the consumer level. I. Fluid milk and fluid milk by-products. J. Milk Food Technol. 23, 280-284.

STAPHYLOCOCCUS AUREUS 109

FOOD PROTECTION COMMITTEE, FOOD NUTRITION BOARD. 1971. Reference methods for the microbiological examination of foods. Nat. Res. Council—Natl. Acad. Sci., Washington, D.C.
FORSGREN, A., FORSUM, V., and HALLANDER, H. O. 1972. Failure to detect cell-associated enterotoxin B in *Staphylococcus aureus* by immunofluorescence. Appl. Microbiol. *23*, 559–564.
FRIEDMAN, M.,E. 1968. Inhibition of staphylococcal enterotoxin B formation by cell wall blocking agents and other compounds. J. Bacteriol. *95*, 1051–1055.
FRIEDMAN, M. E., and WHITE, J. D. 1965. Immunofluorescent demonstration of cell-associated staphylococcal enterotoxin B. J. Bacteriol. *89*, 1155.
FROST, A. J. 1967. Phage typing of *Staphylococcus aureus* from dairy cattle in Australia. J. Hyg. *65*, 311–319.
FUNG, D. Y. C., and WAGNER, J. 1971. Capillary tube assay for staphylococcal enterotoxins, A, B, and C. Appl. Microbiol. *21*, 559–561.
GALTON, M. M. *et al.* 1961. A six months survey of staphylococcal flora in the milk from a large dairy herd. U.S. Livestock Sanitary Assoc. *65*, 251–263.
GANDHI, N. R., and RICHARDSON, G. H. 1971. Capillary tube immunological assay for staphylococcal enterotoxins. Appl. Microbiol. *21*, 626–627.
GEIGER, J. C., CROWLEY, A. B., and GRAY, J. P. 1935. Food poisoning from ice cream on ships. J. Am. Med. Assoc. *105*, 1980–1981.
GENIGEORGIS, C. A. 1974. Recent developments on staphylococcal food poisoning. *In* Proc. 16th Ann. Food Hygiene Symp. Teachers Food Hygiene Colleges of Vet. Med., Pullman, Wash.
GENIGEORGIS, C., MARTIN, S., FRANTI, C. E., and RIEMANN, H. 1971A. Inhibition of staphylococcal growth in laboratory media. Appl. Microbiol. *21*, 934–939.
GENIGEORGIS, C., and PRUCHA, J. 1971. Production of enterotoxin C in processed meats. Bacteriol. Proc. p. 18, Am. Soc. Microbiology, Washington, D.C.
GENIGEORGIS, C., RIEMANN, H., and SADLER, W. W. 1969. Production of enterotoxin B in cured meats. J. Food Sci. *34*, 62–68.
GENIGEORGIS, C., and SADLER, W. W. 1966A. Immunofluorescent detection of staphylococcal enterotoxin B. I. Detection in media. J. Food Sci. *31*, 441–449.
GENIGEORGIS, C., and SADLER, W. W. 1966B. Immunofluorescent detection of staphylococcal enterotoxin B. II. Detection in foods. J. Food Sci. *31*, 605–609.
GENIGEORGIS, C., and SADLER, W. W. 1966C. Effect of sodium chloride and pH on enterotoxin B production. J. Bacteriol. *92*, 1383–1387.
GENIGEORGIS, C., and SADLER, W. W. 1966D. Characteristics of strains of *Staphylococcus aureus* isolated from livers of commercially slaughtered poultry. Poultry Sci. *45*, 973–980.
GENIGEORGIS, C., SAVOUKIDIS, M., and MARTIN, S. 1971B. Initiation of staphylococcal growth in processed meat environments. Appl. Microbiol. *21*, 940–942.
GEORGE, E., JR., and OLSON, J. C., JR. 1960. Growth of staphylococci in condensed skim milk at low and moderate temperatures. J. Dairy Sci. *43*, 852.
GERSHENFELD, L. 1968. Iodine. *In* Disinfection, Sterilization, and Preservation, C. A. Lawrence, and S. S. Block (Editors). Lea and Febiger, Philadelphia.
GIBSON, T., and ABD-EL-MALEK, Y. 1957. The development of bacterial populations in milk. Can. J. Microbiol. *3*, 203–213.
GILBERT, R. J. 1969. Cross contamination by cooked meat slicing machines and cleaning cloths. J. Hyg. *67*, 249–254.
GILBERT, R. J., KENDALL, M., and HOBBS, B. C. 1969. Media for the

isolation and enumeration of coagulase-positive staphylococci from foods. *In* Isolation Methods for Microbiologists, D. A. Shapton, and G. W. Gould (Editors). Academic Press, New York.

GILBERT, R. J., and MAURER, I. M. 1968. The hygiene of slicing machines, carving knives, and can-openers. J. Hyg. *66*, 439-450.

GILBERT, R. J., and WATSON, H. M. 1971. Some laboratory experiments on various meat preparation surfaces with regard to surface contamination and cleaning. J. Food Technol. *6*, 163-170.

GILBERT, R. J., and WIENEKA, A. A. 1973. Staphylococcal food poisoning with special reference to the detection of enterotoxin in food. *In* The Microbiological Safety of Food, B. C. Hobbs, and J. H. B. Christian (Editors). Academic Press, New York.

GILCREAS, F. W., and COLEMAN, M. B. 1941. Studies of rebaking cream-filled pastry. Am. J. Public Health *31*, 956-958.

GILDEN, M. M., BAER, E. F., and FRANKLIN, M. K. 1966. Comparative evaluation of a direct plating procedure and an enrichment isolation procedure for detecting coagulase-positive staphylococci in foods. Assoc. Offic. Anal. Chem. *49*, 273-275.

GILLESPIE, W. A., and ALDER, V. G. 1952. Production of opacity in egg-yolk media by coagulase-positive staphylococci. J. Pathol. Bacteriol. *64*, 187-200.

GIOLITTI, G., and CANTONI, C. 1966. A medium for the isolation of staphylococci from foodstuffs. J. Appl. Bacteriol. *29* 395-398.

GOLDENBERG, N., and EDMONDS, G. 1973. Education in microbiological safety standards. *In* The Microbiological Safety of Food, B. C. Hobbs, and J. H. B. Christian (Editors), Academic Press, New York.

GOOGINS, J. A., COLLINS, J. R., MARSHALL, A. L., Jr., and OFFUTT, A. C. 1961. Two gastroenteritis outbreaks from ham in picnic fare. Public Health Rept. (U.S.) *76*, 945-954.

GONZAGA, A. J., MORTIMER, E. A., JR., WOLINSKY, E., and RAMMELKAMP, C. H., JR., 1964. Transmission of staphylococci by fomites. J. Am. Med. Assoc. *189*, 711-715.

GOULD, J. C. 1955. The effect of local antibiotic on nasal carriage of *Staphylococcus pyogenes.* J. Hyg. *53*, 379-385.

GOULD, J. C., and ALLAN, W. S. H. 1954. *Staphylococcus pyogenes* cross-infection: Prevention by treatment of carriers. Lancet *2*, 988-989.

GOULD, J. C., and CRUIKSHANK, J. D. 1957. Staphylococcal infection in general practice. Lancet *2*, 1157-1161.

GOULD, J. C., and McKILLOP, E. J. 1954. The carriage of *Staphylococcus pyogenes* var. *aureus* in the human nose. J. Hyg. *52*, 304-310.

GRAVES, R. R., and FRAZIER, W. C. 1963. Food microorganisms influencing the growth of *Staphylococcus aureus*. Appl. Microbiol. *11*, 513-516.

GREENBERG, B. 1971. Flies and Disease, Vol. 1. Princeton Univ. Press, New Jersey.

GRETLER, A. C., MUCCIOLO, P., EVANS, J. B., and NIVEN, C. F., JR. 1954. Vitamin nutrition of the staphylococci with special reference to their biotin requirements. J. Bacteriol. *70*, 44-49.

GROSS, C. E., and VINTON, C. 1947. Thermal death time of a strain of *Staphylococcus* in meat. Food Res. *12*, 188-202.

GUNDERSON, M. F. 1960. Microbiological standards for frozen foods. *In* Conference on Frozen Food Quality. U.S. Dept. Agr. ARS 74-77, 74-78.

GUNDERSON, M. F. 1963. Food microbiological problems from the standpoint of industry. *In* Microbiological Quality of Foods, L. W. Slanetz, C. O. Chichester, A. R. Garrbin, and Z. J. Ordal (Editors). Academic Press, New York.

GUNDERSON, M. F., and PETERSON, A. C. 1964. A consideration of the microbiology of frozen foods. Assoc. Food Drug Official U.S. Quart. Bull. *28*, 47-61.

GUNDERSON, M. F., and ROSE, K. D. 1948. Survival of bacteria in a precooked fresh frozen food. Food Res. *13*, 254–263.

HACKLER, J. F. 1939. Outbreak of Staphylococcus milk poisoning in pasteurized milk. Am. J. Public Health *29*, 1247–1249.

HAINES, R. B. 1938. The effect of freezing on bacteria. Proc. Royal Soc. (London) Ser. B *124*, 451–463.

HAINES, W. C., and HARMON, L. G. 1973A. Effect of variations in conditions of incubation upon inhibition of *S. aureus* by *Pediococcus cerevisiae* and *Streptococcus lactis*. Appl. Microbiol. *25*, 169–172.

HAINES, W. C., and HARMON, L. G. 1973B. Effect of selected lactic acid bacteria on growth of *Staphyloccus aureus* and production of enterotoxin. Appl. Microbiol. *25*, 436–441.

HAJEK, V., and MARSALEK, E. 1969. A study of staphylococci isolated from the upper respiratory tract of different animal species. Zentr. Bakteriol. Parasitenk. I. Abstr. Orig. *212*, 60–73. *Cited by* Excerpta Med. Sect. 4, 3511, 1970. (German)

HALL, H. E., ANGELOTTI, R., and LEWIS, K. H. 1963. Quantitive detection of staphylococcal enterotoxin B in food in food by gel-diffusion methods. Public Health Rept. (U.S.) *78*, 1089–1098.

HALL, H. E., ANGELOTTI, R., and LEWIS, K. H. 1965. Detection of the staphylococcal enterotoxins in food. Health Lab. Sci. *2*, 179–191.

HALLANDER, H. O., and KORLOF, B. 1967. Enterotoxin producing staphylococci. Acta Pathol. Microbiol. Scand. *71*, 359–375.

HAMDY, M. K., and BARTON, N. D. 1965. Fate of *Staphylococcus aureus* in bruised tissue. Appl. Microbiol. *13*, 15–21.

HAMDY, M. K., BARTON, N. D., and BROWN, W. E. 1965. Sources and portal of entry of bacteria found in bruised poultry tissue. Appl. Microbiol. *12*, 464–469.

HAMMON, W. McD. 1941. Staphylococcus enterotoxins an improved cat test, chemical and immunological studies. Am. J. Public Health *31*, 1191–1198.

HARE, R. 1967. The antiquity of diseases caused by bacteria and viruses, a review of the problem from a bacteriologist's point of view. *In* Diseases in Antiquity, D. Brothwell, and A. T. Sandison (Editors). Charles C. Thomas, Springfield, Ill.

HARE, R., and COOKE, E. M. 1961. Self-contamination of patients with staphylococcal lesions. Brit. Med. J. *2*, 333–336.

HARE, R., and RIDLEY, M. 1958. Further studies on the transmission of *Staphy. aureus*. Brit. Med. J. *1*, 69–73.

HARE, R., and THOMAS, C. G. A. 1956. Transmission of *Staphylococcus aureus*. Brit. Med. J. *2*, 840–844.

HARRY, E. G. 1967A. Some characteristics of *Staphylococcus aureus* isolated from the skin and upper respiratory tract of domestic and wild (feral) birds. Res. Vet. Sci. *8*, 490–499.

HARRY, E. G. 1967B. The characteristics of *Staphylococcus aureus* isolated from cases of staphylococcosis in poultry. Res. Vet. Sci. *8*, 479–489.

HARTSELL, S. E. 1951. The longevity and behavior of pathogenic bacteria in frozen foods. The influence of plating media. Am. J. Public Health *41*, 1072–1077.

HAUSLER, W. J. JR. 1972. Standard Methods for the Examination of Dairy Products, 13th Edition. Am. Public Health Assoc., Washington, D.C.

HAYAMA, T., and SUGIYAMA, H. 1964. Diarrhea in cecectomized rabbits induced by staphylococcal enterotoxin. Proc. Soc. Exptl. Biol. Med. *117*, 115–118.

HEINEMANN, B. 1957. Growth and thermal destruction of *Micrococcus pyogenes* var. *aureus* in heated and raw milk. J. Dairy Sci. *40*, 1585–1589.

HEMPEL, J., and MALESZEWSKI, J. 1968. Preliminary observations on the behavior of a staphylococci in salted herrings. Roczn. Pom. Akad. Med.

Swierczewskiego. Suppl. *2*, 371–374. *Cited by* Excerpta Med. Sect. *17*, 6521, 1969. (Polish)

HENDRICKS, S. L., BELKNAP, R. A., and HAUSLER, W. J., JR. 1959. Staphylococcal food intoxication due to cheddar cheese. I. Epidemiology. J. Milk Food Technol. *22*, 313–317.

HERMAN, L. G., and MORELLI, T. A. 1960. The growth and isolation of coagulase-positive staphylococci on medium no. 110 fortified with egg yolk. Bacteriol. Proc. p. 102, Am. Soc. Microbiology, Washington, D.C.

HEWITT, C. G. 1914. The Housefly. Cambridge Zoological Series. Univ. Press, Cambridge, U.K.

HEWITT, L. F. 1957. Influence of hydrogen-ion concentration and oxidation-reduction conditions on bacterial behavior. *In* Microbial Ecology, R. E. O. Williams, and C. C. Spicer (Editors), Cambridge Univ. Press, London.

HILKER, J. S. *et al.* 1968. Heat inactivation of enterotoxin A for *Staphylococcus aureus* in veronal buffer. Appl. Microbiol. *16*, 308–310.

HILL, I. R., and KENWORTHY, R. 1970. Microbiology of pigs and their environment in relation to weaning. J. Appl. Microbiol. *33*, 299–316.

HILL, W. M. 1972. The significance of staphylococci in meats. *In* Proc. 25th Ann. Reciprocal Meat Conf., Ames, Iowa.

HINTON, N. A., MATTMAN, J. R., and ORR, J. H. 1960. The effect of desiccation on the ability of *Staphylococcus pyogenes* to produce disease in mice. Am. J. Hyg. *72*, 343–350.

HOBBS, B. C. 1964. Food poisoning: Observations on sources of salmonellae, *Clostridium welchii*, and staphylococci. Ann. Inst. Pasteure Lille *15*, 31–41.

HOBBS, B. C. 1968. Food Poisoning and Food Hygiene, 2nd Edition. Arnold, London.

HOBBS, B. C., and FREEMAN, V. 1949. Food poisoning due to pressed beef. Monthly Bull. Min. Health, Public Health Lab. Serv. (British) *8*, 63–67.

HOBBS, B. C., KENDALL, M., and GILBERT, R. J. 1968. Use of phenolphthalein diphosphate agar with polymyxin as a selective medium for the isolation and enumeration of coagulase-positive staphylococci. Appl. Microbiol. *16*, 535.

HOBBS, B. C., and SMITH, M. E. 1954. The control of infection spread by synthetic cream. J. Hyg. *52*, 230–246.

HOBBS, B. C., and THOMAS, M. E. M. 1948. Staphylococcal food poisoning from infected lamb's tongue. Monthly Bull. Min. Health, Public Health Lab. Serv. (British) *7*, 261–266.

HODGE, B. E. 1960. Control of staphylococcal food poisoning. Public Health Rept. (U.S.) *75*, 355–361.

HODOVAL, L. F., RAPAPORT, M. I., and BEISEL, W. R. 1966. A radiobioassay for staphylococcal enterotoxin B antitoxin. J. Lab. Clin. Med. *68*, 678–685.

HOJVAT, S. A., and JACKSON, H. 1969. Effects of sodium chloride and temperature on the growth and production of enterotoxin B by *Staphylococcus aureus*. Can. Inst. Food Technol. *2*, 56–59.

HOPF, E. A., SCHUPPERT, L., ROOP, D. J., and GARBER, H. J. 1971. Gastroenteritis attributed to Hormel San Remo Stick Genoa Salami—Maryland. Morbidity Mortality Weekly Rept. (U.S.) *20*, No. 40, 370.

HOPPER, S. H. 1963. Detection of staphylococcus enterotoxin. I. Flotation antigen-antibody system. J. Food Sci. *28*, 572–577.

HOPTON, J. 1961. A selective medium for the isolation and enumeration of coagulase-positive staphylococci from foods. J. Appl. Bacteriol. *24*, 121–124.

HORWOOD, M. P., and MINCH, V. A. 1951. The number and types of bacteria found on the hands of food handlers. Food Res. *16*, 133–136.

HUANG, I-Y., and BERGDOLL, M. S. 1970A. The primary structure of staphylococcal enterotoxin B: I. Isolation, composition, and sequence of tryptic peptides from oxidized enterotoxin B. J. Biol. Chem. *245*,

STAPHYLOCOCCUS AUREUS 113

3493-3510.
HUANG, I-Y., and BERGDOLL, M. S. 1970B. The primary structure of staphylococcal enterotoxin B: II. Isolation, composition, and sequence of chymotryptic peptides. J. Biol. Chem. 245, 3511-3517.
HUANG, I-Y., and BERGDOLL, M. S. 1970C. The primary structure of staphylococcal enterotoxin B: III. The cyanogen bromide peptides of reduced and aminoethylated enterotoxin B and the complete amino acid sequence. J. Biol. Chem. 245, 3518-3525.
HUBER, D. A., ZABOROWSKI, H., and RAYMAN, M. M. 1958. Studies on the microbiological quality of precooked frozen meals. Food Technol. 12, 190-194.
HUCKER, G. J., and HAYNES, W. C. 1937. Certain factors affecting the growth of food poisoning micrococci. Am. J. Public Health 27, 590-594.
HUSSEMANN, D. L., and TANNER, F. W. 1947. Relation of certain cooking procedures to staphylococcus food poisoning. Am. J. Public Health 37, 1407-1414.
HUTCHISON, J. G. P., GREEN, C. A., and GRIMSON, T. A. 1957. Nasal carriage of Staphylococcus aureus in nurses. J. Clin. Pathol. 10, 92-95.
IANDOLO, J. J., CLARK, C. W., BLUHM, L., and ORDAL, Z. J. 1965. Repression of staphylococcus aureus in associative culture. Appl. Microbiol. 13, 646-649.
IANDOLO, J. J., and ORDAL, Z. J. 1966. Repair of thermal injury of Staphylococcus aureus. J. Bacteriol. 91, 134-142.
IANDOLO, J. J., ORDAL, Z. J., and WITTER, L. D. 1964. The effect of incubation temperature and controlled pH on the growth of Staphylococcus aureus MF31 at various concentrations of NaCl. Can. J. Microbiol. 10, 808-811.
INNES, A. G. 1960. Tellurite-egg agar, a selective and differential medium for the isolation of coagulase-positive staphylococci. J. Appl. Bacteriol. 23, 108-113.
IWASAWA, Y., and ISHIHARA, K. 1967. Resistance of Staphylococcus aureus to desiccation, heat, and ultraviolet rays in relation to phage pattern. Japan J. Microbiol. 11, 305-309.
JACKSON, H., and WOODBINE, M. 1963. The effect of sublethal heat treatment on the growth of Staphylococcus aureus. J. Appl. Bacteriol. 26, 152-158.
JACOBS, S. I., WILLIS, A. T., GOODBURN, G. M. 1963. Significance of deoxyribonuclease production by staphylococci. Nature 200, 709-710.
JAY, J. M. 1961. Incidence and properties of coagulase-positive staphylococci in certain market meats as determined on three selective media. Appl. Microbiol. 9, 228-232.
JAY, J. M. 1962A. Further studies on staphylococci in meats. III. Occurrence and characteristics of coagulase-positive strains from a variety of non-frozen market cuts. Appl. Microbiol. 10, 247-251.
JAY, J. M. 1962B. Further studies on staphylococci in meats. IV. The bacteriophage pattern and antibiotic sensitivity of isolates from non-frozen meats. Appl. Microbiol. 10, 252-257.
JAY, J. M. 1963. The relative efficacy of selective media in isolating coagulase positive staphylococci from meats. J. Appl. Bacteriol. 26, 69-74.
JAY, J. M. 1966. Production of lysozyme by staphylococci and its correlation with three other extracellular substances. J. Bacteriol. 91, 1804-1810.
JAYAKAR, P. A., and BHASKARAN, C. S. 1968. Incidence of Staphylococcus aureus in the staff of a hospital and a study of the biological characters of the strains isolated. Indian J. Med. Sci. 22, 858-862.
JEFFRIES, C. D., HOLTMAN, D. F., and GUSE, D. G. 1957. Rapid method for determining the activity of microorganisms on nucleic acids. J. Bacteriol. 73, 590-591.

114 FOOD MICROBIOLOGY

JENSEN, L. B. 1945. Microbiology of Meats. Garrard Press, Champaign, Ill.
JOHNSON, H. M., BUKOVIC, J. A., KAUFFMAN, P. E., and PEELER, J. T. 1971. Staphylococcal enterotoxin B: Solid-phase radioimmunoassay. Appl. Microbiol. 22, 837–841.
JOHNSON, H. M., HALL, H. E., and SIMON, M. 1967. Enterotoxin B: Serological assay in cultures by passive hemagglutination. Appl. Microbiol. 15, 815–818.
JONES, D., and NIVEN, C. F., JR. 1964. Comments on the species Staphylococcus epidermidis (Winslow and Winslow) Evans and Staphylococcus saprophyticus Fairbrother. Intern. Bull. Bacteriol. Nomenclature Taxonomy 14, 45–51.
JONES, H. D., KING, G. J. G., FENNELL, H., and STONE, D. 1957. The growth of Staphylococcus aureus in milk with special reference to food poisoning. Monthly Bull. Min. Health, Public Health Lab. Serv. (British) 16, 109–122.
JONES, R. H., and BENNETT, F. W. 1965. Bacteriophage types and antibiotic sensitivity of staphylococci from bovine milk and human nares. Appl. Microbiol. 13, 725–731.
JORDAN, E. O., DACK, G. M., and WOOLPERT, O. 1931. The effect of heat, storage and chlorination on the toxicity of staphylococcus filtrates. J. Prevent. Med. 5, 383–386.
JUNGHERR, E., and PLASTRIDGE, W. N. 1941. Avian Staphylococcosis. J. Am. Vet. Med. Assoc. 98, 27–32.
KADAN, R. S., MARTIN, W. H., and MICKELSEN, R. 1963. Effects of ingredients used in condensed and frozen dairy products on thermal resistance of potentially pathogenic staphylococci. Appl. Microbiol. 11, 45–49.
KALLANDER, A. 1953. Examination of cooks and serving personnel with reference to enterotoxin producing bacteria. Nord. Hyg. Tidskr. 34, 1–8 (Danish). Cited by Elek, 1959.
KAO, C. T., and FRAZIER, W. C. 1966. Effect of lactic acid bacteria on growth of Staphylococcus aureus. Appl. Microbiol. 14, 251–255.
KATO, E., KHAN, M., KUJOVICH, L., and BERGDOLL, M. S. 1966. Production of enterotoxin A. Appl. Microbiol. 14, 966–972.
KAY, C. R. 1962. Sepsis in the home. Brit. Med. J. 1, 1048–1052.
KERELUK, K., and GUNDERSON, M. F. 1961. Survival of bacteria in artifically contaminated frozen meat pies after baking. Appl. Microbiol. 9, 6–10.
KIENITZ, M. 1964. Food intoxication caused by staphylococcus. Aerztl. Praxis 16, 872. Cited by Bergdoll, 1970.
KINTNER, T. C., and MANGEL, M. 1953A. Survival of staphylococci and salmonellae experimentally inoculated into salad dressing prepared with dried eggs. Food Res. 18, 6–10.
KINTNER, T. C., and MANGEL, M. 1953B. Survival of staphylococci and salmonellae in puddings and custards prepared with experimentally inoculated dried eggs. Food Res. 18, 492–496.
KLEMPERER, R., and HAUGHTON, G. 1957. A medium for the rapid recognition of penicillin-resistant coagulase-positive staphylococci. J. Clin. Pathol. 10, 96–99.
KODAMA, T., HATA, M., and SIBUYA, Y. 1940. Outbreak of staphylococcal food-poisoning in Yokosvka City. Kitasato Arch. Exper. Med. 17, 115–126. Cited by Dolman, 1943.
KRAFT, A. A. et al. 1963. Effect of method of freezing on survival of microorganisms on turkey. Poultry Sci. 42, 128–137.
LABUZA, T. P., CASSIL, S., and SINSKEY, A. J. 1972. Stability of intermediate moisture foods. 2. Microbiology. J. Food Sci. 37, 160–162.
LACHICA, R. V. F., and DEIBEL, R. H. 1969. Detection of nuclease activity in semisolid and broth cultures. Appl. Microbiol. 18, 174–176.

LACHICA, R. V. F., HOEPRICH, P. D., and GENIGEORGIS, C. 1972. Metachromatic agar-diffusion microslide technique for detecting staphylococcal nuclease in foods. Appl. Microbiol. 23, 168–169.

LACHICA, V. R., WEISS, K. F., and DEIBEL, R. H. 1969. Relationships among coagulase, enterotoxin, and heat stable deoxyribonuclease production of Staphylococcus aureus. Appl. Microbiol. 18, 126–127.

LASMANIS, J., and SPENCER, G. R. 1953. The action of hypochlorite and other disinfectants on Micrococci with and without milk. Am. J. Vet. Res. 14, 514–516.

LAWRENCE, C. A. 1968. Quarternary ammonium surface active disinfectants. In Disinfection, Sterilization, and Preservation, C. A. Lawrence, and S. S. Block (Editors). Lea and Febiger, Philadelphia.

LECHOWICH, R. V., EVANS, J. B., and NIVEN, C. F., JR. 1956. Effect of curing ingredients and procedures on the survival and growth of staphylococci in and on cured meats. Appl. Microbiol. 4, 360–363.

LEPPER, M. H., JACKSON, G. G., and DOWLING, H. F. 1955. Characteristics of the micrococcal nasal carrier state among hospital personnel. J. Lab. Clin. Med. 45, 935–942.

LEWIS, M. N., WEISER, H. H., and WINTER, A. R. 1953. Bacterial growth in chicken salad. J. Am. Dietet. Assoc. 29, 1094–1099.

LINDBORN, G., and LAURELL, G. 1967. Studies on the epidemiology of staphylococcal infections. IV. Effect of nasal chemotherapy on carrier state in patients and on postoperative sepsis. Acta Pathol. Microbiol. Scand. 69, 237–245.

LISTON, J., and RAJ, H. D. 1962. Food poisoning problems of frozen seafoods. J. Environ. Health 25, 194–198.

LOEB, L. 1903. The influence of certain bacteria on the coagulation of the blood. J. Med. Res. 10, 407–419.

LOH, W. P., and ABIOG, A. 1957. Staphylococci in a community hospital. I. Nasopharyngeal-carrier rate of hospital personnel on maternity wards and antiobiotic sensitivity of the staphylococci isolated. New Engl. J. Med. 256, 177–179.

LOKEN, K. I., and HOYT, H. H. 1962. Studies on bovine staphylococci mastitis. I. Characterization of staphylococci. Am. J. Vet. Res. 23, 534–540.

LONGREE, K. 1972. Quantity Food Sanitation, 2nd Edition. John Wiley & Sons, New York.

LONGREE, K., MORAGNE, L., and WHITE, J. C. 1960. Cooling starch-thickened food items with cold tube agitation. J. Milk Food Technol. 23, 330–336.

LONGREE, K., MORAGNE, L., and WHITE, J. C. 1963. Cooling menu items by agitation under refrigeration. J. Milk Food Technol. 26, 317–322.

LONGREE, K., PADGHAM, R. F., WHITE, J. C., and WEISMAN, B. A. 1957. Effect of ingredients on bacterial growth in soups. J. Milk Food Technol. 20, 170–177.

LONGREE, K., and WHITE, J. C. 1955. Cooling rates and bacterial growth in food prepared and stored in quantity. I. Broth and white sauce. J. Am. Dietet. Assoc. 31, 124–132.

LONGREE, K. et al. 1958. Bacterial growth in poultry stuffings. J. Am. Dietet. Assoc. 34, 50–57.

LONGREE, K., WHITE, J. C., CUTLAR, K. L., and WILLMAN, A. R. 1959A. Bacterial growth in potato salad and turkey salads. J. Am. Dietet. Assoc. 35, 38–44.

LONGREE, K., WHITE, J. C., and LYNCH, C. W. 1959B. Bacterial growth in protein-base sandwich fillings. J. Am. Dietet. Assoc. 35, 131–138.

LOTTER, L. P., MEYER HORSTMAN, B. S., and REHG, V. 1968. Distribution of staphylococci and micrococci among humans and their physical environment during simulated aerospace confinement. Am. J. Clin. Pathol. 49, 414–422.

LOWBURY, E. J. L. 1960. Infection in burns. Brit. Med. J. *1*, 994–1001.

LOWBURY, E. J. L. 1961. Skin disinfection. J. Clin. Pathol. *14*, 85–90.

LOWBURY, E. J. L., LILLY, H. A., and BULL, J. P. 1963. Disinfection of hands: Removal of resident bacteria. Brit. Med. J. *1*, 1251–1256.

LOWBURY, E. J. L., LILLY, H. A., and BULL, J. P. 1964A. Methods for disinfection of hands and operation sites. Brit. Med. J. *2*, 531–536.

LOWBURY, E. J. L., LILLY, H. A., and BULL, J. P. 1964B. Disinfection of hands: Removal of transient organisms. Brit. Med. J. *2*, 230–233.

LUDLAM, G. B. 1949. A selective medium for the isolation of *Staphylococcus aureus* from heavily contaminated materials. Monthly Bull. Min. Health, Public Health Lab. Serv. (British) *8*, 15–20.

LUNDBECK, H., and TIRUNARAYANAN, M. O. 1966. Investigation on the enzymes and toxins of staphylococci. Study of the 'egg yolk reaction' using an agar plate assay method. Acta Pathol. Microbiol. Scand. *68*, 123–134.

LYONS, P. R., and MALLMANN, W. L. 1954. A bacteriological study of cottage cheese with particular reference to public health hazards. J. Milk Food Technol. *17*, 373–376.

MacDONALD, A. 1944. Staphylococcal food poisoning caused by cheese. Monthly Bull. Min. Health, Public Health Lab. Serv. (British) *3*, 121–122.

MacDONALD, A. 1946. *Staphylococcus aureus* in cows' milk: The results of phage-typing. Monthly Bull. Min. Health, Emergency Public Health Lab. Serv. (British) *5*, 230–233.

MAH, R. A., FUNG, D. Y. C., and MORSE, S. A. 1967. Nutritional requirements of *Staphylococcus aureus* S-6. Appl. Microbiol. *15*, 866–870.

MALLMANN, W. L., and SCHALM, O. 1932. The influence of the hydroxyl ion on the germicidal action of chlorine in dilute solution. Mich. Eng. Expt. Sta. Bull *44*.

MALTMAN, J. R., ORR, J. H., and HINTON, N. A. 1960. The effect of desiccation on *Staphylococcus pyogenes* with special reference to implications concerning virulence. Am. J. Hyg. *72*, 335–342.

MANN, P. H. 1959. Antibiotic sensitivity testing and phage typing of staphylococci found in the nostrils of dogs and cats. J. Am. Vet. Med. Assoc. *134*, 469–470.

MANUFACTURED MEAT PRODUCTS WORKING PARTY. 1950. Rept. Min. Food, London.

MARKUS, Z., and SILVERMAN, G. J. 1968. Enterotoxin B production by nongrowing cells of *Staphylococcus aureus*. J. Bacteriol. *96*, 1446–1447.

MARKUS, Z., and SILVERMAN, G. J. 1969. Enterotoxin B synthesis by replicating and non-replicating cells of *Staphylococcus aureus*. J. Bacteriol. *97*, 506–512.

MARKUS, Z., and SILVERMAN, G. J. 1970. Factors affecting the secretion of enterotoxin A. Appl. Microbiol. *20*, 492–496.

MARTIN, D. M., and WHITEHEAD, J. E. M. 1949. Carriage of penicillin-resistant *Staphylococcus pyogenes* in healthy adults. Brit. Med. J. *1*, 173–175.

MARTIN, N. H. 1942. The relation of pyogenic skin infections to skin carrier rate. Brit. Med. J. *2*, 245–246.

MARTIN, R. R., and WHITE, A. 1968. The reacquisition of staphylococci by treated carriers: A demonstration of bacterial interference. J. Lab. Clin. Med. *71*, 791–797.

MARTYN, G. 1949. Staphylococci in the newborn: Their coagulase production and resistance to penicillin and streptomycin. Brit. Med. J. *1*, 710–712.

MATTHIAS, J. Q., SHOOTER, R. A., and WILLIAMS, R. E. O. 1957. *Staphylococcus aureus* in the faeces of hospital patients. Lancet *1*, 1172–1173.

MAY, K. N., and KELLY, L. E. 1965. Fate of bacteria in chicken meat during freeze-dehydration, rehydration, and storage. Appl. Microbiol. *13*, 340–344.

McBURNEY, R. 1933. Food poisoning due to staphylococci: Report of an outbreak. J. Am. Med. Assoc. *100*, 1999–2001.

McCARTHY, P. A., BROWN, W., and HAMDY, M. K. 1963. Microbiological studies of bruised tissues. J. Food Sci. 28, 245-253.
McCLESKEY, C. S., and CHRISTOPHER, W. N. 1941. Some factors influencing the survival of bacteria in cold-pack strawberries. Food Res. 6, 327-333.
McCOY, D. W., and FABER, J. E. 1966. Influence of food microorganisms on staphylococcal growth and enterotoxin production in meat. Appl. Microbiol. 14, 372-377.
McCROAN, J. E., McKINLEY, T. W., BRIM, A., and HENNING, W. C. 1964. Staphylococci and salmonellae in commercial wrapped sandwiches. Public Health Rept. (U.S.) 79, 997-1004.
McDADE, J. J., and HALL, L. B. 1963A. An experimental method to measure the influence of environmental factors on the viability and the pathogenicity of Staphylococcus aureus. Am. J. Hyg. 77, 98-108.
McDADE, J. J., and HALL, L. B. 1963B. Survival of Staphylococcus aureus in the environment. I. Exposure on surfaces. Am. J. Hyg. 78, 330-337.
McDIVITT, M. E., and HAMMER, M. L. 1958. Cooling rates and bacterial growth in cornstarch pudding. J. Am. Dietet. Assoc. 34, 1190-1194.
McDIVITT, M. E., and HUSSEMANN, D. L. 1954. Comparison of three media for the isolation of enterotoxigenic micrococci. Am. J. Public Health 44, 1455-1459.
McGIRR, O. 1969. Screening of food handlers. Proc. Roy. Soc. Med. 62, 601-602.
McKINLEY, T. W., and CLARKE, E. J. 1964. Imitation cream fillings as a vehicle of staphylococcal food poisoning. J. Milk Food Technol. 27, 302-304.
McLEAN, R. A., LILLY, H. D., and ALFORD, J. A. 1968. Effects of meat-curing salts and temperature on production of staphylococcal enterotoxin B. J. Bacteriol. 95, 1207-1211.
McLEOD, R. W., ROUGHLEY, R. J., and RICHARDS, J. 1962. Staphylococcus aureus in cheddar cheese. Effect of numbers on survival. Australian J. Dairy Technol. 17, 54-56.
McNAMARA, M. J., MARSTON, J., and WATSON, K. A. 1966. The influence of respiratory disease on the nasal carriage of staphylococci. Am. J. Epidemiol. 84, 431-438.
MESHALOVA, A. N., and MIKHAILOVA, I. F. 1964. Laboratory Diagnosis of Infectious Diseases: Methodological Manual, 2nd Edition. Biuro Nauchnoi Informatsii, Moscow. Cited by Thatcher and Clark, 1968. (Russian)
MESSER, J. W. et al. 1970. Microbiological quality survey of some market foods in two socioeconomic areas. Bacteriol. Proc., 12, Am. Soc. Microbiology, Washington, D.C.
MEYER, K. F. 1953. Food poisoning. New Engl. J. Med. 249, 765-773, 804-812, 843-852.
MICKELSEN, R., FOLTZ, V. D., MARTIN, W. H., and HUNTER, C. A. 1963. Staphylococci in cottage cheese. J. Milk Food Technol. 26, 74-77.
MILLER, W. A. 1955. Effect of freezing ground pork and subsequent storing above 32°F upon the bacterial flora. Food Technol. 9, 332-334.
MILLER, W. A., and SMULL, M. L. 1955. Efficiency of cooling practices in preventing growth of micrococci. J. Am. Dietet. Assoc. 31, 469-473.
MILLIAN, S. J., BALDWIN, J. N., RHEINS, M. S., and WEISER, H. H. 1960. Studies on the incidence of coagulase-positive staphylococci in a normal unconfined population. Am. J. Public Health 50, 791-798.
MINER, M. L., SMART, R. A., and OLSON, A. E. 1968. Pathogenesis of staphylococcal synovitis in turkeys: Pathogenic changes. Avian Diseases 12, 46-60.
MINETT, F. C. 1938. Experiments on Staphylococcus food poisoning. J. Hyg. 38, 623-637.
MOORE, T. D., and NELSON, F. E. 1962. The enumeration of Staphylococcus aureus on several tellurite-glycine media. J. Milk Food Technol. 25, 124-127.

MOOREHEAD, S., and WEISER, H. H. 1946. The survival of staphylococci food poisoning strain in the gut and excreta of the housefly. J. Milk Food Technol. 9, 253–259.

MORAGNE, L., LONGREE, K., and WHITE, J. C. 1959. Heat transfer in white sauce cooled in flowing water. J. Am. Dietet. Assoc. 35, 1275–1282.

MORAGNE, L., LONGREE, K., and WHITE, J. C. 1960. The effect of some selected factors on the cooling of food under refrigeration. J. Milk Food Technol. 23, 142–150.

MORAGNE, L., LONGREE, K., and WHITE, J. C. 1961. Cooling custards and puddings with cold tube agitation. J. Milk Food Technol. 24, 207–210.

MORAGNE, L., LONGREE, K., and WHITE, J. C. 1963. Effect of a scraper-lifter agitation on cooling time of food. J. Milk Food Technol. 26, 182–184.

MORRISON, S. M., FAIR, J. F., and KENNEDY, K. K. 1961. Staphylococcus aureus in domestic animals. Public Health Rept. (U.S.) 76, 673–677.

MORSE, S. A., and MAH, R. A. 1967. Microtiter hemagglutination-inhibition assay for staphylococcal enterotoxin B. Appl. Microbiol. 15, 58–61.

MORSE, S. A., MAH, R. A., and DOBROGOSZ, W. J. 1969. Regulation of staphylococcal enterotoxin B. J. Bacteriol. 98, 4–9.

MORTON, H. E. 1968. Alcohols. In Disinfection, Sterilization, and Preservation, C. A. Lawrence, and S. S. Block (Editors). Lea and Febiger, Philadelphia.

MORTON, H. E., and KLAUDER, J. V. 1944. Germicidal soaps: Report of Council on Pharmacy and Chemistry. J. Am. Med. Assoc. 124, 1195–1201.

MUCH, H. 1908. About a precursor of fibrin ferment in cultures of Staphylococcus aureus. Biochem. Z. 14, 143–155. (German)

MULLER, E. 1967. The ecology of Staphylococcus aureus on the surface of human skin. I. The phenomenon of the so-called self disinfection ability of the skin surface. Arch. Klin. Exptl. Dermatol. 30, 371–382. Cited by Excerpta Med. Sect. 4, 5303, 1968. (German)

MUNCH-PETERSEN, E. 1961. Staphylococcal Carriage in Man. Bull. World Health Organ. 24, 761–769.

MUNCH-PETERSEN, E. 1963. Staphylococci in food and food intoxication. A review and an appraisal of phage typing results. J. Food Sci. 28, 692–719.

MURAKAMI, M., and ASAKAWA, Y. 1966. Studies on pathogenic staphylococci. II. Isolation of Staphylococcus aureus from the feces of healthy humans. Ann. Rep. Shizvoka Perfect. Hyg. Res. Lab. 14, 30–36. Cited by Excerpta Med. Sect. 4, 5534, 1967. (Japanese)

NAGAKI, S. et al. 1955. Acute staphylococcal gastro-enteritis with rash, the complication induced by antibiotic treatment. Japan J. Med. Sci. Biol. 8, 149–169.

NAHMIAS, A. J., LEPPER, M. A., HURST, V., and MUDD, S. 1962. Epidemiology and treatment of chronic staphylococcal infections in the household. Am. J. Public Health 52, 1828–1843.

NEAVE, F. K., DODD, F. H., and KINGWILL, R. G. 1962. A method for controlling udder disease. Vet. Rec. 78, 521–523.

NEAVE, F. K., and HOY, W. A. 1947. The disinfection of contaminated metal surfaces with hypochlorite solutions. J. Dairy Res. 15, 25–54.

NEWBOULD, F. H. 1968. Epizootiology of mastitis due to Staphylococcus aureus. J. Am. Vet. Med. Assoc. 153, 1683–1687.

NEWCOMER, J. L., RAMSEY, E. W., and EATON, H. D. 1962. Effect of air flow in a refrigerator on cooling rate. J. Am. Dietet. Assoc. 40, 39–40. N.Y. State Dept. Health. 1940. Gastroenteritis in Oswego County traced to homemade ice cream. Health News 17, 104.

NIVEN, C. F., JR., and EVANS, J. B. 1955. Popular misconceptions concerning staphylococcus food poisoning. Proc. 7th Res. Conf. Am. Meat Inst., Univ. Chicago, Chicago.

NOBLE, W. C. 1966. Staphylococcus aureus on the hair. J. Clin. Pathol. 19, 570–572.

NOBLE, W. C., and DAVIES, R. R. 1965. Studies on the dispersal of staphylococci. J. Clin. Pathol. *18*, 16–19.
NOBLE, W. C., VALKENBURG, H. A., and WOLTERS, C. H. L. 1967. Carriage of *Staphylococcus aureus* in random samples of a normal population. J. Hyg. *65*, 567–573.
NOONE, P., GRIFFITHS, R. J., and TAYLOR, C. E. D. 1970. Hexachlorphane for treating carriers of *Staphylococcus aureus.* Lancet *1*, 1202–1203.
NUNHEIMER, T. D., and FABIAN, F. W. 1940. Influence of organic acid, sugar, and sodium chloride upon strains of food poisoning staphylococci. Am. J. Public Health *30*, 1040–1049.
NYHAN, J. F., and COWHIG, M. J. 1967. Inadequate milking machine vacuum reserve and mastitis. Vet. Rec. *81*, 122–124.
OAKLEY, C. L., and FULTHROPE, A. J. 1953. Antigenic analysis by diffusion. J. Pathol. Bacteriol. *65*, 49–60.
OBERHOFER, T. R., and FRAZIER, W. C. 1961. Competition of *Staphylococcus aureus* with other organisms. J. Milk Food Technol. *24*, 172–175.
O'GRADY, F., and WITTSTADT, F. 1963. Nasal carriage of *Staphylococcus pyogenes.* II. Bacterial ecology of the nose. Am. J. Hyg. *77*, 187–194.
OGSTON, A. 1880. About Abscesses. Arch. Klin. Chir. *25*, 588–600. (German)
OGSTON, A. 1881. Report upon micro-organisms in surgical diseases. Brit. Med. J. *1*, 369–375.
OGSTON, A. 1882. Micrococcus poisoning. J. Anat. *17*, 24.
OLLIVIER, M. 1830. On the poisonous effects of certain spoiled articles of food (Abstr.) Lancet *2*, 838–839.
OMORI, G., and KATO, J. 1959. A staphylococcal food-poisoning caused by a coagulase-negative strain. Biken's J. *2*, 92.
ORTH, D. S., and ANDERSON, A. W. 1970A. Polymyxin-coagulase-deoxyribonuclease agar. I. A selective isolation medium for coagulase-positive staphylococci. Appl. Microbiol. *20*, 508–509.
ORTH, D. S., and ANDERSON, A. W. 1970B. Polymyxin-coagulase-mannitol-agar. I. A selective isolation medium for coagulase-positive staphylococci. Appl. Microbiol. *19*, 73–75.
OUCHTERLONY, O. 1949. Antigen antibody reactions in gels. Acta Pathol. Microbiol. Scand. *25*, 507–515.
OUCHTERLONY, O. 1953. Antigen antibody reactions in gels. IV. Types of reactions in coordinated systems of diffusion. Acta Pathol. Microbiol. Scand. *32*, 231–240.
OUDIN, J. 1952. Specific precipitation in gels and its application to immunochemical analysis. Methods Med. Res. *5*, 335–378.
OWEN, R. 1907. The bacteriology of meat poisoning. Physician Surgeon *29*, 289–298.
PARFENTJEV, I. A., and CATELLI, A. R. 1964. Tolerance of *Staphylococcus aureus* to sodium chloride. J. Bacteriol. *88*, 1–3.
PARKER, M. T., et al. 1971. Report (1966–1970) of the subcommittee on phage-typing of staphylococci to the International Committee on Nomenclature of Bacteria. Intern. J. Syst. Bacteriol. *21*, 167–170.
PASSET, J. 1885. About microorganisms of purelent tissue inflamation of man. Fortschr. Med. *3*, 33–43. (German)
PASTEUR, L. 1880. On the extension of the germ theory to the etiology of some common diseases. Compt. Rend. Acad. Sci. (Paris) *90*, 1033–1044. (French)
PATTERSON, J. T. 1963. Salt tolerance and nitrate reduction by micrococci from fresh pork, curing pickles, and bacon. J. Appl. Bacteriol. *26*, 80–85.
PEAVY, J. E., DICKERSON, M. S., and GONZALEZ, J. L. 1968. Food poisoning—Laredo, Texas. Morbidity Mortality Weekly Rept. (U.S.) *17*, 109–110.
PETERS, H. A. 1966. Nutritional requirements of enterotoxigenic strains of

120   FOOD MICROBIOLOGY

Staphylococcus aureus. Disertation Abstr. *26*, 4971-4972.
PETERSON, A. C., BLACK, J. J., and GUNDERSON, M. F. 1962A. Staphylococci in competition. I. Growth of naturally occurring mixed populations in precooked frozen foods during defrost. Appl. Microbiol. *10*, 16-30.
PETERSON, A. C., BLACK, J. J., and GUNDERSON, M. F. 1962B. Staphylococci in competition. II. Effect of total numbers and proportion of staphylococci in mixed cultures on growth in artificial culture medium. Appl. Microbiol. *10*, 23-30.
PETERSON, A. C., BLACK, J. J., and GUNDERSON, M. F. 1964A. Staphylococci in competition. III. Influence of pH and salt on staphylococcal growth in mixed populations. Appl. Microbiol. *12*, 70-76.
PETERSON, A. C., BLACK, J. J., and GUNDERSON, M. F. 1964B. Staphylococci in competition. IV. Effect of starch and kind and concentration of sugar on staphylococcal growth in mixed populations. Appl. Microbiol. *12*, 77-82.
PETERSON, A. C., BLACK, J. J., and GUNDERSON, M. F. 1964C. Staphylococci in competition. IV. Effect of eggs, egg plus carbohydrates, and lipids on staphylococcal growth. Appl. Microbiol. *12*, 83-86.
PHILLIPS, A. W., JR., and PROCTOR, B. E. 1947. Growth and survival of an experimentally inoculated *Staphylococcus aureus* in frozen precooked foods. J. Bacteriol. *54*, 49.
PILLSBURY, D. M., and REBELL, G. 1952. The bacterial flora of the skin. Factors influencing the growth of resident and transient organisms. J. Invest. Derm. *18*, 173-186.
PREONAS, D. L. *et al.* 1969. Growth of *Staphylococcus aureus* MF31 on the top and cut surfaces of southern custard pies. Appl. Microbiol. *18*, 68-75.
PRICE, P., NEAVE, F. K., RIPPON, J. E., and WILLIAMS, R. E. O. 1954. The use of phage typing and penicillin sensitivity tests in studies of staphylococci from bovine mastitis. J. Dairy Res. *21*, 342-353.
PRICE, P. B. 1938. The bacteriology of normal skin: A new quantitative test applied to a study of the bacterial flora and the disinfectant action of mechanical cleansing. J. Infect. Diseases *63*, 301-318.
PRICE, P. B. 1951. Fallacy of a current surgical fad—the 3 minute scrub with hexachlorophene soap. Ann. Surg. *134*, 476-485.
PRICE, P. B. 1968. Surgical antiseptics. *In* Disinfection, Sterilization and Preservation, C. A. Lawrence, and S. S. Block (Editors). Lea and Febiger, Philadelphia.
PRINDLE, R. F., and WRIGHT, E. S. 1968. Phenolic compounds. *In* Disinfection, Sterilization and Preservation, C. A. Lawrence, and S. S. Block (Editors). Lea and Febiger, Philadelphia.
QUEVEDO, F., and CARRANZA, N. 1966. The role of flies in the contamination of food in Peru. Ann. Inst. Pasteur Lille *17*, 199-202.
RADICH, R., SANDINE, W. E., and ELLIKER, P. R. 1967. Inhibition of *Staphylococcus aureus* by *Streptococcus diacetilactis* in milk and cream filling. J. Dairy Sci. *52*, 880.
RAJ, H. 1966. A new procedure for the detection and enumeration of coagulase-positive staphylococci from frozen seafoods. Can. J. Microbiol. *12*, 191-198.
RAJ, H., and LISTON, J. 1961. Detection and enumeration of coagulase-positive staphylococci. Bacteriol. Proc. p. 68, Am. Soc. Microbiology, Washington, D.C.
RAJ, H. D., and BERGDOLL, M. S. 1969. Effect of enterotoxin B on human volunteers. J. Bacteriol. *98*, 833-834.
RAJULU, P. S., FOLTZ, V. D., and LORD, T. H. 1960. The canine as a reservoir of pathogenic staphylococci. Am. J. Public Health Suppl. *50*, 74-78.
RAMMELL, C. G., and HOWICK, J. M. 1967. Enumeration of coagulase-positive staphylococci in cheese. J. Appl. Bacteriol. *30*, 382-388.

RAVENHOLT, R. T., EELKEMA, R. C., MULHERN, M., and WATKIN, R. B. 1961. Staphylococcal infection in meat animals and meat workers. Public Health Rept. (U.S.) 76, 879-888.

READ, R. B., JR., and BRADSHAW, J. G. 1966A. Thermal inactivation of staphylococcal enterotoxin B in veronal buffer. Appl. Microbiol. 14, 130-134.

READ, R. B., JR., and BRADSHAW, J. G. 1966B. Staphylococcal enterotoxin B thermal inactivation in milk. J. Dairy Sci. 49, 202-203.

READ, R. B., JR., and BRADSHAW, J. G. 1967. α-Irradiation of staphylococcal enterotoxin B. Appl. Microbiol. 15, 603-605.

READ, R. B., JR., BRADSHAW, J., PRITCHARD, W. L., and BLACK, L. A. 1965B. Assay of staphylococcal enterotoxin from cheese. J. Dairy Sci. 48, 420-424.

READ, R. B., JR., PRITCHARD, W. L., BRADSHAW, J., and BLACK, L. A. 1965A. In vitro assay of staphylococcal enterotoxins A and B from milk. J. Dairy Sci. 48, 411-419.

REISER, R. F., and WEISS, K. F. 1969. Production of staphylococcal enterotoxins A, B, and C in various media. Appl. Microbiol. 18, 1041-1043.

RICCIARDI, G. 1967. A search for carriers of staphylococcus among workers in food industries. Igiena Mod. 60, 719-735. Cited by Abstr. Hyg. 4331, 1968. (Italian)

RIDLEY, M. 1959. Perineal carriage of Staphylococcus aureus. Brit. Med. J. 1, 270-273.

RIVAS, V. M. T. et al. 1965. Statistical study from the bacteriological aspects of cheeses suspected of having caused food poisoning. Salud Publica Mexico 7, 243-249. (Spanish)

ROBINSON, J., and THATCHER, F. S. 1965. Determination of staphylococcal enterotoxin by an indirect hemagglutination inhibition procedure. Bacteriol. Proc. p. 72, Am. Soc. Microbiology, Washington, D.C.

ROGERS, R. E., and GUNDERSON, M. F. 1958. Roasting of frozen stuffed turkeys. II. Survival of Micrococcus pyogenes var. aureus in inoculated stuffing. Food Res. 23, 96-102.

ROODYN, L. 1960. Recurrent staphylococcal infections and the duration of the carrier state. J. Hyg. 58, 11-19.

ROSENBACH, F. J. 1884. Microorganisms in wound infection in man. J. F. Bergmann, Wiesbaden. (German)

ROTH, L. M., and WILLIS, E. R. 1957. The medical and veterinary importance of cockroaches. Smithsonian Inst. Misc. Collections 134, No. 10, Publ. 4299.

ROUGHLY, R. J., and McLEOD, R. W. 1961. Longevity of Staphylococcus aureus in Australian cheddar cheese. Australian J. Dairy Technol. 6, 110-112.

ROUNDTREE, P. M. 1963. The effect of desiccation on the viability of Staphylococcus aureus. J. Hyg. 61, 265-272.

ROUNDTREE, P. M., and BARBOUR, R. G. H. 1951. Nasal carrier rates of Staphylococcus pyogenes in hospital nurses. J. Pathol. Bacteriol. 63, 313-324.

ROUNDTREE, P. M., FREEMAN, B. M., and JOHNSTON, K. G. 1956. Nasal carriage of Staphylococcus aureus by various domestic and laboratory animals. J. Pathol. Bacteriol. 72, 319-324.

RYBERG, R. E., and CATHCART, W. H. 1942. Staphylococci and salmonella control in foods. II. Effect of pure fruit fillings. Food Res. 7, 10-15.

SAHU, S. P., and MUNRO, D. A. 1969. Observations of systemic and localized infection associated with the isolation of Staphylococcus in chickens in North Carolina. Avian Diseases 13, 684-689.

SALOMON, L. L., and TEW, R. 1968. Assay of staphylococcal enterotoxin B by latex agglutination. Proc. Soc. Exptl. Biol. Med. 129, 539-542.

SALZER, R. H., KRAFT, A. A., and AYRES, J. C. 1967. Microorganisms isolated from turkey giblets. Poultry Sci. 46, 612-615.

SCHANTZ, E. J. *et al.* 1965. Purification of staphylococcal enterotoxin B. Biochem. *4*, 1011–1016.

SCHEUSNER, D. L., HOOD, L. L., and HARMON, L. G. 1973. Effect of temperature and pH on growth and enterotoxin production by *Staphylococcus aureus*. J. Milk Food Technol. *36*, 249–252.

SCHNEIERSON, S. S. 1965. Antibiotic susceptibility of pathogenic microorganisms isolated in 1963. N.Y. State J. Med. *65*, 542–547.

SCHWABACHER, H., CUNLIFFE, A. C., WILLIAMS, R. E. O., and HARPER, G. J. 1945. Hyaluronidase production by staphylococci. Brit. J. Exptl. Pathol. *26*, 124–129.

SCOTT, W. J. 1953. Water relations of *Staphylococcus aureus* at 30°C. Australia J. Biol. Sci. *6*, 549–564.

SCOTT, W. J. 1957. Water activity of food spoilage microorganisms. Adv. Food Res. *7*, 83–127.

SEGALOVE, M. 1947. The effect of penicillin on growth and toxin production by enterotoxin staphylococci. J. Infect. Diseases *81*, 228–243.

SEGALOVE, M., and DACK, G. M. 1941. Relation of time and temperature to growth and enterotoxin production of staphylococci. Food Res. *6*, 127–133.

SEGALOVE, M., and DACK, G. M. 1951. Growth of bacteria associated with food poisoning experimentally inoculated into dehydrated meat. Food. Res. *16*, 118–125.

SEGALOVE, M., DAVISON, E., and DACK, G. M. 1943. Growth of food-poisoning strain of *Staphylococcus aureus* experimentally inoculated into canned foods. Food Res. *8*, 54–57.

SELWYN, S., VERMA, B. S., and VAISHNAV, V. P. 1967. Factors in the bacterial colonization and infection of the human skin. Indian J. Med. Res. *55*, 652–656.

SEMINIANO, E. N., and FRAZIER, W. C. 1966. Effect of pseudomonads and Achromobacteraceae on growth of *Staphylococcus aureus*. J. Milk Food Technol. *29*, 161–164.

SESSOMS, A. R., and MERCURI, A. J. 1969. Efficiency of five selective media for recovering coagulase-positive staphylococci from mixed cultures. Poultry Sci. *48*, 1637–1639.

SHAFFER, C. H., and STUART, L. S. 1968. Methods of testing sanitizers and bacteriostatic substances. *In* Disinfection, Sterilization and Preservation, C. A. Lawrence, and S. S. Block (Editors). Lea and Febiger, Philadelphia.

SHAH, D. B., and WILSON, J. B. 1963. Egg yolk factor of *Staphylococcus aureus*. I. Nature of the substrate and enzyme involved in the egg yolk opacity reaction. J. Bacteriol. *85*, 516–521.

SHARF, J. M. 1966. Recommended Methods for the Microbiological Examination of Foods, 2nd Edition. Am. Public Health Assoc., New York.

SHARPE, M. E., FEWINS, B. G., REITER, B., and CUTHBERT, W. A. 1965. A survey of the incidence of coagulase-positive staphylococci in market milk and cheese in England and Wales. J. Dairy Res. *32*, 187–192.

SHARPE, M. E., NEAVE, F. K., and REITER, B. 1962. Staphylococci and micrococci associated with dairying. J. Appl. Bacteriol. *25*, 403–415.

SHAW, C., STITT, J. M., and COWAN, S. T. 1951. Staphylococci and their classification. J. Gen. Microbiol. *5*, 1010–1023.

SHELTON, L. R., *et al.* 1960. A bacteriological survey of the frozen precooked food industry. U.S. Dept. Health, Education, and Welfare, Food and Drug Admin., Washington, D.C.

SHEMANO, I., HITCHENS, J. T., and BEILER, J. M. 1967. Paradoxical intestinal inhibitory effects of staphylococcal enterotoxin. Gastroenterology *53*, 71–77.

SHEWAN, J. M. 1962. Food poisoning caused by fish and fishery products. *In* Fish as Foods, Vol. III, G. Borgstrom (Editor). Academic Press, New York.

SHOOTER, R. A., and WYATT, H. V. 1955. Mineral requirements for growth of

Staphylococcus pyogenes. Effect of magnesium and calcium ions. Brit. J. Exptl. Pathol. *36*, 341–350.
SHOOTER, R. A., and WYATT, H. V. 1956. Mineral requirements for growth of *Staphylococcus pyogenes.* Effect of potassium ions. Brit. J. Exptl. Pathol. *37*, 311–317.
SIEMS, H., KUSCH, D., SINELL, H. J., and UNTERMANN, F. 1971. Occurrence and characteristics of staphylococci in different production stages of meat processing. Fleischwirschaft *10*, 1529–1533. (German)
SILBERG, S. L., NESER, W. B., and BLENDER, D. C. 1967. Predisposition of the host to staphylococcal reinfection. Can. J. Public Health *58*, 411–413.
SILLIKER, J. H. 1969. Some guidelines for the safe use of fillings, toppings, and icings. Baker's Dig. *43*, No. 1, 51–54.
SILLIKER, J. H., JANSEN, C. E., VOEGELI, M. M., and CHMURA, N. W. 1962. Studies on the fate of staphylococci during the processing of hams. J. Food Sci. *27*, 50–56.
SILLIKER, J. H., and McHUGH, S. A. 1967. Factors influencing microbial stability of butter-cream-type fillings. Cereal Sci. Today *12*, No. 1, 63–65, 73–74.
SILVERMAN, G. J, et al. 1961. Microbial analysis of frozen raw and cooked shrimp. I. General results. Food Technol. *15*, 455–458.
SILVERMAN, S. J. 1963. Serological assay of culture filtrates for *Staphylococcus* enterotoxin. J. Bacteriol. *85*, 955–956.
SILVERMAN, S. J., KNOTT, A. R., and HOWARD, M. 1968. Rapid, sensitive assay for staphylococcal enterotoxin and a comparison of serological methods. Appl. Microbiol. *16*, 1019–1023.
SINELL, H. J., and KUSCH, D. 1969. Selective culturing of coagulase-positive staphylococci from chopped meat. Arch. Hyg. *153*, 56–66. *Cited by* Excerpta Med. Sect. 4, 2332, 1969. (German)
SLANETZ, L. W., and BARTLEY, C. H. 1953. The diagnosis of staphylococcal mastitis, with special reference to the characteristics of mastitis staphylococci. J. Infect. Diseases *92*, 139–151.
SLOCUM, G. G., and LINDEN, B. A. 1939. Food poisoning due to staphylococci with special reference to staphylococcus agglutination by normal horse serum. Am. J. Public Health *29*, 1326–1330.
SMART, R. A., MINER, M. L., and DAVIS, R. V. 1968. Pathogenesis of staphylococcal synovitis in turkeys: Cultural retrieval in experimental infection. Avian Diseases *12*, 37–46.
SMITH, H. B. H., and FARKAS-HIMSLEY, H. 1969. The relationship of pathogenic coagulase-negative staphylococci to *Staphylococcus aureus.* Can. J. Microbiol. *15*, 879–890.
SMITH, H. W. 1957. The multiplication of *Staphylococcus aureus* in cows' milk. Monthly Bull. Min. Health, Public Health Lab. Serv. (British) *16*, 39–52.
SMITH, H. W., and CRABB, W. E. 1960. The effect of diets containing tetracyclines and penicillin on the *Staphylococcus aureus* flora of the nose and skin of pigs and chickens and their human attendants. J. Pathol. Bacteriol. *79*, 243–249.
SMITH, P. B. 1970. Bacteriophage typing. *In* Manual of Clinical Microbiology, J. E. Blair, E. H. Lennette, and J. P. Truant (Editors). American Society of Microbiology, Bethesda, Md.
SMITH, P. B., HANCOCK, G. A., and RHODEN, D. L. 1969. Improved medium for detecting deoxyribonuclease-producing bacteria. Appl. Microbiol. *18*, 991–993.
SMITH, P. B., McCOY, E., and WILSON, J. B. 1962. Identification of staphylococci in nonfat dry milk by the fluorescent antibody technique. J. Dairy Sci. *45*, 729–734.
SMITH, P. B. 1972. Bacteriophage typing of *Staphyloccus aureus.* In The Staphylococci, J. O. Cohen (Editor). Wiley-Interscience, New York.
SPEARE, G. S. 1954. Staphylococcus pseudomembranous enterocolitis, a complication of antibiotic therapy. Am. J. Surg. *88*, 523–534.

124   FOOD MICROBIOLOGY

SPLITTSTOESSER, D. F., HERVEY, G. E. R., II, and WETTERGREEN, W. P. 1965. Contamination of frozen vegetables by coagulase-positive staphylococci. J. Milk Food Technol. 28, 149-151.
STABYJ, B. M., DOLLAR, A. M., and LISTON, J. 1965. Post-irradiation survival of Staphylococcus aureus in sea foods. J. Food Sci. 30, 344-350.
STARK, R. L., and MIDDAUGH, P. R. 1969. Immunofluorescent detection of enterotoxin B in food and a culture medium. Appl. Microbiol. 18, 631-635.
STEWART, G. T. 1965. The lipases and pigments of staphylococci. Ann. N.Y. Acad. Sci. 128, 132-151.
ST. GEORGE, C., RUSSELL, K. E., and WILSON, J. B. 1962. Characteristics of staphylococci from bovine milk. J. Infect. Diseases 110, 75-79.
STICKLER, D. J., and FREESTONE, M. 1971. Coagulase and deoxyribonuclease tests for staphylococci. Med. Lab. Technol. 28, 96-97.
STILES, M. E., and WITTER, L. D. 1965. Thermal inactivation, heat injury, and recovery of Staphylococcus aureus. J. Dairy Sci. 48, 677-681.
STONE, R. V. 1943. Staphylococcic food-poisoning and dairy products. J. Milk Technol. 6, 7-16.
STRAKA, R. P., and COMBES, F. M. 1952. Survival and multiplication of Micrococcus pyogenes var. aureus in creamed chicken under various holding, storage, and defrosting condition. Food Res. 17, 448-455.
STRATFORD, B., RUBBO, S. C., CHRISTIE, R., and DIXSON, S. 1960. Treatment of nasal carrier of Staphylococcus aureus with framycetin and other antibacterials. Lancet 2, 1225-1227.
STRAUSS, J. S., and KLIGMAN, A. M. 1956. The bacteria responsible for apocrine odour. J. Invest. Derm. 27, 67-69.
STRITAR, J. E. 1941. Studies in meat canning problems. Proc. Chem. Operating Sec., 36th Ann. Conv. Am. Meat Inst., Chicago.
STRITAR, J., DACK, G. M., and JUNGEWALTER, F. G. 1936. The control of staphylococci in custard-filled puffs and eclairs. Food Res. 1, 237-246.
SUGIYAMA, H., CHOW, K. L., and DRAGSHEDT, L. K., II. 1961. Study of emetic receptor sites for staphylococcal enterotoxin in monkeys. Proc. Soc. Exptl. Biol. Med. 108, 92-95.
SUGIYAMA, H., and HAYAMA, T. 1965. Abdominal viscera as site of emetic action for staphylococcal enterotoxin in the monkey. J. Infect. Diseases 115, 330-336.
SUMMERS, M. M., LYNCH, P. F., and BLACK, T. 1965. Hair as a reservoir of staphylococci. J. Clin. Pathol. 18, 13-15.
SURGALLA, M. J., BERGDOLL, M. S., and DACK, G. M. 1952. Use of antigen-antibody reactions in agar to follow the progress of fractionation of antigenic mixtures: Application to purification of staphylococcal enterotoxin. J. Immunol. 69, 357-365.
SURGALLA, M. J., BERGDOLL, M. S., and DACK, G. M. 1953. Some observations on the assay of staphylococcal enterotoxin by the monkey-feeding test. J. Lab. Clin. Med. 41, 782-788.
SURGALLA, M. J., BERGDOLL, M. S., and DACK, G. M. 1954. Staphylococcal enterotoxin: Neutralization by rabbit antiserum. J. Immunol. 72, 398-403.
SURGALLA, M. J., and DACK, G. M. 1945. Growth of Staphylococcus aureus, Salmonella enteritidis, and alpha-type streptococcus experimentally inoculated into canned meat products. Food Res. 10, 108-113.
SURKIEWICZ, B. F. 1966. Bacteriological survey of the frozen prepared food industry. I. Frozen cream-type pies. Appl. Microbiol. 14, 21-26.
SURKIEWICZ, B. F., GROOMES, R. J., and PADRON, A. P. 1967A. Bacteriological survey of the frozen prepared food industry. III. Potato products. Appl. Microbiol. 15, 1324-1331.
SURKIEWICZ, B. F., GROOMES, R. J., and SHELTON, L. R., JR. 1968. Bacteriological survey of the frozen prepared food industry. IV. Frozen breaded fish. Appl. Microbiol. 16, 147-150.

SURKIEWICZ, B. F., HARRIS, M. E., and JOHNSTON, R. W. 1973. Bacteriological survey of frozen meat and gravy produced at establishments under federal inspection. Appl. Microbiol. 26, 574-576.

SURKIEWICZ, B. F., HYNDMAN, J. B., and YANCEY, M. V. 1967B. Bacteriological survey of the frozen prepared food industry. II. Frozen breaded raw shrimp. Appl. Microbiol. 15, 1-9.

SURKIEWICZ, B. F., JOHNSTON, R. W., ELLIOTT, R. P., and SIMMONS, E. R. 1972. Bacteriological survey of fresh pork sausage produced at establishments under federal inspection. Appl. Microbiol. 23, 515-520.

SURKIEWICZ, B. F., JOHNSTON, R. W., MORAN, A. B., and KRUMM, G. W. 1969. A bacteriological survey of chicken eviscerating plants. Food Technol. 23, 1066-1069.

TAGER, M., and DRUMMOND, M. C. 1965. Staphylocoagulase. Ann. N.Y. Acad. Sci. 128, 92-111.

TAKAHASHI, I., and JOHNS, G. K. 1959. Staphylococcus aureus in cheddar cheese. J. Dairy Sci. 42, 1032-1037.

TANNER, F. W., and TANNER, L. P. 1953. Food-borne Infections and Intoxications, 2nd Edition. Garrard Press, Champaign, Ill.

TARDIO, J. L., and BAER, E. F. 1971. Comparative efficiency of two methods and two plating media for isolation of Staphylococcus aureus from foods. J. Assoc. Offic. Anal. Chem. 54, 728-731.

TARR, H. L. A. 1942. The action of nitrites on bacteria: Further experiments. J. Fisheries Res. Board Can. 6, 74-89.

TATINI, S. R. 1973. Influence of food environments on growth of Staphylococcus aureus and production of various enterotoxins. J. Milk Food Technol. 36, 559-563.

TATINI, S. R., CORDS, B. R., McKAY, L. L., and BENNETT, R. W. 1971. Effect of growth temperature higher than optimal on production of staphylococcal enterotoxins and deoxyribonuclease. Bacteriol. Proc. p. 17, Am. Soc. Microbiology, Washington, D.C.

TERPLAN, G., and ZAADHOF, K. J. 1969. Diagnostic and food hygiene importance of Staphylococcus aureus in cows' milk. Deut. Tieraerztl. Wochscht. 76 217-221. Cited by Excerpta Med. Sect. 4, 2336, 1969. (German)

THATCHER, F. S. 1963. The microbiology of specific frozen foods in relation to public health: Report of an international committee. J. Appl. Bacteriol. 26, 266-285.

THATCHER, F. S., and CLARK, D. S. 1968. Microorganisms in Foods: Their Significance and Methods of Enumeration. Univ. of Toronto Press, Canada.

THATCHER, F. S., COMTOIS, R. D., ROSS, D., and ERDMAN, I. E. 1959. Staphylococci in cheese: Some public health aspects. Can. J. Public Health 50, 497-503.

THATCHER, F. S., ROBINSON, J., and ERDMAN, I. 1962. The vacuum pack method of packaging foods in relation to the formation of the botulinum and staphylococcal toxins. J. Appl. Bacteriol. 25, 120-124.

THATCHER, F. S., and ROSS, D. 1960. Multiplication of staphylococci during cheese making: The influence of milk storage temperatures and of antibiotics. Can. J. Public Health 51, 226-234.

THATCHER, F. S., and SIMON, W. 1955. The resistance of staphylococci and streptococci isolated from cheese to various antibiotics. Can. J. Public Health 46, 407-409.

THATCHER, F. S., and SIMON, W. 1956. A comparative appraisal of the properties of "Staphylococci" isolated from clinical sites and dairy products. Can. J. Microbiol. 2, 703-714.

THOMAS, C. T., WHITE, J. C., and LONGREE, K. 1966. Thermal resistance of salmonellae and staphylococci in foods. Appl. Microbiol. 14, 815-820.

THOMPSON, J. D., and JOHNSON, D. 1963. Food temperature preferences of surgical patients. J. Am. Dietet. Assoc. *43*, 209–211.

TOMPKIN, R. B., AMBROSINO, J. M., and STOZEK, S. K. 1973. Effect of pH and sodium chloride on enterotoxin A production. Appl. Microbiol. *26*, 833–837.

TROLLER, J. A. 1971. Effect of water activity on enterotoxin B production and growth of Staphylococcus aureus. Appl. Microbiol. *21*, 435–439.

TROLLER, J. A. 1972. Effect of water activity on enterotoxin A production and growth of Staphylococcus aureus. Appl. Microbiol. *24*, 440–443.

TROLLER, J. A., and FRAZIER, W. C. 1963A. Repression of Staphylococcus aureus by food bacteria. I. Effect of environmental factors on inhibition. Appl. Microbiol. *11*, 11–14.

TROLLER, J. A., and FRAZIER, W. C. 1963B. Repression of Staphylococcus aureus by food bacteria. II. Causes of inhibition. Appl. Microbiol. *11*, 163–165.

TUCKEY, S. L., STILES, M. E., ORDAL, Z. L., and WITTER, L. D. 1964. Relation of cheese-making operations to survival and growth of Staphylococcus aureus in different varieties of cheese. J. Dairy Sci. *47*, 604–611.

UNTERMANN, F. 1972. Contribution to the occurrence of enterotoxin producing staphylococci in man. Zbl. Bakt. Hyg. I. Abstr. Orig. A *222*, 18–26. (German)

U.S. DEPT. AGR., CONSUMER AND MARKETING SERV. 1966. Accuracy of attribute sampling: A guide for inspection personnel. U.S. Dept. of Agriculture, Washington, D.C.

U.S. DEPT. HEALTH, EDUCATION, AND WELFARE, FOOD AND DRUG ADMIN. 1968. Human foods: Current good manufacturing practices (Sanitation) in manufacturing, processing, packing, or holding. Federal Register [21 CFR part 128] *33*, No. 247, 19023–19026.

U.S. DEPT. HEALTH, EDUCATION, AND WELFARE. 1971. Hexachlorophene and Newborns. FDA Drug Bull. Dec., Food and Drug Admin., Rockville, Md.

U.S. DEPT. HEALTH, EDUCATION, AND WELFARE, PUBLIC HEALTH SERV. 1962. Food service sanitation manual. Public Health Publ. *934*.

U.S. DEPT. HEALTH, EDUCATION, AND WELFARE, PUBLIC HEALTH SERV. 1969. Sanitary food service: Instructor's guide. Public Health Serv. Publ. *90*.

U.S. DEPT. HEALTH, EDUCATION, AND WELFARE, PUBLIC HEALTH SERV. 1967. Reported Foodborne Outbreaks 1966. National Communicable Disease Center, Atlanta, Ga.

U.S. DEPT. HEALTH, EDUCATION, AND WELFARE, PUBLIC HEALTH SERV. 1968. Foodborne Outbreaks Status Report for 1967. National Communicable Disease Center, Atlanta, Ga.

U.S. DEPT. HEALTH, EDUCATION, AND WELFARE, PUBLIC HEALTH SERV. 1969. Foodborne Outbreaks Annual Summary 1968. National Communicable Disease Center, Atlanta, Ga.

U.S. DEPT. HEALTH, EDUCATION, and WELFARE, PUBLIC HEALTH SERV. 1970. Foodborne Outbreaks Annual Summary 1969. National Communicable Disease Center, Atlanta, Ga.

U.S. DEPT. HEALTH, EDUCATION, AND WELFARE, PUBLIC HEALTH SERV. 1971. Foodborne Outbreaks Annual Summary 1970. Center for Disease Control, Atlanta, Ga.

U.S. DEPT. HEALTH, EDUCATION, AND WELFARE, PUBLIC HEALTH SERV. 1972. Foodborne Outbreaks Annual Summary 1971. Center for Disease Control, Atlanta, Ga.

U.S. DEPT. HEALTH, EDUCATION, AND WELFARE, PUBLIC HEALTH SERV. 1973. Foodborne Outbreaks Annual Summary 1972. Center for Disease Control, Atlanta, Ga.

U.S. DEPT. HEALTH, EDUCATION, AND WELFARE, PUBLIC HEALTH SERV. 1974. Foodborne and Waterborne Outbreaks Annual Summary 1973. Center for Disease Control, Atlanta, Ga.

VANDERBOSCH, L. L., FUNG, D. Y. C., and WIDOMSKI, M. 1973. Optimum temperature for enterotoxin production by *Staphylococcus aureus* S-6 and 137 in liquid medium. Appl. Microbiol. *25*, 498–500.

VANDERZANT, C., and NICKELSON, R. 1969. A microbiological examination of muscle tissue of beef, pork, and lamb carcasses. J. Milk Food Technol. *32*, 357–361.

VARGA, D. T., and WHITE, A. 1961. Suppression of nasal skin and aerial staphylococci by nasal application of methicillin. J. Clin. Invest. *40*, 2209–2214.

VIRGILIO, R. *et al.* 1970. Bacteriological analysis of frozen shrimp. 2. Staphylococci in precooked frozen Chilean shrimp. J. Food Sci. *35*, 845–848.

VOGEL, R. A., and JOHNSON, M. 1960. A modification of the tellurite-glycine medium for use in the identification of *Staphylococcus aureus*. Public Health Lab. *18*, 131–133.

WADSWORTH, C. A. 1957. Slide microtechnique for the analysis of immune precipitates in gel. Intern. Arch. Allergy Appl. Immunol. *10*, 355–360.

WAIN, H., and BLACKSTONE, P. A. 1956. Staphylococcal gastroenteritis: Report of a major outbreak. Am. J. Digest. Diseases *1*, 424–429.

WALKER, G. C., and HARMON, L. G. 1966. Thermal resistance of *Staphylococcus aureus* in milk, whey, and phosphate buffer. Appl. Microbiol. *14*, 584–590.

WALKER, G. C., HARMON, L. G., and STINE, C. M. 1961. Staphylococci in colby cheese. J. Dairy Sci. *44*, 1272–1282.

WEATHERALL, J. A. C., and WINNER, H. I. 1963. The intermittent use of hexachlorophene soap—a controlled trial. J. Hyg. *61*, 443–449.

WEBSTER, R. C., and ESSELEN, W. B. 1956. Thermal resistance of food poisoning organisms in poultry stuffing. J. Milk Food Technol. *19*, 209–212.

WEED, L. A., MICHAEL, A. C., and HARGER, R. N. 1943. Fatal staphylococcus intoxication from goat milk. Am. J. Public Health *33*, 1314–1318.

WEIMANN, O. *et al.* 1971. Staphylococcal gastroenteritis associated with salami: United States. Morbidity Mortality Weekly Rept. (U.S.) *20*, No. 28, 253, 258.

WEIRETHER, F. J., LEWIS, E. E., ROSENWALD, A. J., and LINCOLN, R. E. 1966.. Rapid quantitative serological assay of staphylococcal enterotoxin B. Appl. Microbiol. *14*, 284–291.

WEISER, H. H., WINTER, A. R., and LEWIS, M. A. 1954. The control of bacteria in chicken salad. I. *Micrococcus pyogenes* var. *aureus*. Food Res. *19*, 465–471.

WENZEL, R. P. 1974. Food poisoning. *In* Current Therapy 1974, H. F. Conn (Editor). Saunders, Philadelphia.

WEST, L. S. 1951. The Housefly. Comstock Publishing Associates, Ithaca, N.Y.

WETHINGTON, M. C., and FABIAN, F. W. 1950. Viability of food poisoning staphylococci and salmonellae in salad dressing and mayonnaise. Food Res. *15*, 125–134.

WIEDEMAN, K., WATSON, M. A., MAYFIELD, H., and WALTER, W. G. 1956. Effect of delayed cooking on bacteria in chicken casserole. J. Am. Dietet. Assoc. *33*, 37–41.

WIEDEMAN, K., WATSON, M. A., NEILL, J., and WALTER, W. G. 1957. Effect of hot holding time on bacteria in meat loaf. J. Am. Dietet. Assoc. *32*, 935–940.

WILKINSON, R. J. *et al.* 1965. Effective heat processing for the destruction of pathogenic bacteria in turkey rolls. Poultry Sci. *44*, 131–136.

WILKOFF, L. J., WESTBROOK, L., and DIXON, G. J. 1969. Factors affecting the persistence of *Staphylococcus aureus* on fabrics. Appl. Microbiol. *17*, 268–274.

WILLIAMS, G. C., SWIFT, O. B. E., VOLLUM, R. L., and WILSON, G. S. 1946. Three outbreaks of staphylococcal food poisoning due to ice cream. Monthly Bull. Min. Health, Emergency Public Health Lab. Serv. (British) *5*, 17–25.

WILLIAMS, J. D., WALTHO, C. A., AYLIFHE, G. A. J., and LOWBURY, E. J. L. 1967. Trials of fine antibacterial creams in the control of nasal carriage of *Staphylococcus aureus*. Lancet *2*, 390-392.

WILLIAMS, R. E. O. 1946. Skin and nose carriage of bacteriophage types of *Staphylococcus aureus*. J. Pathol. Bacteriol. *58*, 259-268.

WILLIAMS, R. E. O. 1963. Healthy carriage of *Staphylococcus aureus*: Its prevalence and importance. Bacteriol. Rev. *27*, 56-71.

WILLIAMS, R. E. O. 1965. Pathogenic bacteria on the skin. *In* Skin bacteria and their role in infection, H. I. Maibach, and G. Hildick-Smith (Editors). McGraw-Hill, New York.

WILLIAMS, R. E. O. 1966. Epidemiology of airborne staphylococcal infection. Bacteriol. Rev. *30*, 660-672.

WILLIAMS, R. E. O. 1967. *Staphylococcus aureus* as commensal and pathogen. J. Pathol. Microbiol. *30*, 932-945.

WILLIAMS, R. E. O., and MILES, A. A. 1945. The bacterial flora of wounds and septic lesions of the hand. J. Pathol. Bacteriol. *57*, 27-36.

WILLIAMS, R. E. O., RIPPON, J. E., and DOWSETT, L. M. 1953. Bacteriophage typing of strains of *Staphylococcus aureus* from various sources. Lancet *1*, 510-514.

WILLIAMS-SMITH, W. 1957. The multiplication of *Staphylococcus aureus* in cows' milk. Monthly Bull. Min. Health Public, Health Lab. Serv. (British) *16*, 39-52.

WILSON, E., FOTER, M. J., and LEWIS, K. H. 1959. A rapid test for detecting *Staphylococcus aureus* in food. Appl. Microbiol. 7, 22-26.

WILSSENS, A. T. E., and VANDE CASTEELE, J. C. 1967. Occurrence of micrococci and staphylococci in the intestines of piglets and pigs. J. Appl. Bacteriol. *30*, 336-339.

WISEMAN, G. M. 1970. The beta- and delta-toxins of *Staphylococcus aureus*. *In* Microbial Toxins, Vol. 3, T. C. Montie, S. Kadis, and S. J. Ajl (Editors). Academic Press, New York.

WOLF, F. S. *et al.* 1970. Staphylococcal food poisoning traced to butter—Alabama. Morbidity Mortality Weekly Rept. (U.S.) *19*, No. 28, 271.

WOODARD, W. G. 1968. The Survival and Growth of Bacteria in Turkey Rolls During Processing and Storage. M. S. Thesis. Iowa State Univ., Ames, Iowa.

WOODBURN, M., BENNION, M., and VAIL, G. E. 1962. Destruction of salmonellae and staphylococci in precooked poultry products by heat treatment before freezing. Food Technol. *16*, 98-100.

WOODBURN, M. J., and STRONG, D. H. 1960. Survival of *Salmonella typhimurium, Staphylococcus aureus*, and *Streptococcus faecalis* frozen in simplified food substrates. Appl. Microbiol. 8, 109-113.

WOODIN, A. M. 1965. Staphylococcal leukocidin. Ann. N.Y. Acad. Sci. *128*, 152-164.

WOODIN, A. M. 1970. Staphylococcal leukocidin. *In* Microbial Toxins, Vol. 3, T. C. Montie, S. Kadis, and S. J. Ajl (Editors). Academic Press, New York.

ZEBOVITZ, E., EVANS, J. G., and NIVEN, C. F. 1955. Tellurite-glycine agar: A selective plating medium for the quantitative detection of coagulase-positive staphylococci. J. Bacteriol. *70*, 686-690.

ZEHREN, V. L., and ZEHREN, V. F. 1968A. Examination of large quantities of cheese for staphylococcal enterotoxin A. J. Dairy Sci. *51*, 635-644.

ZEHREN, V. L., and ZEHREN, V. F. 1968B. Relation of acid-development during cheesemaking to development of staphylococcal enterotoxin A. J. Dairy Sci. *51*, 645-649.

ZOTTOLA, E. A. 1968. Staphylococcus food poisoning. Univ. Minnesota Agr. Ext. Serv. Bull. *354*.

ZOTTOLA, E. A., and JEZESKI, J. J. 1963. Effect of heat on *Staphylococcus aureus* in milk and cheese during commercial manufacture. J. Dairy Sci. *46*, 600.

CHAPTER 3

John A. Troller | Salmonella and Shigella

Much of the current interest in salmonellosis, the food-borne enteric disease caused by members of the genus *Salmonella*, stems from an increased awareness of this disease which commenced early in the nineteen sixties. At that time researchers could look back on comparative incidence histories of typhoid fever and salmonellosis and note that, paradoxically, as the incidence of the typhoid fever declined, the incidence of salmonellosis increased. This pattern has been essentially maintained over the ensuing 10–13 years.

The reason for this continual upsurge in the numbers of reported cases of salmonellosis is not known with any degree of certainty; however, improved detection methods and more thorough awareness and reporting of this disease are probably involved. In addition, however, there can be little doubt that a real increase in salmonellosis is occurring throughout the world.

SALMONELLA CLASSIFICATION

The genus *Salmonella* is found in the family *Enterobacteriaccae*, a group of Gram-negative, non-spore-forming bacilli commonly cultured from the contents of vertebrate intestines. Salmonellae are usually motile with peritrichous flagellae, produce $H_2 S$, produce acid from glucose, maltose, mannitol and sorbitol, utilize citrate, but do not ferment salicin, sucrose and lactose. They are indole-negative and reduce nitrate to nitrite. Like many of the enterobacter group, they produce mucoid colonies, especially if incubated at low temperatures.

The seventh edition of Bergey's "Manual of Determinative Bacteriology" describes ten *Salmonella* species, distinctions between species being primarily based on host specificity, morphology and to a limited extent carbohydrate fermentation. A number of other classification schemes for this genus have been proposed; however, none have received universal acceptance. Kauffmann and Edwards (1952) suggested that the genus *Salmonella* be divided into three species, *S. typhosa*, *S. choleraesuis* and *S. enterica*; with the latter species serving as a repository for the approximately 1,500 serotypes currently in existence. A similar three species system of nomenclature was proposed by Ewing (1968) who designated the three species as: *S. typhi*, *S. cholerae-suis*, and *S. enteritides*. A more general but somewhat similar grouping of the salmonellae is based on their host

TABLE 3.1
GROUPINGS OF SALMONELLAE BASED ON HOST SPECIFICITY

Host Preference	Disease Symptoms	Organism
Group I Primarily adapted to man.	Blood stream invasion. Enteric Fever. Carrier rate high. Long incubation period.	*S. typhi* *S. enteritidis* serotype *paratyphi A & C*
Group II Primarily adapted to animals	Abortion (cattle, horses, sheep) Salmonellosis (poultry)	*S. enteritidis* serotypes, *dublin*, *abortusequi & abortusovis* serotype *pullorum*
Group III Not adapted.	Gastroenteritis (man) Short incubation period	Most serotypes

preference (Table 3.1). The unadapted group, which contains organisms with little or no host preference, contains most of the isolates of importance to food microbiologists and most of the currently known serotypes.

Complicating the problem of obtaining a suitable system of nomenclature are the wide variety of media and methods employed by investigators in this area, the inherently close interrelationships of various genera comprising the *Enterobacteriaceae*, and the ability of many genera, including *Salmonella* to undergo genetic changes which result in altered biochemical characteristics and antibiotic resistance.

As a practical matter, food microbiologists should recognize that serological and biochemical alterations can rarely occur and thus result in, for example, a lactose-fermenting strain of *Salmonella*. In addition, while identification of species within this genus is of importance to those investigating human or animal enteric diseases, biochemical tests to determine the presence of *Salmonella* in a food, followed by serotyping through use of the Kauffmann and White (Kauffmann 1966) schema are the principle tools on which the food microbiologist relies for *Salmonella* identification.

Serology

Three types of antigens are responsible for the serological reactions of the salmonellae; K (capsular), O (somatic) and H (flagellar).

The K or capsular antigens are possessed by some types of salmonellae and are thought to be a component of the envelope surrounding this organism. The principle significance of K antigens, from a diagnostic standpoint, lies in their ability to mask O antigens, thus making the latter unavailable for agglulutination with the

SALMONELLA AND SHIGELLA   131

corresponding O antibodies. The K antigens are heat labile and following their destruction by boiling, the heat stable O antigens are available for antibody agglutination. The Vi antigen of *Salmonella typhi* is an example of a capsular or K antigen which must be destroyed before this organism will react with group D somatic antisera.

Somatic (O) antigens are those antigens associated with the cell surface. There exist more than 40 different O antigens among the various *Salmonella* serotypes which are organized into lettered groups according to content of one or more of these antigens. The specificity of the various O antigens is largely attributed to the presence of terminal sugars, such as glucose, paratose, abequose and colitose, found at the end of cell wall polysaccharides. In laboratory investigations of food-borne salmonellae and their source, commercial preparations of combined O antisera groups, so-called polyvalent sera are commonly used. This procedure eliminates the tedious testing of the many different individual O antigens.

The H antigens located in the flagella of salmonellae are typically proteinaceous in composition and thus relatively heat labile. These H antigens exist as two forms; a specific, or phase one form, and a group, or phase two form. A given strain of *Salmonella* may contain one (monophasic) or both (biphasic) forms. The determination of both phases is essential to the final assignation of a serotype to a foodborne salmonella isolate.

Because the salmonellae possess more than forty O antigens and numerous H antigens, the combinations and permutations of these various antigens are extensive, thus accounting for the more than 1,500 serotypes mentioned earlier in this chapter. Fortunately for the food microbiologist, most of the *Salmonella* strains isolated during outbreaks of foodborne disease are of relatively few serotypes (Table 3.2), thus simplifying the serotyping of food isolates. Kauffmann (1966) has organized the salmonellae into subgroups based on O antigenicity and types based on H antigenicity. This system of classification, the Kauffmann-White schema, while serving as the basis for the differentiation of the hundreds of serotypes, has also served to organize the salmonellae into epidemiologically useful groups. An additional factor which simplifies serological identification in food laboratories is the presence of commercially available and standardized antisera, plus published procedures for their use. As noted earlier, the principle purpose of serotyping food isolates is twofold; to assist in epidemiological investigations and to provide final and positive confirmation (only necessary to polyvalent O grouping) of the genus itself.

TABLE 3.2
THE TEN MOST FREQUENTLY ISOLATED SEROTYPES
OF SALMONELLA FROM HUMAN SOURCES IN 1972*

Frequency Rank	Serotype	Percent of all Serotypes	Somatic Group
1	*typhimurium*	25.8	B
2	*newport*	8.4	C_2
3	*enteritidus*	6.5	D
4	*infantis*	6.3	C_1
5	*heidelberg*	5.6	B
6	*saint-paul*	3.9	B
7	*thompson*	2.6	C_1
8	*derby*	2.4	B
9	*oranienburg*	2.4	C_1
10	*javiana*	2.2	D
Total		66.1	

*In part from: Center for Disease Control: *Salmonella* Surveillance Annual Summary, 1972.

The relationship of individual serotypes to foodborne illness is uncertain at the present time. Table 3.2 shows that of the ten most frequently encountered (by the CDC) serotypes in 1972, somatic groups B and C, are most common. Whether or not all serotypes are potential food pathogens is not currently known; however, until more definitive data is available, it must be assumed that a salmonella isolation from a given food carries the implication that this food, if consumed, could produce the symptoms of salmonellosis.

Detection

The detection and enumeration of salmonellae is a complex subject with numerous variations in sampling procedures, media and cultural techniques. In addition, some procedures favor certain serotypes or are more useful for specific types of foods. Before selecting a given analytical scheme for use in the laboratory, the food microbiologist is encouraged to consider his specific needs. Sensitivity, that is the inherent ability of the test method to detect small numbers of salmonella in large sample volumes, should be considered an important criterion of any procedure or analytical scheme.

Sampling.—In addition to a great many proprietory sampling plans currently existing in the food industry, there are a number of additional sampling procedures which have appeared in the literature. Obviously the number of samples analyzed will influence the degree of reliability which can ultimately be put on a *Salmonella* sampling program. Basically, any sampling scheme should meet three criteria:
   1) Sensitivity—The plan should be able to detect low numbers of salmonellae even in the presence of high numbers of other bacterial species.

2) Economic feasibility—Minimal risk factors as related to sampling and analysis costs plus an economically acceptable level of false positive results.

3) Broad applicability—Program should be adaptable to the special needs of a variety of foods processed under a variety of conditions. Differences in age and health of consumers with regard to potential susceptibility should be considered as well as recognition that some foods are inherently more hazardous, from the standpoint of *Salmonella* contamination, than others.

A sampling plan meeting the above criteria was established by the National Academy of Science, Committee on Salmonella, a committee of the National Research Council (1969). This plan in a slightly revised form has been adopted by the FDA Bureau of Foods and approved for agency use.

Basically this plan takes into account the ultimate manner in which the food is to be used, such as final cooking, etc., the abuse potential of the product, the presence of sensitive ingredients and finally, the population at risk. Using these factors, a system of five food categories was established to classify foods:

I.   Food products for consumption by infants, the aged and infirm.

II.   Food products subject to the three hazard characteristics:
      A. High risk or sensitive ingredients.
      B. Absence of a pasteurizing process step.
      C. A potential for microbial growth if mishandled or abused.

III. Food products subject to two of the above hazard characteristics.

IV.   Food products subject to one of the above hazard characteristics.

V.   Food products subject to none of the above hazard characteristics.

The FDA Compliance Program Guidance manual groups categories III, IV and V into a single category III and lists various food products in each category.

The number of sample units recommended for collection in the FDA manual (above) is shown in Table 3.3 and varies slightly (categories II and III) from those recommended by the Committee on Salmonella. One difference between the FDA plan and the plan recommended by this committee is that the latter avoids rejection on the basis of a single positive result if an additional number of samples are found to be negative. For example, in Category I, 60 units tested and found negative is equivalent to testing 92 units with one positive; thus additional units could be tested if a positive were found in an

TABLE 3.3
COMPARISON OF NAS AND FDA *SALMONELLA* SAMPLING PLANS

	NAS COMMITTEE*		FDA*	
	Sample Frequency Units	Samples With 1 Positive Units	Sample Frequency Units	Samples With 1 Positive Units
Category I	60	92	60	None
Category II	29	48	30	None
Category III	13	22	15	None
Category IV	13	22	No Category	
Category V	13	22	No Category	

*All sample units are 25 g.

original, smaller sample group. The FDA plan, on the other hand, does not permit additional sampling in the case of a positive result and thus requires all samples to be negative.

As mentioned above there are other sampling plans in existence with varying degrees of merit including a USDA plan which, while similar, puts greater emphasis on monitoring food processes and plant environments rather than each specific lot of product. The extent to which each food processor avails himself of these or other sampling schemes is a matter of personal preference; however, it is recommended that quality control managers at least be familiar with one or both of the above plans. Once an adequate sampling plan has been established, food samples are analyzed by a series of tests which involve one or more of the following steps: sample preparation, preenrichment, elective enrichment, plating, biochemical tests and serology. The exact make-up and sequence of the analysis scheme for individual applications is a matter of individual judgement which should be based on the type of food to be analyzed, likelihood of contamination and laboratory capabilities. Because individual needs vary quite widely from laboratory to laboratory, basic principles involved in each analysis step will be discussed in this section. This discussion will then be followed by a description of three commonly used *Salmonella* analysis schemes.

Sample Preparation.—Prior to analysis, the food sample is maintained in the same condition in which it exists prior to use or consumption. When analyzed, special care must be taken to assure that microorganisms in the sample are not destroyed or adversely affected by the treatment the sample receives during preparation. In addition, prolonged storage of samples in any condition is to be avoided because a gradual and consistent decrease in numbers of organisms in the food may occur. Gram-negative bacteria such as *Salmonella* appear to be especially susceptible to declines in counts during storage. Also, *Salmonella* identification procedures may be long and complex, involving numerous transfers from medium to medium; therefore, special care must be taken to assure that the identity of each sample is carefully recorded and maintained throughout the analysis and until the final report is completed.

Normally, samples are blended in commercially available, sterile, blender jars. Usually sterile pre-enrichment medium is the blending medium. Blending times vary with the type of food to be analyzed, but normally range from 1 to 3 minutes. The purpose of blending is to remove salmonellae from the food it is embedded in without injuring the organism to the extent that its subsequent growth is jeopardized.

Cultural Procedure

Preenrichment.—The general purpose of preenrichment is to aid in the recovery of "injured" or attenuated salmonellae to assure that greater numbers of these organisms are available for inoculation into selective media. Although preenrichment is now widely used, there are certain circumstances, such as the cultivation of heavily contaminated hydrated samples, when this step should be omitted. In such instances, the preenrichment procedure, because it is nonselective, may favor the growth of other organisms in a food to the detriment of the salmonellae present.

There are a number of types of media that have been recommended for preenrichment; however, lactose broth, as originally described by North (1961) appears to be the most commonly used and most effective from the standpoint of good recovery rates. Taylor (1961) suggested enrichment media containing carbohydrates other than lactose; however, subsequent reports surprisingly indicated that the choice of preenrichment medium did not seem to influence the eventual recovery of salmonellae. Although various procedures such as antibody flocculation, selective heat treatments, and centrifugation of extracted foods have been tested in the past, the lactose broth procedure of North continues to be widely used for the preenrichment of dried and/or heated foods. The recommended length of incubation for lactose broth preenrichments is usually 18 to 24 hours.

One theory for the apparent effectiveness of lactose-containing preenrichment broths in the recovery of salmonellae states that the fermentation of lactose by competitors tends to lower the pH of the preenrichment broth to a level (5.8 to 6.3) at which the salmonellae are selectively favored. Another possibility is that by favoring competitors during preenrichment through increasing their rate of growth, their subsequent inhibition in the more inhibitory enrichment media is enhanced.

Selective Enrichment.—The purpose of the selective enrichment phase of *Salmonella* analysis is basically to inhibit competing organisms while enhancing the growth of salmonellae. The recent review of Litchfield (1973) lists numerous types of enrichment media; however, the two media most commonly used are tetrathionate and selenite-cystine broths. The former contains compounds such as bile salts, tetrathionate, sodium thiosulfate and brilliant green that are inhibitory to organisms other than salmonellae. Selenite-cystine broth contains the inhibitor, sodium selenite which is reduced to selenium by microorganisms. This compound reacts with sulfur-containing amino acids to inhibit certain types of bacteria. In addition, the fermentation of lactose in this medium by competing bacteria tends to reduce pH levels to ranges in which *Salmonella* species are favored.

Each of the selective media tends to inhibit salmonellae to some extent; therefore many laboratories prefer to use both selenite-cystine and tetrathionate broths, thus improving the ability to detect a variety of different types and species of these organisms.

Plating.—Enrichment broths, after incubation, commonly contain mixed populations of bacteria which must be further selected for salmonellae. To accomplish this task, a number of selective plating media are available. Like the enrichment broths described above, these media contain a number of nutrients favorable for the growth of salmonellae plus various inhibitors and indicating compounds which prevent the growth of organisms other than *Salmonella* species or, by virtue of specific reactions, indicate the presence of such foreign species.

No attempt will be made in this chapter to list the numerous types of plating media that have been described in the literature. An abbreviated listing of some of the commonly used plating media and the appearance of *Salmonella* colonies growing on them are described by Litchfield (1973). This author prefers to use five media, Hektoen's agar, brilliant green, S-S, bismuth sulfite and XLD. The FDA, on the other hand, specifies plating on bismuth sulfite, S-S and brilliant green agar. The Food Protection Committee of the National Academy of Science (1970) recommends only two selective plating media; brilliant green and bismuth sulfite. Obviously there is considerable variability between laboratories in the number and types of media used for plating; however, since two selective enrichment broths are usually employed, with a separate set of plates for each enrichment, the amount and variety of media used can be substantial.

Biochemicals.—Biochemical testing involves a series of confirmatory media in which salmonellae produce recognizable changes. Litchfield (1973) lists a number of media used in biochemical testing by various laboratories. Most of the biochemical tests involve one or more of the following tests in each medium: motility, glucose, sucrose, lactose, mannitol or salicin fermentation, decarboxylation of lysine, and $H_2 S$ production. The basic purpose of biochemical testing is to confirm that characteristic colonies appearing during plating are, in fact, *Salmonella* species.

Serotyping.—Frequently, confirmed cultures of salmonellae are serotyped to establish their identity. Data relative to the serological identity of salmonellae is especially useful for determining the specific source of the strain in question.

The basic serological characteristics of these organisms are discussed earlier in this chapter; however, a few additional points relating specifically to the serological investigation of food isolates should be mentioned.

Both O and H antigens are determined during serotyping. In the case of the former, a small amount of culture material from one of the biochemical slants, preferably the TSI slant, is used. However, in the case of H or flagellar antigens, the suspect culture must first be subcultured in a relatively complete medium such as tryptone broth, motility broth or brain-heart or veal infusion broth. The purpose of this procedure is to assure the formation of flagella which, of course, bear H antigens.

For the initital phases of serotyping, polyvalent O and H antisera are normally reacted with the *Salmonella* cultures. These polyvalent sera are representative of all serotypes reported to be encountered in foods. After obtaining positive polyvalent reactions, group specific O and H reactions are carried out with O grouping and H pooled antisera arranged in accordance with the Kauffmann-White scheme of identification. A commonly used, pooled H antiserum system is that of Spicer-Edwards which provides an antiserum mixture of the serotype groups most usually found in foods. Some workers advocate the use of the Spicer-Edwards system as a screening tool by subculturing suspect or lactose-negative colonies obtained from plating media. By this technique biochemical tests and some polyvalent testing are omitted and considerable analysis time is saved.

Rapid Analysis Procedures

The rapid analysis of foods for salmonellae assumes considerable importance in industry where finished product must either be warehoused pending analytical results or must be shipped before such data are available. In the latter case, a positive laboratory finding would necessitate an expensive and potentially embarrassing recall from the market or food distribution channels, with the possibility that some of this food could be consumed. Thus, procedures which allow for the rapid yet sensitive analyses of *Salmonella* assume considerable public health and economic importance.

Enrichment Serology.—One procedure, requiring 50 hours for a positive result versus 5 days for conventional procedures (described above) was proposed by Sperber and Deibel (1969). In this procedure, samples are pre-enriched and enriched as in the longer, cultural schemes; however, following enrichment, transfer is made to M-broth. After incubation (six hours) in this medium, the culture is serologically typed with Spicer-Edwards polyvalent H antiserum. A similar, but faster modification of this procedure, utilizing an 8 hour preenrichment and 16 hour enrichment culture, has also been developed. The principal advantages of this scheme are that it is even more rapid than the regular enrichment serology technique, requiring only 32 hours for completion, and it is adaptable to an 8 hour working day.

Both procedures appear to give results comparable to the cultural technique; however, some problems have been encountered with specific types of foods and in foods highly contaminated with competing organisms. Egg albumen, particularly, appears to be difficult to analyze by enrichment serology procedures.

Fluorescent Antibody Technique (FAT).—The FAT procedure, as it is presently constituted, is primarily a screening tool. It involves preenrichment, enrichment and elective enrichment in modified M-broth (M broth minus mannose). Following elective enrichment, smears are stained with a fluorescent-tagged antibody (described below) and examined by fluorescence microscopy. In the case of positive smears, retained M-broth cultures are analyzed by conventional cultural techniques for confirmation.

Staining and the interpretation of stained slides are the keys to success with this technique. The staining procedure most commonly used is the direct method in which fluorescein-isothiocyanate dye is conjugated with a commercially available antiserum containing H and O antibodies for salmonellae of significance in foods. This fluorescein-labelled antiserum is placed directly on fixed smears of modified M broth culture. After exposure for 30 minutes, the slides are rinsed, dried and examined under UV light with a fluorescence microscope. *Salmonella* cells will normally show a green or chartreuse fluorescence which, because of somatic or cell wall

SALMONELLA AND SHIGELLA 139

staining, will resemble an elongated doughnut. If the serum used for staining contained H antibodies, as many of them do, fluorescing flagella may also be seen, although the presence of flagellar staining is not necessary to declare a given sample positive.

The recognition of a salmonella-positive strain is a subjective judgement on the part of the technician performing the test and thus is subject to human error. Background material may at times stain or salmonellae may go unrecognized in debris-laden fields. An overly liberal interpretation of the stained slides can result in a relatively high percentage of false positive results or an overly conservative approach may result in a high incidence of false negatives. In addition, certain food samples may tend to give a greater percentage of false positive or negative results than others. Generally, a thorough and lengthy period of training for technicians, in which routine as well as known, naturally contaminated samples are tested, is advisable.

Various laboratories have compared the FAT procedure with the conventional, cultural procedures. Some laboratories have reported false negative (FAT negative, reference method positive) results in as many as 6% of all samples tested whereas others have reported no false negatives in thousands of samples. Variations in the foods tested, differences in preenrichment, elective and selective procedures, type of sera used and technician capability have been suggested as the reasons for the wide variability in results obtained with this technique. In this author's experience, false negatives have not been encountered with a variety of foods and food related materials (Tables 3.4 and 3.5) tested over an eight month period.

TABLE 3.4
COMPARISON[1] OF RESULTS OBTAINED WITH
VARIOUS COMMODITIES USING FAT AND
CONVENTIONAL (BAM) PROCEDURES

	No. Samples	% FAT Pos.	% BAM Pos.	% FALSE Pos.
Nut Meats	36	15	0	15
Albumen	290	5	1	4
Peanut Butter	49	10	0	10
Spices	39	0	0	0
Cocoa	218	4.5	0	4.5
Non-Fat Dry Milk	124	12	7	5
Environmental	63	15	0	15
Seed Meals	138	7	4	3
Cake Mix	31	8	0	0
Flour	6	16	0	16
Coconut	3	0	0	0

[1] Personal data of the author.

TABLE 3.5
SUMMARY DATA — FAT — BAM COMPARISON[1]

	%
FAT Positive	= 7.7
BAM Positive	= 1.6
False Positive (BAM Neg., FAT Pos.)	= 6.0
False Negative (BAM Pos., FAT Neg.)	= 0
False Positive — Technician A	= 7.3
False Positive — Technician B	= 4.3

[1] Personal data of the author.

While not as rapid as the QES or ES procedures, the FAT technique is more rapid than conventional analysis, depending on the cultural procedures utilized. Litchfield (1973) has listed a number of published FAT procedures which have appeared over the years. One of the most useful of these schemes is that proposed by Insalata *et al.* (1972) which adapts this procedure to working-day time periods through utilization of shaken pre-enrichment and enrichment cultures. As a result, the elapsed time required to declare a given sample negative has been reduced to only 29 hours.

EPIDEMIOLOGY OF SALMONELLOSIS

Symptoms

As noted earlier in this chapter, salmonellae have the capability of causing two separate, pathological reactions in humans; a febrile disease characterized by typhoid and paratyphoid fevers, and gastroenteritis characteristic of food-borne disease. It is the latter that we are concerned with in this chapter.

The symptoms of salmonellosis are vomiting, diarrhea, nausea, abdominal pain and usually a moderate fever. These symptoms appear within 6 to 24 hours after ingestion of the infected food and usually last for 1 to 5 days. The number of salmonellae required to produce a pathogenic response appears to vary widely, depending on the general health of the patient, the serotype involved, the food ingested and its processing history. Actual figures cited have ranged from 6.5 to as high as 1.6×10^{10} cells per gram of food consumed; however, the latter number referred to a host-adapted (poultry) strain of *S. pullorum* which might be expected to have a reduced infectivity for humans.

The manner in which food-borne, host-nonspecific salmonellae produce salmonellosis is not clearly defined. Some workers refer to a "joint action" of the bacterial cell and an endotoxin elaborated by the organism whereas others implicate only an endotoxin produced by salmonellae in the small intestine. Although there are few reports on this subject in the literature, it seems probable that invasion of the bloodstream and/or establishment of localized infections in specific organs are probably rare. Prost and Riemann (1967) have indicated that the disease symptoms occur when an endotoxin is released from *Salmonella* cells as a result of the lytic activity of the stomach contents. Thus, relatively large numbers of bacteria would be required to produce a response and the cause and intensity of the disease would therefore be related to the number of ingested salmonellae. The fact that these organisms are readily cultivated from faecal specimens of infected individuals suggests that they survive and perhaps grow in the alimentary tract. Although some growth may occur in the intestines, the above authors feel that intestinal growth is not etiologically essential. This theory has been questioned by a number of workers who regard salmonellosis as strictly an infectious disease and who feel that toxins found inside or outside the *Salmonella* cell are not factors in production of disease symptoms.

Salmonellosis attacks with greater severity and frequency individuals who are physiologically stressed in some manner, as, for example, the aged, the ill and the very young. In these groups the infection can become generalized and highly invasive and in these instances may be characterized by a relatively high mortality rate.

Incidence

As with many other types of food-borne illnesses, it is extremely difficult to estimate, with any degree of assurance, the actual attack rate of salmonellosis in the U.S. because many cases (some experts have estimated as high as 99%) go unrecognized and unreported. There can be little doubt however, that this disease is increasing in incidence in many parts of the world.

Several characteristic disease patterns can be discerned: 1) the frequency of human isolations is greatest during late summer and early autumn, 2) as noted above, attack rates are greatest in the very young and remain static from the age of 15 on, 3) the number of serotypes involved may be very great; however, the 10 most frequently isolated serotypes caused from 60 to 70% of salmonellosis, 4) the case fatality rate is normally less than 1% of those affected.

Distribution in Animals

Because food animals are regarded to be at the "beginning" of the *Salmonella* infection cycle, much attention has been directed toward infected animals and animals which serve as a reservoir for this organism. The exact number of animals thus infected is unknown; however, the incidence is probably high. The most probable source of salmonellae in animals is the food that they eat. Frequently, animal feeds contain animal by-products such as bone meal and dehydrated meat scraps which become contaminated with *Salmonella* species during and after processing. Some reports have shown that contamination rates in feeds may be as high as 86% and seldom drop below the 20% level. In addition, feed ingredients frequently may be imported, thus aiding the widespread dissemination of various serotypes into areas of the world where they previously had been rare.

Much effort has been expended to break the "cycle of infection" at the point of animal feeds, but only indifferent success has been obtained. Recent reports from Europe indicate that despite efforts to reduce the occurrence of salmonellae in animal feeds, the incidence of this disease is increasing. In this country, studies conducted by the USDA have shown that it may be commmercially feasible to produce *Salmonella*-free animals; however, maintenance of *Salmonella*-free feed is a continuing problem in such instances. Recently, animal feeds have been brought under the Food, Drug and Cosmetic Act (Section 402 (a)) which considers these materials to be adulterated and thus unfit for interstate shipment if they contain salmonellae.

It has been said that poultry probably constitute the largest, individual reservoir of salmonellae. In addition to serving as carriers for this organism, poultry are susceptible to acute disease as a result of *Salmonella* infection which can result in reduced egg production, lowered fertility and high mortality rates in chicks. Many states have recognized the importance of poultry as a reservoir of *Salmonella* and have instituted intensive efforts to maintain *Salmonella*-free flocks. In some instances these efforts have been successful, but whether a significant decrease will occur in the incidence of the human disease as a result of the effort is, as yet, undetermined.

Eggs have gained a degree of notoriety as sources of salmonellae through incrimination in large outbreaks of salmonellosis. As noted above, poultry constitute one of the major carriers of this disease, and thus it is not too surprising that eggs will be similarly contaminated. Eggs become contaminated in the oviducts of hens or after laying, if moist conditions permit the growth of salmonellae on the egg surface through the shell and membrane. Contamination of egg products can occur during breaking, pouring, or drying of commercially processed eggs. The source of these organisms may be contaminated fecal matter from hens carrying this organism.

During preparation by the housewife, eggs are normally cooked at temperatures that will destroy contaminating salmonellae. Thus, unless recontamination occurs, fresh eggs are seldom a food-borne disease problem. Processed eggs, however, are another matter. In commercially processed egg products, shelled eggs may be held for various periods at room temperature which could easily allow the growth of salmonellae. In addition, normal processing procedures, such as freezing or drying, are seldom sufficient to destroy this organism. Pasteurization of egg products has been required by the USDA since 1965; however, functionality requirements often limit the extent of heat treatments that can be used. Dried eggs may be heated at 135-140°F for 7-14 days to destroy salmonellae. With liquid eggs the problem is somewhat more difficult. Procedures involving heating in the presence of various chemicals such as hydrogen peroxide or aluminum or iron salts have been to some extent used, but functionality is a problem. Some types of this organism, for example *S. senftenberg 775W*, are reported to be more resistant to moist heat than others. Generally, the most effective means of preventing or eliminating *Salmonella* from commercial egg products are: 1) encouraging the establishment of salmonella-free flocks of laying hens, 2) use of only high quality eggs, 3) proper sanitation of equipment and plant environment, and 4) pasteurization of the finished product to whatever extent is feasible.

Salmonellosis as a result of the consumption of contaminated meat from swine and cattle, while frequent in the U.S., apparently is somewhat less common than in foreign countries. A number of reasons have been proposed for this disparity, such as better refrigeration, more widespread consumption of thoroughly cooked meat or more effective sanitation in meat packing houses. Whatever the reason, salmonellae in meats and in the meat-producing environment continue to be a serious problem.

Domestic meat animals normally serve as reservoirs for many *Salmonella* serotypes of food significance, the most common of which is the relatively non-host specific *S. typhimurium*. Salmonellae probably enter the animals from contaminated feeds, made from processed livestock offal. Galton *et al.* (1954), in an excellent study of the meat processing chain, found rather high percentages of *Salmonella* contamination from various sites in both live and slaughtered swine. The frequency of contamination appeared to increase as the animal moved to and through the abbatoir despite the fact that the scalding operation reduced surface counts. Swabs taken from dehairing machines were positive from 20 to 68% of the time and the evisceration area yielded 74% positives. High levels of contamination were also found on knives and similar pieces of apparatus used throughout the slaughtering operation. Considerable attention has also been devoted to the "shedding" of salmonellae by swine and cattle during transport to the abbatoir and in holding pens. Anderson (1961), studying this effect in England, found that the infection rate increased from 0.5% in calves on the farm to 35.6% in calves after holding for several days at the abbatoir. Contamination of cut-up beef and pork varies widely with reports ranging from 5% to greater than 10% of examined samples. Red meat and meat products were reported to cause 25% of all food-related salmonellosis during 1972. These figures included sausages and other types of processed meats in which salmonellae may survive.

Various aspects of the contamination of milk and dairy products have been discussed by Marth (1969) in his extensive review of this subject. Probably some of the largest salmonellosis outbreaks, from the standpoint of numbers of persons affected, have involved milk and other dairy products; however, seldom has fresh, pasteurized milk been involved unless, of course, post pasteurization contamination has occurred. Cows normally do not excrete salmonellae in their milk, but these organisms may enter the milk stream from fecal contamination, contaminated utensils or animal or human carriers.

Dried milk has been implicated in a number of outbreaks of salmonellosis both in the U.S. and in Europe. An outbreak occurring in 1966 caused by the consumption of non-fat dried milk containing S. new-brunswick was attributed by Schroeder (1967) to poorly designed and constructed equipment which resulted in product accumulations and incomplete sanitation. Various surveys of dried milk and the plants in which it was produced showed that 0.2% of the milk samples tested and 13% of the plant environmental samples were contaminated with salmonellae.

In addition to dried milk, salmonellae have been found in ice cream, butter and cheese although in the latter two food products survival of salmonellae is not extended, probably as a result of the bactericidal effects of fatty acids formed during the ripening process.

A number of food products of plant origin have been implicated as causes of salmonellosis with dried coconut being most frequently mentioned in this regard. The most likely sources of this contamination arise from contact with human carriers or contaminated water during processing. In recent years, the frequency of isolation of salmonellae from coconut has decreased dramatically with the advent of improved sanitation during processing and probably most importantly, the employment of a terminal pasteurization step. Potato salad, chocolate candy and cocoa have also been reported to have been involved in outbreaks. Although the number of outbreaks of salmonellosis attributed to foods of plant origin are relatively few, the food microbiologist is well advised to be aware of the possibility that salmonellae may on occasion enter such products either directly or indirectly via combination with contaminated ingredients.

CONTROL OF SALMONELLOSIS

The basic principles for the control of salmonellae in foods and in the food environment are those that are familiar to all students of food microbiology: a) Prevent the entry of this organism in foods, b) Prevent the growth of the organism, c) Destroy the organism.

Contamination Prevention

Ideally, one of the most effective and perhaps controversial means of breaking the cycle of *Salmonella* contamination of foods is to assure that animal feeds are free of this organism. Once this is done, stringent measures to detect salmonellae and eliminate the organism from animals during transport, in holding pens and in slaughter houses may be undertaken.

As a practical matter, many of the above measures are virtually impossible to achieve although efforts have, for some time, been underway to maintain *Salmonella*-free poultry flocks and to achieve improved processing of cattle feed. There is much more that can and must be done, however, to achieve a reduction in salmonellae in our food supplies and thereby the incidence of salmonellosis.

A task force of the APHIS dealing specifically with this disease made the following recommendations:

(1) Increase and improve *Salmonella* surveillance of cooked foods with prompt removal of *Salmonella*-contaminated foods from the market.

(2) Reduce cross contamination in processing facilities by physical means and by control of air flow patterns.

(3) Improve water treatment.

(4) More effective monitoring of food production facilities and equipment.

(5) Improve reporting of salmonellosis.

(6) Extend existing programs for the control of salmonellae in some types of feeds to all animal feeds.

(7) Improve training of regulatory agency inspectors to provide better understanding of the *Salmonella* problem.

In addition, a number of proposals dealing with specific industries were listed by this group. A similar task force, sponsored by the FDA also considered measures by which the incidence of salmonellae could be curtailed and made the following recommendations:

(1) Increase educational efforts to reduce the incidence of salmonellosis caused by mishandling and careless preparation of food.

(2) Encourage voluntary programs designed to eliminate salmonellae from animal feeds and increase surveillance of feed blending operations.

(3) Give some priority to measures that reduce the risk of *Salmonella* contamination in the distribution and preparation of human foods.

The elimination of contamination by salmonellae in human foods is not a simple matter. Isolated attempts to eliminate the organism from one portion of the food supply to "break the chain of infection" are not only difficult and expensive to achieve but probably of limited value unless accompanied by more broadly applied control measures.

Prevention of *Salmonella* Growth

Probably the most widely used method for the prevention of *Salmonella* growth in foods is by storage at low temperatures. Other means by which growth of this organism can be curtailed or totally prevented will become evident later in this chapter.

Salmonella Destruction in Foods

Chemical Treatment.—Included in the list of various chemicals suggested as *Salmonella*-cidal agents for use in foods are peroxides, beta-propiolactone, and ethylene and propylene oxides. In most cases, these chemicals suffer from such disadvantages as inducing off-flavors (especially in egg products), and lack of approval from regulatory agencies. Other chemicals that have been tested at one time or another are quaternary ammonium compounds, sorbic acid, sodium nitrite, and various antibiotics.

Radiation Treatment.—Salmonellae are relatively susceptible to gamma irradiation although some radiation-resistant strains have been reported. Comer *et al.* (1963) reported that a dose range of 0.36 to 0.54 Mrad was required to produce a 10^7 reduction in counts of 18 *Salmonella* strains in frozen whole egg. Following irradiation at the maximal dose, these authors could not detect flavor or functionality alterations in this product. However, Ingram *et al.* (1961) did note flavor alterations in a similar product exposed to a dose level of 0.50 Mrad. These organisms are also destroyed by ultraviolet irradiation in thin films of liquid egg products; however, as in the case of chemical treatments, off-flavor development and lack of regulatory approval have curtailed the application of these procedures.

Heat Treatment.—The most effective and widely used treatment to achieve destruction of salmonellae in foods is heat. These organisms are eliminated during normal household cooking, provided the food has reached a uniform temperature of 70-75°C for a reasonable period of time (usually 3-7 minutes). Decimal reduction times or D values (time required to kill 90% of the organisms present in a given food) vary from 0.6 minutes to 31 minutes at 134.6°F. The most heat resistant strains thus far tested have belonged to the serotype *S. senftenberg*, the most well known being *S. senftenberg* strain 775W described by Winter *et al.* (1946) as a heat resistant mutant. Mutants of this type are probably the exception rather than the rule in the genus *Salmonella*. However, until further studies have been done to determine the extent of heat resistance, it is probably well to note that resistance may occasionally occur.

Heat treatments for liquid whole egg have been devised in which functionality is not adversely affected. Extensive thermal destruction studies have shown that liquid egg must be heated to at least 140°F for at least 3-1/2 minutes, a treatment which destroys all but the most heat-resistant *Salmonella* strains. Other egg products, for example dried egg albumen, can be rendered virtually free of salmonellae by exposing this product to temperatures of 125-130°F for 7-10 days without adversely affecting the functionality of the albumen.

Thermal destruction of salmonellae in milk and milk products has also been investigated. Read *et al.* (1968) found that a number of *Salmonella* serotypes isolated from dry milk powders were inactivated by pasteurization processes. *S. senftenberg* 775W, however, when inoculated into dry milk did survive these heat treatments. Other studies using strains of *S. typhi* indicated that these organisms did not survive pasteurization in continuous flow, plate-type pasteurizers.

The thermal destruction of salmonellae, like many organisms, can be influenced by other factors such as the presence of certain chemicals, pH, and water activity (a_w). In the latter case, it has been demonstrated (Goepfert and Biggie, 1968) that reduced a_w levels markedly increase the heat resistance of salmonellae in milk chocolate. Other workers have reported that low numbers of *S. anatum* in milk chocolate survived 16 hours but not 24 hours of heating at 71°C. However, additional studies (Barrile and Cone, 1970) show dramatic decreases in heat resistance if 1 to 4% water was added to the chocolate before heating.

Low Temperature Storage and Freezing.—Freezing and refrigerated storage of foods, while effective in preventing growth of *Salmonella* species, seldom result in killing of these organisms. A phenomenon described as "cold injury" may result following freezing in which growth of the organism following restoration of optimal growth conditions is curtailed for a time. A similar effect, "heat injury", is observed with sublethal heating.

In summary, probably the most effective means of controlling salmonellae is to keep these organisms from entering foods. As noted, the accomplishment of this task is not an easy matter, and a coordinated effort involving most of the following would probably be required: 1) The use of *Salmonella*-free feeds, 2) Careful screening of carrier animals, 3) Removal of the organism from food processing environments, and 4) Effective consumer education programs to prevent cross contamination in the home.

An effort of the magnitude required to obtain these goals would be expensive in terms of money and manpower expended. The tools to accomplish these tasks are available, but the piecemeal commitment of these tools to the objective of eliminating food-borne salmonellosis is probably of doubtful value.

FACTORS AFFECTING GROWTH OF SALMONELLAE

Temperature

Under normal laboratory conditions, salmonellae will grow well if incubated at 37°C; however, the complete range of growth temperatures lies between 5 and 47°C. Minimal growth temperatures have been emphasized in the literature, primarily because many foods rely for their stability and safety on refrigerated storage. Lowest growth temperatures are normally observed if the pH level of the food or medium is poised at pH 7.0 to 8.0 and incubated for extended periods of time. Under these conditions, Matches and Liston (1967) showed that *S. heidelberg* would grow at temperatures between 4.0 and 5.7°C and several other serotypes grew in the range 6.6 to 8.2°C. The fact that 15 days of incubation were required to produce appreciable growth in these studies indicates that growth rates were extremely low. Thus, one can reasonably expect to prevent growth of most salmonellae at refrigeration temperatures providing that this temperature does not exceed 10°C. At the other end of the temperature scale, if growth of salmonellae is to be prevented, food should not be exposed to temperatures of less than 50°C. The range between 10 and 49°C has been referred to as the "incubation danger zone" (Bowmer, 1965).

pH

Under optimal conditions in artificial media, salmonellae will grow in the pH range of 4.0 to 9.0. However, considerable differences in minimal values exist with regard to the type of acidulant employed. Various workers (Levine and Fellers 1940; Goepfert and Hicks 1969; and Chung and Goepfert 1970) have extensively examined the effect of minimal pH levels on the growth of these organisms and have found that, in addition to the type of acid employed, water activity, oxygen supply, level of inoculum and nutritional aspects interact with pH to play a part in the growth response of salmonellae. Optimal pH levels for growth of this organism are in the range 6.0 to 8.0 according to Banwart and Ayres (1953) and most media for the isolation and identification of this organism are adjusted to pH 6.8 to 7.2.

Organic acids, when present in sufficient concentration, have some degree of bactericidal activity. Acetic acid appears to be especially effective in this regard, although a number of other acids are also effective in killing this organism. Although research thus far has been somewhat limited, it has been suggested that organic acids normally found in the rumen of cattle may kill ingested salmonellae and could, thus, influence the ability of these animals to shed the organism with faecal matter. In humans, however, the presence of viable salmonellae in the feces of afflicted individuals indicates that at least some of the organisms are able to survive the acid conditions present in the stomach . . . perhaps the greater resistance of these organisms to hydrochloric acid (normally found in the stomach) accounts for this fact.

Nutritional Requirements

Members of the genus *Salmonella* are not normally regarded as fastidious organisms with respect to nutritional requirements. In most instances, laboratory media for the selection and cultivation of this organism do not require supplementation, although concern has been expressed by some researchers concerning the depletion of certain thermolabile nutrients, for example, thiamine, in autoclaved media. In addition, there exists a wide variety of *Salmonella* strains genetically adapted to require specific nutrients, usually amino acids. These strains are quite distinct from the "wild type" food-borne salmonellae and are not normally encountered in the food environment.

Oxygen Requirements

Members of the genus *Salmonella* are facultative aerobes and grow well when incubated under normal aerobic conditions. Growth stimulation can be obtained by aeration, as, for example, the use of shaking to obtain acceleration of growth in enrichment media. The minimal oxidation-reduction potential (ORP) reached in culture media by six *Salmonella* serotypes was studied by Oblinger and Kraft (1973) who found that this value varied from +80 to -315 mv at 15°C and from -384 to -422 at 37°C. A rapid drop in the ORP was observed during the mid- to late exponential growth phase, followed by a partial recovery during early maximum stationery phase. No attempt was made to determine the initial, optimal ORP for growth of the serotypes tested.

Moisture Requirements

Moisture requirements for the growth of microorganisms are commonly expressed in terms of water activity (a_w) which is equal to the equilibrium relative humidity of a given solution divided by 100. A relatively thorough definition of the a_w requirements of the salmonellae exists in the literature. Much of this work has been conducted by Scott, Christian and their colleagues at the CSIRO laboratories in Australia.

Christian and Scott (1953) found that 15 strains of *Salmonella* grew at a_w levels between 0.999 and 0.945 in laboratory media. However, in meat and soup, growth appeared as low as 0.93 a_w. The salmonellae appeared to respond similarly when incubated aerobically or anaerobically. Additional work (Christian, 1955A) showed that the minimal a_w level for growth was raised in nutritionally incomplete media, but the addition of various amino acids and vitamins to the media again reduced minimal growth a_w levels to those of nutritionally complete media. Maximal growth rates of *S. oranienburg* were achieved at a water activity of 0.99 irrespective of the solute employed to obtain this a_w level. In addition to influencing the growth of salmonellae, the moisture condition of a given system can greatly affect the influence of heat on this organism. Generally, as the a_w is reduced, the heat resistance of salmonellae increases; however, a number of workers (Goepfert *et al.* 1970; Baird-Parker *et al.* 1970) have indicated that it is difficult to predict with reliability the heat resistance of salmonellae in low a_w foods. Obviously, additional work is needed on the interaction of thermal treatments and a_w.

Associative Growth

Salmonellae found in foods are seldom present as pure cultures, and in fact, some foods, for example cheese and other dairy products, may be protected from the growth of these pathogens by competition from lactic acid starter bacteria. Studies with cheese made from milk inoculated with *Salmonella* have indicated that these organisms grow during the initial phases of cheese making followed by a steady decline during the curing portion of the process. A number of mechanisms have been proposed as the cause of this phenomenon, such as antibiotic production by cheese curing organisms, hydrogen peroxide production, or the production of volatile fatty acids, especially acetic acid, during ripening. The reduction in pH level which occurs during cheese making probably is not a factor in this suppression.

The effect of the growth of naturally present and inoculated bacteria on salmonellae in a number of food products has also been studied. A decline in inoculated salmonellae in Thuringer sausage has been attributed to acidity developed by lactic acid starter bacteria or sodium chloride present in the sausage. Various pigment-producing species of the genus *Pseudomonas* have been reported (Oblinger and Kraft, 1970) to inhibit ten salmonellae strains in poultry. Similarly, Splittstoesser and Segen (1970) suggested that salmonellae were inhibited in frozen food by the normal microflora of these products. Other workers also have attributed the demise of *Salmonella* populations in thawed frozen foods to competition from the normal flora.

Much remains to be learned about the response of salmonellae to other bacteria in foods. In certain products, although a marked repression occurs, it is doubtful if total destruction of all salmonellae present in a given food is achieved.

Effect of Chemicals

As noted above, competing microflora, such as that existing in cheese and dairy products, may produce substances that inhibit salmonellae. These substances may be short chain length, volatile fatty acids produced in the food. Usually, reduced pH levels favor this action, probably because these acids are most effective in the undissociated form. Other chemicals that have been investigated for their ability to inhibit the growth of salmonellae are glycerin, sodium nitrite, diethyl pyrocarbonate, various antioxidants such as propyl gallate and nordihydroguairetic acid, ethylene oxide and phenylmercuric nitrate. In addition, a number of spices such as nutmeg, cinnamon, onion and garlic extracts have been found to have some growth-inhibiting properties.

From the standpoint of food sanitation and hygiene, a number of chlorine-containing compounds are effective in the control of salmonellae. Many of these agents are combined with detergents in commercially available preparations. Iodine preparations and anti-bacterial-containing soaps have also been found to be effective as hand rinsing and washing agents.

SHIGELLA

The Disease

The disease caused by members of the genus *Shigella* is termed shigellosis or bacillary dysentery. It is characterized by diarrhea (often with bloody stools), fever, nausea and abdominal cramps. The time of onset following ingestion of contaminated food varies from 7 to 36 hours with symptoms persisting from 1 to 8 days. Longer incubation periods, however, have been reported. The mechanism by which the disease symptoms are elicited is not well-defined, although endotoxin release in the alimentary tract which then causes intestinal inflammation has been proposed by some researchers. In addition, *S. dysenteriae* type 1 produces a potent exotoxin which is heat stable and distinct from the endotoxin. Antitoxin production is stimulated in experimental animals by this toxin. Thus, it is usually neutralized, but it is believed that the "type 1" exotoxin may contribute to the severity and duration of shigellosis in humans. Relatively little is known with regard to the infective dose of the causative organism, though at least one investigator (Dack 1956) has indicated that 1 X 10^8 to 5 X 10^{10} of this organism must be ingested to induce characteristic symptoms. Following infection and recovery, individuals may continue to be carriers of this organism for several weeks or longer.

The Organism

The genus *Shigella* is found in the family *Enterobacteriaceae* and contains relatively few species. *S. sonnei* and *S. flexneri* are the most frequently encountered genera in food outbreaks in the U.S.

Shigellae grow well, aerobically, on commonly used laboratory media, are non-motile and ferment a variety of carbohydrates such as glucose, fructose and galactose, but usually do not ferment lactose and do not produce H_2S. Interspecies differentiation is accomplished principally by biochemical reactions. Unlike the salmonellae, little reliance can be placed on serological identification methods, principally because many species of this genus share common O antigens, with other enteric bacilli. Colicin typing, however, has shown some promise in recent epidemiological investigations.

The optimal temperature for growth of the shigellae is 37° C, but growth of one common species, *S. sonnei* may occur at 45.5° C. Generally refrigeration temperatures are adequate to prevent growth of shigellae in food, although survival may be extended under these conditions (Taylor and Nakamura, 1964). Little additional data exists relative to the effect of other environmental factors on the growth of shigellae.

Epidemiology

Until relatively recently, shigellosis was commonly thought to be caused by the consumption of water contaminated at some point with sewage. Food-borne shigellosis has, since 1965, received increasing emphasis as a result of intensified surveillance on the part of the U.S. Public Health Service.

Epidemiological investigations of shigellosis outbreaks are often difficult to accomplish because *Shigella* species are frequently spread by direct contact; thus secondary spread of the organisms may confuse any data obtained. Small family outbreaks appear to be more common in instances in which water-borne shigellosis is implicated, although exceptions to this generalization exist. Usually these outbreaks occur in isolated homes with poor hygienic conditions. Incidence levels appear to be especially high in the south-western U.S.

Food-borne shigellosis outbreaks may involve greater numbers of people and are usually characterized by the involvement of infected food handlers at some point in the chain of infection. Contaminated fomites such as water glasses and utensils may also be involved. Foods reported to be implicated as a source of the disease include vegetable salad, chocolate pudding, poi, frozen egg albumen, shrimp and tuna salad, improperly pasteurized milk and potato salad.

Control

There are a number of means by which some degree of control of this disease may be obtained:
1) Periodic physical examinations of food handlers to detect carriers of shigellosis. Skin tests may eventually prove useful in identifying individuals most likely to harbor this organism.
2) Educational programs to encourage water supply safety, such as improved location of wells and chemical treatment of water obtained from wells.
3) Storage and handling of food in such a manner as to preclude contamination and/or growth of shigellae.
4) The application of an effective vaccine to high risk groups; for example, institutionalized, mentally retarded children.

154   FOOD MICROBIOLOGY

Summary

Unlike salmonellosis, the health threat posed by shigellosis is largely an unknown quantity. For many years, U.S. health authorities regarded this disease as primarily a water-borne hazard of infrequent occurrence. However, as improved reporting and diagnostic techniques have evolved, it has become apparent that shigellosis is an important food-borne disease as well.

BIBLIOGRAPHY

ANDERSON, E. S., GALBRAITH, N. S., and TAYLOR, C. E. D. 1961. An outbreak of human infection due to *Salmonella typhimurium* phage type 20a associated with infection in calves. Lancet *1*, 854–858.
BAIRD-PARKER, A. C., BOOTHROYD, M., and JONES, E. 1970. The effect of water activity on the heat resistant strains of salmonellae. J. Appl. Bacteriol. *33*, 515–522.
BANWART, G. J., and AYRES, J. C. 1953. Effect of various enrichment broths and selective agars upon the growth of several species of *Salmonella*. Appl. Microbiol. *1*, 296–301.
BARRILE, J. C., and CONE, J. F. 1970. Effect of added moisture on the heat resistance of *Salmonella anatum* in milk chocolate. Appl. Microbiol. *19*, 177–178.
BOWMER, E. J. 1965. *Salmonella* in food—A review. J. Milk Food Technol. *28*, 74–86.
BREED, R. S., MURRAY, E. G. D., and SMITH, N. R. 1957. Bergey's Manual of Determinative Bacteriology, 7th Edition. The Williams & Wilkins Co., Baltimore, Md.
CHRISTIAN, J. H. B. 1955A. The influence of nitrition of the water relations of *Salmonella oranienburg*. Australian J. Biol. Sci. *8*, 75–82.
CHRISTIAN, J. H. B. 1955B. The water relations of growth and respiration of *Salmonella oranienburg*. Australian J. Biol. Sci. *8*, 490–497.
CHRISTIAN, J. H. B., and SCOTT, W. J. 1953. Water relations of salmonellae at 30°C. Australian J. Biol. Sci. *6*, 565–573.
CHUNG, K. C., and GOEPFERT, J. M. 1970. Growth of *Salmonella* at low pH. J. Food Sci. *35*, 326–328.
COMER, A. G., ANDERSON, G. W., and GARRARD, E. H. 1963. Gamma irradiation of *Salmonella* species in frozen whole egg. Can. J. Microbiol. *9*, 321–327.
COMMITTEE ON SALMONELLA. 1969. An evaluation of the *Salmonella* problem. Rept. U.S. Dept. Agr. and Food and Drug Admin., Natl. Acad. Sci., Washington, D.C.
DACK, G. M. 1956. Food Poisoning, 3rd Edition. Univ. Chicago Press, Chicago.
EWING, W. H. 1968. Differentiation of Enterobacteriacea by Biochemical Reactions. National Communicable Disease Center, Atlanta, Ga.
FOOD PROTECTION COMMITTEE. 1970. Reference methods for the examination of foods. Natl. Acad. Sci., Washington, D.C.
GALTON, M. M., SMITH, W. V., McELRATH, H. B., and HARDY, A. V. 1954. *Salmonella* in swine, cattle and the environment of abattoirs. J. Infect. Diseases *95*, 236–245.
GOEPFERT, J. M., and BIGGIE, R. A. 1968. Heat resistance of *Salmonella typhimurium* and *Salmonella senftenberg* 775W in milk chocolate. Appl. Microbiol. *16*, 1939–1940.
GOEPFERT, J. M., and HICKS, R. 1969. Effect of volatile fatty acids on *Salmonella typhimurium*. J. Bacteriol. *97*, 956–958.
GOEPFERT, J. M., ISKANDER, I. K., and AMUNDSEN, C. H. 1970. Relation

of the heat resistance of salmonellae to the water activity of the environment. Appl. Microbiol. *19*, 429–433.

INGRAM, M., RHODES, D. N., and LEY, F. V. 1961. The use of ionizing radiation for the elimination of salmonellae from frozen whole egg. U.K. Atomic Energy Res. Estab. Bull. R*381*, 1–30.

INSALATA, N. F., MAHNKE, C. W., and DUNLAP, W. G. 1972. Rapid, direct fluorescent-antibody method for the detection of salmonellae in food and feeds. Appl. Microbiol. *24*, 645–649.

KAUFFMAN, F., and EDWARDS, P. R. 1952, Classification and nomenclature of *Enterobacteriaceae*. Intern. Bull. Bacteriol. Nomenclature Taxonomy *2*, 2–8.

KAUFFMAN, F. 1966. The Bacteriology of the *Enterobacteriaceae*. The Williams & Wilkins Co., Baltimore, Md.

LEVINE, A. S., and FELLERS, C. R. 1940. Action of acetic acid on food-spoilage microorganisms. J. Bacteriol. *39*, 499–514.

LITCHFIELD, J. H. 1973. *Salmonella* and the food industry—methods for isolation, identification and enumeration. Crit. Rev. Food Technol. *3*, 415–456.

MARTH, E. H. 1969. Salmonellae and salmonellosis associated with milk and milk products. A Review. J. Dairy Sci. *52*, 283–315.

MATCHES, J. and LISTON, J. 1967. Low temperature growth of *Salmonella*. Proc. Inst. Food Technol. p. 82.

NORTH, W. R., JR. 1961. Lactose preenrichment method for isolation of *Salmonella* from dried egg albumen. Appl. Microbiol. *9*, 188–195.

OBLINGER, J. L., and KRAFT, A. A. 1970. Inhibitory effects of *Pseudomonas* on selected *Salmonella* and bacteria isolated from poultry. J. Food Sci. *35*, 30–32.

OBLINGER, J. L., and KRAFT, A. A. 1973. Oxidation-reduction potential and growth of *Salmonella* and *Pseudomonas fluorescens*. J. Food Sci. *38*, 1108–1112.

PROST, E., and RIEMANN, H. 1967. Food-borne salmonellosis. Ann. Rev. Microbiol. *21*, 495–528.

READ, R. B., JR., BRADSHAW, J. G., DICKERSON, R. W., JR., and PEELER, J. T. 1968. Thermal resistance of salmonellae isolated from dry milk. Bacteriol. Proc. p. 9, Am. Soc. Microbiology, Washington, D.C.

SCHROEDER, S. A. 1967. What the sanitarian should know about salmonellae and staphylococci in milk and milk products. J. Milk Food Technol. *30*, 376–380.

SPERBER, W. H., and DEIBEL, R. H. 1969. Accelerated procedure for *Salmonella* detection in dried foods and feeds involving only broth cultures and serological reactions. Appl. Microbiol. *17*, 533–539.

SPLITTSTOESSER, D. F. and SEGEN, B. 1970. Examination of frozen vegetables for *Salmonella*. J. Milk Food Technol. *33*, 111–113.

TAYLOR, B. and NAKAMURA, M. 1964. Survival of shigellae in food. J. Hyg. *62*, 303–311.

TAYLOR, W. I. 1961. Isolation of *Salmonella* from food samples. V. Determination of the method of choice for enumeration of *Salmonella*. Appl. Microbiol. *9*, 487–490.

WINTER, A. R., STEWART, G. F., McFARLANE, V. H., and SOLOWEY, M. 1946. Pasteurization of liquid egg products. III. Destruction of *Salmonella* in liquid whole egg. Am. J. Public Health *36*, 451–459.

CHAPTER 4

R. Bruce Tompkin
and
Lee N. Christiansen

Clostridium Botulinum

BOTULISM

Botulism is caused by the action of a neurotoxin produced by six types of *Clostridium botulinum*. The types (A through F) are distinguished by the serological specificity of the toxin produced by the different strains. The disease normally occurs following consumption of a food in which the organism has grown and produced toxin. There are rare reports of "wound botulism." In this latter form the disease is similar to tetanus, that is, the botulinum organism gains access to a wound, multiplies in the wound, and produces the toxin which affects the individual.

Collectively, the six *C. botulinum* toxins affect a large variety of animals including fish, birds, and most mammals. Well documented outbreaks of human botulism have been associated with A, B, E and F toxin types, with A, B and E being the most common. Type F was isolated in 1958 and has been responsible for only two documented human botulism outbreaks. The type C organism reportedly has caused a few outbreaks of human botulism. Types C and D are most important as causes of botulism in livestock, mink, and fowl and are sometimes called limberneck, western duck sickness, or forage poisoning, depending upon the circumstances.

Sausages of various types were the first recognized source of botulism. Several outbreaks of "sausage poisoning" occurred in Germany and elsewhere in Europe in the eighteenth and nineteenth centuries. Although later experience has shown sausages are not the only source of the disease, the term botulism, adopted in 1870 from the Latin *botulus* for sausage, has been retained.

Symptoms of botulism vary somewhat depending upon the type of toxin involved and the affected individual. The most frequently reported symptoms include a generalized feeling of uneasiness or weakness: blurred and double vision, dry throat, and difficulties in swallowing, pronouncing words, and breathing. Nausea and vomiting generally are encountered in cases of types B and E botulism but are not uncommon in type A outbreaks. Gastrointestinal upset is often the first symptom to appear in cases of type E botulism. Symptoms usually are noted within 18 to 36 hours after ingestion of food containing the toxin. However, the time may vary from a few hours to eight days. Normally, the earlier the onset of symptoms the more severe the illness. Death, if it occurs, usually follows within 3 to 6 days after ingestion of the toxin.

Characteristics of Botulinal Toxins

Botulinum toxin is released from the cell during cell lysis (Bonventre and Kemp, 1960). Type A toxin has been prepared in crystalline form and the remaining toxins have been purified to some extent. All are simple heat labile proteins. The toxins are denatured within minutes of heating at temperatures above $150°F$.

The unique characteristic of the botulinal toxins is their extreme toxicity. It has been estimated that a minimal lethal oral dose for man is about 5×10^{-9} to 5×10^{-8} g (Zacks and Sheff 1970 or, 1 g of toxin properly diluted could kill more than 500 million people.

The toxins of type E and nonproteolytic strains of B and F are activated by brief exposure to proteolytic enzymes such as trypsin. The toxicity of a type E culture, for example, can be increased more than one hundred fold by trypsinization. Trypsinization has been shown to somewhat enhance toxicity of proteolytic type B cultures, but has little effect on type A cultures (Iida 1970A).

Morbidity and Mortality of Botulism

The morbidity rate in botulism outbreaks is normally high and is often 100%. There are examples of persons remaining asymptomatic after consumption of a food which caused botulism in fellow diners. The most probable explanation for these occurrences is the uneven distribution of toxin in a food. This particularly would be true for solid foods where growth of C. botulinum could be confined to a small area. The possibility of natural immunity to botulinum toxin has been suggested (Rogers et al. 1964). Type E toxin was detected in the blood stream of an asymptomatic patient. Antibodies to the toxin were not demonstrable in the serum. Thus, immunity apparently did not result from a prior subclinical exposure to the toxin. Until additional information is available, it must be assumed that natural resistance to botulinum toxin in man is extremely rare.

The mortality rate of botulism in the U.S. has averaged approximately 57% over the past 70 years. Prior to 1949 this figure approached 65%. Since 1949 the average death-to-case ratio has been approximately 30%. This decrease in mortality rate has been attributed to improved treatment of acute respiratory failure and more rapid and accurate diagnosis of the disease coupled with the use of botulinum antitoxin.

Treatment of Botulism

Botulism therapy essentially involves two steps: (1) treatment of symptoms, particularly those associated with respiratory impairment, and (2) the use of therapeutic antitoxin. Treatment of patients with antitoxin has proven particularly effective in cases of type E botulism. In Japan, for example, the mortality rate was reduced from 28.5% for patients not given antitoxin to 4.3% in cases where antitoxin was administered (Iida 1970B).

Characteristics of C. Botulinum

C. botulinum is an anaerobic, spore-forming, rod-shaped, Gram-positive organism; characteristics common to all members of the genus *Clostridium*. The unique feature and principal reason for the classification of the six *C. botulinum* types as a single species, of course, is the toxin each produces. Although serologically distinct, the mode of action and the pharmacological effect of the six toxins appear to be the same. The types of *C. botulinum*, however, have been divided into three groups on the basis of physiological and serological characteristics (Smith and Holdeman 1968, and Holdeman and Brooks 1970). Group I includes the proteolytic *C. botulinum* types A, B and F. Group II consists of nonproteolytic types E, B and F strains, and Group III includes types C and D *C. botulinum*.

More recent studies have shown little genetic relationship between organisms in the different groups, whereas organisms within a group proved to be closely related genetically (Lee and Rieman 1970A and B). It also is interesting that organisms of Group I showed a close genetic tie to *Clostridium sporogenes*. Except for the inability to produce toxin, *C. sporogenes* also is physiologically similar to the Group I organisms. Group II organisms are genetically and physiologically indistinguishable from certain nontoxic clostridia which have been termed "E like" organisms. These close relationships between the various *C. botulinum* types and other nontoxic clostridia emphasize the difficulties which arise when trying to isolate *C. botulinum* and the necessity of determining that a suspect isolate is toxigenic before labeling it as *C. botulinum*.

Occurrence of C. Botulinum in Nature

The six types of *C. botulinum* have been detected in soil, and less frequently in water, from different parts of the world. Early surveys (Meyer and Dubovsky 1922) showed type A was the most common *C. botulinum* type in soils of the western U.S. In the midwest and eastern states type B predominated. Type B also was the principal type detected in soil samples from western Europe.

More recent surveys have shown the widespread distribution of type E *C. botulinum*. The organism has been detected in marine sediments and in the intestinal content of fish from waters throughout the northern hemisphere. Particularly heavy concentrations of type E exist in the Baltic Sea (Johannsen 1963) and the Green Bay area of Lake Michigan (Bott *et al.* 1968). Type E *C. botulinum* also has been detected in inland soil samples. For example, type E was the predominate *C. botulinum* type detected in soils of Russia (Kravchenko and Shishulina 1967) and Japan (Kanzawa *et al.* 1970). Thus, although the species, *C. botulinum*, is widespread throughout the world, it appears that certain types of *C. botulinum* predominate in different geographical regions. The surveys conducted to date have been limited to regions of temperate climates. Information on botulism and distribution of *C. botulinum* throughout the tropical regions of the world remains to be developed.

Foods Implicated in Botulism

The types of *C. botulinum* (A-F) responsible for human botulism closely follow the geographical distribution of the spore types throughout the world. The other major factor determining the incidence of botulism and the type of *C. botulinum* responsible is the diet and eating habits of the population.

Outbreaks in Europe generally have been associated with the growth of type B *C. botulinum* in meat products. The principal cause of botulism in the Soviet Union has been salted or smoked fish. Both types A and E *C. botulinum* have been isolated from the incriminated foods (Dolman 1964; Matveev *et al.* 1967). Nearly all botulism outbreaks in Japan have been caused by type E growth in fermented fish, rice, and a vegetable dish called izushi (Dolman 1964; Iida 1970B).

Information on botulism outbreaks in the U.S. has been tabulated in two reports. One summarizes outbreaks from 1899 to 1964 and includes Canadian outbreaks during this period (Meyer and Eddie 1965). A second report covers U.S. outbreaks between 1899 and 1967 (Anon. NCDC). During the periods covered by the latter report there were approximately nine botulism outbreaks in the United States per year with 2 to 3 cases per outbreak. Nearly 70% of the total outbreaks have been caused by processed vegetables and fruits (Table 4.1). Meat products, fish and seafoods, major causes of botulism in some other parts of the world, each have caused only seven per cent of the total number of outbreaks in the U.S. Dairy products and miscellaneous foods combined have accounted for only two per cent of the outbreaks. The responsible food was not determined in 12% of the outbreaks.

The responsible toxin type of *C. botulinum* was not determined in

TABLE 4.1
PERCENTAGE OF BOTULISM OUTBREAKS IN THE UNITED STATES
ATTRIBUTED TO VARIOUS FOODS, 1899-1967

Type of Food	Total %	Percentage Caused by Toxin Types					
		A	B	E	F	A&B	Unknown
Vegetables	60	33	8	1	0	1	59
Fruits	8	41	6	0	0	1	33
Meat Products	7	16	9	0	1	0	75
Fish & Seafoods	7	7	5	42	0	0	51
Milk & Milk Products	1	25	25	0	0	0	50
Miscellaneous	1	50	0	0	0	0	50
Unknown	12	-	-	-	-	-	-

a majority of the U.S. outbreaks (Table 4.1). The table shows, however, that type A has predominated as the causative organism in outbreaks where the toxin type was identified. The only type of food in which type A has not been responsible for fifty per cent or more of the outbreaks is fish and seafoods, where type E has caused a majority of the outbreaks. The single type E outbreak associated with mushrooms (Table 4.1) is one of only a few throughout the world where this type of botulism has occurred in foods not of marine origin.

The role of home-processed foods as the major source of botulism in the U.S. is evident from Table 4.2. During the period 1899-1967 there were 640 outbreaks involving 1669 individuals of which 948 died. Improvements were made in commercial processing requirements in the 1920s which reduced the number of outbreaks caused by commercially processed foods. The result has been that from 1930-1967 only 19 outbreaks resulted from commercially processed foods compared with 336 outbreaks from home processed foods. The records (Meyer and Eddie, 1965 and Anon. NCDC) indicate that

TABLE 4.2
OUTBREAKS OF BOTULISM ATTRIBUTED TO COMMERCIALLY
PROCESSED OR HOME PROCESSED FOODS IN THE U.S.

Source of Food	1899	1900 1909	1910 1919	1920 1929	1930 1939	1940 1949	1950 1959	1960 1967	Total
Home Processed	1	1	48	77	135	120	50	31	463
Commercially Processed	0	1	14	26	6	1	3	9	60
Unknown	0	0	8	13	13	13	50	20	117
Total	1	2	70	116	154	134	103	60	640

Source: Anon (NCDC)

TABLE 4.3
ROLE OF IMPORTED FOODS AS A SOURCE OF BOTULISM
IN THE U.S. BETWEEN 1930 AND 1967

Year	Product	Country
1931	Antipasto	Italy
1932	Smoked Salmon	Canada
1933	Crab	Japan
1934	Sprats	Germany
1936	Clams	Japan
1963	Liver paste	Canada
1966	Ham (?)	Germany

7 of the 19 outbreaks involved products which had been produced commercially outside the U.S. (Table 4.3). In terms of the number of cans or units of food produced, it is obvious that the likelihood of botulism from commercially processed foods is extremely rare. However, when a commercial product does become implicated as the cause of an outbreak very few processors can survive the impact. It is possible that the mortality of food processing companies whose product is responsible for botulism is as great as the mortality of the individuals consuming the product. This, in itself, is incentive to avoid the risk of producing hazardous product.

Methods For Preventing Botulism

Historically, raw agricultural commodities such as raw meats, fish, and poultry; fresh vegetables and fruits; or raw milk and cream have not been the foods which caused botulism. It is after these foods have been processed in some manner that they have become hazardous.

The factors which are most important for preventing botulism are knowledge that (1) the bacterium produces a heat resistant spore and, (2) the bacterium must multiply in the food in order to produce sufficient toxin to cause illness. This information forms the basis for four general methods for preventing food-borne botulism. These methods are:

1) Heat process foods at a high enough temperature for a sufficient length of time to destroy the vegetative and spore forms of the bacterium.
2) Apply mild heat processing in combination with inhibitory compounds which prevent growth of germinating spores surviving the mild heat process.
3) Formulate the food in a manner which will prevent growth of the bacterium.
4) Control the product temperature to 45°F or lower and in

certain cases 38° F or lower to control type E growth. Labeling and the design of the product container is an important consideration when temperature control is to be relied upon.

Thermal Processing of Low Acid Foods.—An example of the first technique is the use of pressure cooker for home canning and the retort for commercial canning of low acid foods. The safety of shelf stable low acid foods, such as canned vegetables, stews, and soups, is dependent upon an adequate thermal process. The parameters for safe thermal processing were developed in the 1920s and are based upon studies of the resistance of *C. botulinum* spores. The heat resistance data are used to calculate the time required to destroy a large population of spores at various temperatures of processing. The level of spores offering a margin of safety was selected to be 1×10^{12} and the process has since been referred to as the "12D process" or "botulinal cook." This is sufficient heat to reduce a billion spores in each of one thousand cans to only one spore in a thousand cans.

Due to differences in foods and containers it is necessary that each product be considered a separate entity. Heat penetration tests must be conducted for each food and container size. Subsequent calculations then can be used to determine the thermal process to achieve a 12D process. An alternative to this method is to inoculate product with spore suspensions and determine the rate of spore destruction in the product. This can be done using botulinal spores in special laboratory facilities by experienced, immunized personnel. Alternatively, it is more common to use nonhazardous spores of known heat resistance such as the PA3679 strain of *C. sporogenes.*

It is important to consider modifications in the product formula which may alter the rate of heat penetration. Thus, additional tests may be necessary as certain ingredients are replaced or concentrations of ingredients are changed. Although a single change may not be important, a series of slight changes occurring over several years must not be ignored and may necessitate a reevaluation of the safety of the thermal process.

All aspects of canning, sealing, and processing of shelf stable low acid foods must properly be controlled. Regulations have been proposed which are designed to assure a minimum standard of compliance for thermally processed foods in hermetically sealed containers and to minimize the possibility of food-borne botulism from such products. The reader is referred to the Federal Register *38*: 2398- 2410 (1973) for this information.

Mild Heat Plus Inhibitory Compounds.—Canned shelf stable cured meat is a unique class of food which depends upon a combination of four factors for shelf stability. The factors include mild heating to an

F_o value of 0.1 to 0.7, sodium chloride, sodium nitrite and a low level of spores (Riemann 1963A; Silliker 1959). It is generally believed that this system leads to inhibition of outgrowth of surviving spores. This is apparent from tests with inoculated product wherein low levels of viable bacterial spores can be demonstrated after processing by subculturing into suitable bacteriological media.

Despite the favorable public health record for shelf stable canned cured meats, it must be recognized that any abnormality, such as underprocessing or unusually low salt or nitrite levels, presents the possibility of an unstable and potentially hazardous product. Attempts to clearly define the minimum levels of nitrite or salt necessary for safety have been unsuccessful to date. Studies involving artificially contaminated raw materials have demonstrated that the complex system of protection afforded by salt and nitrite is related to the initial level of botulinal spores prior to cooking.

Control of Product Formulation.—Foods may be formulated to prevent the growth of C. botulinum. Two methods by which this can be accomplished are acidification and adjustment of the food's water activity. It is now generally believed that foods having a pH of 4.5 or lower will not support the growth of C. botulinum (Ingram and Robinson 1951, Riemann 1963B). Townsend et al. (1954) and Ohye and Christian (1967) concluded that pH 4.6 provides an adequate safety factor for acidified foods. It appears unlikely that growth and toxin production by type E will occur below pH 4.8 (Segner et al. 1966). The regulation promulgated by the U.S. Food and Drug Administration uses a finished equilibrium pH value greater than 4.6 as the criterion for classifying foods as "low acid" (Edwards 1973). Exceptions to this pH are granted to pears, pineapples, and tomatoes, or the juices thereof, having a pH of less than 4.7 and figs having a pH of 4.9 or below.

The manufacturer must be certain that the product's pH is stable and will not rise above 4.6 during storage. Normally, a mild heat process is applied to acidified shelf stable foods to destroy yeasts, molds and other common spoilage flora. This not only protects the quality of the product but also prevents growth of spoilage microorganisms which might alter the product's pH through their metabolic activity during storage. As an added precaution it may be desirable that the food be formulated with a strong buffering capacity to help maintain a stable low pH.

Considering the importance of the pH not being greater than 4.6, it is essential that the product be monitored with an accurate, sensitive pH meter. Futhermore, the individual measuring the pH must understand the factors which enter into a valid pH determination. This would appear so elementary that it is tempting, but erroneous, to assume that everyone can make an accurate pH determination.

It is also possible to formulate stable products by controlling the amount of water which is available for microbial growth. From a microbiological viewpoint the total moisture content of a food is composed of free and bound water. Free moisture is that portion of the water in a food which is free and available to microorganisms for growth. Bound moisture is water which is tied up by solutes such as inorganic salts (*e.g.*, sodium chloride) and sugar (*e.g.*, glucose, sucrose) and thus not available for microbial activity. Each type of microorganism has a minimum level of free moisture below which it cannot multiply. The amount of free moisture in a food can be determined and expressed as water activity (A_W). Distilled water, for example, has an A_W value of 1.0. It has been reported that foods formulated to contain 50% sugar or 10% salt are inhibitory to *C. botulinum*. This corresponds to an A_W value of 0.935 (Schmidt 1964). Although germination may occur at lower A_W values, the available information indicates *C. botulinum* will not multiply at an A_W below 0.93 (Marshall *et al.* 1971; Baird-Parker and Freame 1967; Ohye and Christian 1967). This information is of practical value for certain shelf stable foods such as canned bread and canned cake.

In addition to formulating foods with an A_W below 0.93, there is wide use of dehydration for food preservation. For example, dried foods such as fruits, cereals, beans, are sufficiently dry that their A_W is well below 0.93.

Certain cured meats traditionally accepted by consumers as having a salty flavor such as country cured meats, glass packed sliced dried beef, and fermented dry sausages are manufactured by processes which combine the advantages of adjusting A_W by formulation control and dehydration. These products have an A_W below 0.93 when they are shipped from the plant. Fermented dry sausages as referred to here exclude the semi-dry or summer sausages which require refrigeration.

Type E *C. botulinum* presents a unique problem to the smoked fish industry. Regulations for commercial processing of smoked fish have included the following: (1) the finished smoked product must have a brine content of 3.5% in the loin muscle, (2) the sodium nitrite content must fall between 100 and 200 ppm in the loin muscle, (3) the fish must be cooked to a minimum of 160°F for no less than 30 minutes, (4) the product must be cooled to 50°F or below within three hours after smoking, and (5) further cooled to 38°F or below within 12 hours after smoking. In addition, all

CLOSTRIDIUM BOTULINUM65

shipping containers, retail packages, and shipping records must be marked to assure the product is held under refrigeration (at 38°F or below) until consumed. The use of nitrite in smoked fish has been questioned and these regulations are under review and are subject to change.

In certain cases it may be prudent to conduct experiments wherein the product is inoculated with *C. botulinum*. This may be necessary to confirm that *C. botulinum* will not multiply in the product or to determine the margin of safety relative to pH and or A_W. Such tests should be conducted only in specialized laboratories.

Refrigeration or Freezing.—A wide variety of perishable foods is capable of supporting growth of *C. botulinum* but these foods are rarely implicated as the cause of botulism. This broad class of foods is dependent upon refrigeration or freezing to protect the quality of the product as well as avert food poisoning. Perishable meats, fish, poultry, dairy products and certain vegetables may be included in this category.

The temperature growth range of the six *C. botulinum* types is between 38° and 122°F. Type E (Schmidt *et al.* 1961) and nonproteolytic strains of type B and F (Eklund *et al.* 1967) can grow at 38°F. The Langlande strain of type F which is proteolytic grew at 39°F (Walls 1967). Proteolytic strains of types A and B have a minimum growth temperature of 50°F (Ohye and Scott 1953). Maximum toxin production usually occurs between 77° and 82.4°F for nonproteolytic strains and 95° to 98.6°F for proteolytic strains (Kautter *et al.* 1970).

The U.S. Public Health Service recommended in 1962 that all potentially hazardous foods be kept at 45°F or below, or 140°F or above, except during necessary periods of preparation and service. Potentially hazardous foods included any perishable food capable of supporting rapid and progressive growth of infectious or toxigenic microorganisms (Anon. 1962). Many states, counties, and municipalities have incorporated the USPHS recommendations into food service regulations. This has lead to an increased awareness of temperature control and should help minimize the possibility of botulism from mishandled foods.

Importance of Food Container and Labeling

Manufacturers must select with care the container into which their products will be packed. This is most important for pasteurized or pre-cooked ready-to-eat products which are dependent upon refrigeration for safety. The container must be of such design and so labeled as to prevent confusion of the customer. It is critical that consumers know which products must be refrigerated for microbiological safety.

Analytical Methods

A food suspected of causing botulism must be tested to determine if it contains botulinal toxin or the presence of *C. botulinum.* One of the many procedures described for this purpose is basically as follows (Anon. 1971).

1) *Sample Preparation*

A clarified extract of the food is prepared by first making a food-buffer slurry, followed by centrifugation and filtration of the supernatant. A portion of the supernatant is trypsinized, particularly if the food is of marine origin giving cause to suspect the presence of type E toxin. A second aliquot is boiled for 10 minutes to destroy any botulinum toxin which may be present.

2) *Toxin Assay*

Presumptive test: Mice are injected with 0.2 - 0.5 ml. of the untreated, trypsinized and heated aliquots of the supernatant and observed for 96 hours. Death of mice injected with untreated and/or trypsinized material and survival of mice receiving the heated sample is presumptive evidence for the presence of botulinum toxin.

Confirmitory test: Presumptively positive samples are confirmed by means of a mouse protection test. For this test mice are injected as follows:

Type A protected — 0.5 ml of type A antitoxin and
0.5 ml of "suspect" supernatant.

Type B protected — 0.5 of type B antitoxin and
0.5 ml of "suspect" supernatant.

Type E protected — 0.5 ml of type E antitoxin and
0.5 ml of "suspect" supernatant.

Unprotected Control — 0.5 ml of "suspect" supernatant.

Only mice which receive antitoxin corresponding to the toxin type in the sample should survive. Thus, death of all mice except those protected with type A antitoxin would show the suspect sample contained type A toxin.

In the normal procedure for isolation and identification of *C. botulinum*: (1) subcultures or enrichments of the test sample are prepared; (2) the subcultures are assayed for botulinal toxin; (3) toxic subcultures are plated to obtain isolated colonies; (4) the various isolates are tested for ability to produce toxin, and (5) the toxin type is identified using mouse protection tests. The original samples and/or the toxic subcultures from step 2 are generally treated to destroy vegetative cells, leaving bacterial spores including those of *C. botulinum*. When trying to isolate type A and proteolytic strains of types B and F, the subculture can be heated at 176° F. for 10 to 30 minutes. These spores will survive such a heat process. Spores of nonproteolytic strains would be killed by a heat process of this nature. Thus, when trying to isolate type E and nonproteolytic type B and F strains, subcultures are treated with an equal volume of absolute ethanol. Both the heating and the alcohol procedures are most effective when extraneous organisms are present as vegetative cells. The spores, particularly those of other anaerobes which survive these treatments, present a problem in the isolation of *C. botulinum*. These organisms compete with *C. botulinum* and some produce substances which inhibit *C. botulinum* growth. Many of these anaerobic contaminants produce colonies which are indistinguishable from *C. botulinum*.

Thus, during the isolation step (3) it may be necessary to pick a very large number of colonies, many of which would not be *C. botulinum*.

BIBLIOGRAPHY

ANON. (No Date). Botulism in the United States. Review of cases, 1899-1967 and handbook for epidemiologists, clinicians, and laboratory workers. U.S. Dept. Health, Education, and Welfare, Public Health Serv. National Center for Disease Control, Atlanta, Ga.

ANON. 1962. Food Service Sanitation Manual, Including a Model Food Service Sanitation Ordinance and Code, 1962 Recommendations of the Public Health Service. U.S. Dept. Health, Education, and Welfare, Public Health Serv., Div. Environ. Eng. Food Protec., Washington, D.C. Publ. *934*.

ANON. 1971. Reference Methods for the Microbiological Examination of Foods. Food Protec. Comm., Food Nutr. Board, Natl. Res. Council—Natl. Acad. Sci., Washington, D.C.

BAIRD-PARKER, A. C., and FREAME, B. 1967. Combined effect of water activity, pH and temperature on the growth of *Clostridium botulinum* from spore and vegetative cell inocula. J. Appl. Bacteriol. *30*, 420–429.

BONVENTRE, P. F., and KEMPE, L. L. 1960. Physiology of toxin production by *Clostridium botulinum* types A and B. IV. Activation of the toxin. J. Bacteriol. *79*, 18–23.

BOTT, T. L., JOHNSON, J., FOSTER, E. M., and SUGIYAMA, H. 1968. Possible origin of the high incidence of *Clostridium botulinum* type E in an inland bay (Green Bay of Lake Michigan). J. Bacteriol. *95*, 1542–1547.

DOLMAN, C. E. 1964. Botulism as a world health problem. *In* Botulism, Proceedings of a Symposium, K. H. Lewis, and K. Cassel (Editors). U.S. Dept. Health, Education, and Welfare, Public Health Serv. Publ. *999-FP-1*.

EDWARDS, C. E. 1973. Thermally processed low-acid foods packaged in hermetically sealed containers. Federal Register *38*, 2398–2410.

EKLUND, M. W., POYSKY, F. T., and WIELER, D. I. 1967. Characteristics of *Clostridium botulinum* type F isolated from the Pacific coast of the United States. Appl. Microbiol. *15*, 1316–1323.

EKLUND, M. W., WIELER, D. I., and POYSKY, F. T. 1967. Outgrowth and

168   FOOD MICROBIOLOGY

toxin production of nonproteolytic type B *Clostridium botulinum* at 3.3 to
5.6 C. J. Bacteriol. *93*, 1461-1462.
HOLDEMAN, L. V. 1967. Growth and toxin production of *Cl. botulinum* type
F. *In* Botulism. 1966. Proceedings of the Fifth International Symposium on
Food Microbiology, M. Ingram, and T. A. Roberts (Editors). Chapman and
Hall, London.
HOLDEMAN, L. V., and BROOKS, J. B. 1970. Variation among strains of
Clostridium botulinum and related clostridia. *In* Proceedings of the First
U.S.-Japan Conference on Toxic Microorganisms, M. Herzberg (Editor).
UJNR Joint Panels on Toxic Microorganisms and U.S. Dept. Interior.
IIDA, H. 1970A. Activation of *Clostridium botulinum* toxin by trypsin. *In*
Proceedings of the First U.S.-Japan Conference on Toxic Microorganisms, M.
Herzberg (Editor). UJNR Joint Panels on Toxic Microorganisms and U.S.
Dept. Interior.
IIDA, H. 1970B. Epidemiology and clinical observations of botulism outbreaks
in Japan. *In* Proceedings of the First U.S.-Japan Conference on Toxic
Microorganisms, M. Herzberg (Editor). UJNR Joint Panels on Toxic Micro-
organisms and U.S. Dept. Interior.
INGRAM, M., and ROBINSON, R. H. M. 1951. A discussion of the literature on
botulism in relation to acid foods. Proc. Soc. Appl. Bacteriol. *14*, 73-84.
JOHANNSEN, A. 1963. *Clostridium botulinum* type E in Sweden and the
adjacent waters. J. Appl. Bacteriol. *26*, 43-47.
KANZAWA, K., ONO, T., KARASHIMADA, T., and IIDA, H. 1970. Distribu-
tion of *Clostridium botulinum* type E in Hokkaido, Japan. *In* Proceedings of
the First U.S.-Japan Conference on Toxic Microorganisms, M. Herzberg
(Editor). UJNR Joint Panels on Toxic Microorganisms and U.S. Dept Interior.
KAUTTER, D. A. *et al.* 1970. The detection, identification, and isolation of
Clostridium botulinum. In Proceedings of the First U.S.-Japan Conference on
Toxic Microorganisms, M. Herzberg (Editor). UJNR Joint Panels on Toxic
Microorganisms and U.S. Dept. Interior.
KRAVCHENKO, A. T., and SHISHULINA, L. M. 1967. Distribution of *Cl.
botulinum* in soil and water in the U.S.S.R. *In* Botulism. 1966. Proceedings of
the Fifth International Symposium on Food Microbiology, M. Ingram, and
T. A. Roberts (Editors). Chapman and Hall, London.
LEE, W. H., and RIEMANN, H. 1970A. Correlation of toxic and nontoxic
strains of *Clostridium botulinum* by DNA composition and homology. J.
Gen. Microbiol. *60*, 117-123.
LEE, W. H., and RIEMANN, H. 1970B. The genetic relatedness of proteolytic
Clostridium botulinum strains. J. Gen. Microbiol. *64*, 85-90.
MARSHALL, B. J., OHYE, D. F., and CHRISTIAN, J. H. B. 1971. Tolerance of
bacteria to high concentrations of NaCl and glycerol in the growth medium.
Appl. Microbiol. *21*, 363-364.
MATVEEV, K. I., NEFEDJEVA, N. P., BULATOVA, T. I., and SOKOLOV, I. S.
1967. Epidemiology of botulism in the U.S.S.R. P. 1-10. *In* Botulism. 1966.
Proceedings of the Fifth International Symposium on Food Microbiology, M.
Ingram, and T. A. Roberts (Editors). Chapman and Hall, London.
MEYER, K. F., and DUBOVSKY, B. J. 1922. The distribution of the spores of
B. botulinus in the United States. IV. J. Infect. Diseases *31*, 559-594.
MEYER, K. F., and EDDIE, B. 1965. Sixty-five years of human botulism in the
United States and Canada. George Williams Hooper Foundation, San
Francisco Medical Center, Univ. of California, San Francisco.
OHYE, D. F., and CHRISTIAN, J. H. B. 1967. Combined effects of tempera-
ture, pH and water activity on growth and toxin production by *Cl. botulinum*
types A, B and E. *In* Botulism. 1966. Proceedings of the Fifth International
Symposium on Food Microbiology, M. Ingram, and T. A. Roberts (Editors).
Chapman and Hall, London.

OHYE, D. F., and SCOTT, W. J. 1953. The temperature relations of *Clostridium botulinum*, types A and B. Australian J. Biol. Sci. *6*, 178–189.
RIEMANN, H. 1963A. Safe heat processing of canned cured meats with regard to bacterial spores. Food Technol. *17*, 39–49.
RIEMANN, H. 1963B. Anaerobe toxins. *In* Chemical and Biological Hazards in Foods, J. C. Ayres, A. A. Kraft, H. E. Snyder, and H. W. Walker (Editors). Iowa State Univ. Press, Ames, Iowa.
ROGERS, D. E., KOENING, M. G., and SPICKARD, A. 1964. Clinical and laboratory manifestations of type E botulism in man. *In* Botulism, Proceedings of a Symposium, K. H. Lewis and K. Cassel (Editors). U.S. Dept. Health, Education, and Welfare, Public Health Serv. Publ. *999-FP-1*.
SCHMIDT, C. F. 1964. Spores of *C. botulinum*: formation, resistance, germination. *In* Botulism, Proceedings of a Symposium, K. H. Lewis, and K. Cassel (Editors). U.S. Dept. Health, Education, and Welfare, Public Health Serv. Publ. *999-FP-1*.
SEGNER, W. P., SCHMIDT, C. F., and BOLTZ, J. K. 1966. Effect of sodium chloride and pH on the outgrowth of spores of type E *Clostridium botulinum* at optimal and suboptimal temperatures. Appl. Microbiol. *14*, 49–54.
SHANTZ, E. J. 1964. Purification and characterization of *C. botulinum* toxins. *In* Botulism, Proceedings of a Symposium, K. H. Lewis, and K. Cassel (Editors). Public Health Serv., U.S. Dept. Health, Education, and Welfare, Cincinnati, Ohio.
SILLIKER, J. H. 1959. The effect of curing salts on bacterial spores. *In* Proceedings of the Eleventh Research Conference, American Meat Institute Foundation, Chicago.
SMITH, L. D. S. and HOLDEMAN, L. V. 1968. The Pathogenic Anaerobic Bacteria. Charles C. Thomas, Springfield, Ill.
TOWNSEND, C. T., YEE, L., and MERCER, W. A. 1954. Inhibition of the growth of *Clostridium botulinum* by acidification. Food Res. *19*, 536–542.
WALLS, N. W. 1967. Physiological studies on *Cl. botulinum*, type F. *In* Botulism. 1966. Proceedings of the Fifth International Symposium on Food Microbiology, M. Ingram, and T. A. Roberts (Editors). Chapman and Hall, London.
ZACKS, S. L., and SHEFF, M. F. 1970. Studies on botulinus toxin. *In* Proceedings of the First U.S.-Japan Conference on Toxic Microorganisms, M. Herzberg (Editor). UJNR Joint Panels on Toxic Microorganisms and U.S. Dept. Interior.

CHAPTER 5

Charles L. Duncan | # Clostridium Perfringens

CLOSTRIDIUM PERFRINGENS

Clostridium perfringens is usually thought of as the most important gas gangrene organism. However, from a public health standpoint, the type A strain of this anaerobic bacterium is one of the most common etiological agents of human food poisoning. Its significance as a major food poisoning organism was not fully accepted in the U.S. until about the mid-1960s. Recognition of the importance of perfringens food poisoning resulted from improvement in isolation and identification techniques and greater emphasis in testing for this organism by public health laboratories. This type of food poisoning has been most important in food service establishments where large quantities of food are served. Incrimination of *C. perfringens* in food poisoning incidents occurring in the home has been minor. This is likely due to the short duration and relative mildness of the disease and failure of the individuals involved to report the poisoning to health officials.

Considerable progress in understanding the nature of this type of food poisoning has been made in recent years. We now know that type A perfringens poisoning occurs following ingestion of large numbers of viable vegetative cells; yet, the symptoms of the poisoning are induced by the action of perfringens enterotoxin that is synthesized and released when the ingested vegetative cells undergo sporulation in the intestinal tract.

Although *C. perfringens* is quite demanding nutritionally, several kinds of foods will support good growth of the organism. Good sanitation and strict adherence to proper food preparation procedures for these high-food-poisoning-potential foods are necessary to prevent perfringens poisoning.

MORPHOLOGICAL CHARACTERISTICS

The occurrence of variation in clostridia is well known. When working with *C. perfringens* in the laboratory, it should be recognized that it is notorious in this respect.

Cellular Morphology

Actively growing vegetative cells of *C. perfringens* usually appear as straight, plump rods with blunt ends. The cells may vary in size

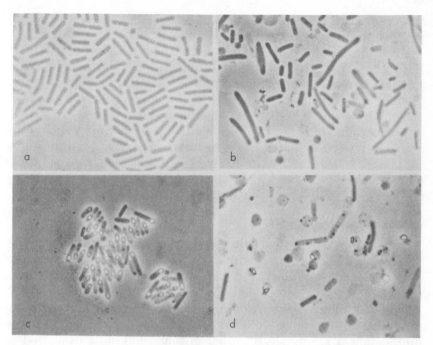

FIG. 5.1. PHASE CONTRAST MICROGRAPHS OF CELLS FROM PURE CULTURES
OF *CLOSTRIDIUM PERFRINGENS* TYPE A SHOWING VARIATIONS IN CELL
MORPHOLOGY THAT MAY OCCUR.
(a) Exponential cells in a growth medium (b) 24 hr old cells in a sporulation medium (c) 8
hr old cells in a sporulation medium (d) 24 hr old cells showing protoplasts in a
sporulation medium.

from 2 to 6 microns long and 0.8 to 1.5 microns wide depending on
the strain, culture age and nature of the growth medium (Figure 5.1).
During the exponential phase of growth, cells may be very short and
appear cubical. The cells may also appear shorter and fatter when
examined in strained direct smears of food. Cells usually occur singly
but may be seen as short chains of 2 or 4 cells in actively growing
cultures.

Spores are usually not seen in smears from foods. They are
produced in vitro in culture media designed for sporulation and
generally are not seen in glucose-containing media used for vegetative
proliferation. Spores are subterminal but their position in different
cells in a given culture may not be consistent. Sporulating cultures
less than about 12 hr old are normally morphologically similar.
However, older cultures tend to undergo recycling, resulting in a new
population of vegetative cells. In addition, some strains tend to form
protoplasts which appear as large spherical structures in the sporulating culture (Figure 5.1). The culture may appear contaminated to

the uninitiated microbiologist due to the variety of cellular forms that may be present in the older culture.

All strains of C. perfringens tested have been nonmotile.

The presence of capsules is difficult to demonstrate in many strains, whereas others produce capsules of considerable size. Capsule production is usually best when a strain is grown with a fermentable carbohydrate under slightly alkaline conditions. With many strains, capsule production may be demonstrated using early exponential phase cultures. Capsules from different strains have been shown to contain a weakly acidic or neutral polysaccharide and an acidic polysaccharide resembling dermatan sulfate (Darby et al. 1970). The polysaccharides appear to be complexed with protein.

Colonial Morphology

Both smooth and rough colonial forms of this organism occur, the smooth form being more typical. Smooth colonies resulting from overnight incubation are about 1 to 3 mm in diameter, low convex, opaque or grayish white, glossy and with entire margins. On the second day of incubation, transparent sectors in, or protuberances extending from, these colonies may appear spontaneously. The protuberances are usually irregular in shape and quite flat. Mutants with altered toxigenicity and sporulating ability may be isolated from these transparent sectors, especially if the culture has been previously treated with a mutagen.

Rough colonies are 3 to 5 mm in diameter with a slightly raised surface, translucent and with lobate margins. Mucoid strains are raised, opaque and glistening with regular margins. Cells from these colonies are heavily encapsulated. Both mucoid and rough variants may be obtained from the typical smooth strains. Some of these variants are stable.

A close relationship exists between phage susceptibility and colonial morphology. Smooth and rough strains are usually phage-susceptible; mucoid strains are usually phage-resistant (Smith 1959).

BIOCHEMICAL CHARACTERISTICS

Nutritional Requirements

The array of amino acids and growth factors found in minimal growth media for C. perfringens indicate its exacting nutritional requirements. The requirements for certain amino acids, i.e. serine, methionine or alanine, and the balance of amino acids in a growth medium may vary from strain to strain. A synthetic minimal medium containing 13 to 14 amino acids, adenine, biotin, calcium pantothenate, pyridoxine, ammonium chloride, magnesium chloride, ferrous chloride, sodium-potassium buffer and glucose has been described for growth of *C. perfringens* (Fuchs and Bonde 1957). This medium may require large inoculum levels of certain strains to initiate growth. A defined medium which supports growth at inoculum levels as low as 10 cells per ml has been described (Riha and Solberg 1971). The nitrogen requirements of *C. perfringens* can be supplied by acid-hydrolyzed casein, provided that tryptophan and an ammonium salt are added. Minimal media which support the growth of this organism are usually not sufficient for toxin synthesis.

A variety of foods are nutritionally adequate to support excellent growth of *C. perfringens*. These include meats and meat dishes, gravies, fish, milk, and legumes such as peas and beans. A polysaccharide isolated from the common dry bean has been shown to stimulate voluminous gas production by *C. perfringens* and thus may be the flatulence factor associated with these legumes (Kurtzman and Halbrook 1970). Protein fractions from sources such as soy protein that are used to fabricate imitation synthetic foods or to supplement standard meat formulations may be stimulatory or inhibitory to growth of *C. perfringens*. For example, the addition of protein additives to turkey meat loaves significantly enhanced the growth rate of *C. perfringens* (Busta and Schroder 1971). Food processors using these protein fractions should realize their possible stimulatory effect on growth of this organism.

Carbohydrate Degradation

C. perfringens is saccharolytic and thus dependent on carbohydrates for energy production. The organism is heterofermentative and produces lactate, acetate, butyrate, ethyl alcohol, carbon dioxide and hydrogen from glucose. In addition to all the enzymes of the Embden-Meyerhof pathway, both lactic acid dehydrogenase and the pyruvate-clastic system have been demonstrated in this organism (Groves and Gronlund 1969A). The conventional hexose-monophosphate pathway is apparently not involved as a major pathway of glycolysis. Pyruvate may serve as an energy source but it will not support growth of *C. perfringens* in a semidefined medium.

A variety of carbohydrates are fermented by *C. perfringens* including glucose, fructose, galactose, mannose, maltose, lactose, sucrose, ribose, xylose, trehalose, starch, dextrin, and glycogen. Fermentation of salicin and glycerol is variable. Mannitol is not fermented. Growth is most rapid in the presence of glucose, mannose or ribose. Both glucose and mannose have been shown to be actively transported via a common mechanism (Groves and Gronlund

1969B). None of the other carbohydrates, including ribose are actively transported.

Biochemical Reactions

Many strains produce stormy fermentation in litmus milk. However, this reaction may not occur with some strains. Since several other clostridia also produce this reaction, it is by no means diagnostic for *C. perfringens*. Hydrogen sulfide is produced in most media. Its production is an important characteristic in selective differential media used for quantitation of this organism. Indole is not produced. The reduction of nitrates to nitrites is a diagnostic test often used. Even though most strains do reduce nitrates some have been isolated that will not. Therefore precaution should be taken in applying this test. Nitrate reduction is not always consistent with a given strain. More consistent results may be obtained by the addition of 0.5% each of galactose and glycerol to the nitrate test medium. Most strains hydrolyze gelatin but not hemoglobin, casein, or coagulated albumen or serum. Some strains produce a collagenase (kappa toxin) that can hydrolyze native collagen.

Phospholipase C (alpha toxin), also called lecithinase, hydrolyzes lecithin to yield phosphoryl choline and a diglyceride. It is a diagnostically important enzyme and may be demonstrated by growing the organism on an egg yolk containing agar. An opaque halo precipitates around colonies on such agar as a result of the lecithinase action.

Hemolysins

Blood agar plates may be used for visualizing the hemolysins produced by *C. perfringens*. The hemolytic patterns may vary depending on the toxigenic type of the strain being tested and the type of blood used in preparing the agar plates. Either partial or complete hemolysis of both may be produced by a given strain due to the activity of the alpha (phospholipase C), delta or theta toxins produced by the strain. Alpha toxin produces partial hemolysis of the red blood cells of cattle, mice, rabbits, sheep and humans, whereas horse and goat cells are resistant. Delta toxin also produces partial hemolysis of sheep, goat, cattle and swine red blood cells, but it is not lytic to horse, rabbit, and human cells. Theta toxin produces complete lysis of the red blood cells of all species investigated except the mouse (Ispolatovskaya 1971). Theta toxin is oxygen-labile and if inactivated by oxygen it can be reactivated by reducing agents such as cysteine. In addition to the hemolytic activity of the recognized toxins, some strains of *C. perfringens* produce an hemolysin serologically distinct from the alpha, delta, or theta toxins. This hemolysin produces complete lysis of red blood cells and is referred to as the non-alpha-delta-theta hemolysin (Brooks *et al.* 1957).

Bacteriocin

Bacteriocins are substances produced by one organism and that have antibiotic-like activity against other members of the same species. Bacteriocin production may occur spontaneously or following ultra-violet light induction in some strains of *C. perfringens* (Mahony and Butler 1971; Mahony *et al.* 1971). The perfringens bacteriocin is heat labile and may be inactivated by treatment with trypsin. It apparently acts on sensitive indicator strains by either inhibition of cell wall synthesis or removal of existing wall material. Protein, DNA and RNA synthesis are not affected in cells treated with bacteriocin.

CULTURAL REQUIREMENTS

Temperature and pH

Growth of *C. perfringens* may occur in a temperature range of about 15°C to 50°C. The optimum temperature for most rapid growth is approximately 45°C. However, at this temperature lysis of many strains occurs soon after the end of exponential growth. Excellent growth may be obtained at 37°C following a lag period of 2 to 4 hr, when low inoculum levels are used; at 46°C little or no lag may be present before exponential growth. A temperature of 46°C is often used in selective isolation techniques. At a temperature of about 50°C, a marked decrease in the initial number of viable cells inoculated into a culture medium may occur during the first 3 to 4 hr of incubation, followed by a sudden increase in growth to a maximum level at about 6 hr. Between 5° and 15°C a stabilization of the inoculum level or slow death may occur. At temperatures between 15° and about 20°C lag times in the neighborhood of 4 to 24 hr may precede vegetative proliferation. Considerably more variation in growth rate exists among different strains at lower temperatures than at higher temperatures. Under optimum conditions the growth rate of *C. perfringens* is extremely rapid. The generation time at pH 7.0 has been shown to range from 10 min at 45°C to 100 min at 25°C (Smith 1971).

The generation time at a given temperature may vary with the pH of the growth medium. In cooked meat broth at 20°C, variable growth was reported to occur at pH 5.8 but rapid growth occurred at pH 7.2 (Barnes *et al.* 1963). The generation time at 37°C was shown to vary from 100 min at pH 5.0 to 21 min at pH 6.5 and 48 min at pH 8.0 (Smith 1971). Growth will occur from about pH 5.0 to 8.0. There is no clear cut pH optimum for growth of *C. perfringens*. The organism grows rapidly between the limits of pH 6.0 and 7.5. The extent of growth is often slightly greater at pH 6.5 than pH 7.0.

An actively growing culture of *C. perfringens* in a glucose-containing medium such as fluid thioglycollate medium, may reduce the pH to less than 5.0 on overnight incubation. Some strains will not survive over 4 to 5 days at such a low pH. Therefore this medium is used primarily for activation of cultures and not for culture carriage. The stability of various strains is considerably better in a meat-containing medium such as cooked meat medium in which buffering action of the meat proteins exits. This kind of medium is often used for carrying stock cultures. However, strains that sporulate at low frequency may be non-viable after a few months storage at room temperature in this medium. Culture preservation may be best accomplished by lyophilization of cells of overnight broth cultures that have been centrifuged and resuspended in a small volume of skim milk.

Anaerobiosis

C. perfringens is an anaerobe that will not produce surface colonies on agar containing media exposed to the air. Liquid media for the growth of this organism are usually poised at a favorable oxidation-reduction potential level by the addition of reducing agents. Low concentrations of agar (0.1 to 0.3%) are often added to the liquid media to prevent diffusion of oxygen into the media. In addition, liquid media are routinely steamed for 15 to 20 min to remove dissolved oxygen, followed by cooling to the desired incubation temperature before inoculation.

The spore-forming anaerobes are generally not as sensitive to oxygen as are the non-spore-forming anaerobes, with *C. perfringens* reportedly being less sensitive than most. The lack of stringent anaerobic requirements for growth of this organism may explain its frequency of isolation from nature and its ability to grow luxuriantly in foods that are not appreciably anaerobic. *C. perfringens* cells will withstand exposure to the air for several hours without appreciable decrease in the viable cell count. An actively metabolizing culture of this organism is not inhibited by atmospheric oxygen even if air is bubbled through the culture. The ability to initiate growth is very much dependent on the oxidation-reduction potential of the medium. The limiting potential at which growth will be initiated is dependent on the pH of the growth menstruum. The limiting

potential for *C. perfringens* may range from +31 mv at pH 7.74
(Hanke and Bailey 1945) to +230 to +250 mv at pH 6.0 (Barnes and
Ingram 1956). One strain was shown to grow more luxuriantly at pH
7.0 at a potential of +200 mv in the presence of oxygen than at +40
mv in the absence of oxygen (Tabatabai and Walker 1970). The
limiting oxidation-reduction potential may vary depending not only
on the strain being tested but also on the inoculum size and the
metabolic state of the inoculated cells. Many anaerobes, including
C. perfringens, can maintain a low oxidation-reduction potential
during active growth by means of the reducing metabolites that are
produced.

Pour plates for colony counts of *C. perfringens* are incubated
under anaerobic conditions in which the air has been replaced by a
different gaseous environment. The most commonly used environment has been a mixture of 90% nitrogen and 10% carbon dioxide.
However, it has been shown that as little as 1% carbon dioxide may
cause a reduction in the viable count of control or irradiated spores
(Futter and Richardson 1970). Control spores grew equally well in
100% nitrogen as in 100% hydrogen but the recovery of irradiated
spores was three times greater in nitrogen than in hydrogen. These
results would indicate that 100% nitrogen is the preferred atmosphere for establishing anaerobic conditions for growth of *C.
perfringens*. In a broth culture, the atmopshere in the headspace is of
less importance. The generation time of eight strains in fluid
thioglycollate medium with an initial 100% nitrogen head space
varied from 12.9 to 16.9 min at 43°C and with an initial 100%
carbon dioxide head space from 12.9 to 17.2 min (Parekh and
Solberg 1970). Normally, broth cultures of *C. perfringens* are
incubated in air with a reducing agent incorporated into the medium.

Water Activity

Water activity (a_w) is an expression of the chemical potential of
water in a solution relative to that of pure water at the same
temperature and pressure. The a_w is numerically equal to the
corresponding relative humidity expressed as a fraction. Since the
metabolism of microorganisms requires an aqueous environment, for
each organism there exists an a_w level below which the cell cannot
properly function under the specified environmental conditions and
growth ceases. The effect of a_w on the growth of *C. perfringens*
varies with the controlling solute used to adjust the a_w, the pH,
temperature and oxidation-reduction potential of the growth
medium (Strong *et al.* 1970; Mead 1969). The limiting a_w for the
growth of *C. perfringens* was shown to be in the range of 0.95 to

0.96 at pH values ranging from 5.5 to 7.0 with glucose as the solute for a_w adjustment. When NaCl or KCl were used for a_w control the limiting a_w for growth was about 0.97. These a_w values compare to a common fluid thioglycollate broth used for growth of *C. perfringens* which has an a_w of 0.995. In growth media at high a_w levels, the effect of decreasing pH is more influential in limiting growth than in media at lower a_w levels. At a_w levels that are near or at the growth limiting level the effect of pH on the growth of *C. perfringens* is considerably diminished.

Although the optimum temperature for growth of *C. perfringens* is in the range of 45°C, when the a_w of the growth medium is decreased, better growth may be obtained at temperatures below optimum. For example, growth of several strains was better at 37°C than at 46°C at all a_w levels below 0.995 regardless of the pH of the medium (Strong *et al.* 1970). A lower temperature limit may also be reached below which higher a_w values become limiting for growth. Expressed in terms of per cent salt, the limiting concentration for growth of *C. perfringens* may vary from 2 to 4% at 20°C, 2 to 5% at 25° and 30°C, 4 to 6% at 37°C and about 4% at 45°C (Smith 1971; Mead 1969).

As the salt concentration is increased, resulting in increasingly lower a_w levels, the oxidation-reduction potential required for growth of *C. perfringens* is decreased (Mead 1969). This is of practical importance in the use of food packaging materials varying in their permeability to oxygen. Materials with low permeability may increase the inhibitory action of NaCl or other preservative agents above that obtained with oxygen impermeable packaging materials.

Obviously many variables are to be considered when investigating the water activity requirements of anaerobes whether it be in the academic research laboratory or in the food processing facility.

Effect of Refrigeration or Freezing on Viability

The viable count of populations of *C. perfringens* cells will decrease considerably following freezing or prolonged refrigerator storage. The decrease in count after exposure to low temperature may be as much as 2 to 3 logs. As would be expected, vegetative cells are more sensitive than are spores. The temperature of exposure may also affect the viable count after storage. A temperature of -5°C has been shown to be more harmful to the cell than -20°C (Barnes *et al.* 1963). Storage at 5°C for 10 days or longer was more lethal to both vegetative cells and spores than storage at -17.7°C (Strong *et al.* 1966). This is not unexpected, since slow lysis of the vegetative cells and germination of the spores may occur at 5°C. However, for shorter storage periods, storage at 5°C may be less deleterious than storage at -17.7°C. It is generally recommended that for short term holding of food samples that are to be tested quantitatively for *C. perfringens*, the samples be refrigerated rather than frozen.

Although vegetative cells of *C. perfringens* at all stages of the growth cycle are sensitive to cold shock, induced by squirting cells into a menstruum at 4°C, the most sensitive cells are those in midexponential growth phase.

SPORES

Sporulation

C. perfringens is notorious for the difficulty one has in obtaining a satisfactory degree of sporulation. Good growth of this organism requires the presence of a readily fermentable carbohydrate. Glucose is used in most growth media to satisfy this requirement. However, under these conditions sporulation and probably sporulation specific activities are repressed. Many media have been reported for sporulation of *C. perfringens*. In some of these a fermentable carbohydrate is not incorporated, and thus very low yield occurs. In others abnormal growth and sporulation may occur due to the high concentration of buffer salts employed. One of the more satisfactory of the reported media is the DS medium (Duncan and Strong 1968). The carbohydrate used in this medium is soluble starch. With starch as the carbohydrate source, the growth rate is restricted, and the enzymes required for sporulation are apparently derepressed resulting in the formation of mature heat resistant spores.

The frequency of sporulation obtained may vary tremendously from strain to strain. In DS medium the per cent sporulation may range from less than 1% to greater than 90% with different strains. Mature, heat-resistant spores of this organism are produced within 10 hr in this medium at 37°C. Continued incubation for 24 hr or longer may result in recycling of many of the spores due to their germination and outgrowth.

Germination Requirements

Germination of *C. perfringens* spores results in the conversion of the resistant and dormant spore into a sensitive and metabolically active form. A variety of amino acids, ribosides or carbohydrates or combinations of these are known to germinate bacterial spores. Generally speaking, the clostridial spores do not require nucleosides for germination. *C. perfringens* spores will germinate in a wide variety of complex media. They may or may not require heat activation for germination. Optimal activation of heat-sensitive spores has been shown to occur by heating at 60° to 70°C for 20 min; whereas, heating at 80°C for 10 or 20 min was optimal for heat resistant spores (Duncan and Strong 1968). Some heat sensitive spores of this organism will germinate without heat shock. When attempting to quantitate spores of *C. perfringens* of unknown heat sensitivity, whether they be in a food or bacteriological culture medium, a heat activation treatment of 20 min at 75°C may be used to obtain an estimate of the spore population.

The optimum germination requirements for one heat-resistant strain of *C. perfringens* were heat activation for 20 min at 75°C, followed by incubation in the presence of glucose, cystine and sodium chloride at pH 6.0 and a temperature of 30°C (Ahmed and Walker 1971). Obviously, these germination requirements may not hold true for spores of a different strain.

Heat Resistance

Spores of different strains of *C. perfringens* may vary considerably in their thermal resistance. Some spores may survive up to 6 hr at 100°C, while heat sensitive spores may survive only a few minutes at this temperature. Quantitatively, the D_{90} °C value for spores of a selected group of heat sensitive spore producing strains was 3 to 5 min; the D_{90} °C value of heat resistant spores was 15 to 145 min (Roberts 1968). This difference in spore heat resistance of various strains was at one time used as a major characteristic to distinguish between food poisoning and non-food poisoning strains of *C. perfringens*. However, it is now known that both heat-sensitive and heat-resistant spore-producing strains may cause food poisoning. At the present time, the heat resistance difference is important from the standpoint of the ability of the spores to survive the cooking process to which foods may be subjected.

Lysozyme Dependent Germination.—Normal physiological germination of the bacterial spore is mediated via a spore-associated lytic enzyme that apparently disrupts the spore cortex by attacking and disrupting the mucopeptide polymer present in the cortical layer of the spore. When spores of *C. perfringens* are heat-injured, the activity of the lytic enzyme is reduced by either specific inactivation of the enzyme or inactivation of some mechanism functioning in its release from a binding site. These spores will not germinate unless a lytic enzyme such as lysozyme is present in the germination medium. In the presence of lysozyme, the heat-injured spores germinate normally (Cassier and Sebald 1969; Duncan *et al.* 1972). This phenomenon is referred to as lysozyme-dependent germination. The phenomenon is very important from a practical standpoint, and is easily demonstrated using heat sensitive spores of *C. perfringens*. If such spores are heat injured just to the point that viable colonies cannot be obtained following plating of the injured spores in normal recovery media, germination and outgrowth of the "non-viable" spores will occur if as little as 1 μg of lysozyme per ml is added to the recovery medium.

The proportion of lysozyme-dependent spores in a population of viable spores depends on the intensity of the heat treatment to which they are subjected; it may be as little as zero for intact spores or up to 100% for heat-sensitive spores that have been heated for 5 min at 95°C (Cassier and Sebald 1969). Heat-resistant spores of *C. perfringens* may also be rendered lysozyme-dependent. However, a critical time-temperature relationship exists in heat injuring the spores sufficiently to produce a dependence on lysozyme for germination without also inactivating enzymes that are associated with outgrowth and that are located in the core of the spore.

Inactivation of the lytic system is an initial manifestation of heat injury in spores of *C. perfringens* and probably spores of other organisms. Precaution must be taken in deciding whether spores of *C. perfringens* that may be present in a food product have been heated sufficiently to cause them to be non-viable. Obviously, if lysozyme or some other lytic-like enzyme is present, the "apparently dead" spores will germinate and grow with little difficulty. The inclusion of 1 to 2 μg of lysozyme per ml of recovery medium is most advantageous in determining viability or non-viability of heated perfringens spores, regardless of whether the spores are from a broth culture or a food sample. This concentration of lysozyme will not affect vegetative growth of *C. perfringens*.

Curing Salts

Sodium nitrite plays an important role in the preservation of cured meats; yet, the actual mechanism by which it prevents spoilage is obscure. Nitrite concentrations far in excess of the 0.02% level commercially acceptable in foods will cause germination of *C. perfringens* spores at a low pH. However, this activity of nitrite is apparently unrelated to its preservative action. In preventing growth from spores in canned cured meats, nitrate apparently functions by preventing outgrowth of the germinated, heat-injured spores. Under laboratory conditions, concentrations of 0.02 and 0.01% sodium nitrate at pH 6.0 inhibited outgrowth of heat damaged *C. perfringens* spores of a heat sensitive and a heat resistant strain, respectively (Labbe and Duncan 1970). Unheated spores required 0.04 and 0.02%, respectively.

The ability of nitrite to prevent growth of *C. perfringens* spores may depend on the nitrite concentration, the concentration of spores present, the degree of heat injury to which the spores were subjected, the heat resistance of the spores, the pH of the menstruum, incubation temperature, and the concentration of sodium chloride. Obviously, the same kind of variables may affect the ability of sodium chloride to inhibit growth of perfringens spores. The variables controlling the effectiveness of curing salts are many and the interactions are complex and certainly not completely understood. Experimental testing will provide the most reliable answer to whether or not perfringens spores will grow in a food system containing curing salts.

Radiation Resistance

C. perfringens spores may differ in their sensitivity to gamma radiation depending upon the substrate in which they are irradiated. For example, resistance has been shown to be greater in cooked meat broth than in phosphate buffer (Midure *et al.* 1965). Spores of heat-resistant strains may be more resistant to radiation than those of heat-sensitive strains. The D value of heat-resistant strains irradiated in aqueous suspension ranged from 0.26 to 0.34 Mrad, whereas those of heat-sensitive strains ranged from 0.12 to 0.21 Mrad (Roberts 1968).

CLASSIFICATION

C. perfringens has been referred to by various binomials since it was first isolated. These include *Bacillus aerogenes capsulatus*, *Bacillus phlegmones emphysematosae*, *Bacillus emphysematosus*, *Bacillus perfringens*, *Bacillus welchii* and *Clostridium welchii*. *C. perfringens* is often referred to as *C. welchii* in medical literature as well as other current literature. However, *C. perfringens* is presently accepted as the valid binomial.

Toxigenic Types

C. perfringens strains may be divided into five types, designated A, B, C, D, and E, on the basis of their ability to produce four major lethal toxins. These toxins are alpha, beta, epsilon and iota. Of these four toxins, type A produces only alpha, type B produces alpha, beta and epsilon, type C produces alpha and beta, type D produces alpha and epsilon, and type E produces alpha and iota toxins. The toxigenic type may easily be determined by testing the mouse lethality of culture filtrates in the presence of commercially available type-specific antiserum. Several other soluble antigens are produced by the various types of *C. perfringens*. These are also designated as toxins and when necessary may be used to subtype the principal types described above (Smith and Holdeman 1968; Ispolatovskaya 1971; Hauschild 1971A). A sixth type, designated type F, is no longer recognized. Instead, strains that were formerly in that type are now considered as type C.

At the present time, it is not known if the enterotoxin active in human food poisoning that is produced by certain type A and type C strains is specific only for these strains or if other types may also produce this toxin.

In North America, only type A has been associated with outbreaks of human food poisoning. Type C strains have been the cause of a very severe and sometimes lethal food poisoning. However, type C food poisoning has been reported in only a few countries, including Germany, New Guinea, Australia and Uganda.

The toxigenic types of *C. perfringens* have a rather limited host range. Type B is associated primarily with lamb dysentery, type C with enterotoxemia of sheep, calves and piglets, type D with pulpy kidney disease of sheep, and type E is occasionally found as a saprophyte in the calf intestine.

Serological Types

The number of different serological types of *C. perfringens* may be as great as several hundred. This diversification creates a problem in serotyping this organism unless a large battery of antisera are available. Such antisera currently are not commercially available. This organism has been serotyped in England, the United States, Japan and Australia. The range of antisera used in England and Australia are very similar, but those in the other two countries differ from each other and from that used in England. There is an obvious need for standardization of the different serotypes.

Antisera are prepared in rabbits by the injection of whole, formalinized cells. Serotyping is done using the slide agglutination technique.

At the present time, no known correlation exists between a specific serotype and its ability to synthesize the enterotoxin associated with human food poisoning.

ECOLOGICAL DISTRIBUTION

C. perfringens is a ubiquitous organism and is perhaps more widely distributed in nature than is any other pathogenic bacterium. Type A is the more prevalent type and may be easily isolated from soil or the intestinal contents of man and animals. Types B, C, D, and E are primarily associated with the intestinal contents of animals and apparently do not survive long in soil. The human carrier rate for heat-sensitive strains of type A is 100%. The actual number of heat sensitive types in the normal human feces may range from about 10^1 to 10^9 per g of fecal material. The concentration of cells in the human feces as well as the serological types may vary from week to week, depending on the quality and quantity of food and drink consumed as well as the general health of the individual (Sutton 1966B). The incidence of heat-resistant type A strains in the intestinal contents is much lower than the heat-sensitive types. The carrier rate may vary from 0% to as much as 88% (Akama and Otani 1970). The incidence of heat-resistant type A strains in the general population is usually low, but persons associated with communal feeding and poor hygienic conditions invariably have a higher carrier rate (Sutton 1966A).

Following an attack of perfringens food poisoning, the concentration of both vegetative cells and spores of this organism increases in the intestinal contents. In an experiment in which human beings were fed known concentrations of different strains of *C. perfringens* type A, an increase in the vegetative and spore concentrations in the feces occurred within 24 hr following ingestion of the cells (Strong *et al.* 1971). This increase occurred whether or not the individuals had food poisoning symptoms and whether an enterotoxin or nonenterotoxin producing strain was fed. Obviously, non-food poisoning strains can multiply in the intestinal tract as easily as can food poisoning strains.

The common occurrence of this organism in the soil results in it being associated with dust as well as being air-borne. Thus, the organism may have ready access to food preparation surfaces.

C. perfringens type A has been isolated from a great variety of foods. Its incidence in American foods has been studied by several laboratories. The reported incidence of the organism is invariably higher if an enrichment procedure is used in isolating the organism than if a direct quantitation of a food is made. The incidence of *C. perfringens* in a variety of foods examined in Wisconsin, Ohio and Montana (Table 5.1) shows that the organism is normally present in our food supply. The incidence is routinely higher in meat and poultry products than in other foods. Various commercially prepared dehydrated sauce and gravy mixes also are contaminated with *C. perfringens*. Some of these products require heating for less than 10 min before serving. If such foods are allowed to cool and are left at a temperature suitable for growth of *C. perfringens* before serving, they may become potential sources of food poisoning.

TABLE 5.1
INCIDENCE OF *CLOSTRIDIUM PERFRINGENS* IN AMERICAN FOODS

	No. of samples tested	Per cent positive for *C. perfringens*
Determined by Direct Quantitation:		
Commercially prepared frozen foods	111	2.7
Raw fruits and vegetables	52	3.8
Spices	60	5.0
Home-prepared foods	165	1.8
Meat, poultry, and fish	122	16.4
Determined Following Enrichment:		
Veal	17	82.0
Beef	50	70.0
Chicken	26	58.0
Lamb	27	52.0
Pork	41	37.0
Spaghetti sauce and mixes	13	53.8
Sauce and gravy mixes	8	12.5
Soup mixes	28	3.6
Cheese and cheese sauce	6	16.7

From Strong *et al.* (1963), Hall and Angelotti (1965), and Nakamura and Kelly (1968).

ISOLATION AND QUANTITATION

The mere isolation of low numbers of *C. perfringens* from a food product does not necessarily mean that a food poisoning hazard exists if the food is consumed. As discussed in the preceeding section, this organism is normally present in our food supply. For a food poisoning hazard to exist, the organism must multiply to relatively high levels and be injested. The exact concentration of organisms required to induce food poisoning is unknown and in fact may vary from strain to strain. However, it may be useful to monitor the population level of this organism in foods that are known to harbor the organism and that may be subject to mishandling before human consumption.

Demonstration of large numbers, usually hundreds of thousands or more per gram, of *C. perfringens* in a suspected food poisoning vehicle supports a diagnosis of perfringens poisoning based on clinical and epidemiological evidence. Further support is provided if the same serotype of the organism is recovered from the food and the stools of patients involved.

A variety of selective, differential media have been devised for the isolation and quantitation of *C. perfringens*. The medium of choice may vary in different countries as well as within a given country. In England, a commonly used medium is horse blood agar which is made selective by spreading a solution of the antibiotic neomycin sulfate over the surface of the agar. Some strains, but not all, produce hemolysis on this blood agar. The presence of C. perfringens colonies must be confirmed by other differential tests.

In the U.S., the media commonly used for the quantitation of this organism make use of its ability to produce hydrogen sulfide from sulfite. Colonies of the organism are produced in a medium containing sulfite and an iron salt. The sulfide formed precipitates as black iron sulfide resulting in typically black C. perfringens colonies. The various media are made selective by the incorporation of antibiotics which inhibit many sulfite reducing as well as non-sulfite reducing anaerobes or facultative anaerobes. These media include sulfite-polymyxin-sulfadiazine (SPS) agar which contains the antibiotics polymyxin B and sulfadiazine (Angelotti et al. 1962), tryptone-sulfite-neomycin (TSN) agar containing the antibiotics polymyxin B and neomycin (Marshall et al. 1965), Shahidi-Ferguson-perfringens (SFP) agar containing kanamycin and polymyxin B (Shahidi and Ferguson 1971), and tryptose-sulfite-cycloserine (TSC) containing D-cycloserine (Harmon et al. 1971). The TSN agar is incubated at 46°C which limits the growth of other sulfite-reducing clostridia, including C. bifermentans which grows equally as well as C. perfringens in the various media. The other selective, differential media are incubated at 35° to 37°C. All media are incubated anaerobically. The diluent of choice in quantitation of C. perfringens is 0.1% peptone in distilled water. Phosphate buffer or saline diluents may result in decreased viable counts as compared to 0.1% peptone water.

In addition to the differential characteristic of black colony formation, some of the media (SFP and TSC) contain egg yolk which allows the detection of lecithinase (alpha toxin) production by C. perfringens strains. On these media the black colonies are surrounded by a white zone of opaque precipitate resulting from lecithinase activity. Egg yolk free TSC medium has been used successfully for quantitation (Hauschild and Hilsheimer 1974).

Spores of C. perfringens may be quantitated using the same plating media as used for vegetative cells. Spore containing samples are heated (usually 75°C for 20 min) to inactivate vegetative cells and to heat shock spores that may have this requirement for germination. The spore population quantitated by this procedure is only an estimate unless the exact heat shock requirements are known.

Presumptive C. perfringens colonies on any of these media should be confirmed by subsequent tests. Tests that are routinely used are motility, nitrate reduction and lactose fermentation.

Some of the selective media are more inhibitory to strains of C.

perfringens than are others and may not allow quantitative recovery of all strains, but they will prevent growth of many unwanted organisms. The SFP agar is the least selective of all the above mentioned media and may allow growth to some extent of a great many facultative anaerobes. In choosing a medium for isolation and quantitation of this organism, the expected ratio of *C. perfringens* cells to other organisms should be considered. When the perfringens level is high, a medium with low selectivity may be used; however, when the perfringens level is low it may be desirable to use a more selective medium that will inhibit unwanted organisms even at the risk of lower recovery of *C. perfringens*.

It may be desirable to enrich for this organism if the cell population is too low for direct isolation by plating procedures. The organism may be readily enriched in a glucose-containing broth such as fluid thioglycollate broth incubated at a temperature of 46°C. Growth of the organism from very low cell levels may be apparent in as little as 4 hr at this temperature.

TYPE A FOOD POISONING

As early as 1945, McClung described an outbreak of human food poisoning in the U.S. resulting from the ingestion of boiled chicken contaminated with *C. perfringens*. However, it was not until 1959 that official epidemiological reports were received of food poisoning outbreaks in which *C. perfringens* type A was the etiological agent. Belated recognition was at last made of a food poisoning organism that had been well established as a food poisoner in England since the early fifties. During the period from 1959 to 1964, there were relatively few reported outbreaks of perfringens food poisoning in the U.S. After 1964, the number of reported outbreaks began to increase substantially each year, with 29 outbreaks being reported in 1967, 56 in 1968, 65 in 1969, and 53 in 1970. During 1968, 1969, and 1970, *C. perfringens* ranked second in frequency as the etiological agent in outbreaks of foodborne disease. However, in these same time periods, this organism was responsible for more individual cases of foodborne disease (34, 64.9, and 29.7%, respectively) than any other agent.

The increase in the number of reported outbreaks in the late sixties resulted from an increased recognition of the organism as a food poisoning agent. Once public health laboratories became aware of the potential of *C. perfringens* to cause food poisoning, the organism was looked for and found to be the causative agent in many outbreaks that previously would have been considered as cause unknown. It is still not known how many outbreaks occur in the

U.S. each year that involve *C. perfringens*. In the future, this organism may be found to surpass both *Staphylococcus* and *Salmonella* as the leading causative agent of foodborne illness.

Conditions Leading to Outbreaks

Perfringens food poisoning may occur when the organism grows in a food product prior to consumption. The organism may multiply over the broad temperature range of about 15° to 50°C and over a pH range of about 5.5 to 8.0. Under optimum conditions, large numbers of cells may be produced in 2 to 3 hr. Therefore, the danger of this type of food poisoning lies in keeping food at a temperature that is apparently neither hot nor cold but is one at which growth of the organism will occur. This hazard may often be present in mass feeding establishments that rely on steam tables or cooling tables for maintaining foods at "safe" temperatures. The hazard may be even greater if foods have been improperly cooled or inadequately reheated before placing on the steam or cold table.

Place of Acquisition.—In the U.S., it has been estimated that 400,000 food service establishments serve about 100,000,000 meals daily. Eighty-two per cent of these meals are served in public eating establishments; the remainder are served in institutions. The great majority of perfringens food poisoning incidents that are reported by health authorities occur in food service establishments that are concerned with mass feeding. For this reason, large numbers of persons are usually associated with perfringens food poisoning outbreaks. Although restaurants normally account for the largest number of outbreaks, the largest number of cases are usually associated with banquets or cafeterias where large numbers of persons are eating a common meal. This common association facilitates recognition of a food poisoning outbreak when it occurs. However, with a restaurant or other commercial establishment that serves a transient customer, food poisoning incidents may go undetected. Thus, the number of outbreaks occurring in restaurants and caused by *C. perfringens* may be substantially higher than is now reported. The same may be true for outbreaks that occur in the home.

Foods Involved.—Beef or beef-containing dishes are the most common vehicles associated with outbreaks of perfringens food poisoning. Poultry products, including turkey and chicken, are the second most common vehicles. Other meats such as pork and lamb have been the vehicles in outbreaks, though to a much lesser extent than beef and poultry products. A variety of other foods have been incriminated in this type of food poisoning including fish, shrimp, crab, beans, potato salad, macaroni and cheese, and olives.

In many of the outbreaks associated with beef or poultry products, the foods often were combined with gravy or dressing or a combination of the two. These are the types of foods commonly served in mass feeding establishments. They generally require low temperature cooking and may be cooked in advance and reheated prior to serving. Improper cooling of such foods after the initial cooking may allow growth of C. perfringens spores that survived the heat treatment. Also, the reheating may not be sufficient to inactivate vegetative cells or spores which may be present, and the temperature may not be maintained high enough to prevent subsequent multiplication of the cells during the actual serving of the food. Such mishandling of foods may be disastrous to those in charge of the food service establishment. Aside from the individual loss, such establishments may suffer from loss of public confidence and even legal action.

Incubation, Duration and Symptoms

The symptoms of perfringens poisoning usually appear after an incubation period of 6 to 22 hr. The symptoms may appear earlier than this and, in experimentally induced food poisoning, diarrhea was reported within 2.5 hr following ingestion of sporulating viable cells of a food poisoning strain (Strong et al. 1971). The cells used in this feeding experiment had been grown in a sporulation medium and were approaching the stage of sporulation at which enterotoxin would be released from the cells. The incubation period of this type of poisoning compares to 12 to 24 hr for Salmonella and 2 to 6 hr for Staphylococcus food poisoning.

The normal duration of perfringens poisoning is about 12 to 24 hr, whereas Salmonella may last from 1 to 14 days and staphylococcal poisoning from 6 to 24 hr.

The clinical illness of perfringens food poisoning is characterized primarily by diarrhea and abdominal cramps. Most patients have acute abdominal pain, while about one-third are affected by nausea and headache. Vomiting, which is common in Staphylococcus food poisoning is infrequent though it may occur. Vomiting was induced in human and monkey subjects experimentally fed enterotoxin-containing preparations (Strong et al. 1971; Duncan and Strong 1971; Hauschild et al. 1971). Although the toxin may induce emesis, in normal cases of food poisoning it probably does not come into contact with the stomach since the poisoning is induced by ingesting viable cells. This may explain the low incidence of vomiting reported in most outbreaks. Fever, which is common in Salmonella food poisoning, rarely occurs in patients suffering from perfringens food poisoning. In fact, the clinical and epidemiologic pattern of perfringens food poisoning is sufficiently characteristic as to be nearly diagnostic.

Although death from perfringens type A food poisoning is rare, several cases have been reported in other countries. These usually have been associated with elderly or debilitated persons. The lethal potential of this organism should not be underestimated, especially in outbreaks that may occur in hospitals or nursing home type facilities.

Perfringens Enterotoxin

For several years after *C. perfringens* became accepted as a food poisoning organism, the mechanism by which the illness was produced was not clear. The requirement that viable cells be ingested for production of symptoms in humans suggested that the mode of action was that of infection. However, the clinical symptoms themselves, such as the lack of fever, high primary attack rate, lack of secondary person to person spread and lack of clinical immunity suggested an intoxication. It is now known that the enterotoxin responsible for the symptoms of perfringens food poisoning is produced in the intestinal tract following ingestion of large numbers of viable cells. The cells undergo multiplication and sporulation in the intestine and release the toxin on lysis of the sporangia to release the free bacterial spore. Thus, this type of food poisoning may be considered a "special" kind of infection in that bacterial invasion of the tissue with the corresponding host responses does not occur; yet, multiplication of the organism in the intestine followed by elaboration of the enterotoxin does occur.

The enterotoxin was first discovered in 1969 on the basis of its ability to induce fluid accumulation in ileal loops of the ligated rabbit intestine (Duncan and Strong 1969). Such an ileal loop test has been used in both rabbits and in lambs to distinguish between enterotoxin and nonenterotoxin producing strains and to assay for the enterotoxin.

Biological Characteristics.—*C. perfringens* type A enterotoxin will induce fluid accumulation in ligated ileal loops of rabbits, lambs, and cattle, diarrhea in these same species, and diarrhea and vomiting in humans and monkeys. It also is lethal in sufficient concentration, induces erythema without necrosis in the skins of guinea pigs and rabbits (Hauschild 1970; Stark and Duncan 1971), and induces a transient increase in capillary permeability in these animal skins following intradermal injection (Stark and Duncan 1972). The specific toxicity of purified toxin has been reported as about 2000 mouse MLD/mg N. (Hauschild and Hilsheimer 1971).

The enterotoxin is antigenic and may be completely neutralized *in vitro* by antiserum prepared in rabbits. Although the toxin is completely neutralized with antiserum *in vitro*, circulating antibody has little or no neutralizing effect on the enterotoxin in the intestinal lumen (Hauschild 1971B). Once the toxin passes from the lumen into the blood stream it may be neutralized by circulating antibody.

Physical and Chemical Properties.—The enterotoxin is heat-sensitive and loses activity when exposed to extremes of pH. Heating a crude toxin preparation for as little as 10 min at 60°C results in loss of biological activity. The enterotoxin is a protein that may be inactivated by the enzyme pronase but not trypsin, chymotrypsin or papain (Duncan and Strong 1969; Hauschild and Hilsheimer 1971). A purified preparation of enterotoxin, apparently free of nucleic acids, lipids or reducing sugars, had a molecular weight of about 36,000, a Strokes radius of 2.6 mμ and an isoelectric point of pH 4.3 (Hauschild and Hilsheimer 1971).

Relationship of Enterotoxin Formation to Sporulation.—The ability of *C. perfringens* type A to produce enterotoxin is directly related to the ability of the organism to sporulate. Genetic studies using mutants with an altered ability to sporulate and revertants of these mutants have shown that the toxin is a unique sporulation-specific gene product (Duncan *et al.* 1972).

Good growth of *C. perfringens* requires the presence of a readily fermentable carbohydrate. Glucose is used in most growth media to satisfy this requirement. However, under these conditions the sporulation and probably sporulation-specific activities are repressed. Early unsuccessful attempts to demonstrate the presence of a cell-free toxin active in perfringens food poisoning were made using such growth media. The toxin was first detected both intracellularly and extracellularly following growth and sporulation of the organism in DS sporulation medium (Duncan and Strong 1969). The carbohydrate used in this medium is soluble starch. With starch as the carbon source, the growth rate is restricted, the enzymes required for sporulation are apparently derepressed resulting in the formation of spores, and enterotoxin is produced. The enterotoxin can be detected neither extracellularly nor intracellularly in a growth medium in which sporulation is repressed. For this reason, the ability of an unknown strain to produce enterotoxin can not be determined *in vitro* unless sporulation of the organism occurs.

The toxin is synthesized by the sporulating cell in association with a late stage of sporulation (Duncan *et al.* 1972). The toxin concentration inside the sporulating cell reaches a peak just before lysis of the sporangium occurs, and is released from the sporangium with its lysis concomitantly with the mature spore release. Toxin cannot be detected in young culture filtrates of sporulating cultures but may be detected after about 10 hr. The enterotoxin appears to be a spore structural protein that is covalently associated with the spore coat (Frieben and Duncan 1973). It may be extracted from spore coats of different strains in more than one molecular weight form.

The ability of only sporulating cells to produce enterotoxin is consistent with the fact that the organism is known to sporulate readily in the intestine and under such conditions would synthesize and release the biologically active toxin. It is not known if under some conditions cell-free toxin could be produced in a food product. Normally, the organism sporulates poorly or not at all in most foods. However, the possibility cannot be ruled out that if *C. perfringens* sporulates in a food product the toxin could be ingested preformed and induce the symptoms of food poisoning.

Among the nonenterotoxin-producing strains are those that sporulate equally as well as the enterotoxin-producing strains. Intracellular accumulation of perfringens enterotoxin in only certain strains may result from loosely regulated synthesis of the toxin protein, whereas in the nonenterotoxin-producing strains, tight regulation of toxin synthesis may prevent enterotoxin detection.

Detection and Quantitation of Enterotoxin.—Enterotoxin may be readily detected by its ability to induce erythema in the skin of guinea pigs or rabbits or by the ligated intestinal loop technique in rabbits and lambs. The skin test is more rapid, sensitive and accurate than the loop technique and therefore superior as a quantitative assay method. When working with preparations of unknown composition or potency, it is most important that the proper antiserum controls be used in these assay procedures to prevent interference from other known perfringens toxins.

The toxin may also be quantitated by electroimmunodiffusion of toxin-containing preparations in flat agar slabs containing specific anti-enterotoxin serum, by counter current immunoelectrophoresis, and by reversed passive hemmaglutination.

TYPE C FOOD POISONING

Strains of *C. perfringens* type C may cause a very serious and frequently lethal food poisoning. This type of food poisoning has not been reported in North America. Two types of the disease have been reported in other countries. One referred to as pig-bel occurs frequently among the Melanesians in the New Guinea highlands and is associated with the traditional pig-feasting that occurs in that area.

The source of the type C strains associated with pig-bel outbreaks is unknown, but the outbreaks apparently result from the unsanitary and crude cooking and handling conditions under which the pigs are prepared and consumed by the natives (Murrell *et al.* 1966). The other disease has been called darmbrand. It is relatively rare, yet it reached pandemic levels in Germany in 1947 and 1948 (Murrell *et al.* 1966). The outbreaks resulted from consumption of canned meat contaminated with heat resistant type C spores.

Type C food poisoning is characterized by severe abdominal cramps, vomiting, diarrhea, acute inflammation of the small intestine with areas of necrosis and gangrene and by a high mortality rate. The principal enterotoxemic agent of *C. perfringens* type C is thought to be beta toxin. The enterotoxin associated with type A strains has also been demonstrated in type C strains isolated from cases of pig-bel (Skjelkvålé and Duncan 1975). The role of enterotoxin in type C poisoning is uncertain.

CONTROL OF PERFRINGENS FOOD POISONING

With the evident contamination of foods with *C. perfringens* and with the ubiquitous nature of the organism, it would be difficult if not impossible to rely on complete elimination of the organism from food as a control procedure. Meat and poultry may be contaminated with this organism during slaughtering and processing operations. This contamination may be carried over to the food preparation equipment and be spread by personnel handling the food and equipment. Foods may also be contaminated by dirty equipment, dust, insects, etc. The ease by which contamination may occur emphasizes the need for proper personal hygiene and stringent sanitary processing techniques to limit the level of *C. perfringens* contamination in foods.

Since spores of some strains of this organism are very heat resistant, it may be expected that cooked food may contain surviving spores. Even heat-sensitive spores will survive many cooking processes, especially those that involve slow roasting, baking or cooking times at low temperatures. This has been amply demonstrated in experiments designed to reproduce normal cookery procedures. Therefore, it is impossible to heat all foods sufficiently to inactivate all *C. perfringens* spores without making the food organoleptically undesirable. Ultimate control lies in maintaining cooked foods at temperatures that will prevent spore germination and/or multiplication of the vegetative cells. This also is true of foods that have become contaminated subsequent to cooking.

In general, to prevent perfringens food poisoning, foods should be either cooked and immediately eaten hot, or cooled rapidly and refrigerated within 1 to 1.5 hr until required. The foods should never be allowed to cool to room temperature before refrigeration as rapid bacterial growth may occur during this slow cooling period. Refrigeration may not be effective if the temperature of the cooling unit is too high, the unit too small, or the containers used to store the foods too large to allow rapid chilling. Shallow, sanitized containers should be used for bulky food. Prechilling of the foods by ice bath or other such means prior to refrigeration may be necessary.

Partial cooking of foods on one day with subsequent reheating the next day should be avoided. If this is impossible, the food should be boiled or re-cooked thoroughly. The practice of holding foods such as warm meat with natural juices or gravy in a warming oven prior to serving should be avoided unless the temperature is maintained above $60°C$ ($140°F$). If cooked foods are to be kept or served cold, they should be maintained at a temperature no higher than $7°C$ (about $45°F$).

Raw foods such as meat and poultry should be kept separate from cooked foods and special attention should be used to insure that different surfaces and equipment are used for processing these foods. Personnel involved in handling raw foods should thoroughly cleanse their hands and if possible wear plastic disposable gloves before handling cooked foods. Common equipment that may be used to process both raw and cooked foods should be thoroughly washed and sanitized after contact with the raw food.

A particular problem exists in cooking large carcasses of poultry that during cooking may not reach an internal temperature sufficient for lethality of spores and many vegetative cells of *C. perfringens*. It has been recommended that turkeys should be cooked until the internal breast temperature reaches at least $73.8°C$ ($165°F$), preferably higher (Bryan *et al.* 1971).

There exists on the American market a great variety of convenience foods. Many of these are pre-cooked and are either frozen, refrigerated, or are hot. In the hands of the consumer they may require only warming or a short cooking time before use. Another convenience food service is that of vended foods. These are now widely used in schools, colleges, hospitals, factories and other such places. Abuse of such foods, especially those containing meat or poultry products by either the producer or the consumer may result in an expected encounter with perfringens food poisoning.

BIBLIOGRAPHY

AHMED, M., and WALKER, H. W. 1971. Germination of spores of *Clostridium perfringens*. J. Milk Food Technol. *34*, 378–384.

CLOSTRIDIUM PERFRINGENS
195

AKAMA, K., and OTANI, S. 1970. *Clostridium perfringens* as the flora in the intestine of healthy persons. Japan. J. Med. Sci. Biol. *23*, 161-175.

ANGELOTTI, R., HALL, H. E., FOTER, M. J., and LEWIS, K. H. 1962. Quantitation of *Clostridium perfringens* in foods. Appl. Microbiol. *10*, 193-199.

BARNES, E. M., DESPAUL, J. E., and INGRAM, M. 1963. The behaviour of a food poisoning strain of *Clostridium welchii* in beef. J. Appl. Bacteriol. *26*, 415-427.

BARNES, E. M., and INGRAM, M. 1956. The effect of redox potential on the growth of *Clostridium welchii* strains isolated from horse muscle. J. Appl. Bacteriol. *19*, 117-128.

BROOKS, M. E., STERNE, M., and WARRACK, G. H. 1957. A reassessment of the criteria used for type differentiation of *Clostridium perfringens* types. J. Pathol. Bacteriol. *74*, 185-195.

BRYAN, F. L., McKINLEY, T. W., and MIXON, B. 1971. Use of time-temperature evaluations in detecting the responsible vehicle and contributing factors of foodborne disease outbreaks. J. Milk Food Technol. *34*, 576-582.

BUSTA, F. F., and SCHRODER, D. J. 1971. Effect of soy proteins on the growth of *Clostridium perfringens*. Appl. Microbiol. *22*, 177-183.

CASSIER, M., and SEBALD, M. 1969. Lysozyme-dependent germination of spores of *Clostridium perfringens* ATCC 3624 after heat treatment. Ann. Inst. Pasteur *117*, 312-324. (French)

DARBY, G. K., JONES, A. S., KENNEDY, J. F., and WALKER, R. T. 1970. Isolation and analysis of the nucleic acids and polysaccharides from *Clostridium welchii*. J. Bacteriol. *103*, 159-165.

DUNCAN, C. L., LABBE, R. G., and REICH, R. R. 1972. Germination of heat and alkali altered spores of *Clostridium perfringens* type A by lysozyme and an initiation protein. J. Bacteriol. *109*, 550-559.

DUNCAN, C. L., and STRONG, D. H. 1968. Improved medium for sporulation of *Clostridium perfringens*. Appl. Microbiol. *16*, 82-89.

DUNCAN, C. L., and STRONG, D. H. 1969. Ileal loop fluid accumulation and production of diarrhea in rabbits by cell-free products of *Clostridium perfringens*. J. Bacteriol. *100*, 86-94.

DUNCAN, C. L., and STRONG, D. H. 1971. *Clostridium perfringens* type A food poisoning. I. Response of the rabbit ileum as an indication of enteropathogenicity of strains of *Clostridium perfringens* in monkeys. Infect. Immun. *3*, 167-170.

DUNCAN, C. L., STRONG, D. H., and SEBALD, M. 1972. Sporulation and enterotoxin production by mutants of *Clostridium perfringens*. J. Bacteriol. *110*, 378-391.

FREIBEN, W. R., and DUNCAN, C. L. 1973. Homology between enterotoxin protein and spore structural protein in *Clostridium perfringens* type A. European J. Biochem. *39*, 393-401.

FUCHS, A. R., and BONDE, G. J. 1957. The nutritional requirements of *Clostridium perfringens*. J. Gen. Microbiol. *16*, 317-329.

FUTTER, B. V., and RICHARDSON, G. 1970. Viability of clostridial spores and the requirements of damaged organisms. II. Gaseous environment and redox potentials. J. Appl. Bacteriol. *33*, 331-341.

GROVES, D. J., and GRONLUND, A. F. 1969A. Glucose degradation in *Clostridium perfringens* type A. J. Bacteriol. *100*, 1420-1423.

GROVES, D. J., and GRONLUND, A. F. 1969B. Carbohydrate transport in *Clostridium perfringens* type A. J. Bacteriol. *100*, 1256-1263.

HALL, H. E., and ANGELOTTI, R. 1965. *Clostridium perfringens* in meat and meat products. Appl. Microbiol. *13*, 352-357.

HANKE, M. E., and BAILEY, J. H. 1945. Oxidation-reduction potential requirements of *Cl. welchii* and other clostridia. Proc. Soc. Exptl. Biol. Med. *59*, 163-166.

HARMON, S. M., KAUTTER, D. A., and PEELER, J. T. 1971. Improved medium for enumeration of *Clostridium perfringens*. Appl. Microbiol. *22*, 688-692.

HAUSCHILD, A. H. W. 1970. Erythemal activity of the cellular enteropathogenic factor of *Clostridium perfringens* type A. Can. J. Microbiol. *16*, 651-654.

HAUSCHILD, A. H. W. 1971A. *Clostridium perfringens* toxins types B, C, D, and E. *In* Microbial Toxins, Vol. IIA, S. Kadis, T. C. Montie, and S. J. Ajl (Editors). Academic Press, New York.

HAUSCHILD, A. H. W. 1971B. *Clostridium perfringens* enterotoxin. J. Milk Food Technol. *34*, 596-599.

HAUSCHILD, A. H. W., and HILSHEIMER, R. 1971. Purification and characteristics of the enterotoxin of *Clostridium perfringens* type A. Can. J. Microbiol. *17*, 1425-1433.

HAUSCHILD, A. H. W., and HILSHEIMER, R. 1974. Enumeration of foodborne *Clostridium perfringens* in egg yolk-free tryptose-sulfite-cycloserine agar. Appl. Microbiol. *27*, 521-526.

HAUSCHILD, A. H. W., WALCROFT, M. J., and CAMPBELL, W. 1971. Emesis and diarrhea induced by enterotoxin of *Clostridium perfringens* type A in monkeys. Can. J. Microbiol. *17*, 1141-1143.

ISPOLATOVSKAYA, M. V. 1971. Type A *Clostridium perfringens* toxin. *In* Microbial Toxins, Vol. IIA, S. Kadis, T. C. Montie, and S. J. Ajl (Editors). Academic Press, New York.

KURTZMAN, R. H., and HALBROOK, W. U. 1970. Polysaccharide from dry navy beans, *Phaseolus vulgaris*: Its isolation and stimulation of *Clostridium perfringens*. Appl. Microbiol. *20*, 715-719.

LABBE, R. G., and DUNCAN, C. L. 1970. Growth from spores of *Clostridium perfringens* in the presence of sodium nitrite. Appl. Microbiol. *19*, 353-359.

MAHONY, D. E., and BUTLER, M. E. 1971. Bacteriocins of *Clostridium perfringens*. 1. Isolation and preliminary studies. Can. J. Microbiol. *17*, 1-6.

MAHONY, D. E., BUTLER, M. E., and LEWIS, R. G. 1971. Bacteriocins of *Clostridium perfringens*. 2. Studies on mode of action. Can. J. Microbiol. *17*, 1435-1442.

MARSHALL, R. S., STEENBERGEN, J. F., and McCLUNG, L. S. 1965. Rapid technique for the enumeration of *Clostridium perfringens*. Appl. Microbiol. *13*, 559-563.

MEAD, G. C. 1969. Combined effect of salt concentration and redox potential of the medium on the initiation of vegetative growth by *Clostridium welchii*. J. Appl. Bacteriol. *32*, 468-475.

MIDURA, T. F., KEMPE, L. L., GRAIKOSKI, J. T., and MILONE, N. A. 1965. Resistance of *Clostridium perfringens* type A spores to γ-radiation. Appl. Microbiol. *13*, 244-247.

MURRELL, T. G. C. *et al.* 1966. The ecology and epidemiology of the pig-bel syndrome in man in New Guinea. J. Hyg. *64*, 375-396.

NAKAMURA, M., and KELLY, K. D. 1968. *Clostridium perfringens* in dehydrated soups and sauces. J. Food Sci. *33*, 424-425.

PAREKH, K. G., and SOLBERG, M. 1970. Comparative growth of *Clostridium perfringens* in carbon dioxide and nitrogen atmospheres. J. Food Sci. *35*, 156-159.

RIHA, W. E., and SOLBERG, M. 1971. Chemically defined medium for the growth of *Clostridium perfringens*. Appl. Microbiol. *22*, 738-739.

ROBERTS, T. A. 1968. Heat and radiation resistance and activation of spores of *Clostridium welchii*. J. Appl. Bacteriol. *31*, 133-144.

SHAHIDI, S. A., and FERGUSON, A. R. 1971. New quantitative, qualitative, and confirmatory media for rapid analysis of food for *Clostridium perfringens*. Appl. Microbiol. *21*, 500-506.

SMITH, H. W. 1959. The bacteriophages of *Clostridium perfringens*. J. Gen. Microbiol. *21*, 622-630.

SMITH, L. Ds. 1971. Factors affecting the growth of *Clostridium perfringens. In* SOS/70 Proc. 3rd Intern. Cong. Food Sci. Technol., Inst. Food Technol., Chicago.

SMITH, L. Ds., and HOLDEMAN, L. V. 1968. The Pathogenic Anaerobic Bacteria. Charles C. Thomas, Springfield, Ill.

STARK, R. L., and DUNCAN, C. L. 1971. Biological characteristics of *Clostridium perfringens* type A enterotoxin. Infec. Immun. *4*, 89–96.

SKJELKVALE, R., and DUNCAN, C. L. 1975. Enterotoxin formation by different toxigenic types of *Clostridium perfringens.* Infect. Immun. *11*, 563–575.

STARK, R. L., and DUNCAN, C. L. 1972. Transient increase in capillary permeability induced by *Clostridium perfringens* type A enterotoxin. Infect. Immun. *5*, 147–150.

STRONG, D. H., CANADA, J. C., and GRIFFITHS, B. B. 1963. Incidence of *Clostridium perfringens* in American foods. Appl. Microbiol. *11*, 42–44.

STRONG, D. H., DUNCAN, C. L., and PERNA, G. 1971. *Clostridium perfringens* type A food poisoning. II. Response of the rabbit ileum as an indication of enteropathogenicity of strains of *Clostridium perfringens* in human beings. Infect. Immun. *3*, 171–178.

STRONG, D. H., FOSTER, E. F., DUNCAN, C. L. 1970. Influence of water activity on the growth of *Clostridium perfringens.* Appl. Microbiol. *19*, 980–987.

STRONG, D. H., WEISS, K. F., and HIGGINS, L. W. 1966. Survival of *Clostridium perfringens* in starch pastes. J. Am. Dietet. Assoc. *49*, 191–195.

SUTTON, R. G. A. 1966A. Distribution of heat-resistant *Clostridium welchii* in a rural area of Australia. J. Hyg. *64*, 65–74.

SUTTON, R. G. A. 1966B. Enumeration of *Clostridium welchii* in the faeces of varying sections of the human population. J. Hyg. *64*, 367–374.

TABATABAI, L. B., and WALKER, H. W. 1970. Oxidation-reduction potential and growth of *Clostridium perfringens* and *Pseudomonas fluorescens.* Appl. Microbiol. *20*, 441–446.

CHAPTER 6

J. Liston | # Vibrio Parahaemolyticus

V. parahaemolyticus is a Gram-negative organism of marine origin which causes gastroenteritis in individuals eating contaminated seafood. The organism is a major cause of bacterial gastroenteritis in Japan and possible in other S.E. Asian countries where consumption of raw or lightly processed fish and shellfish is common. In recent years it has been identified as a cause of food poisoning outbreaks in the U.S. related to the consumption of shellfish and crustaceans. *V. parahaemolyticus* has also been implicated in the U.S. in human septicaemias and soft tissue infections commonly due to wound contamination by seawater.

The organism was not recognized as a pathogen until recently because it is halophilic and does not grow on the media normally used in the investigation of food poisoning outbreaks. Moreover the symptoms in patients are similar to those of salmonellosis and shigellosis, leading frequently to arbitrary diagnosis of these diseases, even though no *Salmonella* or *Shigella* strains were isolated. *V. parahaemolyticus* is now generally recognized as a cause of food poisoning, and methods for its isolation and identification have been developed and should be applied in all cases of food poisoning in which seafoods or salted food products are suspected as the vehicle.

NOMENCLATURE

There is some confusion and considerable controversy in the early Japanese literature concerning the name of this organism, and recently proposals in the U.S. literature have again raised nomenclatural problems. Fujino *et al.* (1953), who originally isolated the organism from an outbreak of food poisoning resulting from the consumption of boiled salted sardines, called their isolates *Pasteurella parahaemolyticus*. Takikawa and Fujisawa (1956) independently isolated a similar organism from an outbreak involving salted cucumbers and called it *Pseudomonas enteritis*. Later Miyamoto *et al.* (1961) put both strains in a new genus *Oceanomonas* as *O. parahaemolytica* and *O. enteritis* and described a third species, *O. alginolytica*. Sakazaki *et al.* (1963) made definitive studies of a large number of strains and finally defined the food poisoning strains as *V. parahaemolyticus*. This nomenclature is now generally accepted. Miyamoto's third species, which is very similar to *V. parahaemolyticus* and has been identified as a biotype, is now known as

V. alginolyticus. Recently, Baumann and Baumann (1973) have proposed that these organisms should be placed in a redefined genus *Beneckea*, but this nomenclature is still in dispute.

BACTERIOLOGY

The confusion over nomenclature arises from the fact that *V. parahaemolyticus* is a member of a group of closely similar *Vibrio* types which are commonly present in inshore marine environments and occur frequently on fish and shellfish. All the organisms in the group are mildly halophilic, but they show different degrees of tolerance to salt in the range 0 to 10% and this, together with different temperature ranges for growth, provides some assistance in distinguishing *V. parahaemolyticus* from the other species. However, the difficulty of separating *V. parahaemolyticus* from other commonly occurring mesophilic vibrios in seafoods or other marine samples should not be underestimated, and it is usually necessary to complete a full series of tests to ensure accurate identification. Samples from stools of patients require less complex testing methods, since interfering vibrios are not usually present in them.

Morphology

V. parahaemolyticus is a facultatively anaerobic Gram-negative rod which rarely shows characteristic vibroid curvature in culture, but does typically produce round bodies after a few days' incubation. It is motile in liquid media by possession of a sheathed polar flagellum, but is reported also to produce unsheathed peritrichous flagella when grown on solid medium. Electron micrograph preparations of thin sections show it to possess the typical morphological features of a Gram-negative rod. Growth is rapid in most commonly used liquid and solid laboratory media containing 2-3% NaCl incubated at suitable temperatures. Colonies on solid media are usually smooth, moist, circular, and opaque, though rough colonies are also produced. Generally, they are indistinguishable from colonies of other common Gram-negative bacteria on nonselective media.

Physiology

The most outstanding physiological feature of this organism is its requirement for salt (Table 6.1). Sakazaki *et al.* (1961) have reported requirements for Na^+, K^+, Mg^{++}, and $PO_4^=$. Growth is optimal in media containing 3% NaCl, and growth occurs over the range 0.5 to 9% NaCl. Some strains show slight growth at 10% NaCl. *V. parahaemolyticus* will grow on Blood Agar, MacConkey's, Wilson-Blair, Bismuth Sulfate, and Aronson's media without added salt.

TABLE 6.1
PHYSIOLOGICAL CHARACTERISTICS OF *V. PARAHAEMOLYTICUS*

	Minimum	Optimum	Maximum
NaCl concentration	0.5%	3%	9%
Temperature	8°C	37°C	44°C
pH	4.5	6.5-9.0	11
Oxygen		Facultative anaerobe	
A_w	<0.94	not known	

There is rapid inactivation of the organism in distilled water or buffered medium which does not contain salt, and all dilution or test media for the organism should therefore contain 2- 3% NaCl.

The temperature range for growth is usually found to be between 8° and 44°C. Some authors have reported growth to as low as 5°C, and this may indicate strain variation in this regard. Growth in foods may be more restricted. Thus Vanderzant and Nickelson (1972) reported no growth at 3, 7, and 19°C in shrimp, and studies in the author's laboratory have indicated no growth in oysters at 5, 8, or 11°C. From a practical point of view, it seems unlikely that significant growth of the organism will occur at temperatures in the normal range of storage of refrigerated foods (*i.e.* 0- 10°C).

The optimum temperature is 37°C. At this temperature the generation time is only 9–12 minutes, while at 25°C and 15°C the generation time is 27 minutes and 120 minutes respectively. Growth occurs rapidly in foods at ambient summer temperatures, as can be seen from the data in Table 6.2.

TABLE 6.2
GROWTH OF *V. PARAHAEMOLYTICUS* IN SEAFOODS

Seafood	Incubation Temperature (°C)	Count in 1000's per g of seafood after			
		0 hour	1 hour	3 hours	5 hours
Raw Shrimp	30	1.8	—	—	3,000
	20	3.0	11.0	120.0	—
Cooked Shrimp	30	1.3	—	—	8,400
	20	5.4	13.0	120.0	—
Crabmeat	30	0.72	4.2	63.0	—
	20	0.72	0.92	13.0	—

V. parahaemolyticus is very sensitive to low temperature, and cultures will die out if held in the refrigerator. Sensitivity appears to be greatest at temperatures close to 0°C and die-off in seafoods at least is less rapid at - 30°C than at - 20°C (Johnson and Liston 1973). In practice, therefore, chilling and holding food at chill temperatures is much more lethal to *V. parahaemolyticus* than freezing and frozen storage. Consequently it is not surprising that *V. parahaemolyticus* has been isolated sometimes in significant numbers from frozen foods (Liston 1973).

V. parahaemolyticus is extremely sensitive to heat and is rapidly destroyed at temperatures above 50°C. A decimal reduction time of less than 4 minutes has been reported at 53°C (Beuchat 1973). Thus pasteurization procedures can be expected to destroy this organism.

In common with other vibrios, *V. parahaemolyticus* shows a marked preference for alkaline conditions, and growth is optimal between pH 6.5 and 9.0. Growth has been reported up to pH 11.0. The minimum pH for growth lies between 4.5 and 5.0. Low pH conditions are apparently lethal for the organism, and Vanderzant and Nickelson (1972) have reported a rapid die-off in shrimp homogenate held at pH 5.0. This suggests that the organism is unlikely to present a serious problem in acid foods.

V. parahaemolyticus is a facultative anaerobe which grows well in the absence of air. Unfortunately there does not appear to be any information available on the minimum Eh which will permit growth of the organism.

Information is not available on the a_w limits for growth of *V. parahaemolyticus* as such. However, since most strains of the organism are unable to grow in simple liquid media containing 10% NaCl, it may be concluded that the minimum a_w for growth is close to 0.94.

Nutrition and Biochemistry

The biochemical characteristics of *V. parahaemolyticus* are quite typical of the *Vibrio* group. A wide range of substrates can be utilized for growth, and the organism is biochemically active, attacking both carbohydrate and protein materials. The range of characteristics is shown in Table 6.3, together with some of the definitive morphological and physiological properties of the organism. The biochemical characteristics are most important in the identification of the organism, and there is no evidence that they are of any practical significance in determining its growth in a particular foodstuff. Physiological characteristics seem to be more important here. Noteworthy biochemical characteristics are the positive oxidase reaction, fermentation of sugars to produce acid but not gas, chitin and starch hydrolysis, and sensitivity to the vibriostatic compound 0-129, all of which are characteristic of the halophilic vibrios. The inability to ferment sucrose or to produce acetoin are characteristics which are useful to distinguish *V. parahaemolyticus* from related vibrios, as discussed below.

TABLE 6.3
GENERAL CHARACTERISTICS OF *V. PARAHAEMOLYTICUS*

Test	Response	Test	Response
Gram stain	−	Lactose acid	−
Motility	+	Sucrose acid	−
Growth in trypticase broth		Cellobiose acid	+
+ 0% NaCl	−	Maltose acid	+
+ 3% NaCl	+	Mannitol acid	+
+ 7% NaCl	+	Starch hydrolysis	+
+ 10% NaCl	−*	Chitin digestion	+*
Growth at 43°C	+	Gelatin liquefied	+
Cytochrome oxidase	+	Sensitive to 0−129	+
Voges Proskauer	−	Sensitive to 0−5 u.	
		penicillin	−
TSI	K/A**	Utilizes as sole source of	
		C and energy	
Arginine dihydrolase	−	L-leucine	+
Lysine decarboxylase	+	L-histidine	+*
NO$_3$ reduced	+	D-galactose	+

*Some strains may show variable response
**Acid butt, alkaline slant, H$_2$S −, no gas

Ability to hemolyze blood is quite common among the mesophilic vibrios, but a unique characteristic of most enteropathogenic strains of *V. parahaemolyticus* is the ability to produce β-hemolysis on a special blood agar called Wagatsuma agar. This is discussed later.

Serology

Vibrio parahaemolyticus strains possess O antigens (somatic), K antigens (capsular), and H antigens (flagellar). All strains share a common H antigen and therefore O antigens and K antigens are used to serologically identify strains. The antigenic schema for *V. parahaemolyticus* is shown in Table 6.4. There are 11 O groups and 53 K types. Each K antigen is specific to an individual O group except for K4, which occurs in O groups 3 and 4 and K30, which occurs in O groups 3 and 5. Practically, typing is carried out using polyvalent K antisera, followed by specific K antisera. Unfortunately the O and K antigens are possessed by other mesophilic vibrios, and it is necessary to identify *V. parahaemolyticus* by physiological and biochemical tests before serological analysis can be usefully applied. Obviously, therefore, this procedure is most valuable in epidemiological studies where strain identification is important.

Bacteriophage

A variety of bacteriophages have been isolated which will cause lysis of *V. parahaemolyticus* strains. Unfortunately, there is again

TABLE 6.4
ANTIGENIC SCHEMA FOR *V. PARAHAEMOLYTICUS*

O group	K-antigen
1	1, 25, 26, 32, 38, 41, 56
2	3, 28
3	4, 5, 6, 7, 29, 30, 31
	33, 37, 43, 45, 48, 54, 57
4	4, 8, 9, 10, 11, 12, 13
	34, 42, 49, 53, 55
5	15, 17, 30, 47
6	18, 46
7	19
8	20, 21, 22, 39
9	23, 44
10	24, 52
11	36, 40, 50, 51

extensive cross lysis in most cases with other mesophilic and even, in some cases, psychrophilic vibrios, so that a satisfactory phage typing system has not yet been developed.

Genetic Characteristics

The DNA base composition of *V. parahaemolyticus* is in the range 43 to 47% G + C. However, other mesophilic marine vibrios also fall within this range. DNA reassociation studies have shown that authentic *V. parahaemolyticus* strains show high DNA homology, usually in excess of 90%. *V. alginolyticus* shows 60-70% homology with *V. parahaemolyticus*, supporting the observed high phenotypic similarity of these two organisms (Anderson and Ordal 1972). All other vibrios tested show low DNA homology with *V. parahaemolyticus* of the order of 30% or less.

Pathogenicity

Not all strains of *V. parahaemolyticus* are enteropathogenic for humans and, indeed, pathogenicity appears to be a variable phenomenon in this species. Pathogenicity appears to be correlated with the possession of a particular hemolysin which is demonstrable as the so-called Kanagawa phenomenon. Kanagawa-positive strains show β-hemolysis when grown on Wagatsuma agar (Table 6.5). Virtually all strains ($<95\%$) of *V. parahaemolyticus* isolated from diarrheal stools are Kanagawa-positive, while less than 1% of strains isolated from fish and shellfish are reported to be positive. Feeding experiments with human volunteers appear to confirm that only Kanagawa-positive strains can cause gastroenteritis (Sakazaki *et al.* 1963). However, the correlation between pathogenicity and the

TABLE 6.5
WAGATSUMA AGAR

Bacto-peptone	10 g
Yeast extract	3 g
Sodium chloride	70 g
Mannitol	10 g
Dipotassium phosphate	5 g
Crystal violet	0.0001 g
Bacto agar	15 g
Distilled water	1,000 ml

Adjust to pH 8. Steam (DO NOT AUTOCLAVE) 30 minutes; cool to 50° C and add 100 ml 20% suspension of washed human red cells. Mix, pour, and dry.

Kanagawa phenomenon may not be an absolute one, since Kanagawa-negative strains have recently been isolated from cases in three outbreaks (Teramoto *et al.* 1971). A remarkable feature of most outbreaks is the finding that only Kanagawa-negative strains can be isolated from the incriminated foodstuff, while Kanagawa-positive strains are isolated from fecal samples. Moreover, it is commonly observed that different serotypes are isolated from foods and from patients' stools. This suggests that pathogenicity may in fact be developed by certain strains within the alimentary canal of the sensitive host. However, in the absence of more definitive information, all authentic *V. parahaemolyticus* strains must be considered to be potentially enteropathogenic.

Animal tests of pathogenicity do not appear to be useful except for the ligated rabbit ileal loop technique. Twedt and Brown (1973) showed a high correlation between Kanagawa-positivity and dilation of ileal loop segments.

There is conflicting information on whether *V. parahaemolyticus* produces a toxin(s). The Kanagawa hemolysis itself does not seem to be toxic, and recent experiments by Twedt and Brown (1973) failed to show unequivocably that either an endotoxin or exotoxin was produced by *V. parahaemolyticus*. Current evidence would suggest that pathogenicity is due to infection characterized by epithelial cell penetration, disturbance of bowel function, and fluid loss. This is supported by the experimental observation that ingestion of about 10^5 living cells is necessary to cause food poisoning symptoms to develop in human volunteers.

EPIDEMIOLOGY AND DISEASE SYNDROME

V. parahaemolyticus is apparently widely distributed in inshore marine environments. However, it only appears in significant numbers during warmer periods of the year. This correlates with both Japanese and U.S. experience of outbreaks of *V. parahaemolyticus* food poisoning which is recognized to be a "summer" disease. In North America, the organism has been isolated from seawater, sediments, plankton, shellfish, crustaceans, and fish. It appears to be most abundant in shellfish and other inshore or estuarine marine animals during summer months. Colwell *et al.* (1973) have postulated that *V. parahaemolyticus* overwinters in sediment and is distributed into the water column by zooplankton during and subsequent to the spring bloom.

Typically the disease occurs in people eating uncooked fish or shellfish contaminated with *V. parahaemolyticus*. The outbreaks in the U.S. have resulted from consumption of precooked crab and shrimp which was recontaminated after cooking. Suspected cases from consumption of steamed clams have also been reported. The U.S. experience is shown in Table 6.6. The very rapid growth rate of *V parahaemolyticus* in the temperature range 20-35°C (Table 6.2) permits multiplication of the population to an infective level within a few hours, before the food is organoleptically "spoiled."

Onset of symptoms occurs most commonly 12-18 hours after ingestion of the contaminated foods. Incubation periods as short as 4 hours and as long as 48 hours have been reported. The principal symptoms are diarrhea, abdominal pain (which may be severe), nausea, and vomiting. Fever, headache, and chills are reported in a significant proportion of cases. The percent incidence of symptoms is shown in Table 6.7 for 205 of 425 cases which occurred in three

TABLE 6.6
OUTBREAKS OF *V. PARAHAEMOLYTICUS* FOOD POISONING IN U.S.

Year	Location	Incriminated Food	Cases	Occasion
1969	Washington State	Shellfish	48[1]	Camp
1969	Washington State	Shellfish	23[1]	Camp
1971	New Jersey	Crab & shrimp	12	Home
1971	Maryland	Crab	320[2]	Picnic
1971	Maryland	Crab	43[2]	Home
1971	Maryland	Crab	15[2]	Private picnic
1971	Maryland	Crab	39[3]	Picnic
1971	Maryland	Crab	8[3]	Home
1971	Maryland	Crab	24	Hospital
1972	Louisiana	Shrimp	600	Party (shrimp boil)

[1] Not fully confirmed bacteriologically
[2] Common source of steamed crabs
[3] Common source of steamed crabs

TABLE 6.7
OCCURRENCE OF SYMPTOMS AMONG 205 CASES
OF *V. PARAHAEMOLYTICUS* FOOD POISONING

Symptom	%
Diarrhea	96.6
Abdominal Cramps	77.6
Nausea	70.2
Vomiting	63.9
Headache	38.5
Fever	26.8
Chills	14.6

Recalculated from data of Dadisman *et al.*, 1973.

outbreaks traced to contaminated crabs which occurred in Maryland in 1971. The attack rate is usually quite high, being 57% in the example quoted here, but may be related to the amount of contaminated food consumed. Thus in the Louisiana shrimp outbreak in 1972 (Center for Disease Control, 1972) the attack rate was 27.3% for those eating less than 10 shrimp and 73% for those eating more than 20 shrimp. Duration of the illness is usually short, from 1-4 days, but cases have been reported where symptoms continued beyond 10 days. Recovery of patients is normally rapid and uneventful. No mortalities have been reported from outbreaks in the U.S. (involving approximately 1,100 cases), and mortality in most Japanese outbreaks has been very low. *V. parahaemolyticus* is present in large numbers in the diarrheal stools and may also be recovered from vomit. However, the organism disappears rapidly from feces after recovery of the patient and cannot be isolated after 5-10 days from onset of symptoms. There is no evidence of secondary person to person spread of *V. parahaemolyticus* food poisoning, and human sources have never been implicated as the primary cause of an outbreak.

ISOLATION AND IDENTIFICATION

A variety of media and methods has been developed for isolation and identification of the organism. Isolation from patients' stools during the diarrheal stage is simple because of the dominance of *V. parahaemolyticus* in feces. Samples may be inoculated directly or after dilution in medium containing 3% NaCl onto thiosulfate-citrate-bile salts-sucrose agar (TCBS). On this medium, *V. parahaemolyticus* produces characteristic round colonies 3-5 mm in diameter, with green or blue centers, after incubation overnight (18 hours) at 35°C. For isolation from seafoods or other environmental samples, it is normally necessary to use an enrichment broth procedure. The comminuted sample suitably diluted in liquid containing 3% NaCl is transferred to nutrient liquid media containing 3% NaCl and incubated at 35-37°C. A good enrichment medium which is also selective for *V. parahaemolyticus* is glucose-salt-Teepol broth (GSTB). An MPN procedure can be used to obtain counts on foods. After overnight incubation in the enrichment medium, the cultures are streaked onto TCBS agar and incubated overnight again at 35-37°C. Difficulties are sometimes experienced in the use of TCBS with environmental samples. In mixed populations where the incidence of *V. parahaemolyticus* is low, relatively large numbers of false positives may be encountered, and it is necessary to confirm the identification by biochemical testing of selected cultures.

Presumptive *V. parahaemolyticus* cultures showing the characteristic colony appearance on TCBS are inoculated into triple sugar iron agar (TSI), motility agar, trypticase soy agar, and trypticase soy broth, all prepared with added NaCl to yield 3% NaCl final concentration. These media should also be incubated at 37°C. Organisms which are Gram-negative, motile, produce an alkaline slant and acid butt in TSI, but no $H_2 S$ or gas production, are then submitted to additional tests. Identification of the isolates as Vibrio depends on: positive cytochrome oxidase test, absence of gas production, anaerobic fermentation of glucose in Hugh & Liefson medium, and sensitivity to the 0-129 vibriostatic compound. Within the vibrio group, the problem is to distinguish *V. parahaemolyticus* from *V. alginolyticus*, *V. anguillarum*, and other unnamed mesophilic vibrios common in seafoods. This is done on the basis of: growth at 42-43°C in trypticase soy broth with 3% NaCl, failure to ferment sucrose, negative Voges-Proskauer test, negative arginine dihydrolase test but positive lysine decarboxylase test, and growth in trypticase soy broth containing 3% and 7% NaCl but no growth in 0% or 10% salt (see Table 6.8). Definitive identification requires completion of the complete range of tests shown. Organisms identified biochemically as *V. parahaemolyticus* may be tested for the Kanagawa phenomenon noted above as a presumptive indication of enteropathogenicity, but this is probably only useful in clinical investigations.

CONTROL OF V. PARAHAEMOLYTICUS IN FOODS

Since *V. parahaemolyticus* is so widespread in natural occurrence, all seafoods should be considered as potential sources of this organism. Japanese experience suggests that other salted foods, particularly brined vegetables, may be contaminated from seafoods

TABLE 6.8
TESTS FOR THE DIFFERENTIATION OF *V. PARAHAEMOLYTICUS*
FROM OTHER MESOPHILIC VIBRIOS

Tests	Response of			
	V. para-haemolyticus	V. algino-lyticus	V. anguil-larum	Other Vibrios
TCBS	+ (green)	− (yellow)	−	±
Growth at 43°C	+	+	−	±
Growth in 8% NaCl	+	+	−	−
10% NaCl	−*	+	−	−
Lysine decarboxylase	+	+	−	±
Arginine dihydrolase	−	−	+	±
Voges-Proskauer	−	+	+	±
Sucrose	−	+	+	±

*Some strains may show variable response

and become a source of potential hazard. Foods consumed hot or shortly after cooking are unlikely to present a hazard because of the extreme heat sensitivity of the organism, but cooked foods which are recontaminated and held for a few hours at temperatures above 15°C may be hazardous. Since the organism is highly sensitive to temperatures in the range 0-5°C, refrigeration of food until consumption is a very effective method of control except perhaps where initial contamination is at a very high level. Freezing, though less lethal than chilling, is also an effective method of control, since it prevents all growth of the organism and does promote die-off.

It is important to recognize that *V. parahaemolyticus* is naturally present on seafoods and is not derived from human or other terrestrial sources in most cases. Control thus must involve the raw seafood itself. Contamination of processed product from raw product or raw product containers seem to have been the principal cause of *V. parahaemolyticus* food poisoning incidents in the U.S. Cross contamination must therefore be avoided. The second important factor in most outbreaks is holding or storage of the foodstuff at a temperature which permits *V. parahaemolyticus* to multiply rapidly. It seems unlikely that fish or shellfish handled and held under good conditions (*i.e.* low temperature) will naturally have sufficiently high counts of *V. parahaemolyticus* at the point of entry to the processing plant to represent a direct hazard. (One exception to this might be molluscan shellfish harvested from enclosed waters during warm weather). Consequently, application of good manufacturing practices should ensure that the product leaving the plant is free from *V. parahaemolyticus* or carries only small numbers of the organism. Maintenance of this safe condition depends on the

VIBRIO PARAHAEMOLYTICUS 209

consistent application of low temperature storage conditions throughout wholesale distribution and retail operations. In the North American situation these procedures should be rigorously applied to crabmeat, lobsters, shrimp, and certain molluscs which may not be cooked by the housewife prior to serving to the family.

Public health authorities and industry concerned with such products should recognize the desirability of regular surveillance through specific bacteriological analysis for *V. parahaemolyticus* of raw material and finished product during the summer or other warm weather periods.

BIBLIOGRAPHY

ANDERSON, R. S., and ORDAL, E. J. 1972. Ribonucleic acid relationship among marine vibrios. J. Bacteriol. *109*, 696–706.
BAROSS, J., and LISTON, J. 1970. Occurrence of *V. parahaemolyticus* and related hemolytic vibrios in marine environments of Washington State. Appl. Microbiol. *20*, 179–186.
BAUMANN, P., and BAUMANN, L. 1973. Phenotypic characterization of *Beneckea parahaemolytica*: A preliminary report. J. Milk Food Technol. *36*, 214–219.
BEUCHAT, L. R. 1973. Interacting effects of pH, temperature, and salt concentration on growth and survival of *Vibrio parahaemolyticus*. Appl. Microbiol. *25*, 844–846.
CENTER FOR DISEASE CONTROL. 1972. *Vibrio parahaemolyticus*—New Jersey. Morbidity and Mortality Weekly Rept. (U.S.) *21*, No. 50, 430.
COLWELL, R. R. *et al.* 1973. *Vibrio parahaemolyticus*—isolation, identification, classification, and ecology. J. Milk Food Technol. *36*, 202–213.
DADISMAN, T. A., NELSON, R., MOLENDA, J. R., and GARBER, H. J. 1973. *Vibrio parahaemolyticus* gastroenteritis in Maryland. 1. Clinical and epidemiologic aspects. Am. J. Epidemiol. *96*, 414–426.
FUJINO, T. *et al.* 1953. On the bacteriological examination of Shirasu food poisoning. Med. J. Osaka Univ. *4*, 299–304.
JOHNSON, H., and LISTON, J. 1973. Sensitivity of *Vibrio parahaemolyticus* to cold in oysters, fish fillets and crabmeat. J. Food Sci. *38*, 437–441.
LISTON, J. 1973. *Vibrio parahaemolyticus*. In Microbial Safety of Fishery Products, C. O. Chichester, and H. D. Graham (Editors). Academic Press, New York.
MIYAMOTO, Y., NAKAMURA, K., and TAKIZAWA, K. 1961. Pathogenic halophiles, Proposals of a new genus "*Oceanomonas*." Japan. J. Microbiol. *5*, 477–486.
SAKAZAKI, R., IWANAMI, S., and FUKUMI, H. 1963. Studies on the enteropathogenic facultatively halophilic bacteria, *Vibrio parahaemolyticus*. 1. Morphological, cultural, and biochemical properties and its taxonomical position. Japan. J. Med. Sci. Biol. *16*, 161–188.
TAKIGAWA, I., and FUJISAWA, T. 1956. An outbreak of food poisoning presumably caused by a marine bacteria. Skokuhin Eisei Kenkyu *6*, 15–19. (Japanese)
TERAMOTO, T., NAKANISHI, H., MAESHIMA, K., and MIWATANI, T. 1971. On the case of food poisoning caused by presumptive Kanagawa phenomenon negative strain. Media Circle *16*, 174–177. (Japanese)
TWEDT, R. M., and BROWN, D. F. 1973. *Vibrio parahaemolyticus*: infection or toxicosis? J. Milk Food Technol. *36*, 129–134.
VANDERZANT, C., and NICKELSON, R. 1972. Survival of *Vibrio parahaemolyticus* in shrimp tissue under various environmental conditions. Appl. Microbiol. *23*, 34–37.

CHAPTER 7

Elmer H. Marth
and
Blanca G. Calanog

Toxigenic Fungi

Benefits man has derived from the use of molds have been known for many years. Molds have not only served to synthesize antibiotics but also to produce some foods. Fermented foods such as some cheese, soysauce, miso, tempeh, and other Oriental delicacies are all prepared with the help of molds.

However, it is also well documented that some molds produce toxic substances. Some fungi elaborate the toxin in large macroscopic fruiting bodies; for example, the toxin produced by certain species of *Amanita*, a poisonous mushroom. Other fungi always grow and sporulate as parasites on living host plants, and sometimes will do so only on a specific host. *Claviceps* is an example of this group of fungi and it produces mycotoxins that have been known for centuries. The ovary of the host cereal is transformed into a hard fungal mass containing ergot alkaloids.

In contrast to fungi that are parasitic on living plants, another group of fungi is saprophytic and causes destruction of dead plants and animal material. Some of these fungi regularly inhabit stored grains, feeds, and similar relatively dry products, and thus, they have been referred to as "storage fungi" (Hesseltine 1969). Kramer *et al.* (1963) reported the abundance in the atmosphere of spores of the following molds: *Cladosporium*, *Fusarium*, *Penicillium*, *Aspergillus*, and *Alternaria*, and added that these molds constitute a considerable part of the airborne biota. Great interest has been afforded these fungi in the last decade because of their wide distribution and potential ability to produce mycotoxins under suitable environmental conditions (Forgacs 1962).

A mycotoxin has been defined as a mold metabolite produced on foodstuffs (or feedstuffs) that causes illness or death when ingested by man or animals (Wilson 1968). This definition eliminates mushroom poisons, although it does include the notorious ergot whose history dates back several centuries. Since ergot and ergotism are well documented, they will not be considered further. Rather, mycotoxins which have attained prominence in recent years will be discussed.

One of the earliest reports of poisoning in cattle induced by fungally contaminated sweet clover was that of Schofield (1924). The Russians also deserve credit for calling attention to the mycotoxicoses that occurred endemically in the Soviet Union. In the

1940's, they observed the occurrence of alimentary toxic aleukia, a serious affliction involving several stages of illness, leucopenia, and hemorrhagic diathesis. Urov disease, a chronic affliction which causes marked skeletal deformities in children and adolescents, also was reported. Wilson (1968) indicated that these diseases were caused by *Fusarium sporotrichoides*, a frequent contaminant of cereal crops. It was also the Russians who reported stachybotrytoxicoses of man and livestock caused by *Stachybotrys atra*, a black fungus that grows on grain and hay (Wilson 1968).

Thornton and Percival (1959) reported the occurrence of facial eczema in sheep and cattle in New Zealand and Australia. They attributed this disease to sporidesmin, a hepatotoxin produced by *Pithomyces chartarum* (once called *Sporidesmium bakeri*). This fungus grows on pasture grass in humid autumn weather. Outbreaks of apparent mycotoxicoses in animals in the U.S. dating back to the last century were cited by Wilson (1968).

Despite these incidents of mycotoxicoses, the etiology of the toxic factors was largely ignored. This was altered drastically in 1960 when outbreaks of an apparently new disease of young turkeys were noted in England (Sargeant *et al.* 1961A). More than 500 outbreaks occurred that resulted in the death of 100,000 birds and a smaller number of calves, pigs, and other domestic animals. Economic losses from these outbreaks prompted extensive research to isolate and identify the cause. Sargeant *et al.* (1961A) reported the causative factor to be a toxic material produced by *Aspergillus flavus*, and they named the substance aflatoxin.

This discovery was significant for two reasons: (a) the aflatoxigenic organism, *A. flavus*, is a common soil fungus throughout the world, and this prompted work the results of which demonstrated that aflatoxin appeared in food in many countries, and (b) aflatoxin was found to be a potent liver poison with marked carcinogenic properties (Sargeant *et al.* 1961A) and hence constituted a hazard to human health. This concern intensified the interest and research efforts of scientists from different disciplines. Because molds can be present on a variety of foods, whether as contaminant or an essential ingredient in its manufacture, food scientists and others have investigated the distribution of toxigenic fungi in foods and the conditions that lead to toxigenesis in such foods.

MYCOTOXINS AND HUMAN HEALTH

Although much information has accumulated on mycotoxins, especially since the discovery of aflatoxin in the early 1960's, effects of these compounds on human health are not well documented. Two possible exceptions are the human diseases, ergotism and alimentary toxic aleukia, caused by toxins produced by fungi in the genera *Claviceps* and *Fusarium*, respectively. Ergotism has been known for centuries and has affected many thousands of people; hence there have been ample opportunities to observe the gangrenous, hallucinogenic, and often fatal consequences of this disease. In contrast, alimentary toxic aleukia was not described in detail until fairly recently. This again was possible only because large numbers of people contracted the disease and hence its effects could be observed readily.

Since most mycotoxins are highly toxic to many experimental animals, it is obviously impossible to experimentally determine their effects on humans. Thus, we need to rely on results obtained from tests with experimental animals and must be alert to make observations on humans when they suffer from known inadvertent intoxication by mycotoxins.

There is considerable variation in the disease processes caused by various mycotoxins. Some information on the principal toxic effects of some mycotoxins produced by aspergilli, penicillia, fusaria, and molds in several other genera is summarized in Tables 7.1, 7.2, 7.3 , and 7.4. It is clearly evident that diseases caused by mycotoxins, at least in experimental animals, take various forms, although the liver and kidneys are often damaged. The toxicity of mycotoxins also is variable, depending on the toxin, test animal, and route of administration. Nevertheless, many of the compounds (Table 7.5) are highly toxic to the usual test animals.

Needless to say, demonstrated occurrence of mycotoxins in foods and feeds and the highly toxic nature of the toxins have raised questions concerning the risk to human health (Goldblatt 1968; Teunisson and Robertson 1967). Answers to such questions are slow in coming but some information on relationships between recently

TABLE 7.1
PRINCIPAL TOXIC EFFECTS OF SOME MYCOTOXINS
PRODUCED BY ASPERGILLI[a]

Aspergillus	Mycotoxin	Principal toxic effect
flavus	Aflatoxins	Liver damage and cancer
parasiticus	Aflatoxins	Liver damage and cancer
ochraceous	Ochratoxins	Fatty liver
clavatus	Patulin	Neurotoxin
oryzae var. *microsporus*	Maltoryzine	Muscular paralysis
terreus	Citrinin	Nephrotoxin
candidus	Citrinin	Nephrotoxin

[a] From Lillehoj *et al.* (1970).

TABLE 7.2
PRINCIPAL TOXIC EFFECTS OF SOME MYCOTOXINS
PRODUCED BY PENICILLIA[a]

Penicillium	Mycotoxin	Principal toxic effect
urticae	Patulin	Neurotoxin
rubrum	Rubratoxins	Generalized hemorrhages
cyclopium	Tremorgen	Tremors, convulsions
islandicum	Islanditoxin, luteoskyrin	Liver damage, cancer, hemorrhages
viridicatum	Hepatotoxin, tremorgen nephrotoxin	Nephroses, tremors, liver damage
toxicarium	Citreoviridin	CNS paralysis, respiratory failure
citrinum	Citrinin	Nephrotoxin

[a] From Lillehoj et al. (1970).

TABLE 7.3
PRINCIPAL TOXIC EFFECTS OF SOME MYCOTOXINS
PRODUCED BY FUSARIA[a]

Fusarium	Mycotoxin	Principal toxic effect
culorum	Unknown	Ataxia
nivale	Nivalenol	Nausea, vomiting, somnolence
tricinctum, scirpi, and equiseti	Diacetoxy-scirpenol, T-2 toxin, butenolide	Diarrhea, weight loss
roseum	Unknown	Nausea, vomiting, death
sporotrichoides	Fusariogenin	Aleukia, bone marrow exhaustion, death

[a] From Lillehoj et al. (1970).

TABLE 7.4
PRINCIPAL TOXIC EFFECTS OF MYCOTOXINS
PRODUCED BY MOLDS IN SEVERAL GENERA[a]

Mold	Mycotoxin	Principal toxic effect
Diplodia zea	Unknown	Salivation, ataxia, death
Rhizoctonia leguminicola	Slaframine	Slobbering, diarrhea, death
Stachybotrys atra	Unknown	Hemorrhage, necrosis of mucus membranes
Pithomyces chartarum	Sporidesmins	Facial eczema, lesions of liver and bile duct
Sclerotinia sclerotiorum	Psoralens	Dermatitis

[a] From Lillehoj et al. (1970).

TABLE 7.5
TOXICITY OF SOME MYCOTOXINS[a]

Mycotoxin	LD_{50} (mg/kg)
Aflatoxin B_1	0.56 - day-old rat (oral)
	5.5 - male rat at weaning (oral)
	1.4 - guinea pig (oral)
Islanditoxin	0.34 - mouse (IV)[b]
Citreoviridin	8-30 - rat (various)
Luteoskyrin	6.6 - mouse (IV)
Ochratoxin A	3 - duckling (oral)
Rubratoxin	3.7 - mouse (IP)[c]

[a] From Feuell (1969).

[b] Intravenous

[c] Intraperitoneal

discovered mycotoxins, especially aflatoxin, and human health is beginning to accumulate.

Several investigators (Alpert and Davidson 1969; Oettle 1964) have reported unusually frequent occurrence of primary liver carcinoma in certain African and Asiatic countries. These workers proposed that the condition may be related to ingestion of hepatotoxic fungal products. Alpert et al. (1968) screened Ugandan food and found that 40% of the samples contained measurable quantities of aflatoxin. Fifteen percent of peanut samples contained levels of aflatoxin B_1 exceeding 1 ppm. Hesseltine (1967) indicated that moldy corn, a major ingredient in the diet of the African Bantu tribe, may be related to hepatomas representing 68% of carcinomas in this tribe.

In humans, most cases of primary hepatoma have had an antecedent of cirrhosis (Kraybill and Shapiro 1969). Robinson (1967) examined samples of breast milk from mothers of cirrhotic children, as well as urine from the children. Using thin-layer chromatographic (TLC) analysis, 7% of the milk samples appeared positive for the presence of aflatoxin B_1.

In Taiwan, three children died after consuming moldy rice which contained 18–22 μg aflatoxin B_1 per 100 g of rice (Detroy et al. 1971). Serck-Hansen (1970) reported observing aflatoxin-induced fatal hepatitis in a 15-year old African boy who had daily eaten fish, beans, meat, and cassava contaminated with aflatoxin. Histological changes in the boy's liver were identical to those observed in aflatoxin-treated monkeys. More recently, Boesenberg (1973) presented the first direct evidence of acute damage to a human after a single dose of aflatoxin. A 45-year old man with liver cirrhosis had consumed an excessive quantity of a mixture of nuts, and died a short while later. Examination of extracts from his liver demonstrated the presence of suspicious blue fluorescing substances on thin-layer chromatograms. UV absorbance and fluorescence showed these substances to have the same properties as aflatoxin B_1 and/or B_2. Several strains of *Aspergillus parasiticus* were isolated from the nuts.

Legator (1969) studied the effect of aflatoxin B_1 on human embryonic lung cells. He observed that 5 ppm of toxin inhibited lung cell growth, while a smaller amount (1 ppm) resulted in a 90% increase in giant cells over the control culture. Gabliks *et al.* (1965) pointed out that there is a correlation between the susceptibility to aflatoxin of cells grown in tissue culture and animals from which the cells were derived.

Although the literature gives few quantitative data concerning the effect of aflatoxin on man, the findings just discussed provide ample evidence that aflatoxin or other mycotoxins may play a role in the etiology of human disease.

MYCOTOXICOSIS IS A WORLDWIDE PROBLEM

The mycotoxin problem, particularly that of aflatoxin, is not confined to a small area; rather, it is universal. Lopez and Crawford (1967) indicated that about 15% of the peanuts sold for human consumption in Uganda contained more than 10 ppm aflatoxin. This is significant since the peanut, because of its high protein content, is an important item in the African diet. Sellschop *et al.*, as cited by Detroy *et al.* (1971), found that 17% of the peanut butter samples obtained in Transvaal, South Africa, contained traces of aflatoxin. A high percentage of the peanuts and peanut cake collected from three districts in India exhibited extensive visible mold damage and about 20 to 40% of the kernels and 82% of the oil cake contained large amounts of aflatoxin (Anon. 1967-1968). Detroy *et al.* (1971) also cited reports to indicate that dried sweet potatoes in Taiwan were heavily contaminated with *A. flavus* and contained 0.01 to 0.18 ppm aflatoxin. Furthermore, aflatoxin was found in peanuts, peanut butter, peanut cakes, and peanut oils. It should be noted that in Taiwan, peanut oil is produced without alkali refining, a procedure used in the U.S. that effectively degrades the toxin.

In the Philippines, Campbell (1969) tested 29 samples of locally produced peanut butter and found that all contained more than 30 ppb aflatoxin. He also detected toxin in peanuts, peanut brittle, other peanut products, corn, beans, cassava, sweet potato, coconut products, cocoa, rice, and some native foods. Five of 21 samples of retail milk in South Africa were positive for aflatoxin M (Purchase and Vorster 1968). Investigators from the Massachusetts Institute of Technology analyzed the food supply available in Thailand, Hong Kong, and Malaysia. Aflatoxin was found in peanuts, corn, dried chili peppers, rice, beans, dried fish, shrimp, garlic, and prepared foods (Shank 1968). Aflatoxin has also been detected in many dehydrated products available for human consumption in Germany (Detroy et al. 1971). In the U.S., Taber and Schroeder (1967) found small amounts (0-91 ppb) of aflatoxin in 78 samples of peanuts that were grown in nine different geographical areas.

Although additional documentation could be provided, these reports are sufficient to indicate that mycotoxins occur with some frequency in areas of the world where humidity and temperature are high and where harvesting, storage, and marketing facilities for susceptible products often are not advanced. Furthermore, mycotoxins also can develop in products even when advanced techniques are used, provided that molds are given the opportunity to grow.

MOLDS THAT PRODUCE TOXINS

After Sargeant et al. (1961A) first reported that aflatoxin was produced in peanuts by A. flavus, numerous investigators directed their interest and efforts toward finding other toxigenic, particularly aflatoxigenic, molds. In 1963, Codner et al. (1963) found that strains of A. parasiticus also could produce aflatoxin. Hodges et al. (1964) reported that aflatoxin was elaborated by Penicillium puberulum.

After these reports were available, Kulik and Holaday (1966) isolated possible toxigenic molds belonging to the genera Aspergillus and Penicillium from corn grains. Extracts from cultures of 107 isolated fungi were assayed for aflatoxin by TLC. Thirty-nine of 107 isolates tested by TLC yielded a positive result. Toxigenic strains included A. flavus, A. parasiticus, Aspergillus ruber, Aspergillus wentii, as well as Penicillium citrinum and Penicillium variabile. Penicillium frequentans and P. puberulum elaborated this toxin only in trace amounts. Duckling bioassays of the extracts showed that only the extracts from A. parasiticus were highly toxic.

An extensive study of toxigenic molds from overwintered cereal was undertaken by Joffe (1960, 1962, 1965). He isolated about 2,600 fungi that included 100 different species of Fusarium, Cladosporium, Alternaria, Penicillium, and Mucor and found 14% of these to be toxigenic. Joffe (1969) also analyzed 1,626 isolates of A. flavus obtained from peanuts and soils in Israel and found that 90% produced aflatoxins B_1 and B_2. Only 8.4% produced the four major aflatoxins (B_1, B_2, G_1, and G_2).

Kurata and Ichinoe (1967A) and Kurata *et al.* (1968A) isolated many strains of *Penicillium* and *Aspergillus* from ground rice and from rice and wheat flours. The same investigators (Kurata and Ichinoe 1967B; Kurata *et al.* 1968B) screened these isolates and found that about 10% were toxic to mice. Austwick and Ayerst (1963) reported that the microflora of groundnuts included *A. flavus, A. parasiticus*, and other species of *Aspergillus* as well as *Penicillium, Fusarium, Rhizopus, Chaetomium, Syncophalastrum,* and *Trichothecium*. Fifty-nine of these isolates were screened for their ability to produce a fluorescent substance when grown on the surface of a defined mineral medium in test tubes. Nineteen of the 29 isolates that exhibited fluorescence were tested for toxin production. Only nine isolates were toxigenic. Culture filtrates from four isolates of *Aspergillus tamarii* and two of *Aspergillus fumigatus* were faintly fluorescent but nontoxic. One isolate of *Rhizopus* was nontoxic and nonfluorescent.

Hesseltine *et al.* (1970) studied 67 strains of the *A. flavus* group. All strains that produced aflatoxin G_1 also formed aflatoxin B_1 but the converse was not true. They also noted that 11 of 14 strains of *A. parasiticus* produced all four aflatoxins.

Parrish *et al.* (1966) examined more than 40 strains of *Aspergillus* which were not members of the *A. flavus* group, and ten strains of *Penicillium* and found none to be toxigenic.

Another extensive survey totalling 260 isolates, and including 43 species of *Penicillium* and seven of *Aspergillus* revealed that only *A. flavus* was able to produce aflatoxin (Mislivec *et al.* 1968). Five isolates, three of *Aspergillus niger* and one each of *P. puberulum* and *P. frequentans*, previously reported as aflatoxigenic (Kulik and Holaday 1966) were found to be nontoxigenic by Mislivec *et al.* (1968). Kulik and Holaday (1966) tried to explain this contradictory finding on the toxigenicity of the same mold isolates in several ways. They indicated that some isolates may lose their aflatoxin-producing ability in culture or that media used by other investigators are more or less conducive to aflatoxin production. They also thought it possible that results from TLC and from duckling assays were misinterpreted.

The numerous but unconfirmed reports that fungi other than *A. flavus* and *A. parasiticus* can synthesize aflatoxin led Wilson *et al.* (1968) to screen 121 mold isolates, including several that had been reported to produce aflatoxin. These latter cultures were supplied by investigators who had reported their strains to be toxigenic. Wilson *et al.* (1968) were unable to confirm the findings of the other researchers. They concluded that aflatoxin synthesis among those cultures surveyed was confined to the *A. flavus* group.

Holker and Underwood (1964) examined 2,000 mutants of *Aspergillus versicolor* that were produced by irradiation of spores with ultraviolet light, and in no instance did TLC reveal any detectable amounts of aflatoxins in the extracts. Scott (1965) studied the microflora of legumes and cereal products in South Africa and reported that 46 strains of 22 species of molds were toxic to ducklings. He was able to isolate five strains of *Aspergillus ochraceus* and three of them were toxigenic. Diener and Davis (1969A) indicated that the overall average of toxigenic isolates from all sources is near 60%, a high percentage considering the other reports that were just discussed.

Some of the apparently contradictory data on the ability of various fungi to produce aflatoxin may result from the appearance on thin-layer chromatoplates of fluorescent substances with an R_f value approximating those of the aflatoxins (Andrellos and Reid 1964; Yokotsuka *et al.* 1967B; 1968). In addition, as pointed out by Wilson *et al.* (1968), possible false-positive findings may also result from "(a) residual contamination of glassware and instrument with aflatoxin solutions, (b) inadvertent use of food material previously

TABLE 7.6

TOXINS PRODUCED BY STRAINS OF MOLDS IN THE GENUS *ASPERGILLUS*.

Mold	Toxin produced	Reference
Aspergillus flavus	Aflatoxin	Kulik and Holaday (1966)
"	Aspertoxin	Kulik and Holaday (1966)
"	Sterigmatocystin	Kulik and Holaday (1966)
"	Hydroxyaspergillic acid	Menzel *et al.* (1945)
"	Aspergillic acid	White and Hill (1943)
"	Kojic Acid	Wilson (1968)
A. parasiticus	Aflatoxin	Kulik and Holaday (1966)
A. niger	Aflatoxin	Kulik and Holaday (1966)
"	Oxalic acid	Kulik and Holaday (1966)
A. ruber	Aflatoxin	Kulik and Holaday (1966)
A. wentii	Aflatoxin	Kulik and Holaday (1966)
A. ochraceus	Ochratoxin	Lai *et al.* (1970)
A. ostianus	Aflatoxin	Scott *et al.* (1967)
A. versicolor	Sterigmatocystin	Bullock *et al.* (1962)
A. oryzae	Aspergillic acid	Wilson (1971A)
"	Kojic acid	Wilson (1971A)
A. sclerotium	Aspergillic acid	Wilson (1971A)
A. chevaliere	Xanthocillin	Saito (1971)
A. sulphureus	Ochratoxin	Lai *et al.* (1970)
A. melleus	Ochratoxin	Lai *et al.* (1970)
A. fumigatus	Fumigatin	Wilson (1971A)
A. rugulosus	Sterigmatocystin	Detroy *et al.* (1971)

TABLE 7.7
TOXINS PRODUCED BY STRAINS OF MOLDS IN THE GENUS *PENICILLIUM.*

Mold	Toxin produced	Reference
Penicillium rubrum	Rubratoxin A,B	Moss (1971)
P. viridicatum	Viridicatin	Wilson (1971B)
P. islandicum	Islanditoxin	Saito *et al.* (1971)
"	Luteoskyrin	Saito *et al.* (1971)
"	Rubroskyrin	Saito *et al.* (1971)
P. cyclopium	Cyclopiazonic acid	Holzapfel (1971)
P. puberulum	Aflatoxin	Kulik and Holaday (1966)
"	Griseofulvin	Wilson (1971)
"	Penicillic acid	Wilson (1971B)
P. rugulosum	Rugulosin	Saito *et al.* (1971)
P. citrinum	Citrinin	Saito *et al.* (1971)
P. notatum	Xanthocillin x	Saito *et al.* (1971)
P. urticae	Patulin	Saito *et al.* (1971)
P. purpurogenum	Rubratoxin B	Saito *et al.* (1971)
P. tardum	Rugulosin	Saito *et al.* (1971)
P. claviforme	Patulin	Saito *et al.* (1971)
P. variabile	Rugulosin	Saito *et al.* (1971)
P. citreo-viride	Citreoviridin	Saito *et al.* (1971)
P. thomii	Penicillic acid	Saito *et al.* (1971)
P. ochrosalmoneum	Citreoviridin	Saito *et al.* (1971)
P. implicatum	Citrinin	Saito *et al.* (1971)
P. expansum	Patulin	Saito *et al.* (1971)
P. brunneum	Rugulosin	Saito *et al.* (1971)

contaminated with aflatoxin as substrates for fungal cultures, and (c) undetected contamination of aflatoxin-negative fungal culture with aflatoxigenic species."

Since aflatoxin produced by *A. flavus* is a very potent carcinogen, most investigations on toxigenic fungi have been related to it. However, some other toxins that have been known for some time also are produced by *A. flavus*. Wilson (1966) reviewed information about these toxic substances which include oxalic acid, kojic acid, tremorgens, and aspergillic acid and related compounds.

Mycotoxins produced by species of molds in other genera have been reported. Tables 7.6, 7.7, and 7.8 list some of the different mycotoxins produced by molds in the genera *Aspergillus, Penicillium, Fusarium, Pithomyces,* and *Byssochlamys.* Table 7.9 gives a classification of mycotoxins according to functional groups. Tables 7.10 and 7.11 provide information on some of the physical properties of aflatoxins and mycotoxins other than the aflatoxins.

VARIOUS MYCOTOXINS AND THEIR PRODUCTION IN FOODS

Numerous studies have shown that toxigenic fungi can be present in or on a variety of foods. Van Walbeek *et al.* (1968) isolated 128

TABLE 7.8
TOXINS PRODUCED BY STRAINS OF MOLDS IN THE GENERA
FUSARIUM, PITHOMYCES AND *BYSSOCHLAMYS*.

Mold	Toxin produced	Reference
Fusarium nivale	Butenolide	Saito *et al.* (1971)
"	Scirpenes T-2	Saito *et al.* (1971)
"	Nivalenol	Saito *et al.* (1971)
"	Fusarenone - x	Saito *et al.* (1971)
F. sporotrichoides	Fusariogenin	Saito *et al.* (1971)
F. poae	Fusariogenin	Saito *et al.* (1971)
F. tricinctum	Scirpenes T-1	Saito *et al.* (1971)
"	Scirpenes T-2	Saito *et al.* (1971)
F. equiste	Scirpenes T-1	Saito *et al.* (1971)
F. sambucinum	Scirpenes T-1	Saito *et al.* (1971)
F. diversisporum	Scirpenes T-1	Saito *et al.* (1971)
F. graminearum	Zearalenone	Ciegler *et al.* (1971)
Pithomyces chartarum	Sporidesmin A,B,C	Ciegler *et al.* (1971)
Byssochlamys fulva	Byssochlamic acid	Moss (1971)

TABLE 7.9
CLASSIFICATION OF MYCOTOXINS ACCORDING TO THEIR FUNCTIONAL GROUPS[a]

Functional group	Mycotoxin
Anthraquinones	Luteoskyrin, Rubroskyrin, Islanditoxin, Rugulosin
Butenolides	Patulin, Penicillic acid
Cyclopeptides	Islanditoxin, Phalloidin
Coumarins	Aflatoxins, Aspertoxins, Sterigmatocystin, Ochratoxin
Nonadrides	Rubratoxin, Glauconic acid, Byssochlamic acid
Piperazine	Gliotoxin, Sporidesmin
Pyrone and pyran	Citrinin, Citreoviridin
Pyrazine	Aspergillic acid and derivatives
Scirpene	T-1 toxin, T-2 toxin, Nivalenol, Fusarenone

[a] From a paper presented by Dr. F. S. Chu at the Food Research Institute Annual Meeting at the University of Wisconsin, Madison, Wisconsin, 1969.

fungi from 74 food samples obtained from households (food apparently linked with illness and food of poor quality), retail stores (samples for routine microbiological assay), and other sources (food in transit and consignments for processing). They found that 16 isolates produced toxins. Of 24 fungi isolated from 20 samples of

TABLE 7.10
PHYSICAL PROPERTIES OF THE AFLATOXINS (DETROY ET AL. 1971)

Aflatoxin	Molecular formula	Molecular weight	Melting point (°C)	362–363 mμ absorption (E)	Fluorescence emission (mμ)	R_f
B_1	$C_{17}H_{12}O_6$	312	268–269	21,800	425	0.56
B_2	$C_{17}H_{14}O_6$	314	286–289	23,400	425	0.53
G_1	$C_{17}H_{12}O_7$	328	244–246	16,100	450	0.48
G_2	$C_{17}H_{14}O_7$	330	237–246	21,100	450	0.46
M_1	$C_{17}H_{12}O_7$	328	299	19,000	425	0.34
M_2	$C_{17}H_{14}O_7$	330	293	21,000	—	0.23
GM_1	$C_{17}H_{12}O_8$	344	276	12,000	—	0.12
B_{2a}	$C_{17}H_{14}O_7$	330	240	20,400	—	0.13
G_{2a}	$C_{17}H_{14}O_8$	346	190	18,000	—	0.10
R_o	$C_{17}H_{16}O_6$	314	230–234	14,100	425	0.54
B_3	$C_{16}H_{14}O_6$	302	217	9,700	—	0.44
$1\text{-}OCH_3B_2$	$C_{18}H_{16}O_6$	360	220–223	—	—	0.81
$1\text{-}OCH_3B_2$	isomer	—	260–262	—	—	—
$2\text{-}OCH_3B_2$	$C_{18}H_{16}O_7$	—	245–247	—	—	—
$1\text{-}C_2H_5B_2$	$C_{19}H_{18}O_7$	358	247	12,580	—	0.77
$1\text{-}C_2H_5G_2$	$C_{19}H_{18}O_8$	374	374	19,500	—	0.66

TABLE 7.11
PROPERTIES OF SOME MYCOTOXINS OTHER THAN AFLATOXINS

Mycotoxin	Molecular formula	Molecular weight	Melting point, °C	Fluorescence under longwave ultraviolet light
Aspergillic acid	$C_{12}H_{20}N_2O_2$	224	97–99	NF
Aspertoxin	$C_{19}H_{14}O_7$	354	—	bright yellow
Butenolide	$C_6H_7N_1O_3$	141	—	—
Citreoviridin	$C_{23}H_{30}O_6$	402	99–101	—
Citrinin	$C_{13}H_{14}O_5$	250	172	yellow
Cyclopiazonic acid	$C_{20}H_{20}N_2O_6$	336	246	—
Fumigatin	$C_8H_{10}O_3$	154	—	—
Griseofulvin	$C_{17}H_{17}O_6Cl$	352	218–221	—
Hydroxyaspergillic acid	$C_{12}H_{20}N_2O_3$	336	150–152	—
Islanditoxin	$C_{24}H_{31}N_5O_7Cl_2$	571	251	—
Kojic acid	$C_6H_6O_4$	142	—	—
Luteoskyrin	$C_{30}H_{22}O_{12}$	574	287	yellow
Ochratoxin A	$C_{20}H_{18}NO_6Cl$	403	—	green
Ochratoxin B	$C_{20}H_{19}NO_6$	369	221	—
Ochratoxin C	$C_{22}H_{22}NO_6Cl$	429	—	—
Patulin	$C_7H_6O_4$	154	—	—
Penicillic acid	$C_8H_{10}O_4$	170	83–85	—
Rubratoxin A	$C_{26}H_{32}O_{11}$	520	210–214	NF
Rubratoxin B	$C_{26}H_{30}O_{11}$	518	168–170	NF
Rugulosin	$C_{30}H_{20}O_{10}$	542	290	—
Sterigmatocystin	$C_{18}H_{12}O_6$	324	—	reddish brown
Viridicatin	$C_{15}H_{11}NO_2$	237	268	violet
Xanthocillin	$C_{18}H_{14}N_2O_2$	290	—	—

NF = nonfluorescent under ultraviolet light
— = no data available

cheese, only one produced aflatoxin. In this investigation, aflatoxins were produced not only by *A. flavus* isolates but also by a *Rhizopus* species and by *A. ochraceus*. Aflatoxin was also produced by six strains of *A. flavus* var *columnaris*; ochratoxin was formed by a *Penicillium* species and by an *A. ochraceus* isolate.

Frank (1968) studied aflatoxin formation in apple juice, prepacked bread, and cheese, the three common domestic foodstuffs that frequently turn moldy under household conditions in Germany. His results indicated that the amount and depth of aflatoxin below the fungal mat in the food depended on the strain of mold growing on the surface, the water content of the food, physical properties of the food, and the storage time and temperature.

The possibility that aspergilli can produce aflatoxin during growth on Cheddar and brick cheese was investigated by Lie and Marth (1967) and Shih and Marth (1972A). Both *A. flavus* and *A. parasiticus* produced aflatoxins during growth on the two types of cheese. Brick cheese was less satisfactory than Cheddar cheese for production of aflatoxin and neither cheese afforded high yields of toxin. *A. parasiticus* produced aflatoxin in brick cheese at 12.8 and 23.9°C but not at 7.2°C, whereas *A. flavus* only formed aflatoxin at 23.9°C during growth on the same cheese. Aflatoxin penetrated into cheese as much as 4 cm from the surface; the amount of toxin decreased as the distance from the surface of cheese increased.

Wildman *et al.* (1967) inoculated a toxigenic strain of *A. flavus* into several solid foods and commercially prepared fruit juices and found evidence that production of aflatoxin was paralleled by formation of mold mycelium. They recovered between 10 and 82 µg aflatoxin per gram of peanuts, grapes, and bread. None or trace amounts were formed on oranges, potatoes, cantaloupe, peaches, beef, and cheese. All the fruit juices, except the cranberry drink, supported formation of aflatoxin. Based on their findings, they hypothesized that poor growth and, consequently, low aflatoxin production were caused by insufficient nitrogen and minerals in the foods.

Alderman and Marth (1974A,B,C) did extensive studies on production of aflatoxin in citrus products. These foods can become contaminated with aspergilli and other molds during harvesting, in transit, and during processing and marketing. Detectable amounts of aflatoxin appeared in peel and juice of grapefruit and lemon after *A. parasiticus* grew in these materials. Concentrated (3:1) grapefruit juice supported production of 35 to 50 times more aflatoxin than did single strength juice. Growth of *A. parasiticus* on intact grapefruit was accompanied by production of aflatoxin that penetrated into the endocarp of the fruit.

Scott *et al.* (1967) obtained an isolate of the *A. ochraceus* group and of *Aspergillus ostianus* Wehmer from a Japanese dried fish product (katsuobushi) and grew them in mycological broth with 0.5% yeast extract. Formation of aflatoxins B_1 and G_1 was observed.

In addition to mold contamination of food, several fungi are used to manufacture certain products. Consequently, these organisms are added to foods as inocula; in other instances, a characteristic flora develops by chance during an aging process (as in the aging of country-cured hams and Italian-type salamis). Examples of the former include use of certain molds to manufacture some types of cheese (Roquefort, Camembert), miso (fermented paste of soybeans, barley and/or rice), soysauce, tempeh, sake, etc. There has been a report that *Aspergillus oryzae*, the mold used to produce miso and soysauce, produces traces of toxin (Detroy *et al.* 1971). However, Hesseltine *et al.* (1966) examined 52 strains of *A. oryzae* obtained from pure culture starters used to manufacture shoyu, miso, and blackbeans and found that none produced aflatoxin. Samples of miso, shoyu, and tempeh also were examined with negative findings. Likewise, Matsuura *et al.* (1970) tested 128 samples of miso, 28 of rice-koji, and 238 strains of industrial koji-molds but all were negative for aflatoxin.

Yokotsuka *et al.* (1967A) obtained 73 industrial strains of *Aspergillus* molds that were used either for food production or found in food preparations. About 30% of these strains, including those used for food manufactured by major Japanese food industries, produced compounds with fluorescence spectra and R_f values on thin-layer chromatoplates that were similar to those of aflatoxins. Further chemical examination of these compounds demonstrated, however, that they were not aflatoxins. Instead, several substances resembling flavacol and aspergillic acid were identified.

Wild rice, a cereal indigenous to northern Wisconsin, Minnesota, and southern Ontario, undergoes a natural fermentation that lasts several weeks. A substantial mold population can develop during the fermentation, especially if the rice is not kept sufficiently moist and turned regularly. Goel *et al.* (1972) recovered aflatoxigenic aspergilli from rice but the toxin was never found in the product.

Aging of country-cured hams is usually done in rooms or attics in which temperature and humidity are not controlled but allowed to fluctuate with the environmental conditions. Heavy mold growth occurs often, particularly on the flesh side of the hams. Ayres *et al.* (1967) reported that the type of mold found is determined by the moisture content of the hams. Thus, during early stages of aging, penicillia predominate; as the available water of the ham decreases, aspergilli begin to grow. When the water activity becomes still less (0.65), *A. ruber* and other xerophilic molds predominate. In the aging of Italian-type salamis (30 to 60 days at 10 to 16°C and a relative humidity of about 75%), profuse mold growth may develop; the predominant molds are either penicillia or penicillia and *Scopulariopsis* (Ayres *et al.* 1967). Bullerman and Ayres (1968) and Bullerman *et al.* (1969) isolated a strain of *A. flavus* from an Italian-type salami and determined the amounts of aflatoxin that could be produced by a known toxigenic strain on salami under simulated conditions of manufacture. They also investigated the amounts of aflatoxins that could be produced by known toxigenic strains of *A. flavus* and *A. parasiticus* on country-cured hams at different stages of aging. They concluded that temperatures below 15°C and humidities of less than 75% prevented aflatoxin development during the aging of salami. Aging of salami for eight weeks and presence of curing ingredients, especially pepper and sodium nitrite, tended to reduce the amount of aflatoxin found. When the temperature approached 30°C, aflatoxin was produced by *A. flavus* and *A. parasiticus* on 6- to 9-month old country-cured hams.

FACTORS AFFECTING MOLD GROWTH AND TOXIN PRODUCTION

The need to produce substantial amounts of mycotoxins for chemical and toxicological studies has led to detailed laboratory investigations of the optimal conditions for toxin production. Some of these conditions include temperature, moisture, pH, aeration and agitation, substrates, and interaction between microorganisms.

Temperature

Agnihotri (1964) and Schroeder and Heins (1967) stated that temperature is the most important factor in the physical environment affecting metabolic activities of fungi. It was this notion that led numerous investigators to study the correlation of various temperatures to growth of and toxin production by some fungi.

Toxins have been generally produced at temperatures that are suboptimal for growth. DiMenna *et al.* (1970) observed that production of sporidesmin by *P. chartarum* is best at 20°C although the optimum growth temperature for the mold is 24°C. Toxins of *Fusarium poae* and *Cladosporium epiphyllum* are produced at temperatures ranging from -15 to 1°C, but no toxin is produced at 23 to 25°C, the optimum growth temperature for these fungi (Joffe 1965). *S. atra* produces toxin over a range of 4 to 36°C and its limits for growth range from 7 to 37°C (Moss and Hill 1970). Rubratoxins are produced by *Penicillium rubrum* on laboratory media at 25 to 30°C (Moss and Hill 1970); however, optimum physical conditions for toxin production have not been established (Jarvis 1971).

Studies to determine temperatures for optimal growth and maximum aflatoxin production by *A. flavus* have resulted in contradictory data. Rabie and Smalley (1965) reported the temperatures for optimal growth and maximum toxin production by *A. flavus* to be 18 and 24°C, respectively. Other investigators (Austwick and Ayerst 1963; Ayerst 1969; Schindler *et al.* 1967) reported 29 to 35°C for optimal growth and 24°C for maximal toxin production. Schroeder and Heins (1967) found a positive correlation between increasing temperature, within the range of 20 to 35°C, and rapid production or accumulation of aflatoxins. Sorenson *et al.* (1967) observed the optimum temperature for production of aflatoxin B_1 and G_1 by *A. flavus* NRRL 2999 on rice to be 28°C. Both aflatoxin B_1 and G_1 were found in lesser amounts at temperatures above 32°C, and the aflatoxin content of rice incubated at 37°C was low even though mold growth was good.

Studies regarding the limiting temperature for aflatoxin production have also been made. Jarvis (1971) reported that aflatoxin is not produced at temperatures below 13°C or above 42°C. However, Sorenson *et al.* (1967) detected small amounts (100 ppb) of aflatoxin B_1 in cultures incubated at 11°C for three weeks. At 8°C, they did not detect any aflatoxin. Ayerst (1969) concluded that the minimum temperature for toxin production does not necessarily agree with the reported minimum growth temperature. That the growth rate of *A. flavus* is not related to aflatoxin production was also reported by Schindler *et al.* (1967) who found that isolates grown at 2.7 and 41°C for 12 weeks did not produce aflatoxins.

Temperature of incubation likewise determines the time required for maximum aflatoxin production. Diener and Davis (1966) obtained maximum yields after 15 days at 20°C or 11 days at 30°C, whereas Jarvis (1971) observed that maximum aflatoxin production occurred between the fourth and seventh day at 24°C (being dependent partially on the carbohydrate content of the medium). Maximum yields of aflatoxin after three to four days at 25°C were reported by Ciegler *et al.* (1966). As the temperature of incubation was increased from 20 to 35°C, Schroeder and Heins (1967) observed that the time required for significant amounts of the toxin to accumulate became progressively shorter.

Temperature also influences the ratio of aflatoxins produced. At lower temperatures, essentially equal amounts of aflatoxin B_1 and G_1 are produced, whereas at 28°C approximately four times as much aflatoxin B_1 as G_1 can be detected. At higher temperatures, relatively less aflatoxin G_1 is formed (Sorenson et al. 1967). Davis and Diener (1970) found the ratio of aflatoxin B_1 and G_1 to be approximately one at 30°C, less than one below 30°C, and more than one above 30°C.

Moisture

Hesseltine et al. (1966) consider temperature and moisture as the two most important factors governing growth of A. flavus in peanuts or in other agricultural commodities. The most commonly used measure of the availability of water to microorganisms is water activity (a_w). This is the ratio of the vapor pressure of water in the substrate to that of pure water at the same temperature and pressure. Ayerst (1969) reported the optimal (0.93 to 0.98) and limiting (0.71 to 0.94) a_w values for some strains of mycotoxic fungi including four species of Aspergillus, two of Penicillium, and one of Stachybotrys. Diener and Davis (1970) stated that aflatoxin formation in stored peanuts generally occurred when the moisture content of the kernels was 10% or higher.

Aeration and Agitation

Hayes and Wilson (1968) reported that patulin and rubratoxin production was as good as or better in stationary cultures than in shake flasks. Mateles and Adye (1965), Shih and Marth (1973A, 1974A), and Shih et al. (1974) also found this to be true for aflatoxin production. However, Davis et al. (1966) and Ciegler et al. (1966) obtained high yields of aflatoxin in shake flasks and in fermenters using high rates of agitation and aeration. They also reported that when yeast extract was added to the medium and cultures were incubated in containers having large surface area to volume ratio, three days were sufficient for surface mycelial mats to produce aflatoxin. Hayes et al. (1966) reported maximum mycelium and aflatoxin production (212.5 mg/l) with an aeration rate of 9,000 ml air/min when A. flavus grew on a yeast extract-sucrose (YES) medium. The amount of aflatoxin produced with substantial aeration and minimal agitation on YES medium was approximately four times that obtained by Mateles and Adye (1965) when they grew the same strain of A. flavus on a synthetic medium.

The effects of carbon dioxide in combination with reduced relative humidity and temperature on growth and aflatoxin production by A. flavus on peanuts were investigated by Sanders et al. (1968). They observed that at a constant temperature, an increase in carbon dioxide concentration caused a decrease in yield of aflatoxin, and at a given carbon dioxide concentration, lowering the temperature decreased the amount of aflatoxin that was produced. At approximately 86% relative humidity, 20% carbon dioxide and 17°C, visible growth and aflatoxin production was inhibited, as well as by 60 or 40% carbon dioxide at 25°C. Landers et al. (1967) found that increasing the carbon dioxide concentration from 0.03 to 20% did not reduce mold development. However, growth and sporulation were inhibited when the concentration of carbon dioxide exceeded 20%. They also observed lower toxin production with concentrations of atmospheric oxygen below 20%.

Shih and Marth (1973A) evaluated several atmospheric conditions for their effects on production of aflatoxin during growth of A. parasiticus in a synthetic medium in a fermenter. They observed that the maximal yield of aflatoxin was obtained when no air was sparged into the quiescent medium. Increasing the rate of aeration enhanced glucose utilization and acid formation but reduced toxin formation by the mold. Replacement of air by various mixtures of O_2 and CO_2 or O_2 and N_2 suppressed toxin formation. Increasing proportions of CO_2 or N_2 in the atmosphere enhanced their inhibitory effect on aflatoxin formation, and complete inhibition of toxin synthesis occurred in atmospheres of 100% CO_2 or 100% N_2. Synthesis of toxin was suppressed more by a high concentration of CO_2 than of N_2.

pH

The effect of pH on production of most mycotoxins other than aflatoxin does not appear to have been investigated, although several workers report the final pH of spent media. The initial pH value of the medium has been reported as not appreciably affecting aflatoxin production except that the organism did not grow well at pH values below 4 (Davis et al. 1966). Unpublished reports cited by Jarvis (1971) indicated that higher yields of aflatoxin were obtained with initial pH values near neutrality. The final pH of media after aflatoxin production has been reported variously as 4 (Davis et al. 1966), 5 (Hayes et al. 1966), and 8 (Diener and Davis 1969), depending on the medium used. Using a synthetic medium in submerged cultures, Mateles and Adye (1965) observed a drop from pH 4 to 2.1 to 2.3 when different carbohydrates were incorporated into the medium as the carbon source for aflatoxin production. These results contradict unpublished findings cited by Jarvis (1971) which showed an initial drop from pH 7 to 4.2 to 4.4 followed by a rise to pH 8.6 to 8.8 soon after the commencement of mycelial autolysis.

The effect of pH on aflatoxin production on a proteinaceous substrate was investigated by Lie and Marth (1968). The pH of slurries of cottage cheese curd was adjusted to 1.0 to 11.35 using 6 N HCl and 6 N NaOH and to 1.8 to 10.7 using 6 N lactic acid and 6 N NH_4OH. One set of slurries treated with HCl and NaOH was inoculated with a spore suspension of a toxigenic strain of A. flavus, and another set with spores of toxigenic A. parasiticus. Spores of these molds were also used to inoculate two sets of cottage cheese slurries whose pH values were adjusted with lactic acid and NH_4OH. The authors observed that when HCl and NaOH were used, A. flavus grew at all initial pH values in the range of 1.7 to 9.34 and that maximum growth occurred at pH values between 3.42 and 5.47. When lactic acid and NH_4OH were used, there was a narrow pH range in which A. flavus could initiate growth (3.13 to 5.90). The authors also noted a pH change toward neutrality as the molds grew. Greatest aflatoxin concentrations were obtained from samples treated with HCl and NaOH which initially were at the extremes, both acidic and alkaline, of the pH values permitting growth. When lactic acid and NH_4OH were used, largest amounts of aflatoxin appeared in samples with the lowest initial pH values which permitted mold growth. Similar observations were made when A. parasiticus spores were used as the inoculum except that this mold was able to initiate growth over a wider pH range than A. flavus did when the substrate was treated with lactic acid and NH_4OH.

Substrate or Medium for Growth

The yield of toxin, particularly aflatoxin, on natural substrates is superior to that obtained on semisynthetic or chemically defined (synthetic) media. Wildman et al. (1967) obtained 1,000 μg aflatoxin per gram of moistened shredded wheat. The same yield was reported by Vogel et al. (1965) using moist crushed peanuts and by Shotwell et al. (1966) with moist rice. Yields up to 4000 μg/g of tissue were obtained by Cucullu et al. (1966) using hearts of peanut kernels. Wildman et al. (1967) concluded that products like moist peanuts, rice, and shredded wheat, all having a large surface area exposed to air and a high ratio of nutrient to moisture are good substrates for aflatoxin production. Diener and Davis (1969) added that, generally, seeds with a high carbohydrate content (rice, wheat, and corn) can support production of relatively larger amounts of aflatoxin than seeds with a low carbohydrate and high oil content such as peanuts, cotton, and soybeans. The same was also reported by Borker et al. (1966). A study cited by Detroy et al. (1971) revealed that strains of A. flavus grown on cheese and other animal products produced less aflatoxin than when the mold was grown on substrates with a high carbohydrate content. It was reasoned that this was probably because the former contain large amounts of oil that are not immediately metabolized by the mold. Lie and Marth (1968) also obtained relatively poor yields of aflatoxin on a proteinaceous (casein) substrate. However, the same investigators obtained substantial amounts of aflatoxin when *A. flavus* and *A. parasiticus* were grown on Cheddar cheese at room temperature (Lie and Marth 1967).

To study the chemistry and toxicology of mycotoxins, it is obviously desirable to have them available in gram quantities. Because chemical synthesis of aflatoxins has not yet been achieved, they must be produced by biological means. Although this may be readily done by growth of *A. flavus* on solid substrates such as ground peanuts or shredded wheat, production of aflatoxins in submerged culture on semisynthetic or synthetic media would be preferred because of ease in scale-up, simplicity of extraction and purification, and the suitability of these media to study the physiology of aflatoxin production. However, Vogel *et al.* (1965) and Kulik and Holaday (1966) found evidence that aflatoxin yields may decrease after repeated transfer of aspergilli on a synthetic medium but can be restored after transfers of the mold on sterilized peanuts.

Although the yield of toxin has frequently been superior on natural substrates rather than in synthetic media, in a few instances (production of toxins by *Penicillium islandicum, Penicillium citreoviride, P. citrinin,* and *Penicillium urticae*) standard Czapek Dox or Raulin-Thom media have proved suitable as substrates. However, frequently it was found essential to supplement mineral salts media with malt extract (Hayes and Wilson 1968), yeast extract (Davis *et al.* 1966; Merwe *et al.* 1965), peptone, cornsteep liquor, casamino acids (*Davis et al.* 1966, 1967, 1968; Mateles and Adye 1965; Schroeder 1966), or other extracts to obtain good toxin production.

In a medium such as Czapek Dox broth, little or no aflatoxin was produced even though the fungus grew well. When 1% yeast extract was added to the medium, the fungus synthesized the toxin (Davis *et al.* 1964). Addition to the medium of 15 elements generally required by plants increased yields of aflatoxin. However, aflatoxin was not formed if minerals were added in the absence of yeast extract. Thus, one or more of the added minerals stimulated production of aflatoxin but did not replace the required factor in yeast extract. Davis *et al.* (1966) reported that isolates of *A. flavus* as stationary cultures produced 0.2 to 63 mg of aflatoxin B_1 and G_1 per 100 ml in a semisynthetic medium consisting of 20% sucrose and 2% yeast extract incubated for 6 days at 25°C. Yeast extract-sucrose (YES) medium has also been used by Scott *et al.* (1970) to produce 14 mycotoxins. Extracts like cornsteep liquor and casamino acids are poor supplements for ochratoxin production.

Eldridge *et al.* (1965) developed a chemically defined liquid medium that supported abundant growth and aflatoxin synthesis by strains of *A. flavus*. The medium consisted of a mixture of carbohydrates, vitamins, minerals, and organic nitrogenous compounds, and yielded 10 to 110 mg toxin per liter. Mateles and Adye (1965) and Parrish *et al.* (1966) reported a maximum total yield of 91 and 4.3 μg aflatoxin B_1 per gram of culture medium, respectively, when a synthetic medium consisting of glucose, ammonia, and salts was used. Wildman *et al.* (1967) grew *A. flavus* in a series of chemically defined liquid media and obtained the highest yield of aflatoxin (350 μg/g) in a medium consisting of Czapek Dox broth, 5.8 to 8.6% glucose, and 0.4% citric acid.

Various carbon sources have been used successfully in synthetic and semisynthetic media. For aflatoxin production by *A. flavus*, glucose, sucrose, or fructose are the preferred carbon sources; however, 14 of the 17 carbon sources tested by Mateles and Adye (1965) supported production of aflatoxin. Davis and Diener (1968) reported that mannose and xylose stimulated aflatoxin production by *A. parasiticus* although these same carbohydrates have been reported to inhibit toxin production by *A. flavus* (Mateles and Adye 1965). Production of ochratoxin by *A. ochraceus* is optimal in the presence of either 3% sucrose or of a combination of lactose and sucrose. Only 75% as much ochratoxin was obtained on galactose. Patulin production by *P. urticae* and *Penicillium patulum* on Czapek Dox medium occurs maximally in the presence of glucose or maltose.

Lactic acid has been reported to increase ochratoxin production on media containing casamino acids and ammonium acetate as nitrogen sources, while citric acid stimulated aflatoxin production on Czapek Dox medium containing glucose (Wildman *et al.* 1967). Davis and Diener (1968) presented data that supported the hypothesis of Adye and Mateles (1964) that a carbon compound, to support both aflatoxin production and mold growth, must be metabolized through both the hexose monophosphate and the classical glycolytic pathways. Maximum concentration of aflatoxin in the medium has been correlated with exhaustion of fermentable carbohydrate and onset of mycelial autolysis (Jarvis 1971).

Inorganic nitrogen sources for semisynthetic and synthetic media vary considerably in their ability to stimulate aflatoxin production, ammonium sulfate and potassium nitrate being the most effective (Mateles and Adye 1965). Poor yields of ochratoxin and sporidesmin have been obtained with inorganic nitrogen sources (Ferreira 1966). Glycine and glutamic acid are the most effective single amino acids to enhance aflatoxin production, followed by alanine, aspartic acid, and glutamine (Davis *et al.* 1967). Asparagine was reported as stimulating sporidesmin production on a glucose-mineral salts medium (Jarvis 1971).

The effect of inorganic ions on aflatoxin production is variable. Zinc is essential, cadmium and iron stimulate aflatoxin production, but iron also depresses mycelial growth (Davis and Diener 1968; Nesbitt *et al.* 1962). Davis *et al.* (1967) found that molybdenum stimulated toxin production. Lee *et al.* (1966) observed that barium inhibited aflatoxin production and that cultivation of a toxigenic strain of *A. flavus* on a barium-containing medium resulted in production of an atoxigenic mutant.

Shih and Marth (1973A, 1974A,B) compounded a synthetic medium that supported production by *A. parasiticus* of approximately 1500 μg aflatoxins (B_1 + B_2 + G_1 + G_2) per 100 ml of substrate. The medium contains, per liter, 50 g glucose, 6 g $(NH_4)_2SO_4$ 5 g KH_2PO_4, 6.4 g K_2HPO_4, 0.5 g $MgSO_4 \cdot 7H_2O$, 2 g glycine, 2 g glutamic acid, 10 mg $FeSO_4 \cdot 7H_2O$, 5 mg $ZnSO_4 \cdot 7H_2O$, and 1 mg $MnSO_4 \cdot H_2O$.

Microbial Interaction

In nature, toxigenic molds seldom exist as pure cultures but are usually associated with other microorganisms. This association may lead to competition between the toxigenic fungi and other microorganisms with resulting restriction of mold growth and toxin production. The converse also could be true; other microorganisms could stimulate growth and production of toxin by molds.

Schroeder and Ashworth (1965) noted smaller amounts of aflatoxin in peanut kernels from parasite-damaged pods than in kernels from broken pods. They concluded that microbial competition or microbial breakdown of toxin might be responsible. Ashworth *et al.* (1965) demonstrated limited development of *A. flavus* when grown competitively with *A. niger* and *Rhizoctonia solani* on peanut kernels. These same workers also reported that *A. niger, R. solani, Fusarium oxysporium*, and *Macrophomina phaseoli* degraded aflatoxin in peanuts and in an aflatoxin-containing liquid medium. Wildman *et al.* (1967) obtained smaller aflatoxin yields when *A. flavus* grew competitively in a liquid medium with *Penicillium* sp. *Rhizopus oryzae*, inoculated into a culture of *A.*

parasiticus, has been reported to inhibit aflatoxin production (Jarvis 1971). *R. oryzae* was also able to metabolize preformed aflatoxin in a yeast-extract sucrose (YES) medium.

Alderman *et al.* (1973) grew *A. parasiticus* and *P. rubrum* separately, associatively, or in combination with *Penicillium italicum* and/or *Lactobacillus plantarum*. A glucose-salts broth and single-strength and concentrated grapefruit juice served as substrates. In the broth medium incubated at 28°C for 7 days, production of aflatoxin by *A. parasiticus* was enhanced when the culture also contained *P. rubrum* plus *L. plantarum*; was unaffected by the presence of *P. italicum*; and was reduced by the presence of *P. rubrum*, *P. rubrum* plus *P. italicum*, or *L. plantarum*. Production of aflatoxin in single strength grapefruit juice held at 28°C for 7 days was enhanced when *A. parasiticus* grew together with *P. rubrum* and *L. plantarum*. Presence of all other cultures or combinations reduced aflatoxin production. Similar results were obtained with concentrated grapefruit juice.

Rubratoxin production in broth was enhanced when *P. rubrum* grew in association with *A. parasiticus*. All other combinations of cultures yielded less rubratoxin than did *P. rubrum* by itself. In single strength grapefruit juice, presence of *A. parasiticus* caused a marked decline in rubratoxin production. Slight increases in yield of rubratoxin were obtained when cultures contained *L. plantarum* or *A. parasiticus* plus *L. plantarum*. Other combinations of cultures neither enhanced nor reduced rubratoxin production. All combinations of cultures enhanced rubratoxin production in concentrated grapefruit juice with the highest yield appearing when *A. parasiticus* and *L. plantarum* grew together with *P. rubrum*. Production of rubratoxin was similar in single strength and concentrated grapefruit juice.

Burmeister and Hesseltine (1966) surveyed 329 microorganisms for aflatoxin sensitivity. Twelve *Bacillus* sp., *Clostridium sporogenes*, and a streptomycete were inhibited when 30 μg of crude aflatoxin (36% pure)/ml was incorporated into the growth substrate. A strain of *Bacillus brevis* and two of *Bacillus megaterium* were found most sensitive, being inhibited by 10 and 15 μg aflatoxin/ml, respectively. Lillehoj *et al.* (1967) reported aflatoxin B_1 to inhibit growth of *Aspergillus awamori*, *Penicillium chrysogenum*, *Penicillium duclauxi*, and four isolates of *A. flavus*. Aflatoxin B_1, as well as crude aflatoxin, has been shown to exhibit antimicrobial activity against various strains of *Streptomyces* sp. and *Nocardia* sp. (Arai *et al.*, 1967).

Ciegler *et al.* (1966) screened approximately 1000 microorganisms for their ability to degrade aflatoxin. Only *Flavobacterium auranti-acum* was able to irreversibly remove aflatoxin B_1 from a nutrient solution. Under experimental conditions, *F. aurantiacum* completely detoxified several aflatoxin-containing foods.

METHODS TO DETECT AND MEASURE MYCOTOXINS

Methods to detect and estimate the concentration of mycotoxins have been developed with particular emphasis on aflatoxins. Basically, however, the general principles involved could apply to assays for other mycotoxins with appropriate variations in methodology. The following discussion will describe methods for aflatoxin assay, since they have been refined more than procedures to measure other mycotoxins.

Physicochemical Analyses

Detroy *et al.* (1971) mentioned the following factors that may complicate physicochemical assays for aflatoxins: (a) aflatoxins are found in agricultural commodities which are complex and hence a preliminary extraction procedure is necessary; (b) extraction procedures adequate for one commodity are inadequate for another; (c) the amount of toxin is small, usually in the microgram per kilogram range; (d) distribution in natural products is usually uneven, requiring large samples for toxin detection; and (e) the assay is nonspecific and requires further confirmatory tests.

Early methods developed for physicochemical assays have been summarized into unit operations by Borker *et al.* (1966) and by Pons and Goldblatt (1969). These operations include: (a) primary extraction, (b) extract purification, and (c) thin-layer chromatographic separation of aflatoxins. The latter investigators (Pons and Goldblatt 1969) went a step further by adding another unit operation which involved estimation of amounts of aflatoxins on TLC plates.

Extraction

Early procedures to remove aflatoxins from mold-damaged products involved long exhaustive Soxhlet extraction with methanol. These procedures had a common disadvantage in that large amounts of polar lipids, pigments, and carbohydrates were present in the primary extracts, and often remained even after subsequent steps to purify the extract. Later, several investigators proposed the use of more efficient extraction solvents, in rapid equilibrium systems, to reduce materially the time of extraction. Extraction procedures for a particular food or feed differ depending on the composition of the product. Shih and Marth (1971) developed a rapid and efficient method to recover aflatoxin from cheese. The method was based on the fact that chloroform and methanol can extract aflatoxin efficiently from toxin-containing substrates and that hexane can be used to remove interfering lipids and pigments from crude extracts. Mixtures of solvents (chloroform, methanol, and water) were prepared and tried. "The procedure involves: (a) blending the sample with a mixture of chloroform, methanol, and water [solvents are used in such proportions that a miscible (monophasic) system is formed], (b) adding more chloroform and water so the mixture becomes biphasic, (c) filtering to remove the food residue, (d) separating the lower chloroform layer which contains virtually all of the aflatoxin, and (e) purification, if necessary, of the material in step (d) after it has been concentrated." This procedure was adapted for extracting toxins from rice, peanut butter, and corn meal and was found suitable for aflatoxin recovery provided the chloroform:methanol:water mixture was kept in proportions of 25:50:20, v/v/v, before and in proportions of 50:50:45, v/v/v, after dilution. Detroy et al. (1971) described different extraction procedures for peanut and peanut products, corn and corn products, cottonseed and cottonseed meal, oats, cocoa, coffee and tea, copra and coconut, milk, meat, and others. Even for the same product, different extraction procedures have been proposed by various workers. Pons and Goldblatt (1969) tabulated several extraction procedures (both exhaustive and equilibrium methods) to recover aflatoxins from peanuts.

Purification

Purification of the extract involves preliminary removal of pigments, polar lipids, and carbohydrates, as well as separation of aflatoxins from residual interfering materials. Pons and Goldblatt (1969) also tabulated representative systems to remove pigments and polar lipids as well as systems to separate aflatoxins. They concluded that chromatographic extract purification systems provide cleaner extracts for subsequent thin-layer chromatographic analysis than those obtained solely by liquid-liquid partition purification systems.

TLC Separation

Separation of aflatoxins by TLC is possible because of the inherent fluorescence properties of the toxins under longwave ultraviolet illumination. Pons and Goldblatt (1969) listed the different conditions proposed by various workers to separate aflatoxins on silica gel; including the type of silica gel and its thickness, development solvent, and solvent path. Satisfactory resolution of the different aflatoxins is often a problem. The degree of resolution as well as the R_f values are markedly influenced by variables which have been examined in detail by Nesheim (1969). These variables included: (a) differences in commercial silica gel-calcium sulfate adsorbent preparations, (b) adsorbent particle size, (c) concentration and nature of the calcium sulfate binder, (d) silica gel layer thickness, (e) moisture content, (f) vapor phase composition in the developing chamber, and (g) the developing solvent.

In an effort to find improved procedures for measurement of aflatoxins with TLC, Shih and Marth (1969B) studied some of these variables. They reported that a 0.25-mm deep layer of Adsorbosil-5 provided better separation than did one with a thickness of 0.50 mm. They also compared 12 different solvent systems, suggested by other investigators, for their ability to separate aflatoxins. When plates were developed in lined and equilibrated tanks, unsatisfactory separation was observed, except when methanol:chloroform, 1:99, served as the solvent system. Marked improvement in separation was obtained with several solvent systems when an unlined and un-equilibrated tank was used for development. Approximately 60 min were required to complete development when the tanks were unlined, whereas approximately 30 min were needed when the tank was lined. They reasoned that the slower movement of the solvent front may have served to improve resolution of toxins when plates were developed in unlined tanks. Additional variables like distance travelled by the solvent front and storage of heat activated TLC plates were also studied. Improved separation of aflatoxin was obtained when the distance travelled by the solvent front was increased from 15 to 18 cm, and when freshly activated (2-hr old) rather than stored (8-hr old) plates were used. Stubblefield *et al.* (1969) suggested addition of water to the solvent system used in TLC to obtain more reproducible results in laboratories where temperature and humidity vary. Shih and Marth (1971) used chloroform:methanol:water, 98:1:1 as a solvent system and obtained reproducible results and good toxin resolution.

Because of these variables, Pons and Goldblatt (1969) recommended that to identify aflatoxin in a sample extract, it is essential to use authentic standards chromatographed under the same conditions as the unknown.

Quantitation

Initially, estimation of the amount of aflatoxin on TLC plates was based on visual comparison of fluorescence intensities under long-wave ultraviolet illumination as exhibited by the unknown sample and a suitable standard. This procedure is properly classified as a semiquantitative method. This visual estimation has a precision of no better than ±20% for a single observation and under operating conditions is probably close to ±28% (Beckwith and Stoloff 1968).

Among the first efforts to improve the accuracy and precision of aflatoxin measurements by use of objective methods was the report by Nabney and Nesbitt (1965). Concentrated chloroform solutions of partially purified extracts from peanut meals were streaked across silica gel plates that were then developed successively with diethyl ether and with chloroform:methanol to separate aflatoxins B_1 and B_2 as bands at R_f values of 0.30 and 0.25, respectively. The material in the bands was removed, eluted with methanol, and the eluate examined by spectrophotometry at 363 mμ, where aflatoxins B_1 and B_2 had similar absorption spectra. Since absorption is much less sensitive than fluorescence, by a factor of 1000 or more, some 3 to 10 μg of isolated aflatoxin were required for accurate measurement.

More sensitive fluorodensitometric measurement of aflatoxins directly on silica gel coated plates was first suggested by Ayres and Sinnhuber (1966) who used a recording densitometer equipped to measure fluorescence emission. Pons et al. (1966) and Stubblefield et al. (1967) evaluated this method and found it to be a very useful tool for measurement of aflatoxin. Presently used densitometric methods employing commercially available fluorodensitometers have proven more accurate than the visual method with an average deviation of about ±2% (Stubbelfield et al. 1967). Beckwith and Stoloff (1968) modified the densitometric procedures as used by most investigators to eliminate inherent inaccuracies.

The comparatively expensive instruments used for these analyses led Peterson et al. (1967) to develop an economical combination of a simple darkroom densitometer connected to an available recorder which gave more reliable data for routine aflatoxin analyses than did visual observation.

The basis for most tests used to measure aflatoxin content of food or feed has been fluorescence under ultraviolet light. Since valuable products may be condemned on the basis of such tests, accuracy is imperative. Compounds with fluorescence resembling that of the aflatoxins are often found in extracts of agricultural products prepared for aflatoxin assay. They may be metabolites of micro-organisms other than those recognized as producers of the toxins or they may be natural constituents of the products. Thus, occurrence of compounds with fluorescent and chromatographic behavior similar to that of aflatoxins has dictated the need for confirmatory tests.

Confirmatory Chemical Tests

Andrellos and Reid (1964) devised a test to confirm the presence of aflatoxin B_1. They treated portions of aflatoxin B_1 with three reagents—formic acid plus thionyl chloride, acetic acid plus thionyl chloride, and trifluoroacetic acid—and then spotted the reaction products on TLC plates. These consistently and characteristically yielded spots on TLC plates with R_f values different from those for authentic aflatoxin B_1. The tests also proved equally applicable to aflatoxin G_1. A modification of the procedure just described was studied collaboratively in 19 different laboratories and Stoloff (1967) recommended its adoption as an official method. Crisan and Grefig, as cited by Pons and Goldblatt (1969), suggested the preparation of oximes and 2,4-dinitrophenylhydrazones of aflatoxin B_1 and B_2 as confirmatory tests. Aflatoxins G_1 and G_2 do not form analogous derivatives. Spraying a TLC plate with 2,4-dinitrohydrazine produced a deep yellow to orange spot, indicating a derivative of aflatoxin B_1.

Besides derivative formation, further assurance that aflatoxin is truly present in a given unknown extract may be obtained by use of internal standards spotted on top of aliquots of the unknown on two or more TLC plates which are then developed using different solvent systems.

Exposure of developed plates to iodine vapors to distinguish interfering materials from aflatoxins was suggested by Mislivec *et al.* (1968). Interfering materials ceased to fluoresce under ultraviolet light and became visible in normal light as maroon spots, whereas the aflatoxin continued to fluoresce unless the treatment was extended beyond 1 min.

Instrumental analyses have also been used to confirm the presence of aflatoxins. Their characteristic ultraviolet, infrared, nuclear magnetic resonance, excitation, and fluorescence spectra have been presented earlier in this review. Gajan *et al.* (1964) proposed oscillographic polarity to identify aflatoxins B_1 and G_1. Purified B_1 and G_1 gave characteristic oscillographic traces having peak potentials at -1.33 and -1.25 ± 0.02 volts versus a silver wire electrode, respectively, in an electrolyte containing tetramethyl ammonium bromide and lithium chloride in aqueous methanol. This has been found suitable for quantitative analysis of fairly pure preparations.

Biological Assay

Lack of a definitive chemical test has necessitated further confirmation of suspected aflatoxin contamination using a biological test. However, biological assays thus far developed also suffer from the same inadequacy as chemical assays, *i.e.*, comparative lack of specificity. In addition, bioassays are not as sensitive as TLC for detection and should only be used as a confirmatory test following chemical identification. Nevertheless, biological assays have certain advantages which are discussed in detail by Brown (1968). Briefly, he stated that "the active compound does not have to be known, or if it is known, the chemical identification is not necessary. The active compound does not have to be in a pure state."

Asplin and Carnaghan (1961) proposed the use of ducklings for aflatoxin assay, and when combined with chemical identification, this is probably the most widely used and accepted procedure to confirm the presence of aflatoxin in various commodities. Sensitivity of ducklings to injury by aflatoxin and the almost immediate induction of bile duct proliferation are the two reasons for widespread use of the technique. The assay is based, not on lethality, but on the degree of hyperplasia of the bile duct epithelium which is roughly related to the amount of toxin fed to the animal. Butler (1964) described the histological changes in the livers of one-day old Khaki Campbell ducklings caused by a 15-μg dose of aflatoxin. These included extensive biliary proliferation in the liver with fatty degeneration of the peripheral parenchymal cells. Other investigators also reported on liver injury in ducklings caused by aflatoxin poisoning. It should be noted that hyperplasia induced by aflatoxin is not specific since other toxic agents like dimethylnitrosamine and cycasin cause a similar response (Butler 1964). Butler (1964), however, also observed that hyperplasia was not seen with such hepatotoxic agents as carbon tetrachloride, ethionine, and thio-acetamide. Sterigmatocystin, a compound closely related to aflatoxin and believed by some to be an intermediate in its biosynthesis, was found to be nontoxic to ducklings and evoked no bile duct hyperplasia (Lillehoj and Ciegler 1968). Hartley *et al.* 1963) separated crude aflatoxin into four fractions and found the following LD_{50} values (single dose to day-old ducklings, 48-hr incubation): aflatoxin B_1, 30 μg and aflatoxin G_1, 60 μg. Aflatoxins B_2 and G_2 were less toxic than B_1 and G_1; the lethal dose for B_2 and G_2 (over 4 days) was greater than 200 μg.

Although the duckling assay is the generally accepted biological test, it is comparatively expensive, time-consuming, and requires highly trained technicians. Also, this assay is certainly not quantitative and should only be considered semiquantitative or qualitative.

Other biological systems have been examined for aflatoxin bioassay. Verrett *et al.* (1964) developed the chicken embryo technique which involves injection of eggs by either the yolk or air cell route followed by incubation of the eggs for 21 days. They found that the air cell injection procedure was more sensitive than the yolk injection technique, that there was a rapid decrease in sensitivity to aflatoxin with increasing age, and the maximum toxic effect was obtained with pre-incubation injections. In addition to mortality, they also found reproducible and significant retardation of growth in the non-surviving embryos. Platt et al. (1962) reported that injection of aflatoxin into the yolk of 5-day old chicken embryos caused death and that the amount of toxin required was only 1/200th of that needed for a positive result in the day-old duckling assay. On the other hand, Gabliks et al. (1965) reported that duck embryos were 4 to 5 times more susceptible to aflatoxin than were chick embryos.

Townsley and Lee (1967) described the inhibition by aflatoxin B_1 of cell cleavage in fertilized mollusk (Bankia setacea) eggs, without preventing fertilization or nuclear division. Thus, in the presence of aflatoxin, fertilized eggs remained unicellular and became multi-nuclear, whereas controls were multicellular. This bioassay requires a minimum of technique and training and is sensitive to 0.05 µg aflatoxin per milliliter. Disadvantages of this assay include the requirement for seawater, the availability of the particular mollusk, and the need to work at low temperatures.

Trout has been reported to be extremely sensitive to aflatoxin (Legator 1969), and this led investigators to study smaller fish which might also be sensitive and could serve as a low-cost test animal. Abedi and McKinley (1968) tested aflatoxin B_1 against developing eggs and larvae of zebra fish (Brachydanio rerio) and found them acutely sensitive to submicrogram quantities of the toxin.

Numerous cell cultures have been described in the literature (Legator 1969). Some of these are human embryonic lung cells, human leukocytes, marsupial kidney, calf kidney cell line, duck and chicken embryo primary culture, HeLa cell line, rat liver slices, duck liver slices, rat kidney slices, rat fibroblasts, and primary explants of human embryo liver cells.

Juhasz and Greczi (1964) reported that addition of aflatoxin to calf kidney monolayer cultures caused destruction of the cytoplasm and nucleus in affected cultures. Legator et al. (1965) found growth of cultured heteroploid human embryonic lung cells was reduced after 48 hr of incubation with 0.05 to 1.0 ppm of aflatoxins; cells did not grow when the concentration of aflatoxin was 5.0 ppm. After an 8- to 12-hr exposure to 1 ppm aflatoxin B_1, there was a 92% increase in giant cells over the control. When Chang cells were used, there were decreases in cell number, protein, RNA, and DNA with increasing aflatoxin B_1 concentrations (Gabliks *et al.* 1965). This suggested enlarged cells since the cell number decreased but the protein, RNA, and DNA content per cell increased with an increasing concentration of aflatoxin. Cultures of chick embryo liver cells showed a characteristic morphological change 12 hr after dosing with aflatoxin (Clements 1968); the nucleoli decreased in volume relative to the nuclei and eventually disappeared. Gabliks *et al.* (1965) described the general effect of aflatoxin B_1 on cell cultures as inhibition of growth followed by progressive granulation, rounding, and finally sloughing of the cells from the glass. Since aflatoxin induces hepatomas in test animals, possible susceptibility may be determined by means of cell culture technique.

Microbiological assay methods for antibiotics based on inhibition of bacterial growth were tested for aflatoxin determination. Burmeister and Hesseltine (1966) surveyed 329 microorganisms (including bacteria, fungi, algae, and one protozoan) for their sensitivity to aflatoxin. They reasoned that a sensitive microorganism might supplement the duckling assay in establishing if contaminated feeds contained sufficient aflatoxin to be toxic. One strain of *B. brevis* and two of *B. megaterium* were most sensitive to aflatoxin, being inhibited by 10 and 15 μg per milliliter, respectively. This sensitivity was subsequently used by Clements (1968) to develop a rapid assay involving an agar diffusion technique. A spore suspension of *B. megaterium* was used to seed agar plates and a zone of inhibition caused by the aflatoxin was measured after 15 to 18 hr of incubation at 35 to 37°C. Clements (1968) noted that aberrant forms consisting of elongated cells could be observed in stained smears of the organism obtained from the margin of the zone of inhibition. The advantages of this procedure include the following: less than 1 hr is required to set up the test, results are obtained after overnight incubation, and the test is sensitive to as little as 1 μg aflatoxin B_1. Burmeister (1967) obtained a patent for a serial dilution method which utilized *B. megaterium* to test for aflatoxin. Growth or lack of growth after 72 hr of incubation at 30°C is the criterion used for the presence or absence of aflatoxin. *B. brevis* could also be used as the assay organism in this procedure.

Smith (1965) suggested use of *Escherichia coli* as the basis of an *in vitro* assay for aflatoxin in foodstuffs to replace the lengthy and tedious duckling feeding method. The basis for this assay is that pure aflatoxin, added *in vitro*, inhibits the amino acid-s-RNA ligase (or the amino acid-activating enzyme) activity in *E. coli* preparations.

Legator (1969) summarized the genetic and nongenetic responses of *E. coli*, *Staphylococcus aureus*, *F. aurantiacum*, 12 *Bacillus* species, *Streptomyces* sp., *C. sporogenes*, *Nocardia* sp., fungi, gram-positive and gram-negative bacteria, and cell cultures. Genetic responses included inhibition of DNA polymerase, phage induction, abnormal cell morphology (filament formation) in microorganisms, inhibition of mitosis and of DNA synthesis, an increase in thymidine kinase activity, and chromosome aberrations. The principal effect on microorganisms was inhibition of growth. In cell cultures, nongenetic effects included cell destruction, growth inhibition, reduction of leucine (^{14}C) incorporation into protein, reduction of leucine activation, and cell degeneration. Legator (1969) also attempted to rate these biological systems as to their sensitivity, ease of operation, and potential significance to man. Because of the numerous biological systems suggested for bioassay, he proposed that further investigations should be directed towards improving currently available methods rather than looking for additional biological indicators.

Screening

It is important to note that occurrence of fungi on a specific product alone does not necessarily indicate the presence of any mycotoxin. Thus, when one is confronted with numerous fungi obtained from samples of food or feed, it is necessary to screen them for toxigenic species or strains and for toxin production without the lengthy and involved physicochemical and biological procedures which have been described. Rapid screening procedures have been developed.

Vogel *et al.* (1965) described a rapid test to recognize aflatoxin-synthesizing strains of the *A. flavus-oryzae* group. Their test involved cultivating the molds on Czapek Dox agar enriched with an aqueous extract of peanuts, and observing a bright blue fluorescence in the medium when placed under an ultraviolet lamp. This fluorescence is specific for aflatoxin B$_1$. Three consecutive subcultures on the medium appear necessary if an unknown strain is negative in the first test.

Mislivec *et al.* (1968) used rice as the substrate when they screened molds in the genera *Aspergillus* and *Penicillium* for aflatoxin production. Chloroform extracts were analyzed by TLC and fluorescence was not used as the final criterion for toxigenic potential of the isolates. Rather, the fluorescent compounds with R$_f$ values similar to those of aflatoxin were separated by two-dimensional chromatography and treated with iodine vapors. Aflatoxins were distinguished from interfering materials by their continued fluorescence in ultraviolet light after the iodine treatment.

Scott *et al.* (1970) described a convenient TLC screening

procedure to detect 18 mycotoxins. With this procedure, they also evaluated their earlier semi-micro culture technique (van Walbeek 1969). Dense spore suspensions from molds grown on fresh slants of potato dextrose agar plus yeast extract were inoculated into 5 ml of yeast extract-sucrose (YES) medium in a 30-ml vial and incubated for 7 days at 25°C. The medium was extracted with hot chloroform and concentrated before TLC analysis. The screening procedure involved TLC with two suitable general solvent systems and one initial spray reagent. The solvent systems used were toluene-ethylacetate-formic acid (6:3:1) and benzene-methanol-acetic acid (24:2:1). Toxins were detected in visible or ultraviolet light before and after spraying the plate with a freshly prepared mixture of 0.5 ml p-anisaldehyde in 85 ml of methanol containing 10 ml glacial acetic acid and 5 ml concentrated sulfuric acid and then heating at 130°C for 8 to 20 min. Although all toxins migrated when acidic solvent systems were used, nivalenol remained close to the origin and citrinin and luteoskyrin formed streaks. Fluorescent colors of certain mycotoxins in ultraviolet light were described by Steyn (1969). The anisaldehyde spray allows detection of nonfluorescent toxins. Scott *et al.* (1970) established the order of migration and color of the different mycotoxins using the two solvent systems and then mixed the toxins and spotted the mixtures. Production of gliotoxin and zearalenone was not obtained in YES medium; however, the value of YES medium for screening toxigenic fungi by the semi-micro technique was established.

Eppley (1968) also introduced a method to screen for zear-alenone, aflatoxin, and ochratoxin. A new developing system to simultaneously separate and detect 11 different mycotoxins was likewise described by Steyn (1969). He removed water-soluble and lipid materials by liquid-liquid partition; neutral material was also removed from mixtures containing acidic mycotoxins (cyclopiazonic acid, secalonic acid D, and ochratoxins A and B). Thin-layer chromatoplates were prepared with a slurry of silica gel and aqueous oxalic acid in a 1:2 ratio. The solvent system he used was chloroform-methylisobutylketone (4:1). Plates were spotted with each mycotoxin in chloroform-methanol solution and allowed to develop in a tank saturated with solvent vapor. Spots were detected by exposure to longwave ultraviolet light and spraying the plate with color reagents. He used two spray reagents: concentrated sulfuric acid and 1% ethanolic ferric chloride. When oxalic acid was omitted from the silica gel slurry, mobility of neutral metabolites was unaffected, whereas acidic compounds did not move.

Jarvis (1971) summed up the principles involved in mycotoxin methodology: "When screening fungi for potential toxicity, it is essential to confirm the results of chemical analysis (*e.g.*, TLC and spectroscopy) with biological tests. Laboratory tests against microorganisms, brine shrimps, etc. are valuable but they still require confirmation by animal toxicity tests, since it is the production of pathological symptoms in animals which is the important criterion of toxicity. However, as in the toxicological screening of food additives, it is always difficult to extrapolate the results of animal tests to man. Experimental procedures such as the use of animal and human cells in tissue culture may provide information on the potential health hazard to man of mycotoxins."

DETOXIFICATION OF MYCOTOXINS

Detoxification is defined in this review as inactivation of toxic materials formed by growth of some fungi at some stage of their history. Most of the investigation on detoxification has been done on aflatoxins; hardly anything has been reported for other mycotoxins. Consequently, methods to detoxify that will be discussed are those which apply to aflatoxins.

In theory, there are three main ways in which a foodstuff containing a toxic substance might be rendered biologically harmless (Feuell 1966): (a) complete removal of the toxin (by means of a selective solvent), (b) conversion of the toxin to a nontoxic derivative (since biological activity is highly structure dependent), and (c) degradation of the toxin to simpler, inactive products by the action of heat, radiation, or chemical attack with acids, alkali, or oxidizing agents.

In practice, the availability of the above mentioned methods may be severely restricted by the nature of the particular foodstuff to be detoxified. To detect and estimate the amount of aflatoxin before and after detoxification treatments, both physicochemical (fluorescence) and biological (duckling bioassay) methods are used. Biological testing of the treated samples is considered essential as the mere disappearance of fluorescence following a treatment cannot unequivocally be taken as evidence of detoxification; slight changes in the aflatoxin molecule may destroy its fluorescent properties without appreciably reducing its toxicity.

Extensive work on detoxification procedures for peanuts containing aflatoxins was started at the Tropical Products Institute in London in 1961 (Feuell 1966). Fairly simple treatments that might be commercially practical and not too expensive were considered. These included radiation, heat, solvent extraction and reactive chemicals (acids, alkali, and gases).

Radiation

Feuell (1966) reported no inactivation when he subjected peanut meal with aflatoxin to ultraviolet light for 8 hr. Similar results were observed when the meal was exposed to gamma rays at a dosage of 2.5 megarads (the maximum practicable). On the other hand, Andrellos and Reid (1967) observed formation of photoproducts from ultraviolet irradiation of aflatoxin B_1 and G_1 on TLC plates or in a methanolic solution. The major photoproduct of B_1 was significantly less toxic than the parent material. Photodegradation of aflatoxin in solutions was thus demonstrated; however, Feuell (1966) concluded that the inability of either ultraviolet or gamma radiation to detoxify contaminated meals negates the method as a practical technique.

Heat

Anhydrous aflatoxins are stable up to their melting points of around 250°C (Coomes et al. 1966; Feuell 1966; Fischbach and Campbell 1965). However, it was shown that autoclaving moist groundnut meal at 15 lb/in^2 (120°C) reduced its aflatoxin content (Feuell 1966). The process was not rapid, 4 hr of autoclaving being necessary to reduce the amount of active aflatoxin from 7000 to 350 µg per kilogram. Further experiments suggested that aflatoxin undergoes hydrolysis or even more extensive degradation (Coomes et al., 1966). Heating as a means to detoxify peanut meals is of little practical value since the appearance and other characteristics of the meal are unfavorably changed and its biological value is probably reduced through degradation of protein. Moreover, since it was found necessary to raise the temperature to over 300°C (Fischbach and Campbell 1965), inactivation by heat treatment is impractical because of inadequate equipment and facilities.

Solvent Extraction

Feuell (1966) stated that petroleum and hexane-type solvents did not remove appreciable quantities of aflatoxins from peanut meal. He selected three common organic solvents to extract toxins: acetone, benzene, and chloroform. He found that aflatoxin extractibility is not the same as solubility; benzene and chloroform are excellent solvents for pure aflatoxins but are unable to extract toxins quantitatively from peanut meal. Sargeant et al. (1961B) showed that the toxin could be completely extracted with methanol. They suggested that only polar solvents would be effective for removing all the aflatoxin from peanut products.

Although aflatoxin is virtually insoluble in water, an appreciable fraction has been removed by aqueous extraction. Lee (1965) showed that if peanut meal is moistened with one-half its weight of water, the aflatoxin can be fully extracted by shaking the mash with chloroform, and a very clean extract is obtained. Robertson et al. (1965) found azeotropic mixtures of hexane with acetone-water or methanol to be effective in extracting aflatoxins. Cucullu et al. (1966) demonstrated the suitability of aqueous acetone as a solvent for extracting aflatoxin from contaminated oilseed products. Moreover, the practicability of this solvent for detoxification was observed in feeding trials, wherein little modification of the nutrient value was apparent. It has been theorized that water and hydroxylated solvents like methanol either break down cell barriers or affect their constituents and so facilitates release of aflatoxins into extracting solvents (Feuell 1966). Detroy et al. (1971) tabulated the various solvent systems that have been used to extract aflatoxins from oilseed meals and other commodities.

Reactive Chemicals

Chemical inactivation of aflatoxin in a contaminated commodity requires a system that can convert the toxin to a nontoxic derivative without deleterious changes in the raw material (Detroy et al. 1971). Feuell (1966) determined the effect of acid, alkali, propylene oxide, sulfur dioxide, and chlorine on toxic solutions of peanut meal extracts. Only the acid and chlorine treatments effectively reduced toxicity. Fischbach and Campbell (1965) reported that 90% of the toxin in toxic meal was destroyed with chlorine gas. However, they also stated that treatment with gaseous chlorine alters the organoleptic quality of the meal, since it leaves a residual odor of chlorine. Feuell (1966) also had reservations about the practicability of the treatment since chlorinated fats and proteins can be highly toxic. Use of 5% sodium hypochlorite has been found effective for inactivating toxins (Fischbach and Campbell 1965; Stoloff and Traeger 1965). Sreenivasamurthy et al. (1967) proposed to destroy aflatoxin in toxic peanut meal by heating the meal at 80°C for 30 min with hydrogen peroxide at a pH of 9.5. Their method was based on the fact that the lactone ring in the toxin molecule opens in an alkaline medium, as evidenced by a change in solubility, and it is susceptible to oxidation when the ring remains open. Hydrogen peroxide has been found most suitable for detoxification because it imparts little residual smell or taste to the product. Although Feuell (1966) reported on the stability of aflatoxins to alkali treatment, Goldblatt (1966) treated oilseed meals with alkaline agents including ammonia gas and ammonium hydroxide and found this procedure to be a promising approach to eliminate aflatoxin from many important protein sources. Dollear *et al.* (1968) verified the efficacy of ammonia, methylamine, and sodium hydroxide as detoxifying agents of aflatoxin-contaminated meal. They also reported that weight gains of animals receiving treated meals were comparable to those receiving aflatoxin-free meals.

Biological Means

Ciegler *et al.* (1966) screened yeasts, molds, bacteria, actinomycetes, algae, and fungi for their ability to degrade aflatoxins. Only one bacterium, *F. aurantiacum* NRRL B-184, took up the toxin irreversibly from the solution. Duckling assays showed the detoxification of aflatoxin by this bacterium to be complete with no new toxic product formed. Toxin-contaminated milk, oil, peanut butter, peanuts, and corn were rapidly and completely detoxified, whereas contaminated soybeans were detoxified 86% after 12 hr of incubation. Some molds and mold spores partially transformed aflatoxin B_1 to new fluorescing compounds. One disadvantage in using molds and mold spores was that long periods of incubation (up to 11 days) were required; no reaction was noted before three days of incubation. Teunisson and Robertson (1967) reported that *Tetrahymena pyriformis* W decreased the concentration of 2 μg aflatoxin B_1 per milliliter by about 67% in 48 hr with production of a bright blue fluorescent substance, whose toxicity was not determined. Ciegler and Peterson (1968) noted a new fluorescing compound was formed with concomitant disappearance of aflatoxin B_1 when toxin was added to acid-producing mold cultures. This compound was identified to be hydroxydihydroaflatoxin B_1 which was less toxic to ducklings than the parent compound. This process has also been carried out efficiently by simply adding acids to slurries of contaminated meals (Feuell 1966).

None of the biological processes just described appear to be satisfactory for industrial use. Although further research is necessary, it should be realized that any process adopted in the U.S. will require approval by the Food and Drug Administration, which requires proof that the nutritive value of the feed or food is unimpaired and that all toxins are destroyed.

IN CONCLUSION

For centuries molds have grown on foods and feeds. The ability of these substrates to spoil the products probably has been recognized for an equally long time. However, concern about toxic substances produced by certain molds during growth on foods and feeds is of recent origin. Serious research efforts to understand the problem did not begin until after aflatoxin was discovered in England early in the 1960's. In spite of this recent beginning much information has accumulated on mycotoxins and mycotoxicoses.

It may well be that worldwide toxic metabolites of molds constitute the single greatest food-borne hazard to human health. The ubiquity of toxigenic molds, the ease with which molds can grow on most foods and feeds, and the stability of most mycotoxins makes the problem of mycotoxicosis more difficult to control than some other forms of food intoxication. Although much has been learned about the problem, its causes, and its control, years of additional research undoubtedly will be needed before all mycotoxins are recognized.

Finally, this review has dealt primarily with those phases of the mycotoxin problem that are most closely associated with the handling of foods. Readers who desire additional information should consult the books edited by Goldblatt (1969), Wogan (1965), Ciegler *et al.* (1971), and Kadis *et al.* (1971, 1972).

BIBLIOGRAPHY

ABEDI, A. H., and MCKINLEY, W. P. 1968. Zebra fish eggs and larvae as aflatoxin test organisms. J. Assoc. Offic. Anal. Chem. *51*, 902–905.

ADYE, J. C., and MATELES, R. I. 1964. Incorporation of labelled compounds into aflatoxins, Biochim. Biophys. Acta *86*, 418–420.

AGNIHOTRI, V. P. 1964. Studies on aspergilli. XVI. Effect of pH, temperature and carbon and nitrogen interaction. Mycopathol. Mycol. Appl. *24*, 305–314.

ALDERMAN, G. G., EMEH, C. O., and MARTH, E. H. 1973. Aflatoxin and rubratoxin produced by *Aspergillus parasiticus* and *Penicillum rubrum* when grown independently, associatively, or with *Penicillium italicum* or *Lactobacillus plantarum*. Z. Lebensm. Unters. Forsch. *153*, 305–311.

ALDERMAN, G. G., and MARTH, E. H. 1974A. Experimental production of aflatoxin in citrus juice and peel. J. Milk Food Technol. *37*, 308–313.

ALDERMAN, G. G., and MARTH, E. H. 1974B. Experimental production of aflatoxin on intact citrus fruit. J. Milk Food Technol. *37*, 451–456.

ALDERMAN, G. G., and MARTH, E. H. 1974C. Production of aflatoxin in concentrated and diluted grapefruit juice. J. Milk Food Technol. *37*, 395–397.

ALPERT, M. E., and DAVIDSON, C. S. 1969. Mycotoxins: A possible cause of primary carcinoma of the liver. Am. J. Med. *46*, 325–329.

ALPERT, M. E., HUTT, M. S. R., and DAVIDSON, C. S. 1968. Hepatoma in Uganda. Lancet *1*, 1265–1267.

ANDRELLOS, P. J., and REID, G. R. 1964. Confirmatory tests for aflatoxin B_1. J. Assoc. Offic. Anal. Chem. *47*, 801–803.

ANDRELLOS, P. J., and Reid, G. R. 1967. Photochemical changes of aflatoxin B_1. J. Assoc. Offic. Anal. Chem. *50*, 346–350.

ANON. 1967-1968. Annual Report For 1967-1968. Regional Research Laboratory, Hyderabad, India.

ARAI, T., ITOL, R., and KOYAMA, Y. 1967. Antimicrobial activity of aflatoxins. J. Bacteriol. *93*, 59–64.

ASHWORTH, L. J., JR., SCHROEDER, H. W., and LANGLEY, B. C. 1965. Aflatoxins: Environmental factors governing occurrence in Spanish peanuts. Science *148*, 1228–1229.

ASPLIN, F. D., and CARNAGHAN, R. B. A. 1961. The toxicity of certain groundnut meals for poultry with special reference to their effect on ducklings and chickens. Vet. Rec. *73*, 1215–1219.

AUSTWICK, P. K. C., and AYERST, G. 1963. Toxic products in groundnuts. Groundnut microflora and toxicity. Chem. Ind. (Brit.), 55–61.

AYERST, G. 1969. The effects of moisture and temperature on growth and spore germination in some fungi. J. Stored Prod. Res. *5*, 127–141.

AYRES, J. C., LILLARD, D. A., and LEISTNER, L. 1967. Mold-ripened meat products. Reciprocal Meat Conf. Proc. *20*, 156.

AYRES, J. L., and SINNHUBER, R. O. 1966. Fluorodensitometry of aflatoxin on thin layer plates. J. Am. Oil Chem. Soc. *43*, 423–424.

BECKWITH, A. C., and STOLOFF, L. 1968. Fluorodensitometric measurement of aflatoxin thin layer chromatograms. J. Assoc. Offic. Anal. Chem. *51*, 602–609.

BOESENBERG, H. 1973. Effect of aflatoxin on animal and man. Z. Lebensm. Unters. Forsch. *151*, 245–249. (German)

BORKER, E., INSALATA, N. F., LEVI, L. P., and WITZEMAN, J. S. 1966. Mycotoxins in feeds and foods. Advan. Appl. Microbiol. *8*, 315–351.

BROADBENT, J. H., CORNELIUS, J. A., and SHONE, G. 1963. The detection and estimation of aflatoxins in groundnuts and groundnut materials. Part II. Thin layer chromatographic method. Analyst *88*, 214–216.

BROWN, R. F. 1968. Proceedings of the First U.S.-Japan Conference on Toxic Microorganisms. U.S. Dept. Interior and UJNR Panels on Toxic Microorganisms, Washington, D.C.

BULLERMAN, L. B., and AYRES, J. C. 1968. Aflatoxin-producing potential of fungi isolated from cured and aged meats. Appl. Microbiol. *16*, 1945–1946.

BULLERMAN, L. B., HARTMAN, P. A., and AYRES, J. C. 1969. Aflatoxin production in meats. II. Aged dry salamis and aged country cured hams. Appl. Microbiol. *18*, 718–722.

BULLOCK, E., ROBERTS, J. C., and UNDERWOOD, J. G. 1962. Studies on mycological chemistry. Part XI. The structure of isosterigmatocystin and an amended structure for sterigmatocystin. J. Chem. Soc. (Brit.), 4179.

BURMEISTER, H. R. 1967. Microbiological screening process for aflatoxin. U.S. Pat. 3,360,441.

BURMEISTER, H. R., and HESSELTINE, C. W. 1966. Survey of the sensitivity of microorganisms to aflatoxins. Appl. Microbiol. *14*, 403–404.

BUTLER, W. H. 1964. Acute liver injury in ducklings as a result of aflatoxin poisoning. J. Pathol. Bacteriol. *88*, 189–196.

CAMPBELL, C. 1969. 83rd Meeting, Association of Official Analytical Chemists, Washington, D.C. In Microbial Toxins, Vol. 7, A. Ciegler, S. Kadis, and S. J. Ajl (Editors). Academic Press, New York.

CIEGLER, A., DETROY, R. W., and LILLEHOJ, E. B. 1971. Carcinogenic lactones. In Microbial Toxins, Vol. 7, A. Ciegler, S. Kadis, and S. J. Ajl (Editors). Academic Press, New York.

CIEGLER, A., KADIS, S., and AJL, S. J. 1971. Microbial Toxins, Vol. 6. Academic Press, New York.

CIEGLER, A., LILLEHOJ, E. B., PETERSON, R. E., and HALL, H. H. 1966. Microbial detoxification of aflatoxin. Appl. Microbiol. *14*, 934–939.

CIEGLER, A., and PETERSON, R. E. 1968. Aflatoxin detoxification: Hydroxydihydroaflatoxin B_1. Appl. Microbiol. *16*, 665–666.

CIEGLER, A., PETERSON, R. E., LAGODA, A. A., and HALL, H. H. 1966. Aflatoxin production and degradation by *A. flavus* in 20 liter fermenters. Appl. Microbiol. *14*, 826–833.

CLEMENTS, N. L. 1968. Note on a microbiological assay for aflatoxin B_1: A rapid confirmatory test by effects on growth of *B. megaterium*. J. Assoc. Offic. Anal. Chem. *51*, 611–612.

CODNER, R. C., SARGEANT, K., and YEO, R. 1963. Production of aflatoxin by the culture of strains of A. flavus-oryzae on sterilized peanuts. Biotechnol. Bioeng. 5, 185-192.
COOMES, T. J., CROWTHER, P. C., FEUELL, A. J., and FRANCIS, B. J. 1966. Experimental detoxification of groundnut meals containing aflatoxins. Nature 209, 406-407.
CUCULLU, A. F., LEO, S. L., MAYNE, R. Y., and GOLDBLATT, L. A. 1966. Determinations of aflatoxins in individual peanuts and peanut sections. J. Am. Oil Chem. Soc. 43, 89-92.
DAVIS, N. D., and DIENER, U. L. 1968. Growth and aflatoxin production by A. parasiticus from various carbon sources. Appl. Microbiol. 16, 158-159.
DAVIS, N. D., and DIENER, U. L. 1970. Environmental factors affecting the production of aflatoxins. Proc. First U.S.-Japan Conf. Toxic Microorganisms. U.S. Govt. Printing Office, Washington, D.C., p. 43.
DAVIS, N. D., DIENER, U. L., and AGNIHOTRI, V. P. 1967. Production of aflatoxins B₁ and G₁ in chemically defined medium. Mycopathol. Mycol. Appl. 31, 251-256.
DAVIS, N. D., DIENER, U. L., and ELDRIDGE, D. W. 1966. Production of aflatoxins B₁ and G₁ by A. flavus in a semi-synthetic medium. Appl. Microbiol. 14, 378-380.
DAVIS, N. D., DIENER, U. L., and LANDERS, K. E. 1964. Factors influencing the production of aflatoxin by A. flavus growing on laboratory media. Third Natl. Peanut Res. Conf. Auburn University, Auburn, Alabama, (July 9-10).
DETROY, R. W., LILLEHOJ, E. B., and CIEGLER, A. 1971. Aflatoxin and related compounds. In Microbial Toxins, Vol. 7, A. Ciegler, S. Kadis, and S. J. Ajl (Editors). Academic Press, New York.
DIENER, U. L., and DAVIS, N. D. 1966. Aflatoxin production by isolates of A. flavus. Phytopathol. 56, 1390-1393.
DIENER, U. L., and DAVIS, N. D. 1969A. Aflatoxin formation by A. flavus. In Aflatoxin, L. A. Goldblatt (Editor). Academic Press, New York.
DIENER, U. L., and DAVIS, N. D. 1969B. Production of aflatoxin on peanuts under control environments. J. Stored Prod. Res. 5, 251-258.
DIENER, U. L., and DAVIS, N. D. 1970. Limiting temperature and relative humidity for aflatoxin production by A. flavus in stored peanuts. J. Am. Oil Chem. Soc. 47, 347-351.
DIMENNA, M. E., CAMPBELL, J., and MORTIMER, P. H. 1970. Sporidesmin production and sporulation in Pithomyces chartarum. J. Gen. Microbiol. 61, 87-96.
DOLLEAR, F. G. et al. 1968. Elimination of aflatoxins from peanut meals. J. Am. Oil Chem. Soc. 45, 862-865.
ELDRIDGE, D. W., DAVIS, N. D., DIENER, U. L., and AGNIHOTRI, V. P. 1965. Aflatoxin production by A. flavus in a chemically defined liquid medium. Phytopathol. 55, 498.
EPPLEY, R. M. 1968. Screening method for zearalenone, aflatoxin and ochratoxin. J. Assoc. Offic. Anal. Chem. 51, 74-78.
FERREIRA, N. P. 1966. Recent advances in research on ochratoxin. 2. Microbiological aspects. In Biochemistry of Some Foodborne Microbial Toxins, R. I. Mateles, and G. N. Wogan (Editors). MIT Press, Cambridge, Mass.
FEUELL, A. J. 1966. Aflatoxin in groundnuts. Part IX. Problems of detoxification. Trop. Sci. 8, 61-70.
FEUELL, A. J. 1969. Types of mycotoxins in foods and feeds. In Aflatoxin, L. A. Goldblatt (Editor). Academic Press, New York.
FISCHBACH, H., and CAMPBELL, A. D. 1965. Note on detoxification of the aflatoxins. J. Assoc. Offic. Agr. Chem. 48, 28.
FORGACS, J. 1962. Mycotoxicoses: Symptoms and general aspects. Proc. Md. Nutr. Conf. Feed Manufacturers, p. 19.

FRANK, H. K. 1968. Diffusion of aflatoxins in foodstuffs. J. Food Sci. 33, 98-100.
GABLIKS, J., SCHAEFFER, W., FRIEDMAN, L., and WOGAN, G. N. 1965. Effect of aflatoxin B₁ on cell cultures. J. Bacteriol. 90, 720-723.
GAJAN, R. J., NESHEIM, S., and CAMPBELL, A. D. 1964. Note on identification of aflatoxin by oscillographic polarography. J. Assoc. Offic. Anal. Chem. 47, 27-28.
GOEL, M. C. et al. 1972. Changes in the microflora of wild rice during curing by fermentation. J. Milk Food Technol. 35, 385-391.
GOLDBLATT, L. A. 1966. Some approaches to the elimination of aflatoxins from protein concentrates. Advan. Chem. Ser. 57, 216-227.
GOLDBLATT, L. A. 1968. Aflatoxin and its control. Econ. Bot. 22, 51-62.
GOLDBLATT, L. A. 1969. Aflatoxin. Academic Press, New York.
HARTLEY, R. D., NESBITT, B. F., and O'KELLY, J. 1963. Toxic metabolites of A. flavus. Nature 198, 1056-1058.
HAYES, A. W., DAVIS, N. D., and DIENER, U. L. 1966. Effect of aeration on growth and aflatoxin production by A. flavus in submerged culture. Appl. Microbiol. 14, 1019-1021.
HAYES, A. W., and WILSON, B. J. 1968. Bioproduction and purification of rubratoxin B. Appl. Microbiol. 16, 1163-1167.
HESSELTINE, C. W. 1967. Aflatoxins and other mycotoxins. Health Lab. Sci. 4, 222-228.
HESSELTINE, C. W. 1969. Mycotoxins. Mycopathol. Mycol. Appl. 39, 371-383.
HESSELTINE, C. W., SHOTWELL, O. L., ELLIS, J. J., and STUBBLEFIELD, R. D. 1966. Aflatoxin formation by A. flavus. Bacteriol. Rev. 30, 795-805.
HESSELTINE, C. W., SORENSON, W. G., and SMITH, M. 1970. Taxonomic studies of the aflatoxin-producing strains in the A. flavus group. Mycologia 62, 123-132.
HODGES, F. A. et al. 1964. Mycotoxins: Aflatoxin isolated from P. puberulum. Science 145, 1439.
HOLKER, J. S. E., and UNDERWOOD, J. G. 1964. A synthesis of a cyclopentenocoumarin structurally related to aflatoxin B₁. Chem. Ind. 45, 1865-1866.
HOLZAPFEL, C. W. 1971. Cyclopiazonic acid and related toxins. In Microbial Toxins, Vol. 7, A. Ciegler, S. Kadis, and S. J. Ajl (Editors). Academic Press, New York.
JARVIS, B. 1971. Factors affecting the production of mycotoxins. J. Appl. Bacteriol. 34, 199-213.
JOFFE, A. Z. 1960. The microflora of overwintered cereals and its toxicity. Bull. Res. Council Israel, Sect. D 9, 101.
JOFFE, A. Z. 1962. Biological properties of some toxic fungi isolated from overwintered cereals. Mycopathol. Mycol. Appl. 16, 201.
JOFFE, A. Z. 1965. Toxin production by cereal fungi causing toxic alimentary aleukia in man. In Mycotoxins in Foodstuffs, G. N. Wogan (Editor). MIT Press, Cambridge, Mass.
JOFFE, A. Z. 1969. Aflatoxin produced by 1,626 isolates of A. flavus from groundnut kernels and soil in Israel. Nature 221, 492.
JUHASZ, S., and GRECZI, E. 1964. Extracts of mould-infected groundnut samples in tissue cultures. Nature 203, 861-862.
KADIS, S., CIEGLER, A., and AJL, S. J. 1971. Microbial Toxins, Vol. 7. Academic Press, New York.
KADIS, S., CIEGLER, A., and AJL, S. J. 1972. Microbial Toxins, Vol. 8. Academic Press, New York.
KRAMER, C. L., PADY, S. M., and WILEY, B. J. 1963. Kansas aeromycology. XIII. Diurnal studies 1959-1960. Mycologia 55, 380-401.
KRAYBILL, H. F., and SHAPIRO, R. E. 1969. Implications of fungal toxicity

to human health. *In* Aflatoxin, L. A. Goldblatt (Editor). Academic Press, New York.

KULIK, M. M., and HOLADAY, C. E. 1966. Aflatoxin: A metabolite product of several fungi. Mycopathol. Mycol. Appl. *30*, 137–140.

KURATA, H., and ICHINOE, M. 1967A. Studies on the population of toxigenic fungi in foodstuffs. I. Fungal flora of flour-type foodstuffs. J. Food Hyg. Soc. Japan *8*, 237–246.

KURATA, H., and ICHINOE, M. 1967B. Studies on the population of toxigenic fungi in foodstuffs. II. Toxigenic determination for the fungal isolates obtained from the flour-type foodstuffs. J. Food Hyg. Soc. Japan *8*, 247–252.

KURATA, H. *et al.* 1968A. Studies on the population of toxigenic fungi in foodstuffs. III. Mycoflora of milled rice harvested in 1965. J. Food Hyg. Soc. Japan *9*, 23–28.

KURATA, H. *et al.* 1968B. Studies on the population of toxigenic fungi in foodstuffs. V. Acute toxicity test for representative species of fungal isolates from milled rice harvested in 1965. J. Food Hyg. Soc. Japan *9*, 379–386.

LAI, M., SEMENIUK, G., and HESSELTINE, C. W. 1970. Conditions for production of ochratoxin by *Aspergillus* species in a synthetic medium. Appl. Microbiol. *19*, 542–544.

LANDERS, K. E., DAVIS, N. D., and DIENER, U. L. 1967. Influence of atmospheric gases on aflatoxin production by *A. flavus* in peanuts. Phytopathol. *57*, 1086–1090.

LEE, E. G. H., TOWNSLEY, P. M., and WALDEN, C. C. 1966. Effect of bivalent metals on the production of aflatoxins in submerged cultures. J. Food Sci. *31*, 432–436.

LEE, W. V. 1965. Quantitative determination of aflatoxin in groundnut products. Analyst *90*, 305–307.

LEGATOR, M. S. 1969. Biological assay for aflatoxin. *In* Aflatoxins, L. A. Goldblatt (Editor). Academic Press, New York.

LEGATOR, M. S., ZUFFANTE, S. M., and HARP, A. R. 1965. Aflatoxin: Effect on cultured heteroploid human embryonic lung cells. Nature *208*, 345–347.

LIE, J., and MARTH, E. H. 1967. Formation of aflatoxin in Cheddar cheese by *A. flavus* and *A. parasiticus*. J. Dairy Sci. *50*, 1708–1710.

LIE, J., and MARTH, E. H. 1968. Aflatoxin formation by *A. flavus* and *A. parasiticus* in a casein substrate at different pH values. J. Dairy Sci. *51*, 1743–1747.

LILLEHOJ, E. B., and CIEGLER, A. 1968. Biological activity of sterigmatocystin. Mycopathol. Mycol. Appl. *35*, 373–376.

LILLEHOJ, E. B., CIEGLER, A., and DETROY, R. W. 1970. Fungal toxins. *In* Essays in Toxicology, Vol. 2, F. R. Blood (Editor). Academic Press, New York.

LILLEHOJ, E. B., CIEGLER, A., and HALL, H. H. 1967. Fungistatic action of aflatoxin B_1. Experientia *23*, 187–188.

LOPEZ, A., and CRAWFORD, M. A. 1967. Aflatoxin content of groundnuts for human consumption in Uganda. Lancet *2*, 1351–1354.

MATELES, R. I., and ADYE, J. C. 1965. Production of aflatoxins in submerged culture. Appl. Microbiol. *13*, 208–211.

MATSUURA, S., MANABE, M., and SATO, T. 1970. Proceedings of the First U.S.-Japan Conference on Toxic Microorganisms. U.S. Dept. Interior and UJNR Panels on Toxic Microorganisms, Washington, D.C.

MENZEL, A. E. O., WINTERSTEINER, O., and RAKE, G. 1945. Note on the antibiotic substances elaborated by an *A. flavus* strain and by an unclassified mold. J. Bacteriol. *46*, 109.

MERWE, K. G. *et al.* 1965. Ochratoxin A, a toxic metabolite produced by *A. ochraceus* Wilh. Nature *205*, 1112–1113.

MISLIVEC, P. B., HUNTER, J. H., and TUITE, J. 1968. Assay for aflatoxin production by the genera *Aspergillus* and *Penicillium*. Appl. Microbiol. *16*, 1053–1055.

MOSS, M. O. 1971. The rubratoxins, toxic metabolites of *P. rubrum*. In Microbial Toxins, Vol. 7, A. Ciegler, S. Kadis, and S. J. Ajl (Editors). Academic Press, New York.

MOSS, M. O., and HILL, I. W. 1970. Strain variation in the production of rubratoxins by *P. rubrum* Stoll. Mycopathol. Mycol. Appl. *40*, 81–88.

NABNEY, J., and NESBITT, B. F. 1965. A spectrophotometric method for determining the aflatoxins. Analyst *90*, 155–160.

NESBITT, B. F., O'KELLY, J., SARGEANT, K., and SHERIDAN, A. 1962. Toxic metabolites of *A. flavus*. Nature *195*, 1062–1063.

NESHEIM, S. 1969. Conditions and techniques for thin layer chromatography of aflatoxins. J. Am. Oil Chem. Soc. *46*, 335–338.

OETTLE, A. G. 1964. Cancer in Africa, especially in regions of the Sahara. J. Natl. Cancer Inst. *33*, 383–436.

PARRISH, F. W., WILEY, B. J., SIMMONS, E. G., and LONG, L. 1966. Production of aflatoxins and kojic acid by species of *Aspergillus* and *Penicillium*. Appl. Microbiol. *14*, 139–140.

PETERSON, R. E., CIEGLER, A., and HALL, H. H. 1967. Densitometric measurement of aflatoxin. J. Chromatog. *27*, 304–307.

PLATT, B. S., STEWART, R. J. C., and GUPTA, S. R. 1962. The chick embryo as a test organism for toxic substances in food. Proc. Nutr. Soc. *21*, 30–31.

PONS, W. A., JR., and GOLDBLATT, L. A. 1969. Physicochemical assay of aflatoxins. In Aflatoxins, L. A. Goldblatt (Editor). Academic Press, New York.

PONS, W. A., JR., ROBERTSON, J. A., and GOLDBLATT, L. A. 1966. Objective fluorometric measurement of aflatoxins on thin layer chromatographic plates. J. Am. Oil Chem. Soc. *43*, 665–669.

PURCHASE, I. F. H., and VORSTER, L. J. 1968. Aflatoxin in commercial milk samples. S. African Med. J. *42*, 219.

RABIE, C. J., and SMALLEY, E. B. 1965. Influence of temperature on the production of aflatoxin by *A. flavus*. Symp. Mycotoxins Foodstuffs, Agr. Aspects, Pretoria, S. Africa. *Cited by* Jarvis, B. 1971. J. Appl. Bacteriol. *34*, 199.

ROBERTSON, J. A., JR., LEE, L. S., CUCULLU, A. F., and GOLDBLATT, L. A. 1965. Assay of aflatoxin in peanuts and peanut products using acetone-hexane-water for extraction. J. Assoc. Oil Chem. Soc. *42*, 467–471.

ROBINSON, P. 1967. Infantile cirrhosis of the liver in India. Clin. Pediatrics *6*, 57–62.

SAITO, M., ENOMOTO, M., and TATSUMO, T. 1971. Yellowed rice toxins. In Microbial Toxins, Vol. 7, A. Ciegler, S. Kadis, and S. J. Ajl (Editors). Academic Press, New York.

SANDERS, T. H., DAVIS, N. D., and DIENER, U. L. 1968. Effect of carbon dioxide, temperature and relative humidity on production of aflatoxin in peanuts. J. Am. Oil Chem. Soc. *45*, 683–685.

SARGEANT, K., SHERIDAN, A., O'KELLY, J., and CARNAGHAN, R. B. A. 1961A. Toxicity associated with certain samples of groundnuts. Nature *192*, 1096–1097.

SARGEANT, K., O'KELLY, J., CARNAGHAN, R. B. A., and ALLCROFT, R. 1961B. The assay of a toxic principle in certain groundnut meals. Vet. Rec. *73*, 1219–1223.

SCHINDLER, A. F., PALMER, J. G., and EISENBERG, W. V. 1967. Aflatoxin production by *A. flavus* as related to various temperatures. Appl. Microbiol. *15*, 1006–1009.

SCHOFIELD, F. W. 1924. Damaged sweet clover. The cause of a new disease in cattle simulating hemorrhagic septicemia and blackleg. J. Am. Vet. Med. Assoc. *64*, 553–575.

SCHROEDER, H. W. 1966. Effect of corn steep liquor on mycelial growth and

254 FOOD MICROBIOLOGY

aflatoxin production in *A. paraciticus*. Appl. Microbiol. *14*, 381–385.
SCHROEDER, H. W., and ASHWORTH, L. S., JR. 1965. Aflatoxins in Spanish peanuts in relation to pod and kernel conditions. Phytopathol. *55*, 464–465.
SCHROEDER, H. W., and HEINS, H., JR. 1967. Aflatoxins: Production of the toxins in vitro in relation to temperature. Appl. Microbiol. *15*, 441–445.
SCOTT, DE, B. 1965. Toxigenic fungi isolated from cereal and legume products. Mycopathol. Mycol. Appl. *25*, 213–222.
SCOTT, P. M., LAWRENCE, J. W., and VAN WALBEEK, W. 1970. Detection of mycotoxins by thin layer chromatography: Application to screening of fungal extracts. Appl. Microbiol. *20*, 839–842.
SCOTT, P. M., VAN WALBEEK, W., and FORGACS, J. 1967. Formation of aflatoxins by *A. ostianus* Wehmer. Appl. Microbiol. *15*, 945.
SELLSCHOP, J. P. F., KRIEK, N. P. J., and DUPREEZ, J. C. G. 1965. Distribution and degree of occurrence of aflatoxin in groundnuts and groundnut products. Symp. Mycotoxins Foodstuffs, Agr. Aspects, Pretoria, S. Africa, p. 9.
SERCK-HANSSEN, A. 1970. Aflatoxin-induced fatal hepatitis. Arch. Environ. Health *20*, 729–731.
SHANK, R. C. 1968. Activity of MIT Thailand Laboratory. Intern. Agency Res. Cancer Working Conf., Studies Role Aflatoxin Human Disease, Lyon, France. *Cited in* Microbial Toxins, Vol. 7, A. Ciegler, S. Kadis, and S. J. Ajl (Editors). Academic Press, New York.
SHIH, C. N., and MARTH, E. H. 1969A. Aflatoxins not recovered from commercial mold-ripened cheeses. J. Dairy Sci. *52*, 1681–1682.
SHIH, C. N., and MARTH, E. H. 1969B. Improved procedures for measurement of aflatoxins with thin layer chromatography and fluorometry. J. Milk Food Technol. *32*, 213–217.
SHIH, C. N., and MARTH, E. H. 1971. A procedure for rapid recovery of aflatoxins from cheese and other foods. J. Milk Food Technol. *34*, 119–123.
SHIH, C. N., and MARTH, E. H. 1972A. Experimental production of aflatoxin on brick cheese. J. Milk Food Technol. *35*, 585–587.
SHIH, C. N., and MARTH, E. H. 1972B. Production of aflatoxin in a medium fortified with sodium chloride. J. Dairy Sci. *55*, 1415–1419.
SHIH, C. N., and MARTH, E. H. 1973A. Aflatoxin produced by *Aspergillus parasiticus* when incubated in the presence of different gases. J. Milk Food Technol. *36*, 421–425.
SHIH, C. N., and MARTH, E. H. 1973B. Release of aflatoxin from the mycelium of *Aspergillus parasiticus* into liquid media. Z. Lebensm. Unters. Forsch. *152*, 336–339.
SHIH, C. N., and MARTH, E. H. 1974A. Aflatoxin formation, lipid synthesis, and glucose metabolism by *Aspergillus parasiticus* during incubation with and without glucose. Biochim. Biophys. Acta *338*, 286–296.
SHIH, C. N., and MARTH, E. H. 1974B. Some cultural conditions that control biosynthesis of lipid and aflatoxin by *Aspergillus parasiticus*. Appl. Microbiol. *27*, 452–456.
SHIH, C. N., McCOY, E., and MARTH, E. H. 1974. Nitrification by aflatoxigenic strains of *Aspergillus flavus* and *Aspergillus parasiticus*. J. Gen. Microbiol. *84*, 357–363.
SHOTWELL, O. L., HESSELTINE, C. W., STUBBLEFIELD, R. D., and SORENSON, W. G. 1966. Production of aflatoxin on rice. Appl. Microbiol. *14*, 425–428.
SMITH, R. H. 1965. The inhibition of amino acid activation in liver and *Escherichia coli* preparations by aflatoxin in vitro. Biochem. J. *95*, 43–44.
SORENSON, W. G., HESSELTINE, C. W., and SHOTWELL, O. L. 1967. Effect of temperature on production of aflatoxin on rice by *A. flavus*. Mycopathol. Mycol. Appl. *33*, 49–55.
SREENIVASAMURTHY, V., PARPIA, H. A. B., SRIKANTA, S., and

SHANKAR, A. 1967. Detoxification of aflatoxin in peanut meal by hydrogen peroxide. J. Assoc. Offic. Anal. Chem. 50, 350-354.
STEYN, P. S. 1969. The separation and detection of several mycotoxins by thin layer chromatography. J. Chromatog. 45, 473-475.
STOLOFF, L. 1967. Collaborative study of a method for the identification of aflatoxin B₁ by derivative formation. J. Assoc. Offic. Anal. Chem. 50, 354-360.
STOLOFF, L., and TRAGER, W. 1965. Recommended decontamination procedures for aflatoxins. J. Assoc. Offic. Agr. Chem. 48, 681.
STUBBLEFIELD, R. D., SHANNON, G. M., and SHOTWELL, O. L. 1969. Aflatoxins: Improved resolution by thin layer chromatography. J. Assoc. Offic. Anal. Chem. 52, 669-672.
STUBBLEFIELD, R. D. et al. 1967. Production of aflatoxin on wheat and oats: Measurement with a recording densitometer. Appl. Microbiol. 15, 186-190.
TABER, R. A., and SCHROEDER, H. W. 1967. Aflatoxin-producing potential of isolates of A. flavus-oryzae group from peanuts (Arachis hypogaea). Appl. Microbiol. 15, 140-144.
TEUNISSON, D. J., and ROBERTSON, J. A. 1967. Degradation of pure aflatoxins by Tetrahymena pyriformis. Appl. Microbiol. 15, 1099-1103.
THORNTON, R. H., and PERCIVAL, J. C. 1959. A hepatotoxin from Sporidesmium bakeri capable of producing facial eczema diseases in sheep. Nature 183, 63.
TOWNSLEY, P. M., and LEE, E. G. H. 1967. Response of fertilized eggs of the mollusk Bankia setacea to aflatoxin. J. Assoc. Offic. Agr. Chem. 50, 361-363.
VAN WALBEEK, W., SCOTT, P. M., and THATCHER, F. S. 1968. Mycotoxins from food-borne fungi. Can. J. Microbiol. 14, 131.
VERRETT, M. J., MARLIAC, J. P., and MCLAUGHLIN, J. 1964. Use of chicken embryo in the assay of aflatoxin toxicity. J. Assoc. Offic. Agr. Chem. 47, 1003-1006.
VOGEL, DE, P., VAN RHEE, R., and KOELENSMID, B. 1965. A rapid screening test for aflatoxin synthesizing aspergilli of the A. flavus-oryzae group. J. Appl. Bacteriol. 28, 213-220.
WHITE, E. C., and HILL, J. J. 1943. Studies on antibacterial products formed by molds. I. Aspergillic acid, a product of a strain of A. flavus. J. Bacteriol. 45, 433-443.
WILDMAN, J. D., STOLOFF, L., and JACOBS, R. 1967. Aflatoxin production by a potent A. flavus link isolate. Biotechnol. Bioeng. 9, 429-437.
WILSON, B. J. 1966. Toxins other than aflatoxins produced by Aspergillus flavus. Bacteriol. Rev. 30, 478-484.
WILSON, B. J. 1968. Mycotoxins. In The Safety of Foods. Avi Publishing Co., Westport, Conn.
WILSON, B. J. 1971A. Miscellaneous Aspergillus toxins. In Microbial Toxins, Vol. 7, A. Ciegler, S. Kadis, and S. J. Ajl (Editors). Academic Press, New York.
WILSON, B. J. 1971B. Miscellaneous Penicillium toxins. In Microbial Toxins, Vol. 7, A. Ciegler, S. Kadis, and S. J. Ajl (Editors). Academic Press, New York.
WILSON, B. J., CAMPBELL, T. C., HAYES, A. W., and HANLIN, R. T. 1968. Investigation of reported aflatoxin production by fungi outside the Aspergillus group. Appl. Microbiol. 16, 819-821.
WOGAN, G. N. 1965. Mycotoxins in Foodstuffs. MIT Press, Cambridge, Mass.
YOKOTSUKA, T. et al. 1967A. Production of fluorescent compounds other than aflatoxins by Japanese industrial molds. In Biochemistry of Some Foodborne Microbial Toxins, R. I. Mateles, and G. N. Wogan (Editors). MIT Press, Cambridge, Mass.
YOKOTSUKA, T. et al. 1967B. Studies on the compounds produced by moulds.

Part I. Fluorescent compounds produced by Japanese industrial moulds. J. Agr. Chem. Soc. Japan *41*, 32–38.

YOKOTSUKA, T., KIKUCHI, T., SASAKI, M., and OSHITA, K. 1968. Aflatoxin G-like compounds with green fluorescence produced by Japanese moulds. J. Agr. Chem. Soc. Japan *42*, 581–585.

CHAPTER 8

Dean O. Cliver | # Viruses

VIRUSES

Viruses are different. They share many properties with living things, and they borrow life from host cells in order to make more viruses, but they are not really alive. During a great portion of the virus' "life cycle," it is a small, inert, relatively simple particle. Depending on the type of virus, this particle might be from 25 to 250 nm (about one to ten one-millionths of an inch) in diameter and be roughly spherical or elongated. Its components include a nucleic acid core (RNA or DNA, but not both) and a protein coat, sometimes with a lipid envelope outermost.

Virus is produced by the infected host cell. The relationship is quite specific: one virus will usually be capable of infecting only a limited variety of cells in a limited number of species. During a portion of the infectious cycle, there are no particles within the host cell, and the virus is said to be "eclipsed."

The cell often loses its specialized function in the body during infection. It may later die, or it may multiply out of control to form a tumor or neoplasm. If enough cells become involved, the body as a whole will be seen to be diseased. Well-known virus diseases include colds, influenza, measles, smallpox, poliomyelitis, and many others; only a few of these seem likely to be transmitted in foods.

Antibiotics have been of little value in treating virus infections; but if the individual survives the early phases of his infection (which is usual), he is likely to recover as a result of his body's own immune processes. Immunity can also be induced by vaccination; however, vaccines are available for only a few of the many known types of viruses. Transmission of the other viruses must be limited by preventing passage of the agent between hosts.

A virus may pass directly from one host to another as a result of immediate physical contact or diffusion through air for short distances. Indirect transmission requires that the virus be carried by a living thing (a "vector"), by an inanimate object (a "fomes"), or by food or water (a "vehicle"). The food vehicle is the present topic, but it should be noted that virus can also be transmitted to the food in any of the ways which are enumerated above. The raw material of the food (vegetable or animal) may have contained virus at the time of harvest or slaughter. Viruses hazardous to man do not infect plants, but some may infect animals. Animal viruses which cannot infect man are of interest if they may be transmitted through foods; they are sometimes a cause of carcass condemnations, they cause barriers to trade in animal food products, and data regarding the persistence of animal viruses in foods may be applicable to food-borne "human" viruses. The agents of foot and mouth disease, rinderpest, African swine fever, and Newcastle disease are in this category.

Diseases transmissible from animals to man (but seldom from man to man) may be called "zoonoses." Zoonotic viruses are sometimes present in foods derived from infected animals, either because of the presence of virus in the food-source tissue or because of later contamination of the food with virus-containing body fluids or digestive tract contents. The virus of tick-borne encephalitis is an important zoonotic agent which will be discussed further below. The viruses associated with leukemic diseases of chickens and cattle have also been suggested as possible food-borne zoonotic agents: much further work will be required before their significance is firmly established.

Viruses may also be present in vegetables or in shellfish (*i.e.*, bivalve molluscs) at the time of harvest. Rather than infection, the presence of virus in these instances generally denotes contact with polluted water. Uptake of virus by vegetables may be relatively inefficient, but shellfish are capable of concentrating virus from their environmental waters (Metcalf and Stiles 1968).

A foodstuff which was free of virus at the time of harvest or slaughter may later become contaminated in processing, storage, or distribution. As was noted above, contamination may be directly of human origin, or it may reach the food by means of a vector, a fomes, or the water vehicle. The human who works with food which is later consumed by others has most frequently been implicated, on epidemiologic grounds, as the source of contaminating virus (Cliver 1971). This conclusion must be drawn from a very limited sample; a great many food-borne disease outbreaks, probably caused by viruses, cannot be traced to their ultimate source.

DISEASES

A disease must generally occur as a common-source outbreak to be recognized as having been food-borne. A common source is seldom suspected in diseases with very delayed or insidious onsets, or when very small numbers of cases are seen. Any of these could be reasons why transmission of a number of viruses through foods has not yet been documented. On the other hand, food will be suspected as a vehicle if a large number of cases occurs among people who have eaten together recently, or if the symptoms of the disease are seen first in the digestive tract, or if the disease is one which has been described previously as food-borne.

Some animal viruses, not significantly hazardous to human health, which are found in foods were listed above. The most studied of these has been the agent of foot and mouth disease. Upon slaughter of an infected animal, this virus may be found in tissues throughout the body. Any of these tissues, distributed and used as food, might carry the virus. Clear evidence has been obtained recently that the virus is also shed in the milk of infected cattle, sometimes before the onset of recognizable symptoms of the disease (Burrows *et al.* 1971). This mode of dissemination may have played a significant role in the 1967-1968 outbreak of foot and mouth disease in England and Wales (Dawson 1970). A respiratory virus of cattle has also been found in the milk of infected animals (Kawakami *et al.* 1966).

Garbage is one of the end products of food processing and consumption (human flesh and excrement are the others). Uncooked pork scraps in garbage fed to swine have been responsible for the transmission of hog cholera and probably of African swine fever. Viruses probably persist in the tissues of other species in garbage; these seem likely to do less harm because the garbage is not fed to other animals of the same species.

The best documented of food-borne zoonoses is tick-borne encephalitis. Lactating goats, infected with the virus by the bite of a tick vector, shed the agent in their milk (Ernek *et al.* 1968). Other dairy species (cattle and sheep) may also be involved at times, but the majority of human illnesses seems to result from consumption of inadequately pasteurized goats' milk. The problem appears to be quite regional; it has been reported only from eastern Europe. A tick-borne encephalitis virus found in North America evidently does not infect dairy animals with significant frequency (Kokernot *et al.* 1969).

Two other zoonoses should be mentioned here, although the causative agents are not true viruses, but specialized bacteria which multiply only inside of appropriate living host cells. While the agents are bacteria, they are more frequently discussed with viruses than with bacteria because they are studied by virologic techniques. Ornithosis is an infection of birds, including the domestic poultry species. It usually infects man by the respiratory route and is seen most frequently in those employed in growing, killing, and processing poultry, rather than in consumers. Q-fever is a rickettsial infection, principally of ruminants, which also infects (by the respiratory route) those who work with infected animals or their carcasses. The agent of Q-fever, *Coxiella burneti*, probably infects consumers only through inadequately pasteurized milk from infected cattle. The agents of both of these diseases are large enough to be seen with a microscope, and (also in contrast to true viruses) both diseases respond to antibiotic treatment.

The remainder of known food-borne viruses belong to the human "enteric complex." These are viruses which infect by the oral route when swallowed and which have a primary site of multiplication in the intestines, with the result that the virus is shed in feces. The enteric complex comprises several groups of viruses. The size, shape, and chemical composition of particles of several of these groups differ significantly; and some have not yet been characterized by electron microscopy or other means. The diseases they cause also differ significantly. Most of these viruses also multiply at other sites than the intestines, and the symptoms caused may show no relationship to the digestive tract.

The human virus disease which has been known to be food-borne for the longest time is poliomyelitis. The first recorded incident seems to have taken place in 1914. Like most of those which followed, the vehicle in this outbreak was thought to have been raw milk. The investigator was unable to learn how the milk had gotten contaminated with virus. Present knowledge of the polioviruses indicates that they do not infect cattle, so the ultimate source was almost certainly human. The record of food-associated poliomyelitis has been reviewed thoroughly elsewhere (Cliver 1967); just a few points will be noted here. First, the vehicle most frequently suspected was milk (usually raw). Second, the vehicle in these outbreaks was thought (where any opinion was ventured) to have been contaminated either by an infected human or by flies which had access to human feces. Third, such outbreaks are no longer being seen in the U.S. and other affluent nations, probably because of the effectiveness of the vaccines against this disease and of improved standards in the production of milk and other foods.

Three of some 65 known types of human enteroviruses have been designated poliomyelitis viruses (or "polioviruses"). The remaining types are called either Coxsackie or ECHO viruses (only the former will usually kill suckling mice inoculated in the laboratory). These are thought less likely to cause paralytic poliomyelitis, but they have been shown to produce several other human illnesses associated principally with the central nervous system (*e.g.*, meningitis) or with the heart or skin. All of the human enteroviruses appear to be transmitted in the same ways. Thus, the few kinds of food which have been tested and found to contain polioviruses have been found also to be contaminated with other human enteroviruses. Published examples have included shellfish and ground beef (Sullivan *et al.* 1970). These virus detections were made by laboratory procedures to be discussed below, and they were not associated with outbreaks of human illness.

There are other viruses of the human enteric complex, which differ in size and composition from the enterovirus group, that one might detect in food by laboratory methods. These include the reovirus and adenovirus groups. Isolation of reovirus from oysters has been reported (Metcalf and Stiles 1968), but human adenoviruses evidently have yet to be detected in any food. It appears that one can properly speak of "human" enteroviruses and adenoviruses (*i.e.*, that the types which infect man seldom or never infect other animals) but that the three types of reoviruses are capable of infecting several warm-blooded species.

There is no generally accepted laboratory host for the virus of infectious hepatitis. Therefore, its association with foods is based entirely upon recognized common-source outbreaks. The virus is spread by a fecal-oral cycle, as is true of other members of the human enteric complex. Though it probably infects the intestine first, the principal disease results from inflammation and impaired function of the liver. The liver is often swollen; there may be jaundice, fever, and loss of appetite; and prolonged weakness is a quite constant finding. The incubation period ranges from 10 or 15 to 40 or 50 days, with an average of ~28 days. Deaths from this disease are relatively rare.

I am now aware of 45 reported outbreaks of this disease in which "food" (sometimes in a broad sense of the word) has been implicated as a common source. Most have been reviewed in detail elsewhere (Cliver 1966, 1967, 1969, 1971). These have ranged in size from 2 to 629 cases, with a grand total greater than 3,600. The vehicles implicated have included milk, meats, vegetables, pastries, salads, and other food and beverage products. Shellfish have been named most frequently. In each case, the food was either eaten raw or with very little heating after it had been contaminated. The immediate source of the virus contaminant has been either an infected human, or water or sewage containing human feces.

The connection of polluted water with shellfish is obvious. Other vehicles which have been contaminated by hepatitis virus in sewage or polluted water have included milk and milk products, cold cuts, soft drinks, watercress (U.S. Public Health Service, 1971), and, as a rather special case, water-filled plastic spheres ("freeze balls") used to cool beverages (Reynolds *et al.* 1968).

Several infected humans have been shown epidemiologically to have contaminated foods resulting in hepatitis outbreaks. Nine such persons were overtly ill (three with jaundice) while still handling food. Five others handled food only during the incubation period of their disease, and three denied ever having been ill. One cannot draw broad conclusions on the basis of just 17 persons, but it seems clear that people will sometimes handle food when ill and that the results are likely to be unfortunate.

A final member of the human "enteric complex" of viruses is the agent, or perhaps several, of epidemic viral gastroenteritis. These have been passed serially among human volunteers in bacteria-free filtrates of fecal suspensions. None, apparently, has been cultivated successfully in laboratory hosts such as tissue cultures or animals. The illnesses caused are not identical, beyond the fact that the digestive tract is always involved; incubation periods are uncertain. These are probably, but not surely, virus diseases. It seems reasonable to suppose that some (perhaps many) of the recorded incidents of "food-borne disease of undetermined etiology" are of this kind. Such a conclusion, if one dare call it that, is the strongest that can be drawn from the information presently available.

PERSISTENCE

The cells of a vertebrate animal normally die within hours after it is killed for food. The cells of shellfish and of vegetables, unless cooked or frozen, remain alive for longer periods or until the food is consumed. There is no significant opportunity for viruses infectious for man to multiply in foods. Viruses multiply only in appropriate living host cells, and cells which occur in foods seem either to die quickly or to be inappropriate. This means that the amount of virus present in a food at harvest or slaughter, or at the time that subsequent contamination occurs, is the greatest quantity there ever will be. Thereafter, the contaminating virus can only persist or be inactivated.

A virus particle is said to have been inactivated when it has lost its ability to cause an infection. This loss of infectivity is irrevocable in most instances. It results from degradation of some essential component of the virus. The component might be the nucleic acid core, the protein coat, or (when present) the lipid envelope. None of the viruses of the enteric complex is known to have a lipid envelope. The components of the virus particle differ quite significantly in susceptibility to adverse factors in the environment.

Thermal inactivation is probably most important where foods are concerned. Thermal inactivation, in the present context, will be taken to mean any inactivation in which thermal energy is the cause of degradation of a virus component. Cooking of a food is an apparently clear-cut process by which to bring about thermal inactivation of virus. This is true if the portion of the food containing the virus is sufficiently heated. An example of inadequate heat penetration permitting virus to persist within a food is afforded by clams steamed just until the shells open. This cooking practice has led to sporadic cases of infectious hepatitis (Koff *et al.* 1967) and to a recent small outbreak of this disease (U.S. Public Health Service, 1972). Similar incidents may be occurring with other foods.

Pasteurization might be expected to entail less rigorous heat treatment than does cooking. This is not necessarily true; most cooking procedures are not nearly so well-defined as the pasteurization processes, so valid comparisons are difficult. A number of studies have been performed on the inactivation of viruses in milk and milk products by pasteurization. These have included a great variety of viruses (enteric and other); several products in addition to raw, whole milk; and both "flash" and holding type pasteurization processes. The results have indicated quite uniformly that any quantity of a virus that one might reasonably expect to find in milk would be inactivated by pasteurization according to U.S. Public Health Service recommendations (Gresiková-Kohútová 1959; Gresiková *et al.* 1961; Sullivan *et al.* 1971B). This is important because viruses do occur in milk, because some of them persist extremely well in milk and milk products under storage conditions, and because improper pasteurization has figured in at least one large outbreak of milk-borne infectious hepatitis (Raska *et al.* 1966).

Many foods are not heat-treated at all on their way to the consumer or before being eaten. Even so, thermal inactivation of viruses may be significant in foods stored and distributed at room temperature. Foods perishable enough to require refrigeration or freezing are another matter. In such situations, viruses of the enteric complex, at least, seem likely to persist beyond the shelf life of the food (Lynt 1966). Inactivation of viruses at room temperature and below is, as yet, poorly characterized. A very elegant study, reported a few years ago, showed that the nucleic acid core of poliovirus, rather than its protein coat, was denatured at these temperatures (Dimmock 1967). Since the nucleic acid is enclosed rather effectively by protein, one might have expected that this kind of inactivation would be little influenced by a food in which the virus was suspended. This has since been found not to be strictly true. Under the mildly acid conditions which are common to many foods, several constituents have a signficant influence upon the rate of inactivation of polioviruses and other, related agents (Cliver *et al.* 1970). This might mean either that the coat protein is made permeable by mild acidity, so that the internal nucleic acid is exposed to the exterior environment, or that the coat protein itself is more labile under these circumstances. The question has practical significance in determining how likely different foods are to harbor viruses.

The chemical environment within a food, then, seems significant principally as it affects the rate, and perhaps the mode, of thermal inactivation. All viruses are also subject to chemical inactivation, defined here as denaturation of an essential virus component by direct chemical reaction with a substance in the environment. Chemical inactivation, at least of viruses of the enteric complex, appears to require harsher reagents than are normal constitutents of foods. Therefore, the principal use of chemical inactivation is in treating water for use in food processing, in disinfecting food surfaces, and in sanitizing food-contact surfaces. Halogens seem generally better suited to these purposes than, say, quaternary ammonium compounds. A great deal of further research is needed, and concise recommendations are few. Chang and Berg (1959) proposed use of trichlormelamine solution containing 250 ppm titrable chlorine and 0.039% KI for at least 10 min at a temperature of at least 5°C for soaking of fruits and vegetables whose surfaces might be virus-contaminated.

Radiant energy can inactivate viruses. Many viruses can be photosensitized under artificial conditions so as to be susceptible to the action of visible light. All viruses which have been tested are rapidly inactivated by intense ultraviolet light such as might be used in water treatment or in surface decontamination. The obvious limitation of ultraviolet light lies in its lack of penetrating ability. Ionizing radiation (*e.g.*, X-rays, cobalt-60 gamma rays), on the other hand, has a great deal of penetrating power. Though its use in relation to foods has been severely restricted, studies of its ability to inactivate viruses which might be food-borne have been reported (Sullivan *et al.* 1971A). The doses required to cause 90% inactivation of a great variety of viruses clustered remarkably close to 0.4 or 0.45 Mrad in suspensions whose radical-scavenging capacity was similar to those of foods. This means that food irradiation processes aimed at producing commercial sterility could be expected to inactivate over 99.999% of any virus which happened to be present. Low-dose processes have been proposed for other purposes; these could be expected to leave a significant proportion of a contaminating virus still infectious. It seemed possible that such treatment might cause mutations of public health significance in any viruses that might be present. Unpublished studies in our laboratory have not supported this.

Drying also affects viruses. Some seem to be stabilized by drying, but few of these are of serious concern in foods. A number of drying studies have been performed using the polioviruses and related agents as models. These viruses are thought to be inactivated by air-drying. Freeze-drying for storage and distribution of the viruses has been unsuccessful until relatively recently (Rightsel and Greiff 1967; Berge et al. 1971). Freeze-drying of experimentally contaminated foods has also caused a great deal of virus inactivation (Heidelbaugh and Giron 1969). Approximately 99% inactivation of poliovirus resulted from freeze-drying in cream-style corn; the virus that persisted was quite stable during subsequent storage of the food at 5°C (Cliver et al. 1970).

Finally, viruses seem to be subject to biological inactivation, which is defined here as inactivation brought about by living things or their products. The action of microorganisms upon viruses is of especial interest because: (1) wherever viruses occur, they are almost certain to be accompanied by bacteria, and (2) many foods have a microflora of their own. In some experiments, microbial decomposition has had little effect upon virus in foods (Lynt 1966; Cliver et al. 1970). In others, viruses have been inactivated more slowly if a food had been autoclaved prior to experimental contamination (Kalitina 1966). Viruses persist well in such products of microbial action as cottage cheese (Kalitina 1969, 1971) and kefir (Kiseleva 1971). Yet-to-be-published studies in our laboratory have shown that bacteria of certain species (e.g., Pseudomonas aeruginosa, Bacillus subtilis) are capable of inactivating viruses under defined conditions. Specific bacteria capable of virus inactivation have yet to be identified in foods.

DETECTION

A great deal of the needed information concerning the incidence and persistence of viruses in foods will be gotten only by detecting viruses in foods. A virus has been detected, in this context, when it has produced a demonstrable infection in some living host system. When testing unknown foods, and for most experimental purposes, the living host system selected has usually been a tissue culture (Cliver and Grindrod 1969). Cells of human or monkey origin, grown in glass or plastic vessels, have found the greatest use in detecting viruses infectious for man. Embryonated eggs, laboratory animals, and even human volunteers might serve as the host system in studies on certain specific viruses. Agents detectable only in human volunteers have been discussed above. In instances where their use is not absolutely necessary, human volunteers have many liabilities and just one advantage: one can test an unknown food simply by feeding it to the test host. A food sample, to be tested in any other host system, must be converted to some kind of a compatible fluid suspension. The procedures by which such fluid suspensions are prepared might be categorized as dilution, extraction, or concentration methods.

In a dilution method, the sample is suspended in added fluid by shaking, grinding, or homogenization. Bacteria and food components which might be injurious to a tissue culture (or other test host) are reduced in concentration simply by diluting the sample suspension. This is a good approach if: (1) there is likely to be a high level of virus in the sample, (2) one must operate with an absolute minimum of laboratory equipment, or (3) one has no idea what kind of virus might be present. The last of these stipulations refers to the fact most manipulations of the sample, other than dilution, carry at least some risk of removing or destroying one kind of virus or another. That is, the more elegant techniques are usually more selective than the dilution methods. Tissue cultures (or any other test host) are also selective, so there is no one technique which enables one to detect all of the viruses infectious for man.

A dilution procedure is basically not very sensitive; if one is looking for a needle in a haystack, he might rather not begin by adding more "hay." An extraction procedure differs in that the bulk of the food solids, bacteria, and other deleterious substances are excluded from the sample suspension. If the separation has been done well, no further dilution is required before the sample suspension (or extract) can be inoculated into tissue cultures. An extraction procedure may begin by using a selected fluid to dissociate virus from the surface of the food or food particles (Sullivan *et al.* 1970), or one might make a food suspension or homogenate and then try to remove the bacteria and food solids from it. Both centrifugal and filter methods of separation are available. There is a certain risk of virus loss in the elution or suspension step and in the separation step, so extraction methods tend to be more selective than dilution methods.

Finally, one might achieve a great deal more sensitivity, with a given quantity of tissue cultures used in testing, by concentrating the food extract. A sample extract may be regarded as a suspension of a small quantity of virus in a relatively large volume of water. In a concentration procedure, one wishes to reduce the volume of the suspension and to lose as little as possible of the virus in the process. A great many techniques have been used for this purpose, including ultracentrifugation, ultrafiltration, phase separation, and others. Several are quite virus-specific, in that they enhance detection of one kind of virus and virtually preclude detection of another. This problem has been discussed in greater depth elsewhere (Cliver 1971).

Sensitivity and precision vary, but the means are now at hand to detect most viruses infectious for man in most foods. The cost per test probably will not be a great deal higher than that for bacterial surveillance. An unhappy result of using infectivity as a basis for virus detection is that the test results are known only after 2 (rarely) to 7 or more days. While this is objectionable, alternate, rapid methods based upon chemical or physical detection of the virus particles are not immediately in prospect. Use of virus antigens for detection by serologic methods has been proposed but not really attempted as yet.

COUNTERMEASURES

The only sure way to prevent viruses from being transmitted in foods, ever, would be to keep the viruses out of the foods in the first place. There are two categories of viruses which are now known to merit concern; each presents its own problems. Some viruses, which are capable of producing disease in man, infect food-source domestic animals and occur in their milk, eggs, or flesh. The presence of these viruses could be avoided by not using infected animals as a source of food. Clearly, this is more easily said than done. Intensive efforts in animal health, veterinary preventive medicine, and meat inspection can accomplish a great deal, however.

The second important category of viruses comprises the enteric agents. Since these are shed principally or exclusively in feces, the way to keep them out of food is by preventing feces (in any quantity or dilution) from contaminating the product. This, too, is more easily said than done. Fecal contamination may occur directly or by way of polluted water and perhaps flies and other insect vectors. The epidemiologic record suggests that this kind of contamination occurs more frequently in food service and in non-commercial food preparation than in the food processing industry. One ought not conclude that food processors do not make such mistakes. They may be less likely to do so, but they are also less likely to be implicated unless a very large number of illnesses result. It is important that all opportunities for fecal contamination be minimized by using a safe and reliable water supply, by excluding insects, and (most important) by using healthy and attentive food handlers. The record cited above for infectious hepatitis makes it clear that some people will handle food while ill. One would hope that supervisors would prevent this, but some of these individuals *were* supervisors.

Testing of foods for virus contamination is feasible and should be done, both for surveillance and for quality assurance purposes. The development of detection methods for food-borne viruses probably will never end, but good methods are available now. These must be applied selectively, however. It is possible now to identify a number of "high risk" foods, with shellfish undoubtedly at the top of the list, which should be tested routinely or frequently. Other foods, especially those which are strongly heat-treated in their final sealed containers, need little or no scrutiny. In identifying high risk foods, one ought not rely on the consumer for self-protection; a product which will only be safe after the consumer has cooked it is an unsafe product. This last statement is meant rather to reflect upon the reliability of the consumer than of heat. Purposeful, controlled heat treatment is by far the most generally useful means of inactivating food-borne virus. Though all virus contamination of food is preventable, not all of it will be prevented. A sufficient quantity of heat (assuming that the food can be heated) will prevent most of the consequences of human error.

SUMMARY

A great many different foods harbor viruses on occasion. Some of these viruses originate in animals. Others are of human origin: presently available information indicates that the most common source of food-borne virus is the human intestines. Virus in foods can be inactivated by heating equivalent to that used in pasteurizing milk and milk products. Since not all food can be heated, it would be better not to let virus contaminate food at all. Viruses will not multiply in foods, and they may be inactivated by several other means than heat before they reach the consumer, but none of these other means presents a very useful alternative to cooking. The incidence of food-borne virus disease could also be reduced by using available methods to monitor foods for virus contamination. Refrigeration or freezing of a food will tend to preserve any virus that it harbors; with this sole exception, the principles of prevention are the same for viruses as for other food-borne disease agents.

BIBLIOGRAPHY

BERGE, T. O., JEWETT, R. L., and BLAIR, W. O. 1971. Preservation of enteroviruses by freeze-drying. Appl. Microbiol. 22, 850-853.

BURROWS, R. et al. 1971. The growth and persistence of foot-and-mouth disease virus in the bovine mammary gland. J. Hyg. 69, 307-321.

CHANG, S. L., and BERG, G. 1959. Chlormelamine and iodized chlormelamine germicidal rinse formulations. U.S. Armed Forces Med. J. 10, 33-49.

CLIVER, D. O. 1966. Implications of food-borne infectious hepatitis. Public Health Rept. 81, 159-165.

CLIVER, D. O. 1967. Food-associated viruses. Health Lab. Sci. 4, 213-221.

CLIVER, D. O. 1969. Viral infections. *In* Food-borne Infections and Intoxications, H. Riemann (Editor). Academic Press, New York.

CLIVER, D. O. 1971. Transmission of viruses through foods. Crit. Rev. Environ. Control *1*, 551–579.

CLIVER, D. O., and GRINDROD, J. 1969. Surveillance methods for viruses in foods. J. Milk Food Technol. *32*, 421–425.

CLIVER, D. O., KOSTENBADER, K. D., JR., and VALLENAS, M. R. 1970. Stability of viruses in low moisture foods. J. Milk Food Technol. *33*, 484–491.

DAWSON, P. S. 1970. The involvement of milk in the spread of foot-and-mouth disease: an epidemiological study. Vet. Rec. *87*, 543–548.

DIMMOCK, N. J. 1967. Differences between the thermal inactivation of picornaviruses at "high" and "low" temperatures. Virology *31*, 338–353.

ERNEK, E., KOZUCH, O., and NOSEK, J. 1968. Isolation of tick-borne encephalitis virus from blood and milk of goats grazing in the Tribec focus zone. J. Hyg. Epidemiol. Microbiol. Immunol. *12*, 32–36.

GRESIKOVA, M., HAVRANEK, I., and GORNER, F. 1961. The effect of pasteurisation on the infectivity of tick-borne encephalitis virus. Acta Virol. *5*, 31–36.

GRESIKOVA-KOHUTOVA, M. 1959. The effect of heat on infectivity of the tick-borne encephalitis virus. Acta Virol. *3*, 215–221.

HEIDELBAUGH, N. D., and GIRON, D. J. 1969. Effect of processing on recovery of polio virus from inoculated foods. J. Food Sci. *34*, 239–241.

KALITINA, T. A. 1966. Persistence of Coxsackie group B serotypes 3 and 5 in mince meat. Voprosy Pitaniya *5*, 74–77. (Russian)

KALITINA, T. A., 1969. Study of the transmissibility of enteroviruses in milk and milk products. Zhurnal Mikrobiol. Epidemiol. Immunobiol. *7*, 61–64. (Russian)

KALITINA, T. A. 1971. Persistence of the poliomyelitis virus and some other enteric viruses in cottage cheese (curd). Voprosy Pitaniya *30*, 78–82. (Russian)

KAWAKAMI, Y. *et al.* 1966. Infection of cattle with parainfluenza 3 virus with special reference to udder infection. I. Virus isolation from milk. Japan. J. Microbiol. *10*, 159–169.

KISELEVA, L. F. 1971. Survival of poliomyelitis, ECHO, and Coxsackie viruses in some food products. Voprosy Pitaniya *30*, 58–61. (Russian)

KOFF, R. S. *et al.* 1967. Viral hepatitis in a group of Boston hospitals. III. Importance of exposure to shellfish in a nonepidemic period. New England J. Med. *276*, 703–710.

KOKERNOT, R. H., RADIVOJEVIC, B., and ANDERSON, R. J. 1969. Susceptibility of wild and domesticated mammals to four arboviruses. Am. J. Vet. Res. *30*, 2197–2203.

LYNT, R. K., JR. 1966. Survival and recovery of enterovirus from foods. Appl. Microbiol. *14*, 218–222.

METCALF, T. G., and STILES, W. C. 1968. Enteroviruses within an estuarine environment. Am. J. Epidemiol. *88*, 379–391.

RASKA, K. *et al.* 1966. A milk-borne infectious hepatitis epidemic. J. Hyg. Epidemiol. Microbiol. Immunol. *10*, 413–428.

REYNOLDS, R. D. *et al.* 1968. Freeze-ball hepatitis. Arch. Intern. Med. *122*, 48–49.

RIGHTSEL, W. A., and GREIFF, D. 1967. Freezing and freeze-drying of viruses. Cryobiol. *3*, 423–431.

SULLIVAN, R., FASSOLITIS, A. C., and READ, R. B., JR. 1970. Method for isolating viruses from ground beef. J. Food Sci. *35*, 624–626.

SULLIVAN, R. *et al.* 1971A. Inactivation of thirty viruses by gamma radiation. Appl. Microbiol. *22*, 61–65.

SULLIVAN, R. *et al.* 1971B. Thermal resistance of certain oncogenic viruses suspended in milk and milk products. Appl. Microbiol. *22*, 315–320.

U.S. PUBLIC HEALTH SERV. 1971. Morbidity Mortality Weekly Rept. *20*, No. 39, 357.

U.S. PUBLIC HEALTH SERV. 1972. Morbidity Mortality Weekly Rept. *21*, No. 2, 20.

CHAPTER 9

Mario P. de Figueiredo
and
James M. Jay

Coliforms, Enterococci, and Other Microbial Indicators

Concern with the safety and sanitary quality of foods is by no means a recent development, although it is of particular import today when tremendous progress has been achieved in the many phases of the food operation. There has been and continues to be a growing dependence on convenience foods and food service outlets, and concern with various complications in storage and refrigeration needs.

In general, the following factors individually or in combination could render a food unsafe for human consumption: (1) Use of contaminated ingredients, (2) Unsanitary conditions in the plant, (3) Process failure, (4) Post-processing contamination, including mishandling during shipment, and (5) Abuses at the retail level and by the consumer. Testing directly for pathogens is a time consuming and costly effort. In order to minimize these factors, it has been the practice for many years to test for certain groups of organisms or "indicator" organisms, whose presence in a food product within certain limits would be reflective of both sanitary quality and conditions that could lead to the entry and proliferation of pathogens. The major groups of organisms commonly employed as indicators are coliforms, enterococci, and more recently the Enterobacteriaceae. Total numbers provide some indication of food sanitary quality, and total viable plate counts along with direct microscopic slide counts are dealt with here in this context.

COLIFORMS

The coliform group of bacteria are members of the family Enterobacteriaceae. Included in this family are pathogenic groups such as *Salmonella*, *Shigella*, and *Yersinia*. The coliforms differentiate themselves from most other members of the family in being capable of fermenting lactose with the production of acid and gas within 48 hr. The two genera of coliforms are *Escherichia* and *Enterobacter* (formerly *Aerobacter*). A few lactose-fermenting strains may be found in several other genera (Hausler 1972).

With respect to growth requirements, the coliforms have been reported to grow in a variety of culture media and under rather diverse physical and chemical conditions. For example, growth has been reported to occur over the temperature range $-2°C$ to $50°C$ and at pH values from 4.4 to 9.0. The presence of an organic carbon source along with a nitrogen source such as $(NH_4)_2SO_4$ and other minerals will enable *E. coli* to grow. The coliforms grow well on nutrient agar where they generally produce visible colonies well within 24 hr at $37°C$, and in the presence of bile salts which inhibit the growth of most gram-positive bacteria. The latter property is utilized for the selective isolation of these organisms and certain other gram-negative bacteria. The incorporation of lactose and an acid-base indicator into a medium containing bile salts enables one to select for and distinguish between coliforms and noncoliforms.

The natural habitat of *E. coli* is the gastrointestinal canal of man and warm blooded animals. *Enterobacter* is found primarily on vegetation but may be found occasionally in the intestine, sometimes alone, but more frequently along with *E. coli*. The omnipresence of man and animals along with fecal wastes accounts for the existence of *E. coli* in water, soil, on vegetation, in the air, and thus throughout the general environment. The presence of *E. coli* has been long used as an indicator of fecal pollution of natural waters. It might be pertinent to note here that the traditional coliform index failed to forewarn of two important water-borne outbreaks of *Salmonella* and *Shigella* infections in Riverside and Madera, California (Greenberg and Ongerth 1966; Browning and Mankin 1966), indicating that alternative or new indicators need to be explored.

The interpretation of the presence of *E. coli* at given levels as being indicative of fecal pollution in water has been extended, sometimes uncritically, to food products. In foods such as milk, cream, soft cheese, etc. (which provide a good medium for the proliferation of these organisms), this hypothesis can lead to absurd conclusions. The occurrence of this organism in large numbers in processed foods does not indicate fecal contamination in the sense of demonstrating recent contact with fecal matter. What often is indicated is poor practice, such as contaminated or inadequately processed raw materials, personnel contamination, improperly cleaned and sanitized equipment or food contact surfaces, all of which suggest that enteric pathogens may have entered the food product through the same route. More important than just merely demonstrating their presence or absence are the relative numbers of coliforms that exist and whether they are of fecal or non-fecal origin.

Isolation and Enumeration

Among the techniques that may be employed to isolate and quantitate coliforms are direct plating methods, the MPN technique, and membrane filter techniques. In direct plating, one may employ violet red bile (VRB) agar, desoxycholate lactose agar, or other similar media recommended by a reputable agency or association. When VRB agar is employed, coliforms appear as red colonies, although sometimes particles of food absorb the dye and thus create artifacts which could lead to erroneous conclusions. Direct plating methods are of limited sensitivity since the limit of detection is dependent upon the dilution factors used.

The MPN technique is of more value in detecting small numbers of coliforms than are direct plating methods. By this technique populations are estimated on the basis of the number of tubes of broth media exhibiting gas following a 48 hr incubation. Generally, three or five tubes are inoculated with a given quantity of sample. Hall (1964) reported that lauryl sulfate tryptose (LST) broth was better than lactose broth for enumerating coliforms by this procedure.

APHA *Standard Methods* or other reputable references should be followed when performing MPN procedures. Results from the primary inoculation of LST tubes are presumptive and should be confirmed by transferring aliquots from gas-positive LST tubes to brilliant green lactose bile (BGLB) broth. The presence of gas in this medium indicates a positive test. If the confirmed test is positive, the completed test is carried out to determine whether the organisms resemble coliforms in terms of microscopic and culture characteristics.

While LST broth is widely recommended and used for the MPN technique, Moussa *et al.* (1973) recently compared this medium to four others, including a lactose-glutamic acid medium and found that the latter medium allowed for the recovery of more coliforms and fecal coliforms from dehydrated and deep frozen foods than did LST, lactose, brilliant green bile, or EE broths.

The membrane filter (MF) technique is recommended as an alternative method for the examination of water or clear beverages. APHA *Standard Methods* or the National Academy report (1971) should be consulted for more specific directions. With respect to types of membranes, Presswood and Brown (1973) found that at temperatures of 35 and 44.5°C Gelman membrane filters gave results more in agreement with plate counts than did the Millipore membranes.

Fecal Coliforms

In order to distinguish between *E. coli* I (generally considered to be fecal in origin) and *E. aerogenes* (nonfecal), the IMViC pattern of the coliform isolates should be determined. By this formula, these two organisms are identified as follows:

$$I\ M\ V\ C$$

E. coli   + + - -

E. aerogenes - - + +

where I = indole production; M = methyl red reaction; V = Voges-Proskauer reaction (acetoin production); and C = citrate utilization. It should be noted that other coliform types do exist and are apt to provide a variety of patterns by the IMViC formula intermediate between the two extremes (Geldreich and Bordner 1971). Also, Adams (1972) found that 28 of 58 isolates of E. aerogenes from hot spring drainage produced gas in EC broth at 44.5°C.

The determination of the incidence of fecal coliforms in a food product is of more value as a sanitary index than the total number of coliforms per se. Fecal coliforms are determined by use of elevated incubation temperatures and the following three techniques have been employed to this end: (1) The MPN method employing EC broth, (2) The membrane filter technique employing M-FC broth-soaked pads, and (3) The incubation of agar plates or pouches.

The MPN method is the most widely used of the elevated temperature methods. Temperatures from 41 to 46°C have been used by various investigators with 44.5°±0.5 being the most commonly employed. By this method, fecal coliforms produce gas in EC broth at 44.5°C in 24 hr. With respect to the different temperatures employed, Geldreich (1966) studied cultures from feces, polluted and presumably nonpolluted soils and recommended a temperature of 44.5°C for fecal coliforms. On the other hand, studies by Fishbein (1962), Fishbein and Surkiewicz (1964), and Fishbein et al. (1967) suggested the use of 45.5°C. The latter authors reported that the rate of recovery of E. coli I and II from EC broth cultures incubated at 45.5°C and 44.5°C were 77.2 and 53.1% respectively; that only 4% of the E. coli cultures which grew at 44.5°C failed to grow at 45.5°C; and that three times more false positives were recorded at 44.5°C than at 45.5°C. Directions for performing fecal coliforms tests employing elevated temperatures may be found in APHA Standard Methods.

The value of making fecal coliform determinations is established. This is true for natural waters and appears to be so for certain foods as well. In examining fecal coliform density relative to the occurrence of salmonellae in streams, Geldreich and Bordner (1971) found that 53.5% of 71 samples were positive for salmonellae when fecal coliforms were present at 1 to 1,000/100 ml of water and 96.4% of 140 samples were positive when the fecal coliform counts exceeded 1,000/100 ml. With respect to the relative incidence of fecal type coliforms, the above authors found that 96.4% of the coliforms in human feces produced gas in lactose broth at 44.5°C and that 93 to 98.7% of those from livestock, poultry, cat, dog, and rodent feces produced gas at the elevated temperature.

The speed of recovery of coliforms from water and foods has always been of concern to microbiologists. The current APHA *Standard Methods* procedure may take up to 72 hr for confirmation of fecal types (24 to 48 hr for LST incubation followed by 24 hr for EC broth incubation). In an effort to reduce the 72 hr incubation time to 24 hr, Fishbein *et al.* (1967) inoculated frozen food homogenates directly into LST, followed by incubation at 44°C for 24 hr, and found that the results for *E. coli* were similar to those obtained by the APHA method. Andrews and Presnell (1972) developed a new medium for the rapid recovery of *E. coli* and showed that this organism could be recovered from raw seawater within 24 hr without significant loss of accuracy. Andrews *et al.* (1975) later showed that this medium was useful in recovering *E. coli* from shellfish. The medium contains lactose, tryptone, NaCl, 0.1% triton X-100, and 0.05% salicin. Mossel and Vega (1973) used MacConkey agar in plastic pouches incubated at 44°C for 24 hr and found that freshly isolated strains of *E. coli* could be recovered by this method just as well as by 37°C incubations. These investigators reported that this is the most reliable procedure for the direct enumeration of *E. coli*.

Other attempts to speed up the recovery and confirmation of fecal coliforms include the agar-pour plate method of Francis *et al.* (1974). This method employs a medium containing proteose peptone no. 3, yeast extract, lactose, NaCl, 0.005% sodium lauryl sulfate, and 0.03% bromthymol blue with pH adjusted to 7.3. These investigators report that fecal coliforms can be detected after only 7 hr at 41.5°C and appear as yellow to orange colonies with yellow haloes against the bluish-green background of the medium. Another of the more recent rapid techniques is that of Bachrach and Bachrach (1974). This is a radiometric method for detecting coliforms based upon the release of $^{14}CO_2$ from (^{14}C) lactose by the coliforms which must be suspended in a liquid growth medium. The evolved $^{14}CO_2$ is trapped and counted in a liquid scintillation spectrometer, and the authors claim that from 1 to 10 coliforms may be detected within 6 hr of incubation at 37°C by the method. While it may prove to be an excellent research method, the radiometric technique appears to be too time consuming for the routine analysis of multiple food samples.

ENTEROCOCCI

The enterococci are members of the genus *Streptococcus*. The streptococci are Gram positive, catalase negative cocci that form long or short chains. They belong to Lancefield's serologic group D streptococci and the group is usually classified as follows (Niven 1963; Hartman *et al.* 1966):

Enterococci	*Others*
S. faecalis	*S. bovis*
var *liquefaciens*	*S. equinus*
var *zymogenes*	
S. faecium	
var *durans*, or *S. durans*	

Sherman (1937) distinguished enterococci from other group D streptococci on the basis of their ability to grow in the presence of 6.5% NaCl; at a pH of 9.6; at temperatures of 10°C and 45°C; and to withstand a temperature of 60°C for 30 min. Deibel (1964) has noted that there could be strains which fail to produce positive results by one or more of these criteria.

The natural habitat of the group D streptococci is the intestinal canal of man and other animals. Studies have shown that *S. faecalis* and its varieties are more commonly associated with the intestinal tract of man than that of other animals (Bartley and Slanetz 1960) while *S. bovis* and *S. faecium* occur predominantly in cattle (Mieth 1962A) and hogs (Mieth 1960). *S. faecium* var *casseliflavus* is a plant epiphyte which displays most features common to other enterococci (Mundt and Graham 1968). Horses were found by Mieth (1962B) to harbor *S. equinus, S. bovis,* and *S. faecium.* In addition to their fecal origins, these organisms occur widely in nature on plants and in soils, being aided in their distribution by insects, wind, and rain (Mallman and Litsky 1951; Mundt *et al.* 1958; Mundt 1961).

Media and Isolation

Methods for the isolation and enumeration of enterococci are numerous, and some of the newer techniques have been compared to the older ones by Hall (1964), Facklam and Moody (1970), Isenberg *et al.* (1970), Sabbaj *et al.* (1971), Pavlova *et al.* (1972), Lee (1972), Efthymiou *et al.* (1974), Oblinger (1975), and Daoust and Litsky

(1975). Most procedures employ presumptive media followed by confirmatory tests, as in the case with coliforms. Azide, tellurite, bile, neomycin, taurocholate, Tween 80, selenite, NaCl, phenylethyl alcohol, and thallium have been used in media as primary selective agents.

The KF agar medium is recommended by the Association of Food & Drug Officials of the U.S. (1966). This medium contains sodium azide to inhibit catalase positive organisms and tetrazolium chloride which provides red color to colonies. Confirmation procedures for the enterococci may be conducted by using "Sherman tests" (Sherman 1937) or by growth in EVA broth (Litsky et al. 1955), or by one of the more recently developed procedures such as tyrosine decarboxylase activity (Lee 1972).

The combination of azide dextrose and EVA broths has been used by a large number of investigators (Hall 1967). Pavlova et al. (1972) compared five media on the recovery of fecal streptococci from various foods (Table 9.1) and found that although the thallous-acetate medium yielded the highest counts, it was the least selective of the five media employed. The Selective Enterococcus agar medium in the hands of these investigators yielded a slightly higher percent of nonfecal streptococci than the KF medium but required only 24 hr incubation while KF required 48 hr.

One of the most rapid techniques reported for the enumeration of fecal streptococci is a membrane filter fluorescent antibody test which Pugsley and Evison (1975) employed to detect these organisms in water in 10-12 hr. Whether or not this method can be adapted to food use or not is uncertain at this time. One of the problems encountered in the use of many of the available media for group D streptococci is a lack of selectivity for these organisms, and a technique such as FA may provide an excellent solution to this problem. The lack of selectivity of many of the older methods is well known, as Splittstoesser et al. (1961) noted in the case of EVA broth.

Incidence in Foods

These organisms are naturally present in many foods such as frozen seafood (Larkin et al. 1956; Raj et al. 1961), dried whole egg powder (Solowey and Watson 1951), raw and pasteurized milk (White and Sherman 1944), commercially frozen fruits, fruit juices, and vegetables (Hucker et al. 1952; Kaplan and Appleman 1952; Larkin et al. 1955A, B). Deibel (1964) demonstrated that the presence of enterococci in processed meats does not necessarily indicate fecal contamination. The enterococci are dubious as

TABLE 9.1
COMPARISON OF PRESUMPTIVE FECAL STREPTOCOCCAL
NUMBERS PER GRAM OF VARIOUS FOODS ON
5 SELECTIVE MEDIA

Selective Media	Mean Number of Bacteria per gram Food								
	Frozen							Non Frozen	
	Scallops	Stuffed Clams	Clam Sticks	Fish Cakes	Seafood Dinner	Shrimp Croquettes	Crab Cakes	Pastrami Luncheon Meat	Veal & Beef Luncheon Meat
N-Enterococcus Agar	0	7.1×10^2	1.2×10^2	2.2×10^2	3.7×10^2	1.5×10^2	1.4×10^3	2.6×10^2	4.3×10^3
XF-Streptococcus Agar	2.6×10^1	2.4×10^3	2.0×10^3	7.1×10^2	3.8×10^2	3.2×10^2	6.8×10^3	5.5×10^2	1.5×10^4
Thallous-Acetate Agar	1.2×10^2	8.8×10^3	9.0×10^3	1.2×10^3	8.8×10^2	7.4×10^2	9.2×10^3	7.0×10^2	9.8×10^4
Azide-Sorbitol Agar	0	6.8×10^2	8.0×10^2	1.0×10^2	2.2×10^2	1.6×10^2	1.1×10^3	1.2×10^2	2.2×10^3
Selective Enterococcus Agar	6.5×10^1	4.5×10^3	1.5×10^3	6.8×10^2	4.7×10^2	4.3×10^2	6.6×10^3	4.3×10^2	2.3×10^4

From Pavlova et al. (1972). Reprinted with permission.

etiologic agents in food poisoning. Deibel and Silliker (1963) failed to demonstrate food poisoning characteristics of 23 strains of *S. faecalis* and *S. faecium* including nine cultures that had allegedly been implicated in food poisoning outbreaks.

It has been suggested that the enterococci provide a better index of food sanitary quality than do coliforms since they are more resistant to adverse environmental conditions. They adapt well to a wide range of environmental conditions such as drying, freezing, salt concentrations, etc., but an accurate assessment of sanitary quality must be taken into consideration in both product and process.

Some differential physiological characteristics of the group D streptococci are presented in Table 9.2. For more detailed discussions of the biochemical characteristics of these organisms, see the reviews by Shattock (1962) and Deibel (1964).

THE ENTEROBACTERIACEAE

While coliforms and enterococci have received the most study and application as indicator organisms, other groups have been suggested and given more limited use. Mossel *et al.* (1963) suggested the collective use of all members of the family Enterobacteriaceae as a means of assessing fecal contamination. This approach has the apparent advantage of providing for greater flexibility since it is possible for the coliform organisms to be destroyed but not necessarily other members of this family. The organisms are detected and enumerated by an MPN procedure employing Enterobacteriaceae enrichment (EE) broth with confirmation on violet red bile glucose agar plates as necessary.

With respect to the utility of employing the Enterobacteriaceae as indicators for foods, Drion and Mossel (1972) reported that when 1 or 2 X 1 g aliquots of dried foods are tested and found negative for the Enterobacteriaceae, the same degree of consumer protection is afforded as when one examines 60 X 25 g amounts of dried foods for salmonellae and accepts the consignment only when no positives are found. The above is based upon their finding Enterobacteriaceae to salmonellae on a minimum order of 10^3. More recently, Mossel (1974A) reiterated his faith in the examination of foods for this group as sanitary indicators in light of an outbreak in the U.S. of enteropathogenic *E. coli* food poisoning caused by *E. coli* 0124 from imported Camembert cheese (Marier *et al.* 1973). This product contained 10^3 to 10^5/g of coliforms but 10^6 to 10^7/g of Enterobacteriaceae. Mossel believes that while some investigators might have accepted the product on the lower level of coliforms (10^3/g), the product would not have been accepted on the higher number of Enterobacteriaceae.

TABLE 9.2
GROUP D STREPTOCOCCI:
SELECTED DIFFERENTIAL PHYSIOLOGICAL CHARACTERS

	Division 1	Division 2		Division 3	
	S faecalis and varieties	S. faecium	S. durans	S. bovis	S. equinus
β Hemolysis	-/+	-	+/-	-	-
Growth 10°	+	+	+	-	-
Growth 45°	+	+	+	+	+
50°	+	+*	-	-	-
pH 9.6	+	+	+/-	-	-
6.5% NaCl	+/-	+/-	+/-	-	-
40% bile	+	+	+	+	+
Resists 60°C for 30 min.	+	+	+/-	-	+
NH₃ from arginine	+	+	+	-	-
Gelatin liquefied	-/+	-	-	-	-
Tolerates 0.04% Pot. tellurite	+	-	-	-	-
Acid from:					
Glycerol (anaerobic)	+*	-	-	-	-

Mannitol	+	+	– *	–/+	–
Sorbitol	+*	–*	–	–/+	–
L-arabinose	–	+*	–	+/–	–
Lactose	+	+	+	+	–
Sucrose	+*	+/–	–	+	+*
Raffinose	–*	–*	–	+	–
Melibiose	–	+*	*	+	–
Melezitose	+*	–	–	–	–
Starch hydrolyzed	–	–	–	+*	–*
Tetrazolium reduced at pH 6.0	+	–	–	+/–	–

+ = positive result.
– = negative result.
+/= variation between strains, majority positive.
–/= variation between strains, majority negative.
* occasional strains atypical.
Division 1 and Division 2 fulfill the criteria for the "enterococcus group" of Sherman (1938).

There is a need for more comparative work before the true position of the Enterobacteriaceae as food sanitary indicators can be determined. The collective use of all members of this family as indicators would seem to place more significance upon the presence of some genera than past experiences would warrant relative to food sanitary quality. For example, the genera *Serratia*, *Proteus*, and *Erwinia* are members of the family which can often be found in high numbers on fresh vegetables and meats in the total absence of *Salmonella* or *Shigella*.

TOTAL COUNTS

Total counts measure all, or more commonly, a portion of the flora in a food product without reference to specific microbial types. When determined by a direct microscopic method, both viable and nonviable cells are included. Only viable cells are represented if the total count is obtained by plating methods and even then only those cells capable of forming colonies under the specific conditions of medium pH, incubation temperature, etc., are enumerated. Total counts at best reflect the sanitary quality of shelf stable foods; foods such as frozen or dried products which do not support microbial growth (Silliker, 1963). In these products, they are used as yardsticks to measure how the foods were handled during manufacturing. High total counts in foods whose production process includes terminal heating or other lethal treatments should be treated with caution. Here a finished product with low viable counts may be generated even though high count, questionable raw materials were used. Prolonged storage, either frozen or dried, could also result in reduction of total viable counts (Angelotti 1964).

Historically, total counts have been used to evaluate the sanitary quality of milk and have proven useful in predicting its shelf life. The criteria used in this case cannot be extended to all other processed foods, since the general perishable nature of milk is well known and its processing and distribution channels are well controlled. The analyst, therefore, must consider a food sample in light of its known history, age, storage conditions and mode of distribution. Products such as fresh meats provide a different situation (Ingram and Dainty 1971). Microbes play an important role in spoilage. Different types of spoilage are caused by different types of organisms, depending upon the conditions of treatment and storage. Total counts estimate total flora without specifying the types of microbes, but in spoilage, the growth patterns of specific organisms are more important than mere measurement of total count. Peterson and Gunderson (1960) have shown that a rather small number of psychrophilic pseudomonads can bring about off flavors in defrosted chicken pies prior to the growth of a larger number of psychrophiles. Similarly, Lerke *et al.* (1965, 1967) showed that less than 10% of the pseudomonads isolated from spoiled fish were capable of causing the typical odors of spoiled fish. Total viable microorganisms determined by incubating plates in the 0° to 5°C range are, therefore, more predictive of the shelf life of refrigerated foods than the use of higher incubation temperatures (Ingram 1965).

Plate Counts

Aerobic plate counts are of limited value in predicting the safety of processed foods in terms of pathogens or toxins. The finding of a low total count does not support the presumption that a food product is necessarily safe from pathogens. Such a conclusion would, indeed, be dangerous since low counts are not always synonymous with safety. Montford and Thatcher (1961) isolated salmonellae from commercial frozen egg preparations with counts as low as 380/g in one case and below 5,000/g in others. In 1968, a lot of dried milk with a total count of only 50/g was responsible for a salmonellae outbreak in Canada. In yet another instance, a lot of dry milk with a total count of only a few thousand/g caused an outbreak of staphylococcal food poisoning (Foster 1968). In products such as sauerkraut, buttermilk, and natural cheeses, bacteria are essential to a proper development of flavor, so that high total counts should be expected.

The pour plate technique employing incubation temperatures of 30-37°C for periods from 48 to 72 hr is generally recommended for mesophilic counts. The usual methods employed to measure total counts in milk or water cannot be universally applied to all foods since foods vary in their physical and chemical properties as well as in their microflora. Hence, what is applicable to one food is not necessarily applicable to another.

A rapid method for the measurement of total counts has been described by Winter *et al.* (1971). By this method, cells are rinsed from food or equipment surfaces with sterile diluent and concentrated on the surface of membrane filters. The filters are incubated for 4 hr on a suitable medium and then examined microscopically. The results were found to be comparable to those observed through standard plate count techniques. The authors claim that this method has been used successfully for in-line control by several processing plants.

The meaningfulness of total plate counts as sanitary indicators would be much improved if similar methods were followed by all investigators. The value of employing standardized and recommended procedures cannot be over emphasized.

Direct Microscopic Counts

This procedure has been used for some time to determine the quality of raw food materials prior to processing. Examples are high count raw milk and milk powder, raw and pre-cooked frozen foods, and powdered eggs and related products. The method consists of examining a stained film of food under the microscope and determining the number of bacteria/ml. A prescribed number of fields are counted. Methylene blue or other specialized stains may be used as staining agents. While the method has proved useful in the case of high-count raw milk, it has certain limitations inherent in its make-up in that it measures both viable and nonviable cells and organisms may not show up clearly, or may be confused with extraneous matter present. On the positive side, the method has the advantage of rapidity and cheapness. Low microscopic counts would certainly suggest that the food was not grossly contaminated. A thorough description of these procedures is found in APHA *Standard Methods* (1971).

THE USE OF INDICATOR ORGANISMS IN ASSESSING THE SANITATION OF SURFACES

The maintenance of an effective sanitation program is essential if food safety is to be assured. Food contact surfaces should be cleaned and sanitized and then kept safe from recontamination. The adequacy and effectiveness of cleaning and sanitizing should be monitored adequately with a bacteriological control program. When the surfaces are those of small utensils such as spoons, pans, knives, and the like, cleaning and sanitization may be brought about by immersing these items in hot water or by steam cleaning. Certain larger contact surfaces cannot be treated in this fashion but must be cleaned and sanitized by scrubbing and other means. The size, location, and texture of some contact surfaces are such that they are difficult to keep sanitized. Improperly cleaned and sanitized surfaces constitute a significant source of microorganisms to foods during their production, processing, and handling.

Methods

The enumeration of microorganisms on surfaces is a difficult task. The many techniques introduced and employed for this purpose can be divided into one of the following three groups: (1) Swab or swab-rinse methods, (2) Agar contact methods, and (3) Surface rinse methods.

Swab or Swab-Rinse Methods.—These are the oldest and most widely used techniques for the examination of surfaces for microorganisms not only in food and dairy plants, but also in restaurants and hospitals. The swab-rinse technique was standardized by a committee of the American Public Health Association (Tiedeman *et al.* 1948), and specific directions for this method may be found in APHA *Standard Methods* (1972). Around 1950, calcium alginate swabs came into use as an improvement over cotton swabs. When cotton swabs are used, not all of the recovered organisms are freed from the cotton before plating of diluent. This problem is not found when calcium alginate swabs are employed since all recovered organisms are released into the diluent upon dissolution of the alginate by sodium hexametaphosphate. In comparing the relative efficacy of these two swabs in recovering microorganisms from surfaces, most investigators have found alginate to be the better (Higgins 1950; Cain and Steele 1953; Fromm 1959). The swab test to determine the adequacy of sanitizing has been used by the dairy industry for many years (Milk Industry Foundation 1964). The standards recommended are 5/ml maximum for total counts and no coliforms. Other guidelines have been suggested by Patterson (1971).

Swab methods have a long history of use in restaurant sanitation. The cotton swab was given extensive use in restaurants by Kleinfeld and Buchbinder (1947) as they sought to evaluate the effectiveness of dishwashing practices. While at least one investigator (Williams 1967) reported that less than 10% of the organisms present on meat surfaces were removed by swabbing, most have found that around 50% of surface organisms are recovered when good swabbing techniques are used. These methods, nevertheless, give good quality control information when applied in a consistent manner. Angelotti *et al.* (1964) found the swab-rinse method to be best for porous, irregular, and greasy film surfaces while the rodac plate was found to be best for smooth, nonporous, flat or slightly rounded surfaces free of greasy films. In comparing alginate swabs with a surface agar and a sticky film method for recovering microorganisms from wooden surfaces in meat and poultry processing, Mossel *et al.* (1966) found that the swab method recovered significantly more microbes than the other methods but these investigators preferred the other methods since they could be used more routinely than swabs. Gilbert (1970) found that alginate swabs always gave higher recoveries than an agar sausage method from stainless steel, plastic, formica, and wooden surfaces, but he preferred the agar method due to its greater utility for routine use.

In studying the relative numbers of indicator organisms to total viable numbers, swab-rinse methods are probably of greater utility

286   FOOD MICROBIOLOGY

than the other existing methods since replicate plates can be prepared from the same rinse and plated with a battery of selective or differential media. The swab-rinse method also allows for a recovery step for metabolically injured indicator organisms prior to the plating step. Since food particles which may be picked up by the surface agar methods can affect the selectivity of media for indicator organisms, the swab-rinse method allows for sample dilution and consequently less chance for false positive or negative medium reactions.

Agar Contact Methods.—In 1955, Litsky (see Walter, 1955) proposed the use of an agar-syringe method for assessing the microbial flora of surfaces and his method was later modified and used by Angelotti *et al.* (1958). By this method, a 100-ml hypodermic syringe from which the end has been cut to obtain a hollow cylinder is sterilized and filled with the agar medium of choice. After hardening, a layer of agar is pushed beyond the end of the barrel by means of a plunger and pressed for 5 seconds against the surface to be examined. The exposed layer of medium is then cut off with a sterile spatula and placed in a Petri dish followed by incubation. This technique appears to have received only limited use. While Angelotti *et al.* (1958) found that its precision was excellent, these investigators found that it produced the lowest average percent recovery of the six methods they compared.

A technique similar to the above was proposed by ten Cate (1963) and labeled an "agar sausage." By this technique, a sterile agar medium is poured into plastic tubing. Prior to sampling, the outside casing is swabbed with alcohol in the area to be cut and the end is cut off with a sterile knife. Approximately 1 cm of the agar column is pushed out followed by pressing the cut end of the agar firmly onto the test surface. Upon removal from the test surface, a 4—6 mm thick slice is cut off and placed inside a sterile Petri dish with the test surface upward. The dish is incubated at the desired temperature followed by counting of colonies that develop over the incubation period. This technique is further described by Bridson (1969). It has been used largely by European investigators and appears to offer no advantages over the agar syringe method except that provided by the more flexible medium enclosure. In comparing to a swab method, Baltzer and Wilson (1965) found the agar sausage method to be the more suitable for assessing bacon slaughter lines for the presence of clostridia.

A direct surface agar plating (DSAP) method was proposed by Angelotti and Foter (1958) as a reference method for assessing surface contamination. By this method, the test area is overlaid with an agar medium and covered with a sterile plate which remains until the agar solidifies. Angelotti and Foter showed that between 88 – 99% of endospore contaminants could be removed by this method in 95 trials out of 100 from experimentally contaminated nonporous surfaces. This technique does not lend itself to routine use nor was it intended to. Employing it as a reference method, Angelotti et al. (1964) found that the swab-rinse method recovered 47% of experimentally contaminated *Bacillus subtilis* spores from stainless steel surfaces and that the rodac plate method recovered 41% of the spores.

The DSAP method was actually preceeded by contact or direct plating methods such as those of Walter and Hucker (1941) and Barton et al. (1954). The latter authors determined the number of bacteria on tableware by pouring melted agar directly onto the items and allowing it to harden before incubating the utensils. By this method, these investigators found 4 to 29 times more bacteria than by swab tests. A related technique is that of Guiteras et al. (1954) where the object to be sampled is poured with agar in a Petri dish. This method was reported to reveal twice the number of experimental contaminants as did swab methods.

The rodac (replicate organism direct agar contact) plate method is the most widely used of the direct agar methods for assessing surfaces. Rodac plates are available commercially and are filled with 15.5 to 16.5 ml of agar medium 18-24 hr in advance of use. To sample a test area, the cover is removed followed by pressing the agar surface with a force of approximately 2 lb for 3 seconds. Upon replacing the cover, these plates are incubated as desired followed by enumeration of colonies by use of a Quebec colony counter. Angelotti et al. (1964) found the rodac plate to be the best for sampling smooth, nonporous, flat or slightly rounded surfaces free of greasy films in contrast to the swab-rinse method which was found by these investigators to be best for opposite conditions. Rodac plates have been used successfully in hospitals to assess the microbial content of hospital corridor floors, surgical gowns, and surgical wound sites. Vesley and Michaelsen (1964) employed over 13,000 rodac plates to assess hospital floor cleaning procedures and found the method to be satisfactory. The use of the rodac plate is recommended by APHA *Standard Methods* (1972) for the dairy industry.

A somewhat indirect agar plate method is the sticky film method of Thomas that has been used by Mossel et al. (1966). The method consists of pressing a sticky film or mending tape against the surface to be examined and then pressing the exposed side against agar plates. Mossel *et al.* found the method to recover less bacteria from wooden surfaces than swabs but felt that it was more subject to routine use than swab methods. To be more effective than the direct agar methods such as the rodac plate, the adherance of organisms to the film or tape would have to be greater than to agar in which case difficulty would be expected in the subsequent removal of organisms from the tape onto agar plates. The latter problem may be overcome by dissolving the tape in a diluent and then enumerating the organisms by conventional serial dilution techniques. Indeed, Silliker *et al.* (1957) suggested a method of this type for sampling meat surfaces whereby contact agar or contact membranes would be employed for picking up surface organisms.

Surface Rinse Methods.—By these methods, small measured volumes of buffered rinse solutions (10 to 20 ml or more) containing sanitizer neutralizer as necessary are pipetted into or onto the test container surface. The solution may be rinsed or flushed over the area, or the area may be scrubbed with a sterile rubber policeman. After the rinse or scrub, the solution may be plated directly or passed through a membrane filter which is then treated by conventional methods. The plating of the rinse allows for the use of various media to determine indicators as well as total viable numbers. In their comparison of six methods for recovering *Staphylococcus aureus* cells and *Bacillus globigii* spores from china surfaces, Angelotti *et al.* (1958) found the surface rinse methods to give high average percentage recoveries with fairly good precision in contrast to cotton and alginate swabs which gave low average percentage recoveries with poor precision. Mallmann *et al.* (1958) compared swab and rinse methods in the recovery of bacteria from poultry surfaces and found the rinse method to be the better. More recently, Yokoya and Zulzke (1975) developed a method for sampling beef carcasses where the portion of carcass to be sampled is cut away, placed in flasks containing water, sand, and Tween 80, shaken on a rotary shaker, and then plated after appropriate serial dilutions are made. This method gave higher counts than cotton swabs applied to corresponding carcasses. Other techniques for sampling animal carcasses have been reviewed and discussed by Patterson (1972). Not all of these methods are suitable for open flat surfaces which permit the rinse solution to spread out too much.

Testing Surfaces for Indicators vs Total Numbers.—While none of the above techniques which permit routine use have been shown to recover all bacteria from different types of surfaces, several may be used with confidence in recovering sanitary indicators such as coliforms and enterococci. Of great importance in the application of any of these techniques is a consistent method of use from one sampling time to the next. Cotton and alginate swabs may be used with confidence if allowance is made for the types of surfaces to be examined and if further allowance is made for the fact that only around 50% of the contaminants will be removed. Coliforms, enterococci, and staphylococci may be assessed by swabbing with a greater degree of confidence than can total viable numbers. The data presented in Table 9.3 from various meat market surfaces show that

TABLE 9.3

RECOVERY OF BACTERIA FROM MEAT MARKET SURFACES
BY USE OF COTTON SWABS

Surfaces Examined	Viable Nos./in.^{2}			
	Total Count	Coliforms	Enterococci	Staph.
Ground beef tub	120,000	<10	270	<10
Cutting block	110,000	<10	1,700	40
Table of saw	18,000	<10	280	16
Cutting knives	16,000	20	210	<10
Meat grinder die	3,000	<10	900	20
Saw blade	2,700	<10	230	10

Values are averages of three determinations taken just prior to use of equipment each morning.
Total viable counts were determined on PCA; presumptive coliform, enterococci, and staphylococci on MacConkey, azide dextrose, and egg yolk agars, respectively.

while total viable counts ranged from 2,700 to 120,000/in.^{2}, coliform counts were consistently below 10/in.^{2} Counts for coagulase-positive staphylococci were slightly higher than coliforms, with presumptive enterococcal counts being even higher. The existence of indicator counts with this low order of magnitude confers greater confidence on the accuracy of their determination than would be the case for total viable counts where spreaders are more likely to appear and render plate counting less accurate. The higher incidence of enterococci over coliforms by the swab-rinse method is consistent with results from the standard plate count method, which is of necessity the reference method for total viable numbers for nonsurface specimen.

While the rodac and other direct agar plating methods give erratic results relative to total viable counts, results for coliforms and enterococci by these methods are of greater reliability. Data presented in Table 9.4 from meat market surfaces assessed by rodac plates show that in five of the ten determinations total viable numbers could not be determined on plate count agar, due either to the colonies being too numerous to count (TNTC) or to the

290   FOOD MICROBIOLOGY

TABLE 9.4
THE RELATIVE EFFICACY OF RODAC PLATES
TO ASSESS MEAT MARKET SURFACES
FOR TOTAL VIABLE NUMBERS, COLIFORMS, AND ENTEROCOCCI

Surfaces Examined	Numbers/plate Total	(average of duplicates) Coliforms	Enterococci
Cutting Block	TNTC*	8	70
„ „	TNTC	1	174
„ „	2,200	4	133
„ „	480	0	320
Saw Bench	TNTC	50	42
„ „	240	1	4
„ „	S**	5	4
„ „	S	0	72
Meat Grinder	40	0	33

*TNTC = too numerous to count; **S = spreaders.

Total numbers were determined on PCA with 2% agar plus lecithin and Tween 80; presumptive coliforms and enterococci by use of MacConkey and azide dextrose agars, respectively.

existence of surface spreaders on these plates. Due to the normal relation of lower numbers of coliforms and enterococci to total viable numbers, rodac plates with selective and differential media for these groups can be counted with much greater accuracy. Also, there is a much lessened chance for spreaders to develop on these type media.

Overall, the examination of surfaces for total viable counts by the existing techniques cannot be carried out with any degree of high accuracy or confidence due to surface plate spreaders, colonies TNTC in the case of the direct agar plate methods, and to the lack of ability of swab methods to recover more than around 50% of surface contaminants. These two techniques along with that of surface rinse may, however, be employed with greater accuracy and confidence if food plant surfaces are assessed for indicator organisms such as coliforms, enterococci, and staphylococci employing appropriate selective and differential media. Due to the generally lower numbers of these groups in relation to total viable counts, the problems of surface spreaders and uncountable plates are avoided.

Another advantage of assessing surfaces for indicator organisms in preference to total viable counts is that elevated temperature incubations can be used for fecal coliforms by direct incubation of selective media plates. When a swab method is employed, a broth recovery procedure for metabolically injured organisms, such as those discussed in this chapter, can be instituted prior to plating, or an overlay procedure such as that suggested by Hartman *et al.* (1975) and Speck *et al.* (1975) may be employed. By this procedure, the rinse is surface plated on 12 ml of trypticase soy agar and allowed to incubate for 2 hr at 35°C followed by an overlay of an equal volume of violet red bile agar (VRBA) with subsequent incubation for 24 hr. Both groups of investigators showed that pre-incubation with the noninhibitory medium allowed more coliforms to be recovered and consequently higher coliform counts were achieved than by direct plating onto VRBA. It is conceivable that this method is adaptable to enterococci and staphylococci as well and perhaps to elevated temperature incubations for fecal coliforms.

METABOLICALLY INJURED INDICATOR ORGANISMS

Prior to the examination of foods for indicator organisms, it is important to consider whether the isolation procedures should include a recovery step for metabolically injured cells. Injured, damaged, or stressed cells may be present as a result of the food products having been subjected to either sublethal heating, freezing, drying, freeze-drying, irradiation, low pH, salting, or certain other food additives. Injured cells very often manifest this state by their lack of ability to reproduce on media containing salts and selective ingredients which allow the growth of nonstressed cells. To more accurately reflect the incidence of injured cells, recovery methods should be conducted along with the standard procedures. Unless recovery or "resuscitation" methods are employed, up to 99% of the cells that develop on recovery media may fail to develop on selective media.

The thermal injury of bacteria has been reviewed by Allwood and Russell (1970). The non-lethal injury of indicator organisms and their recovery from foods has been reviewed and discussed by Maxcy (1970), Mossel and Ratto (1970), and Speck (1970). With respect to *E. coli* cells, Ray and Speck (1973A) found that over 90% of the cells that survived freezing in water at −78°C were freeze-injured and did not form colonies on VRB or deoxycholate-lactose agar. The freeze-injury was repaired when the cells were allowed to recover in trypticase soy broth (TSB) with yeast extract followed by plating and growth on VRB and deoxycholate-lactose agar. They reported that up to 90% of the cells were repaired in TSB within 30 min at 20° to 45°C and began multiplication within 2 hr at 25°C. Freezing of *E. coli* in foods effected injury to 60 to 90% of survivors and TSB was again found to be an excellent recovery medium. In another report, Speck and Ray (1973) found that *E. coli* freeze-injury could be repaired by a simple medium containing phosphorus and magnesium as well as in complex media devoid of inhibitory ingredients. In addition to being more sensitive to bile and deoxycholate, injured cells of this organism were reported to be more sensitive to lauryl sulfate. In addition to freeze-injury, *E. coli* is damaged by sub-lethal heating (Ray and Speck 1973B) and by freeze-drying and radiations (Sinskey and Silverman 1970). The cell injury-repair cycle is known to occur in pathogens such as *Salmonella typhimurium* (Tomlins and Ordal, 1971), *Staphylococcus aureus* (Iandolo and Ordal 1966), and possibly all bacteria of public health importance.

In regards to *S. faecalis*, Clark *et al.* (1968) found that after exposing a strain to sublethal heat treatment at 60°C for 15 min, less than 1% of the viable cells were able to reproduce on media containing 6% NaCl. These authors also found the heat-injured cells to be more sensitive to incubation temperatures, pH, and to 0.01% methylene blue. Beuchat and Lechowich (1968) also found an increased sensitivity to salt concentrations when *S. faecalis* was exposed to sublethal heat treatments. According to Clark *et al.* (1968), the following properties are among the characteristics of heat-injured *S. faecalis* cells: (1) Increased sensitivity to salt, sodium azide, bromcresol purple, and 0.01% methylene blue, (2) extended lag phase of growth, (3) ability of cells to recover in simple rather than complex media, (4) decreased ability to grow at 10 and 45°C, and (5) decreased ability to grow at pH 9.6.

Metabolically injured cells of *S. faecalis* and presumably other enterococci may be recovered in rich media such as TSB, dextrose, or tryptose phosphate broths as well as in simple synthetic media. The recovery times are longer than for *E. coli* cells.

With respect to the Enterobacteriaceae, Mossel and Ratto (1970) found that less than 40% of 167 dried foods and drugs were positive for these organisms when no recovery step was allowed in the enumeration. When a recovery or resuscitation method was employed, over 51% of the same foods were shown to be positive. These investigators employed TSB as their recovery medium and found that it was more effective for 1-6 hr incubations than lactose broth with overnight incubation. Following recovery in TSB, inoculations were made into EE broth. In a more recent report, Mossel (1974B) recommended the following recovery procedure for Enterobacteriaceae: Suspend 10 g sample of food into 100 ml of TSB for 2 hr at *ca.* 20°C and shake the suspension every half hr (to allow for "resuscitation"). Following this step, inoculations are made into heated EE broth. This author also suggested a modified EE broth and designated the new medium EEL. The EEL medium contains 1 g/L of sodium lauryl sulfate (in place of bile salts), and 15 mg/L of certified brilliant green in buffered dextrose broth. The EEL medium is reported to be less inhibitory to stressed cells than the original EE broth.

BIBLIOGRAPHY

ADAMS, J. C. 1972. Unusual organism which gives a positive elevated temperature test for fecal coliforms. Appl. Microbiol. *23*, 172–173.

ALLWOOD, M. C., and RUSSELL, A. D. 1970. Mechanisms of thermal injury in nonsporulating bacteria. Advan. Appl. Microbiol. *12*, 89–119.

AM. PUBLIC HEALTH ASSOC. 1971. Standard Methods for the Examination of Water and Wastewater, 13th Edition. American Public Health Assoc., Washington, D.C.

AM. PUBLIC HEALTH ASSOC. 1972. Standard Methods for the Examination of Dairy Products, 13th Edition, W. J. Hausler (Editor). American Public Health Assoc., Washington, D.C.

ANDREWS, W. H., DIGGS, C. D., and WILSON, C. R. 1975. Evaluation of a medium for the rapid recovery of *Escherichia coli* from shellfish. Appl. Microbiol. *29*, 130–131.

ANDREWS, W. H., and PRESNELL, M. W. 1972. Rapid recovery of *Escherichia coli* from estuarine water. Appl. Microbiol. *23*, 521–523.

ANGELOTTI, R. 1964. Significance of "total counts" in bacteriological examination of foods. *In* Examination of Food for Enteropathogenic and Indicator Bacteria, K. H. Lewis, and R. Angelotti (Editors). Public Health Serv. Pub. *1142*, Washington, D.C.

ANGELOTTI, R., and FOTER, M. J. 1958. A direct surface agar plate laboratory method for quantitatively detecting bacterial contamination on nonporous surfaces. Food Res. *23*, 170–174.

ANGELOTTI, R., FOTER, M. J., BUSCH, K. A., and LEWIS, K. H. 1958. A comparative evaluation of methods for determining the bacterial contamination of surfaces. Food Res. *23*, 175–185.

ANGELOTTI, R., WILSON, J. L., LITSKY, W., and WALTER, W. G. 1964. Comparative evaluation of the cotton swab and rodac methods for the recovery of *Bacillus subtilis* spore contamination from stainless steel surfaces. Health Lab. Sci. *1*, 289–296.

ASSOC. FOOD DRUG OFFICIALS U.S. 1966. Microbiological examination of precooked frozen foods. Assoc. Food Drug Officials U.S. Quart. Bull.

ASSOC. FOOD DRUG OFFICIALS U.S. 1969. Recommended bacterial limits for frozen precooked beef and chicken pot pies. Ad hoc committee on Microbiology of frozen foods. Assoc. Food Drug Officials U.S. Quart. Bull. Suppl.

BACHRACH, U., and BACHRACH, Z. 1974. Radiometric method for the detection of coliform organisms in water. Appl. Microbiol. *28*, 169–171.

BALTZER, J., and WILSON, D. C. 1965. The occurrence of clostridia on bacon slaughter lines. J. Appl. Bacteriol. *28*, 119–124.

BARTLEY, C. H., and SLANETZ, L. W. 1960. Types and sanitary significance of fecal streptococci isolated from feces, sewage and water. Am. J. Public Health *50*, 1545–1552.

BARTON, R. R., GORFIEN, H., and CARLO, R. M. 1954. Determination of bacterial numbers on tableware by means of direct plating. Appl. Microbiol. *2*, 264–266.

BEUCHAT, L. R., and LECHOWICH, R. V. 1968. Effect of salt concentration in the recovery medium on heat-injured *Streptococcus faecalis*. Appl. Microbiol. *16*, 772–776.

BRIDSON, E. Y. 1969. Isolation of surface micro-organisms with the agar slice technique (agar sausage). *In* Isolation Methods for Microbiologists, D. A. Shapton, and G. W. Gould (Editors). Academic Press, New York.

BROWNING, G. E., and MANKIN, J. O. 1966. Gastroenteritis epidemic owing to sewage contamination of public water supply. J. Am. Waterworks Assoc. 58, 1465–1470.

CAIN, R. M., and STEELE, H. 1953. The use of calcium alginate soluble wool for the examination of cleansed eating utensils. Can. J. Public Health 44, 464–467.

CATE, L. ten. 1963. An easy and rapid bacteriological control method in meat processing industries using agar sausage techniques in Rilsan artificial casing. Fleischwertz. 15, 483–486.

CLARK, C. W., WITTER, L. D., and ORDAL, Z. J. 1968. Thermal injury and recovery of Streptococcus faecalis. Appl. Microbiol. 16, 1764–1769.

DAOUST, R. A., and LITSKY, W. 1975. Pfizer selective enterococcus agar overlay method for the enumeration of fecal streptococci by membrane filtration. Appl. Microbiol. 29, 584–589.

DEIBEL, R. H., and SILLIKER, J. H. 1963. Food-poisoning potential of the enterococci. J. Bacteriol. 85, 827–832.

DEIBEL, R. H. 1964. The group D streptococci. Bacteriol. Rev. 28, 330–336.

DRION, E. F., and MOSSEL, D. A. A. 1972. Mathematical-ecological aspects of the examination for Enterobacteriaceae of foods processed for safety. J. Appl. Bacteriol. 35, 233–239.

EFTHYMIOU, C. J., BACCASH, P., LABOMBARDI, V. J., and EPSTEIN, S. E. 1974. Improved isolation and differentiation of enterococci in cheese. Appl. Microbiol. 28, 417–422.

FACKLAM, R. R., and MOODY, M. D. 1970. Presumptive identification of group D streptococci: The bile-esculin test. Appl. Microbiol. 20, 245–250.

FISHBEIN, M. 1962. The aerogenic response of Escherichia coli and strains of Aerobacter in EC broth and selected sugar broths at elevated temperatures. Appl. Microbiol. 10, 79–85.

FISHBEIN, M., and SURKIEWICZ, B. F. 1964. Comparison of the recovery of Escherichia coli from frozen foods and nut meats by confirmatory incubation in EC medium at 44.5 and 45.5°C. Appl. Microbiol. 12, 127–131.

FISHBEIN, M. et al. 1967. Coliform behavior in frozen foods. I. Rapid test for the recovery of Escherichia coli from frozen foods. Appl. Microbiol. 15, 233–238.

FOSTER, E. M. 1968. Bacteriological Standards for Foods—Their Significance. 72nd Ann. Conf. Assoc. Food Drug Officials U.S., Hartford, Conn.

FRANCIS, D. W., PEELER, J. T., and TWEDT, R. M. 1974. Rapid method for detection and enumeration of fecal coliforms in fresh chicken. Appl. Microbiol. 27, 1127–1130.

FROMM, D. 1959. An evaluation of techniques commonly used to quantitatively determine the bacteriological population of chicken carcasses. Poultry Sci. 38, 887–891.

GELDREICH, E. E. 1966. Sanitary significance of fecal coliforms in the environment. Publ. WP 20-3, U.S. Dept. Interior, Washington, D.C.

GELDREICH, E. E., and BORDNER, R. H. 1971. Fecal contamination of fruits and vegetables during cultivation and processing for market. A review. J. Milk Food Technol. 34, 184–195.

GILBERT, R. J. 1970. Comparison of materials used for cleaning equipment in retail food premises, and of two methods for the enumeration of bacteria on cleaned equipment and work surfaces. J. Hyg. 68, 221–232.

GUITERAS, A. F., FLETT, L. H., and SHAPIRO, R. L. 1954. A quantitative method for determining the bacterial contamination of dishes. Appl. Microbiol. 2, 100–101.

GREENBERG, A. E., and ONGERTH, H. J. 1966. Salmonellosis in Riverside, Calif. J. Am. Waterworks Assoc. 58, 1145–1150.

HALL, H. E. 1964. Methods for isolation and enumeration of enterococci. In Examination of Foods for Enteropathogenic and Indicator Bacteria, K. H. Lewis, and R. Angelotti (Editors). Public Health Serv. Publ. 1142.

HARTMAN, P. A., HARTMAN, P. S., and LANZ, W. W. 1975. Violet red bile 2 agar for stressed coliforms. Appl. Microbiol. *29*, 537–539.

HARTMAN, P. A., REINBOLD, G. W., and SARASWAT, D. S. 1966. Indicator organisms—a review. I. Taxonomy of the fecal streptococci. Intern. J. Syst. Bacteriol. *16*, 197–221.

HAUSLER, W. J. 1972. Standard Methods for the Examination of Dairy Products. American Public Health Assoc., Washington, D.C.

HIGGINS, M. 1950. A comparison of the recovery rate of organisms from cotton wool and calcium alginate wool swabs. Public Health Lab. Serv. Bull. (British) *9*, 50–51.

HUCKER, G. J., BROOKS, R. F., and EMERY, A. J. 1952. The source of bacteria in processing and their significance in frozen vegetables. Food Technol. *6*, 147–155.

IANDOLO, J. J., and ORDAL, Z. J. 1966. Repair of thermal injury of *Staphylococcus aureus*. J. Bacteriol. *91*, 134–142.

INGRAM, M. 1965. Psychrophilic and psychrotropic microorganisms. Ann. Inst. Pasteur de Lille *16*, 111–115.

INGRAM, M., and DAINTY, R. H. 1971. Changes caused by microbes in spoilage of meats. J. Appl. Bacteriol. *34*, 21–39.

ISENBERG, H. D., GOLDBERG, D., and SAMPSON, J. 1970. Laboratory studies with a selective enterococcus medium. Appl. Microbiol. *20*, 433–436.

KAPLAN, M. T., and APPLEMAN, M. D. 1952. Microbiology of frozen orange concentrate. III. Studies of enterococci in frozen concentrated orange juice. Food Technol. *6*, 167–170.

KLEINFELD, H. J., and BECHBINDER, I. 1947. Dishwashing practice and effectiveness (swab-rinse test) in a large city as revealed by a survey of 1,000 restaurants. Am. J. Public Health *37*, 379–389.

LARKIN, E. P., LITSKY, W., and FULLER, J. E. 1955A. Fecal streptococci in frozen foods. I. A bacteriological survey of some commercially frozen foods. Appl. Microbiol. *3*, 98–101.

LARKIN, E. P., LITSKY, W., and FULLER, J. E. 1955B. Fecal streptococci in frozen foods. II. Effort of freezing storage on *Eskcherichia coli* and some fecal streptococci inoculated onto green beans. Appl. Microbiol. *3*, 102–104.

LARKIN, E. P., LITSKY, W., and FULLER, J. E. 1956. Incidence of fecal streptococci and coliform bacteria in frozen fish products. Am. J. Public Health *46*, 464–468.

LEE, W. S. 1972. Improved procedure for identification of group D enterococci with two new media. Appl. Microbiol. *24*, 1–3.

LERKE, P., ADAMS, R., and FARBER, L. 1965. Bacteriology of spoilage of fish muscle. III. Characterization of spoilers. Appl. Microbiol. *13*, 625–630.

LERKE, P., FARBER, L., and ADAMS, R. 1967. Bacteriology of spoilage of fish muscle. IV. Role of protein. Appl. Microbiol. *15*, 770–776.

LITSKY, W., MALLMAN, W. L., and FIFIELD, C. W. 1955. Comparison of the most probable numbers of *Escherichia coli* and enterococci in river water. Am. J. Public Health *45*, 1049–1053.

MALLMAN, W. L., DAWSON, L. E., SULTZER, B. M., and WRIGHT, H. S. 1958. Studies on microbiological methods for predicting shelf-life of processed poultry. Food Technol. *12*, 122–127.

MALLMAN, W. L., and LITSKY, W. 1951. Survival of selected enteric organisms in various types of soil. Am. J. Public Health *41*, 38–44.

MARIER, R. *et al.* 1973. An outbreak of enteropathogenic *Escherichia coli* foodborne disease traced to imported French cheese. Lancet 2, 1376–1378.

MAXCY, R. B. 1970. Non-lethal injury and limitations of recovery of coliform organisms on selective media. J. Milk Food Technol. *33*, 445–448.

MIETH, H. 1960. Examination of the occurrence of enterococci in animals and men. I. Communication: Their occurrence in the gut of healthy domesticated hogs. Zentr. Bakteriol. Parasitenk. Abstr. I Orig. *179*, 456–482.

MIETH, H. 1962A. Examination of the occurrence of enterococci in animals and men. III. Communication: Enterococci flora in feces of cattle. Zentr. Bakteriol. Parasitenk. Abstr. I. Orig. *185*, 47-52.

MIETH, H. 1962B. Examination of the occurrence of enterococci in animals and men. IV. Communication: Streptococci flora in feces of horses. Zentr. Bakteriol. Parasitenk. Abstr. I. Orig. *185*, 166-174.

MILK INDUSTRY FOUNDATION. 1964. Methods of Analysis of Milk and its Products, 3rd Edition. Washington, D.C.

MONTFORD, J., and THATCHER, F. S. 1961. Comparison of four methods of isolating salmonellae from foods, and elaboration of a preferred procedure. J. Food Sci. *26*, 510-517.

MOSSEL, D. A. A. 1974A. Bacteriological safety of foods. Lancet *1*, 173.

MOSSEL, D. A. A. 1974B. Standardization of the selective inhibitory effect of surface active compounds used in media for the detection of Enterobacteriaceae in foods and water. Health Lab. Sci. *11*, 260-267.

MOSSEL, D. A. A., KAMPELMACHER, E. H., and VAN NOORLE JANSEN, L. M. 1966. Verification of adequate sanitation of wooden surfaces used in meat and poultry processing. Zent. Bakt., Parasiten, Infek. u. Hyg. Abstr. I. *201*, 91-104.

MOSSEL, D. A. A., and RATTO, M. A. 1970. Rapid detection of sublethally impaired cells of Enterobacteriaceae in dried foods. Appl. Microbiol. *20*, 273-275.

MOSSEL, D. A. A., and VEGA, C. L. 1973. The direct enumeration of *Escherichia coli* in water using MacConkey's agar at 44°C in plastic pouches. Health Lab. Sci. *10*, 303-307.

MOSSEL, D. A. A., VISSER, M., and CORNELISSEN, A. M. R. 1963. The examination of foods for Enterobacteriaceae using a test of the type generally adopted for the detection of salmonellae. J. Appl. Bacteriol. *26*, 444-452.

MOUSSA, R. S., KELLER, N., CURIAT, G., and DeMAN, J. C. 1973. Comparison of five media for the isolation of coliform organisms from dehydrated and deep frozen foods. J. Appl. Bacteriol. *36*, 619-629.

MUNDT, J. O. 1961. Occurrence of enterococci: Bud, blossom and soil studies. Appl. Microbiol. *9*, 541-544.

MUNDT, J. O., and GRAHAM, W. F. 1968. *Streptococcus faecium* var *casseliflavus*, nov var. J. Bacteriol. *95*, 2005-2009.

MUNDT, J. O., JOHNSON, A. H., and KHATCHIKIAN, R. 1958. Incidence and nature of enterococci in plant materials. Food Res. *23*, 186-193.

NATL. ACAD. SCI.—NATL. RES. COUNCIL. 1964. An evaluation of public health hazards from microbiological contamination of foods. Publ. *1195*. Natl. Acad. Sci.—Natl. Res. Council, Washington, D.C.

NIVEN, C. F., JR. 1963. Microbial indexes of food quality: Fecal streptococci. *In* Microbiological Quality of Foods, L. W. Slanetz, C. O. Chichester, A. R. Gaufin, and Z. J. Ordal (Editors). Academic Press, New York.

OBLINGER, J. L. 1975. Recovery of streptococci from a variety of foods: A comparison of several media. J. Milk Food Technol. *38*, 323-326.

PATTERSON, J. T. 1971. Microbiological assessment of surfaces. J. Food Technol. *6*, 63-70.

PATTERSON, J. T. 1972. Microbiological sampling of poultry carcasses. J. Appl. Bacteriol. *35*, 569-575.

PAVLOVA, M. T., BREZENSKI, F. T., and LITSKY, W. 1972. Evaluation of various media for isolation, enumeration and identification of fecal streptococci from natural sources. Health Lab. Sci. *9*, 289-298.

PETERSON, A. C., and GUNDERSON, M. F. 1960. Role of psychrophilic bacteria in frozen food spoilage. Food Technol. *14*, 413-417.

PRESSWOOD, W. G., and BROWN, L. R. 1973. Comparison of Gelman and Millipore membrane filters for enumerating fecal coliform bacteria. Appl. Microbiol. *26*, 332-336.

PUGSLEY, A. P., and EVISON, L. M. 1975. A fluorescent antibody technique for the enumeration of faecal streptococci in water. J. Appl. Bacteriol. *38*, 63-65.

RAJ, H., WIEBE, W. J., and LISTON, J. 1961. Detection and enumeration of fecal indicator organisms in frozen seafoods. Appl. Microbiol. *9*, 295-303.
RAY, B., and SPECK, M. L. 1973A. Enumeration of *Escherichia coli* in frozen samples after recovery from injury. Appl. Microbiol. *25*, 499-503.
RAY, B., and SPECK, M. L. 1973B. Discrepancies in the enumeration of *Escherichia coli*. Appl. Microbiol. *25*, 494-498.
SABBAJ, J., SUTTER, V. L., and FINEGOLD, S. M. 1971. Comparison of selective media for isolation of presumptive group D streptococci from human feces. Appl. Microbiol. *22*, 1008-1011.
SHATTOCK, P. M. 1962. Enterococci. *In* Chemical and Biological Hazards in Foods, J. C. Ayres, A. A. Kraft, H. E. Snyder, and H. W. Walker (Editors). Iowa State Univ. Press, Ames, Iowa.
SHELTON, L. R. *et al.* 1961. A bacteriological survey of the precooked frozen food industry. Food Drug Admin. U.S. Dept. Health Educ. Welfare, Washington, D.C.
SHERMAN, J. M. 1937. The streptococci. Bacteriol. Rev. *1*, 3-97.
SILLIKER, J. H. 1963. Total counts as indexes of food quality. *In* Microbiological Quality of Foods, L. S. Slanetz *et al.* (Editors). Academic Press, New York.
SILLIKER, J. H., ANDRES, H. P., and MURPHY, J. F. 1957. A new non-destructive method for the bacteriological sampling of meats. Food Technol. *11*, 317-320.
SINSKEY, T. J., and SILVERMAN, G. J. 1970. Characterization of injury incurred by *Escherichia coli* upon freeze-drying. J. Bacteriol. *101*, 429-437.
SOLOWEY, M., and WATSON, A. J. 1951. The presence of enterococci in spray-dried whole egg powder. Food Res. *16*, 187-191.
SPECK, M. L. 1970. Selective culture of spoilage and indicator organisms. J. Milk Food Technol. *33*, 163-167.
SPECK, M. L., and RAY, B. 1973. Recovery of *Escherichia coli* after injury from freezing. Bull. Inst. Intern. Froid, Annexe *5*, 37-46.
SPECK, M. L., RAY, B., and READ, I. B., JR. 1975. Repair and enumeration of injured coliforms by a plating procedure. Appl. Microbiol. *29*, 549-550.
SPLITTSTOESSER, D. F., WRIGHT, R., and HUCKER, G. J. 1961. Studies on media for enumerating enterococci in frozen vegetables. Appl. Microbiol. *9*, 303-308.
TIEDEMAN, W. E. *et al.* 1948. Technic for the bacteriological examination of food utensils. (Part II) Am. J. Public Health *38*, 68-70.
TOMLINS, R. I., and ORDAL, Z. J. 1971. Requirements of *Salmonella typhimurium* for recovery from thermal injury. J. Bacteriol. *105*, 512-518.
VESLEY, D., KEENAN, K. M., and HALBERT, M. M. 1966. Effect of time and temperature in assessing microbial contamination on flat surfaces. Appl. Microbiol. *14*, 203-205.
VESLEY, D., and MICHAELSEN, G. S. 1964. Application of a surface sampling technique to the evaluation of bacteriological effectiveness of certain hospital housekeeping procedures. Health Lab. Sci. *1*, 107-113.
WALTER, W. G. 1955. Symposium on methods for determining bacterial contamination on surfaces. Bacteriol. Rev. *19*, 284-287.
WALTER, W. G., and HUCKER, G. J. 1941. Proposed method for the bacteriological examination of flat surfaces. Am. J. Public Health *31*, 487-490.
WHITE, J. C., and SHERMAN, J. M. 1944. Occurrence of enterococci in milk. J. Bacteriol. *48*, 262 (Abstr.).
WILLIAMS, M. L. B. 1967. A new method for evaluating surface contamination of raw meat. J. Appl. Bacteriol. *30*, 498-499.
WINTER, F. H., YORK, G. K., and EL NAKHAL, H. 1971. Quick counting method for estimating the number of viable microbes on food and food processing equipment. Appl. Microbiol. *22*, 89-92.
YOKOYA, F., and ZULZKE, M. L. 1975. Method for sampling meat surfaces. Appl. Microbiol. *29*, 551-552.

CHAPTER 10

Allen A. Kraft | Gram-positive Nonspore-forming Rods

DISTINGUISHING FEATURES OF CORYNEFORMS

It is not uncommon in examining the literature on Gram-positive, nonspore-forming rods which are often club-shaped, to find the quote that forms the first sentence of the review by Jensen (1952), "There are perhaps few groups of bacteria of which the typical representatives are easier to recognize and the aberrant types more numerous and more difficult to separate than those which originally constituted the genus *Corynebacterium*, and which we now may cautiously call the group of coryneform bacteria." In this article, we will not deviate from the norm, but will repeat the statement for emphasis. Jensen (1952, 1966) indicated that the genus was created by Lehmann and Neumann in 1896 to accommodate the diphtheria bacillus.

For many years, the genus was quite simple and homogeneous and included several species characterized mainly as Gram-positive parasitic bacteria that did not form endospores and were aerobic, nonmotile, nonacid fast rods showing a tendency toward irregular staining. Their name was derived from the characteristic club or wedge shape of their cells. Since the time that the genus *Corynebacterium* had been introduced, to the present, many different organisms with *Corynebacterium* features have been discovered or uncovered. The population explosion of coryneforms over the course of time has been characterized by Jensen (1966) as "transgressions of the defined boundaries of the classical corynebacteria," since the organisms have been found in many sources in nature; as saprophytes in soil, water, milk and other foods and as animal and plant parasites. Some members have been demonstrated to be motile, not strictly aerobic or Gram-positive, and even acid fast so as to resemble the tubercle bacillus. In grouping these, it is more proper to refer to the bacteria having certain similar characteristics as "coryneform" organisms, rather than placing them all in the single genus *Corynebacterium*. Whether or not only bacteria forming club-shaped cells should be designated as coryneform bacteria is a moot point. Recently, Davis and Newton (1969) labelled 70 strains as coryneform bacteria with these generic names: *Mycobacterium, Nocardia, Jensenia, Listeria, Erysipelothrix, Kurthia, Brevibacterium, Arthrobacter, Cellulomonas, Microbacterium* and *Corynebacterium*. On the basis of 70 coded features used in numerical taxonomy, Davis and Newton separated *Corynebacterium* species of animal origin from *Arthrobacter, Brevibacterium, Cellulomonas* and *Microbacterium*, although some strains of this last named species grouped with the animal *Corynebacterium*. In discussing different genera of saprophytic coryneform bacteria, Veldkamp (1970) stated that all coryneforms form irregularly shaped cells during lag-phase growth, although the degree of pleomorphy varies widely among different types as well as within the same species and is dependent on cultural conditions. A similar situation exists with regard to the cycle of development in which large pleomorphic rods become shorter or form coccoid cells; another common characteristic of coryneforms.

Jensen (1966) differed with the classification in the seventh edition of Bergey's Manual (Breed *et al*. 1957) in that he believed the coryneform bacteria should consist of *Corynebacterium, Arthrobacter, Microbacterium*, and *Propionibacterium*. These genera were "coherently" grouped and linked with *Brevibacterium*, whose heterogeneity was exemplified by its close resemblance to *Arthrobacter* in the species *B. linens*. These organisms will be discussed in more detail later. *Cellulomonas* also contains coryneform bacteria, with the genus status based largely on ability to attack cellulose. However, some strains may exist which do not attack cellulose; hence, this distinguishing feature detracts from the validity of the organism as a genus.

For purposes of this review, coryneform bacteria are considered as Gram-positive nonspore-forming rods, generally having these properties: nonacid fast, metachromatic granules, definite palisade arrangement of cells, irregular staining tendencies, although possessing the fine structure characteristic of Gram-positive organisms (da Silva and Holt 1965; Veldkamp 1970). Genera included here are: *Corynebacterium, Arthrobacter, Microbacterium, Cellulomonas* and *Brevibacterium*, as described by Veldkamp 1970. The genus *Kurthia* has also been considered to be in the family *Corynebacteriaceae* (da Silva and Holt 1965). Several other studies have been reported on numerical taxonomy of coryneforms; among these, Davis and Newton (1969) showed that *Mycobacterium* and *Nocardia* formed sufficiently distinct clusters that there was little reason to link them with the corynebacteria. They suggested that these not be joined into a common genus as proposed by Harrington (1966) as a result of his numerical taxonomy study. Veldkamp (1970) still preferred to keep the genus *Corynebacterium* separate from *Mycobacterium* and *Nocardia* on the basis of fine structure (centripetal growth of cross walls to form septa in *Corynebacterium*,

Arthrobacter and *Microbacterium*), G-C ratio, and numerical taxonomy. The clusters of Davis and Newton differentiated *Mycobacterium*, *Nocardia* and *Jensenia* from coryneforms as categorized here.

In addition to cell wall analysis (Cummins and Harris 1956) and electrophoresis as means of classifying coryneforms at the generic level (Robinson 1968), nutritional requirements in association with habitat are of value in comparisons of these organisms.

Morphological Changes

Possibly the most striking features of the coryneforms are those of morphology and associated changes. The growth cycle that occurs in all coryneforms is well illustrated by *Arthrobacter*. Coccoid or short rod-shaped cells in the stationary phase become swollen, and during logarithmic growth may form larger and irregular cells with primary branching. *Microbacterium* may continue to produce secondary branching. Near the end of the lag phase, shorter cells appear, either by breaking of the long irregular shaped cells or by gradual shortening of the cells in the process of division (Veldkamp 1970; Mulder *et al.* 1966). "V" shaped cells may appear at the final stages of cell division as a result of "snapping division." The *Arthrobacter* isolated from liquid egg by Torrey (Kraft *et al.* 1966) showed these striking morphological changes within 24 hr, producing irregularly swollen club-shaped cells, with cell length ranging from 1 μ for the small cocci to 25 μ for the large swollen cells. During the early stages of incubation at 30°C, coccoid cells lengthened until a pair of rods was produced prior to the irregular club-shaped cell formation. Some long filaments fragmented to form shorter rods. Other coryneforms characteristically undergo filament formation from coccoid cell germination, and a cycle of development leading to coccoid elements from pleomorphic rods. The degree of development of the full cycle seems to vary among different coryneforms, but has been demonstrated among species of *Corynebacterium*, *Cellulomonas*, *Microbacterium* and *Brevibacterium* in addition to the more dramatic series of changes shown by *Arthrobacter*.

Cell Wall Analysis and Composition

Cell wall analyses serve to distinguish saprophytic coryneforms from *Corynebacterium diphtheriae* and related human and animal parasites. Details of methods of analysis have been described by Robinson (1968) who combined amino acid composition of the cell wall mucopeptide (primary differentiation) with electrophoretic enzyme patterns of cell-free extracts for further identification within

TABLE 10.1
AMINO ACID COMBINATIONS IN CELL WALLS OF CORYNEFORMS

Alanine and Glutamic Acid

I. DL—Diaminopimelic acid
II. Glycine, ornithine
III. Glycine, aspartic acid, lysine
IV. Glycine, aspartic acid, ornithine
V. Lysine
VI. LL—Diaminopimelic acid
VII. Leucine, aspartic acid, DL—Diaminopimelic acid
VIII. Glycine, aspartic acid, component "U," (unidentified ninhydrin positive component), lysine
IX. Alanine, Glycine, 2:4 Diaminobutyric acid

cell wall groups. Nine combinations of amino acids were found among 28 species of coryneforms (Table 10.1). *Arthrobacter* occurred in four of these groups (Groups III, V, VI, and VII).

Organisms may be identified within these groups on the basis of mobility of esterase, catalase, and peroxidase in cell-free extracts.

Of 57 strains of soil and plant isolates, as well as known strains of *Arthrobacter*, *Cellulomonas*, and *Microbacterium*, Keddie *et al.* (1966) placed the majority in one of four groups based on principal amino acids in the cell wall as follows:

Group A: alanine, glutamic acid, lysine

Group B: alanine, glutamic acid, LL-Diaminopimelic acid, glycine

Group C: (subgroup 1): alanine, lysine, aspartic acid, glycine, component "U"

Group C (intermediate): alanine, glutamic acid, lysine, aspartic acid, glycine, trace or minor amount of component "U"

Group C (subgroup 2): similar to C intermediate with no component "U"

Group D: alanine, glutamic acid, DL-Diaminopimelic acid

Arthrobacter occurred in Groups A, C subgroup 1, and C intermediate; *Cellulomonas* in Group A; and *Microbacterium* in Group D. One strain of *M. Flavum* resembled *C. diphtheriae* and animal diphtheroids. No plant forms were characterized in Group A, but they did occur in all other groups. The isolates and named cultures of *Arthrobacter* and *Cellulomonas* had cell wall amino acids typical of Gram-positive bacteria, lending support to the Gram-positive characteristic which may be obscured by variability during various growth stages.

Differences exist in nutritional requirements and cell wall

composition of dominant coryneforms isolated from soil and plant material. Requirements for biotin or biotin and thiamine dominated among the 67 soil isolates; 17 required biotin alone and 18 needed both vitamins. Of the 47 plant isolates, 13 required both vitamins and 18 also needed pantothenate. None of the plant isolates required biotin alone; all had some requirement for vitamins or amino acids. In contrast also, nine soil isolates required no vitamins or amino acids. More than half of the 67 coryneforms from soil fell into the patterns just described for soil isolates or had a Vitamin B_{12} requirement in addition, whereas only one plant isolate was included in this grouping.

TABLE 10.2
CELL WALL COMPOSITION AND VITAMIN REQUIREMENTS

Cell Wall Group	No. of Strains	Dominant Vitamin Requirements
Group A	22 (2 ungrouped)	None, biotin, or biotin plus thiamine
Group B	15	Thiamine or biotin plus thiamine (3 strains also required B_{12})
Group C	13	Biotin plus thiamine, biotin plus thiamine plus pantothenate and/or terregens factor
Group D	7 (3 ungrouped)	Similar to Group C or not tested

Table 10.2 summarizes dominant vitamin requirements in relation to cell wall composition as indicated by the various amino acid groups.

Only those isolates which had requirements for biotin plus thiamine were heterogeneous in cell wall composition. Other organisms which were similar to each other in vitamin requirements also had the same amino acid pattern in their cell walls. *Arthrobacter* and isolates which were morphologically similar to this genus had a variety of patterns of cell wall amino acids; *Cellulomonas*, on the other hand, was remarkably homogeneous in this respect.

Detailed descriptions of cell wall types of various coryneform bacteria have recently been given by Schleifer and Kandler (1972). They distinguished between characteristics of coryneforms such as oxygen relationship and life cycle differences based on amino acid composition of cell walls. Coryneform bacteria show the greatest differences in peptidoglycan type of all "bacterial families." Coryneforms are the only bacteria that demonstrate more than half of the 28 different types.

ISOLATION AND PROPAGATION OF CORYNEFORMS

Specific selection and isolation media for coryneforms have not been extensively developed, owing largely to the lack of unique or specific physiological properties of organisms in the group. Media containing hydrocarbons as enrichment ingredients may favor their growth as the sole carbon and energy source (Veldkamp 1970). Toluene and the herbicide 2;4-dichlorophenoxyacetic acid may allow *Arthrobacter* to become dominant (Alexander 1967).

Arthrobacter has been extensively studied by Mulder and Antheunisse (1963), who demonstrated a life cycle for these organisms by a sequence of selection and inoculation of coccoid cells grown in a "poor" medium; the cells became rod-shaped after transfer and incubation in a "rich" medium. Morphological changes occurring in this manner, as well as "snapping" division are considered to be characteristic of *Arthrobacter* and have been used as a means of isolation. Mulder *et al.* (1966) used a similar system for isolating and identifying soil arthrobacters and cheese arthrobacters. Procedures and media used are given in Table 10.3. Arthrobacters that grow on non-selective media seeded with food samples may be detected and isolated by examination for the characteristic life cycle; coccoid cells are observed in the stationary phase and pleomorphic rods occur in the logarithmic growth phase (Veldkamp 1970). With "rich" media, these changes are more pronounced than with the "poor" media.

Cellulomonas can be recognized for isolation by cellulose digestion of filter paper (Veldkamp 1970) and may show typical "coryneform morphology" with occasional V-formations of rods.

Microbacterium has been isolated on the basis of heat resistance. Pleomorphic rods isolated from meats by Vanderzant and Nickelson (1969) that survived heating at 80°C for 10 min were classified as *Microbacterium*; those pleomorphic rods that resisted 65°C for 10 min were considered to be "coryneforms." Anagnostoupoulos *et al.* (1964) attempted to increase heat resistance of *M. lacticum* by exposure of cells on an agar surface to 72°C two times; the purpose was to determine if mutation resulted in selection of thermoduric *Microbacterium* on dairy equipment. The physiological condition of the cells appears to be the controlling factor, and strain differences are evident. Heat resistance of different species varies also (Robinson 1966A), with some strains of *M. thermosphactum* surviving 63°C for 3 minutes, but none resistant to 63°C for 5 minutes (McLean and Sulzbacher 1953). Differences in heat resistance of *M. lacticum, M. liquefaciens* and *M. flavum* will be discussed later, but this property has been utilized as a means of classification.

TABLE 10.3
ISOLATION OF ARTHROBACTER

	Soil		Cheese	
Sampling	Soil suspensions shaken for 5 min in physiological saline, diluted.		Homogenized surface of cheese in physiological saline, dilutions streaked on plates.	
Primary Plating	"Poor" medium:			
	Ca(H$_2$PO$_4$)$_2$	0.25 g/1	Tryptone	5.0 g/1
	K$_2$HPO$_4$	1.0	Yeast extract	3.0
	MgSO$_4$·7H$_2$O	0.25	glucose	1.0
	(NH$_4$)$_2$SO$_4$	0.25	NaCl	1.0
	Casein	1.0	Agar	10.0
	Yeast extract	0.7	or: 1:2 diluted TGE agar	
	glucose	1.0	with: Lab-Lemco	1.3 g/1
	agar	10.0	tryptone	2.5, glucose 0.5
	pH	6.9	agar	15.0, pH 7.0
			0.5 ml skim milk per plate	
Incubation	5 days at 25°C		5 days at 25°C	
Subculture	Slopes of same medium		Slopes of same medium	
Incubation	7 days at 25°C		7 days at 25°C	
Subculture	Coccoid colonies, transferred to slopes of "Rich" medium:		Coccoid colonies, transferred to slopes of "Rich" medium (g/1)	
	yeast extract	7 g/1	yeast extract	7
	glucose	10	glucose	10
	agar	10	agar	10
Incubation	24 hrs at 25°C		24 hrs at 15°C	
Microscopic examination	Germinating cocci and rods of irregular size and shape		Rod formation, similar to soil forms	

A selective medium for *M. thermosphactum* was developed by Gardner (1966); the medium was termed STAA by virtue of its composition including streptomycin sulfate, thallous acetate and actidione. *M. thermosphactum* and yeasts were isolated from pork sausage with the basal medium and streptomycin sulfate, but addition of actidione and thallous acetate resulted in inhibition of yeast growth. *Pseudomonas* also grew on the medium; these were accounted for by the oxidase reaction with testing done directly on the plates. Remaining colonies were considered to be *M. thermosphactum*. Various meat products were analyzed, and of 820 gram-positive isolates, 358 were examined in further detail. All exhibited characteristics of *M. thermosphactum*. Gardner *et al.* (1967) used STAA medium for isolation of *M. thermosphactum* or organisms resembling it from packaged pork.

Growth of three species of *Microbacterium* on yeastgel-milk agar was used by Robinson (1966A) for differentiation of the pure cultures. *M. lacticum* was gray-white to greenish yellow, opaque and shiny; *M. liquefaciens* was bright yellow, moist and glistening; and *M. flavum* was pale yellow or creamy, opaque and dry. Brownlie (1966) demonstrated that APT medium allowed faster growth of *Microbacterium* species than did tryptone yeast extract, nutrient broth, and Casitone-yeast extract—Casamino acid media.

Several investigations have indicated that *Microbacterium* spp grow on meat products under reduced oxygen pressure, a condition favoring their development over that of *Pseudomonas* (Weidemann 1965; Gardner *et al.* 1967; Gardner and Carson 1967; Barlow and Kitchell 1966). This property may be influential in determining the ease of isolation of *Microbacterium* from meats.

For isolation of *Corynebacterium*, potassium tellurite is often used in media; this follows from diagnostic procedures for *Corynebacterium diphtheriae* and related members of the genus. Potassium tellurite is used in concentrations of 0.03 to 0.04%. *C. diphtheriae* generally grows in blood agar or serum agar with tellurite in the concentrations specified. The reader is referred to any of several texts on clinical microbiology for isolation and propagation of the pathogens. It is not within the purview of this chapter on food microorganisms to expand further on this subject, since we are concerned here with saprophytic coryneforms.

CORYNEBACTERIUM

Most published information on the genus deals with *C. diphtheriae*, which will not be considered here in detail. Possibly one of the most comprehensive reviews in recent years is that of Barksdale (1970) which presents considerable information on morphology, physiological characteristics, and composition of *C. diphtheriae*, the type species of the genus. *Corynebacterium* are Gram-positive rods, nonacid-fast nonspore-forming and usually are nonmotile. The rods may be slightly curved and may exist as typical "coryneform" club-shaped cells. Old cultures may lose their Gram-positive characteristic, but this varies among species. Granules stain Gram-positive. As mentioned earlier, morphology is characterized by groups of cells arranged in palisade formation, or individual cells forming V or L shaped angles. Formation of granules is affected by the medium in which cultures are grown; serum or egg media enhance granule production.

Cell wall composition has been discussed previously; the cell wall of *Corynebacterium* is characterized by meso-diaminopimelic acid and a polysaccharide-containing galactose, arabinose and possibly mannose. Cell wall composition of *Corynebacterium* was studied by Cummins and Harris (1956) for use as a taxonomic character. As Robinson (1966C) found by gel electrophoresis studies, *Corynebacterium* may be separated according to habitat on the basis of enzyme production. Cell free extracts of 24 *Corynebacterium* cultures examined for esterase, catalase, and peroxidase activity showed that esterases were of value in separating strains or serotypes; human pathogens had a single catalase and no peroxidase, while animal pathogens possessed a double catalase, and plant pathogens demonstrated all three types of enzymes. Plant pathogens appear sufficiently variable in cell wall composition and different from the genus in other features so as to be excluded from the *Corynebacterium* (Robinson 1966C; da Silva and Holt 1965).

G & C Content

Most species of *Corynebacterium* have a G & C content of the DNA in the range of about 55–60%, with variations reported possibly between 48 to 68%, even for similar organisms (Hill 1966). As indicated, the most completely described species are pathogens of man and animals; some doubt still exists as to the position of plant pathogens because some are motile and differ from other members of the genus as discussed above. Food-borne saprophytic *Corynebacterium* have not been described to any great extent in the literature. *C. bovis*, which is sometimes classified as a human pathogen, has been found in aseptically drawn cow's milk, and *C. pyogenes*, *C. ulcerans*, *C. lacticum*, *C. liquefaciens* also were recovered from milk, milk products, and dairy equipment (Jayne-Williams and Sherman 1966). *C. pyogenes*, however, may not truly be considered as a member of the genus *Corynebacterium*, because of differences in cell wall composition from other members of the genus, lack of catalase, and other variations.

Physiological Properties

Temperature Limits.—The majority of the 133 strains studied by Jayne-Williams and Skerman (1966) grew better at 30°C than at 37°C, a property that they considered to be general for certain coryneforms isolated from milk, but particularly heat-resistant organisms. Harrigan (1966), in reviewing work on *C. bovis*, indicated that its optimum growth temperature was 37°C. However, *C. bovis* and *C. lacticum* grow well at 30°C; *C. ulcerans*, *C. pyogenes* grow abundantly at 37°C.

Heat Resistance.—Of the organisms mentioned above, the majority of those classified as *C. lacticum* are heat-resistant, withstanding heating to 63°C for 10 minutes. Some also are resistant to 80°C for 10 minutes. *C. pyogenes* is heat-labile. *C. liquefaciens*, which Jayne-Williams and Skerman (1966) thought could be classified in the genus *Microbacterium*, is also heat-resistant. Other workers (Doetsch and Rakosky 1950) would have placed the organism as a variety of *C. lacticum* because of its ability to liquefy gelatin.

Special Nutritional Considerations

C. bovis has been reported to have a nutritional requirement that is satisfied by Tween-80, which was needed as a supplement to most culture media (Jayne-Williams and Skerman 1966, Skerman and Jayne-Williams 1966). Further, *C. bovis* has a requirement for serum, Tween-20, or egg yolk. Biochemical reactions of *C. bovis* and other members of *Corynebacterium* are given in Table 10.4, taken from Harrigan (1966). The use of Tween agar as a test medium for lipolysis was questioned by Harrigan, who distinguished between the action of lipases and esterases. The *C. bovis* studied have a requirement for fatty acid residues, such as provided by phospholipids as lecithin, which has been suggested as a reason for their existence in the cow's udder. *C. bovis* grows in enriched media; nutrient broth alone is insufficient for testing. Oleic acid may provide a nutrient source, other unsaturated fatty acids (palmitoleic, ricinoleic) support varying degrees of growth. The organism utilizes ammonium salts as a nitrogen source, but growth is accelerated by casein hydrolysate. Some strains require nicotinic acid.

One of the more important reactions of *C. bovis* is its ability to cause rancidity in cream and hydrolysis of butterfat. It also produces acetoin and can grow in broth containing 9% NaCl. Most strains ferment glucose, fructose, maltose, and glycerol (Table 10.4). *C. bovis* produces urease, oxidase, and catalase. Action on egg yolk is negative, as is hydrolysis of gelatin, starch, and casein.

TABLE 10.4
SOME BIOCHEMICAL REACTIONS OF *CORYNEBACTERIUM* STRAINS

Organism	*Acid from:				**Growth in:		Reaction in Egg Yolk Agar	Gelatin liquefied
	Glucose	Fructose	Lactose	Sucrose	Tween 20	Tween 80		
C. bovis Isolates	+	+	–	–	++	++	–	–
C. bovis NCTC3224	+	+	–	–	++	++	–	–
C. ovis NCTC3450	+	+	–	–	++	++	–	–
C. renale NCTC7448	+	+	–	–	++	++	–	–
C. hofmanii NCTC231	–	–	–	–	++	–	–	–
C. pyogenes NCTC6448	+	+	+	+	±	–	–	NT

C. xerosis NCTC7243	+	−	−	±	++	++	−	−
C. flavidum NCTC764	+	+	−	+	NT	NT	NT	−
C. diphtheriae NCTC3985	+	+	−	−	NT	NT	NT	NT
M. lacticum NCIB8450 30°C	+	+	+	−	++	++	−	−
37°C	−	−	±	−	−	±	−	−
NCIB8541 30°C	+	+	−	±	++	++	−	−
37°C	+	+	−	−	+	+	−	−

* Acid: +, Final pH 5.0 or lower; ±, final pH 5.0–5.7
** Growth: ++, Good growth; ±, slight growth; −, no growth: NT, not tested
(Adapted from Harrigan, 1966).

Acid can be produced from glucose anaerobically by the Hugh-Leifson test, but oxygen relationships are not well defined.

ARTHROBACTER

The genus *Arthrobacter* is characteristically known as dominating organisms of soil; however their source may also be dairy products, particularly cheese, eggs, and other foods. Most species studied have been isolated from soil; this is the usual source of the type species, *A. globiformis*.

Much of the basis for identification of *Arthrobacter* relies on morphological characteristics, particularly the cycle of development described earlier. Unfortunately, soil forms exist that may not show the transformation of rods to cocci, or at least demonstrate the cycle only to a limited extent (Keddie *et al.* 1966). This property, or lack of it, may also be characteristic of other coryneforms, with graduations of the cycle not uncommon (Veldkamp 1970). A greater range of habitats than soils has been demonstrated for organisms resembling *Arthrobacter* in morphology and staining characteristics (Keddie *et al.* 1966; Mulder *et al.* 1966), and questions still exist as to whether such organisms all belong in the genus *Arthrobacter*.

Growth Cycle

The cycle of development, or morphological change so characteristic of the genus, itself depends on cultural conditions to a great extent. The role of the medium in demonstration of the life cycle was described by Mulder and Antheunisse (1963). They made observations of cultures of coccoid cells after one, two, and three days of incubation at 25°C in solid media. Under the phase contrast microscope, it was observed that in a poor nutrient agar medium, the coccoid cells germinated to form short rods. After each division the length of the bacteria was reduced, so that after three days, only coccoid cells were seen. Strain differences in cell size occurred, but both strains studied behaved in the same manner. In a rich nutrient medium, longer rods were formed, some branching occurred, and characteristic "diphtheroid" shapes were also observed. Cells often divided by snapping to form a V shape or palisade arrangement. On aging, the length of the cells become smaller. At the end of three days, the large coccoid cells known as "cystites" were observed. Such cells have also been considered as specialized germinating structures because they may give rise to slender rods resembling germ tubes similar to those associated with a spore. Stevenson (1963) believed that cystites resulted from nutritional stress conditions, thus leading to aberrant morphological forms, and not spore-like structures.

Growth of soil arthrobacters in a medium containing an excessive amount of glucose (0.5%) in comparison with the nitrogen source resulted in large coccoid cells being formed; when glucose was not added to a yeast extract medium, the cells were much smaller (Mulder *et al.* 1966). Arthrobacters from cheese do not show the characteristic enlargement of coccoid cells as compared with soil arthrobacters. The accumulation of carbohydrate reserve material in the form of an intracellular polysaccharide is believed to be responsible for the cystite structure (Mulder and Zevenhuizen 1967). Presumably, the cheese arthrobacters are not capable of accumulating the reserve material or the soil arthrobacters are incapable of breaking it down after it is formed. In addition to effects of relatively high amounts of glucose and low levels of nitrogen, inadequate phosphorous and sulfur, and low pH also favor cystite formation. All these influences emphasize the significance of composition of medium in development of the large cells. Nitrogen deficiency apparently promotes polysaccharide accumulation by increasing the supply of glucose-1-phosphate; the polysaccharide was identified as a glycogen by Mulder and Zevenhuizen (1967). Soil arthrobacters may lack the "debranching" enzyme needed for breakdown of the glycogen.

Effects of the medium were also noted by Stevenson (1961) with *A. globiformis.* In a simple basal salts medium, the cells doubled in size, whereas in a complex synthetic medium the cells demonstrated an eight to ten fold increase in size prior to division. After successive divisions continued, the cells gradually reverted to normal coccoid form. Salt concentration of the medium affects morphology in that high amounts of salt (12% for cheese organisms and 5% for soil arthrobacters) promote branching and formation of filamentous rods after prolonged incubation (up to 45 days at 30°C).

The cycle of development is also affected by incubation temperature (Mulder *et al.* 1966). At 10-15°C, the rod form is prolonged with less tendency to form coccoid cells; increasing temperature to 30°C produces coccoid forms very quickly from the rods. The type of organism in question obviously also determines its ability to show various morphological changes; this has already been mentioned in connection with the coryneforms in general. For *Arthrobacter*, the orange-pigmented cheese organisms of Mulder *et al.* (1966) more closely resembled soil forms, in showing less tendency to form coccoid cells, than the gray cheese organisms. However, in a comparison of eight strains of *Arthrobacter*, Sundman (1958) noted that chromogenic organisms showed less tendency toward branching that did non-chromogenic *Arthrobacter*.

The cycle to formation of coccoid cells in the stationary phase is characterized by coccoid cell formation arising from: 1) gradual reduction of the rods at each successive division, and/or; 2) sudden disruption of the large rods into smaller fragments.

As indicated earlier, the development cycle described does not separate *Arthrobacter* from other coryneforms, but rather strengthens the similarities among members of genera considered as coryneforms. Changes of pleomorphic cells into coccoid forms is characteristic of cultures of *Corynebacterium, Cellulomonas, Microbacterium* and *Brevibacterium linens* (Veldkamp 1970). Formation of filaments or branched rods from the coccoid structures is characteristic of other coryneforms in addition to *Arthrobacter*. In addition to the cystites, or large coccoid cells of *Arthrobacter*, smaller coccoid cells, more commonly observed have been referred to as arthrospores. However, these terms in themselves do not seem to be significant in differentiating among organism types.

Gram Reaction

In the past, *Arthrobacter* has been classified as Gram variable. More recently, however, the genus is considered as being Gram-positive because of cell wall composition being characteristic of Gram-positive bacteria. Cummins and Harris (1959) stated that *Arthrobacter* had a cell wall composition following the general patterns of Gram-positive organisms. All coryneform bacteria, including *Arthrobacter*, examined by Keddie *et al.* (1966) were Gram-positive at some stage of their growth. These workers studied 114 representative strains from soil and herbage, as well as 11 known species of *Arthrobacter* (*A. globiformis, A. pascens, A. simplex, A. ramosus, A. aurescens, A. atrocyaneus, A. tumescens, A. terregens, A. flavescens, A. duodecadis,* and *A. citreus*). All *Arthrobacter* showed Gram-positive reactions even in the rod form (which has been stated by others to be Gram-negative) or in the branched rod form characteristic of young cultures; such cultures were strongly Gram-positive on agar media containing yeast extract and peptone. These *Arthrobacter* have the cell wall amino acids typical of Gram-positive bacteria. Fifty strains of *Arthrobacter* isolated from cheese and milk by Mulder and Antheunisse (1963) were mostly Gram-positive as rods or cocci. However, soil *Arthrobacter*, as young rods, were generally Gram-variable or Gram-negative.

In summarizing the Gram reaction, *Arthrobacter* are classified as Gram-positive, with the recognition that the rods may become easily decolorized to show Gram-negative cells with Gram-positive granules. Coccoid cells may give only a weak Gram-positive reaction (Keddie *et al.* 1966).

Cell Wall Composition

Cell wall composition of various coryneforms, particularly those studied by Keddie *et al.* (1966) has been described previously. Emphasis has been placed on the fact that the cell walls of *Arthrobacter* do not contain both arabinose and meso-diaminopimelic acid, the combination characteristic of cell walls of *Corynebacterium, Nocardia,* and *Mycobacterium*. *A. globiformis* has cell walls containing alanine, glutamic acid, and lysine. The most commonly occurring sugar is galactose (Cummins and Harris 1959; Keddie *et al.* 1966). The following is a description of the cell wall composition of species of *Arthrobacter* (from Keddie *et al.* 1966):

A. globiformis
A. citreus   alanine, glutamic acid, lysine, galactose

A. simplex   alanine, glutamic acid, glycine, LL-diamino-
A. tumescens   pimelic acid

A. terregens   alanine, glutamic acid, lysine, aspartic acid,
A. flavescens   glycine; unidentified ninhydrin positive com-
                    ponent.

A. duodecadis   alanine, glutamic acid, lysine, glycine, serine,
                    aspartic acid.

G and C Content

The G and C content of the DNA of *Arthrobacter* species ranges from about 60 to 70 moles percent (Tm). For *A. simplex* and *A. tumescens* the range is higher, in the order of 70 moles percent, or more.

Physiological Properties

Temperature Limits.—Most strains of *Arthrobacter* have an optimum in the range of 20-30°C; most also will grow at 10°C but not at 37°C. Psychrophilic *Arthrobacter* may grow at 0°C with an optimum temperature above or below 20°C (Veldkamp 1970).

The development of a hiemal flora in Narragansett Bay was followed by Sieburth (1965). He compared counts, genera, and growth curves of bacteria isolated semimonthly when temperatures ranged from 0° to 36°C. The organisms with optima at 18 to 27°C quickly disappeared as colder temperatures prevailed, and were replaced by a psychrophilic flora with an optimum at 9°C, but capable of growth at 0°C. Of more than 500 isolates with an optimum growth temperature of 9°C, *Arthrobacter* composed about 7% of the flora. Optima at more than one temperature was observed only for *Arthrobacter*. Other genera included *Pseudomonas*, "*Flavobacterium-Cytophaga*," *Vibrio*, and *Achromobacter*.

Growth of *Arthrobacter* isolated from fresh water fish, as well as known species (*A. aurescens* ATCC 13344, *A. citreus* ATCC 11624, *A. globiformis* ATCC 8010, and *A. simplex* ATCC 6946) was studied at 0, 7, 20, 30, and 37°C (Roth and Wheaton 1962). In general, no sharp line of demarcation exists between psychrophilic and mesophilic arthrobacters, with a continuous graduation in ability to initiate and maintain growth at 0°C. Total number of generations is not affected by incubation temperature, but time to reach maximal growth is greatest at low temperatures.

In general, coryneform bacteria, including *Arthrobacter*, undergo selection depending on seasonal or other environmental temperature changes. These determine development of psychrophilic or mesophilic populations (Veldkamp 1970).

Mulder and Antheunisse (1963) subjected rods and coccoid cells of *A. globiformis* to temperatures of 45, 50, 55, 60 and 65°C to determine if there were differences in heat stability between the two morphological types. After 5 min of heating, cocci were recovered at 45 and 50°C, but no rods grew after treatment at 50°C. No viable cells were found at temperatures above 50°C. Greater heat resistance of the coccoid cells is in keeping with the general resistance of cocci to unfavorable environmental conditions. No *Arthrobacter* from soil or herbage survived heat treatment of 63°C for 30 min (Keddie *et al.* 1966).

pH Effects.—Most *Arthrobacter* strains grow well in the range of slight alkalinity to neutral pH.

Drying and Water Activity.—As indicated by Veldkamp (1970), a characteristic of importance for survival of soil bacteria is their ability to resist harmful effects of drying. *Arthrobacter* has a very high degree of resistance to drying, exceeded by only a few other types of soil bacteria, such as *Bacillus*. *Arthrobacter* made up about 1/2 to 1/3 of the bacterial flora of various soils kept at 0.5% humidity for 2 to 10 months, but after about three years, only *Bacillus* survived (Mulder and Antheunisse 1963). Apparently, similar organisms found in foods may also be expected to withstand low moisture conditions, since cheese coryneforms are found in low water activity environments.

Halophilic Nature.—As may be expected from their ability to withstand drying, *Arthrobacter* species or types of coryneforms resembling *Arthrobacter* isolated from cheese are capable of growth in salt concentrations up to 12–15%. Soil forms of *Arthrobacter* are not as tolerant as cheese types. The distinct ability to grow in high salt levels is most closely associated with *Brevibacterium linens*, which has been suggested by da Silva and Holt (1965) to be transferred to the genus *Arthrobacter*. Soil forms of *Arthrobacter* do not need sodium chloride for growth, but cheese types may need salt for growth. With 4% added salt in the medium, a beneficial growth effect occurs for cheese coryneforms and coryneforms from sea fish; this is pronounced at pH values below 6.5, but may not be a specific NaCl effect, since other salts also enhance growth (Mulder *et al.* 1966). With elevation of temperature from 25 to 32 or 35°C, 4% salt produces beneficial growth effects for cheese coryneforms.

Oxygen Requirements.—Like other coryneforms in general, *Arthrobacter* are aerobic; indeed, they have been described as strict aerobes with a respiratory metabolism.

Nutritional and Biochemical Considerations

In the preceding sections, some discussion has been given on the relation of cell wall composition to nutritional requirements and that aspect of coryneforms, particularly *Anthrobacter*, will not be dealt with in detail here.

Nutritional requirements of *Arthrobacter* cover a wide range from those members of the genus that have simple requirements with no added organic growth factors to those that have a demand for amino acids and vitamins in the medium. The original description of *Arthrobacter* by Conn and Dimmick (1947) included the ability of these bacteria to grow in a simple salts medium with an organic carbon and energy source. As more organisms resembling *Arthrobacter* morphologically have been isolated, the nutritional requirements have been found to be more complex, and this basis for differentiation has not been included in the generic definition, but does distinguish species. It also has been used to distinguish between the soil and herbage forms described by Keddie *et al.* (1966) and the cheese types of Mulder *et al.* (1966) discussed earlier. For example, only the plant organisms required biotin, thiamine, and pantothenate (18 of 47 isolates) and only soil isolates required biotin alone (17 to 67 isolates). *A. globiformis* consists of strains that are non-exacting or that require biotin. A requirement for amino acids in addition to vitamins is common for plant arthrobacters but not usual for soil isolates. Of the 112 coryneform bacteria isolated from different soils by Mulder *et al.* (1966), 95% utilized ammonium nitrate as the sole nitrogen source, and 52% of gray cheese arthrobacters also, while all 15 strains isolated from activated sludge have this ability. Glucose was the carbon and energy source. Owens and Keddie (1969) showed that 90% of 55 soil isolates utilized inorganic nitrogen (ammonium sulfate) as sole nitrogen source, but only 30% of 38 herbage isolates did so. No soil isolates had a methionine requirement, but 60% of the herbage organisms did, although they also used ammonium ion as the major nitrogen source. Of the 46 grey-white cheese arthrobacters of Mulder *et al.* (1966), 17 required glutamic acid plus ammonium salt, 3 needed glutamic acid alone, 1 required methionine alone, but 10 needed methionine and the ammonium salt. Only 13 of these organisms grew with inorganic nitrogen as the nitrogen source.

Some species of *Arthrobacter* require the terregens factor or other growth promoting siderochromes known as sideramines. The terregens factor is required by *A. terregens*, produced by *A. pascens*, although other sideramines will replace it. *A. flavescens* also requires the terregens factor, and *A. citreus* has a shortened lag phase with sideramines present.

Ability to utilize various carbon sources differs among *Arthrobacter* types; lactate generally is not used by soil strains while it is a good carbon source for arthrobacters from cheese (Mulder *et al.* 1966). *A. globiformis* can use many carbohydrates and alcohols from sugars and carboxylic acids as carbon and energy sources (Veldkamp 1970). Glucose, sucrose and glycerol are excellent carbon sources for soil forms; *Brevibacterium linens* also uses these compounds.

For *A. globiformis*, possibly at least two distinguishing nutritional features have already been mentioned, the requirement for no added vitamins or for biotin only, and the ability to use inorganic nitrogen as the sole source of nitrogen. Other *Arthrobacter* species, such as *A. tumescens*, have a vitamin requirement for thiamine only. *A. duodecadis* not only requires thiamine, but also Vitamin B_{12}. Various patterns of vitamin requirements and their relation to nitrogen source needed have been described (Keddie *et al.* 1966; Owens and Keddie 1969).

Special Characteristics

Among other significant biochemical considerations is the ability of *Arthrobacter* to synthesize reserve polysaccharide, which has already been described in connection with cystite formation, and which is more commonly associated with soil arthrobacters than with those from cheese (Mulder *et al.* 1966).

In screening 250 strains of *Arthrobacter* for their ability to release amino acids in a minerals-glucose medium, one strain was found by

Veldkamp *et al.* (1966) to produce riboflavin, differing from the typical *A. globiformis* in that respect. A consistent amount of riboflavin, 5 μg/mg of the dry weight of the cells, was produced regardless of nutritive quality of the medium. The riboflavin producer grew at a slower rate than the normal strain of *A. globiformis*.

As indicated by Veldkamp (1970) the humus portion of soils is rich in aromatic hydrocarbons and *Arthrobacter* species have the ability to attack such compounds. Jensen (1966) described decomposition of the herbicide Endothal (disodium endoxohexahydrophthalate) by *A. globiformis*, accompanied by loss of toxicity of the compound to plants. Stevenson (1967) found that 77% of 130 soil arthrobacters could grow on at least 2 aromatic hydrocarbons of about 40 compounds tested. The lower the nutrient requirements of the organisms, the greater their ability to utilize aromatic sources. While *Arthrobacter* species can break down phenol in activated sludge, they do not seem to be able to act on the aromatic hydrocarbons naphthalene, anthracene or phenanthrene. However, their capability to degrade herbicides and aromatic hydrocarbons offers interesting possibilities in removal of residues from food crops.

Mulder *et al.* (1971) performed extensive studies on comparisons of arthrobacters from dairy waste, activated sludge, soil, and cheese. Dairy waste organisms are similar in several characteristics to soil arthrobacters, but quite different from cheese types. All dairy waste organisms (71 tested), and 95% of 112 soil forms, grew with inorganic nitrogen, although about half the dairy waste activated sludge types needed vitamins. These two types also accumulated polysaccharide reserve material to a large degree, whereas no cheese types did. As mentioned earlier, all cheese forms, both pigmented and non-pigmented, could tolerate 8% NaCl; no soil forms and only 7% of the dairy waste arthrobacters tolerated salt at a concentration of 8%.

Proteolytic activity is characteristic of soil arthrobacters, orange strains of cheese organisms similar to or belonging to *B. linens*, and dairy waste organisms. Non-pigmented cheese types are not generally proteolytic. This characteristic is of importance in the activity of the cheese organisms in ripening of the product, a phenomenon still demanding more research. Descriptions of various *Arthrobacter* species usually indicate hydrolysis of gelatin. Veldkamp (1970) reviewed the strong proteolytic ability of proteinase of *A. urefaciens* as a peptidase.

Inhibition of *Arthrobacter* by *Pseudomonas* has been demonstrated for soil and seawater arthrobacters (Sieburth, 1966, 1967). A consistent inverse relationship between the two genera was observed in Narragansett Bay, R. I., during semi-monthly checks over a two year period. A polysaccharide-like heat labile substance produced by the *Pseudomonas* isolates caused growth inhibition, inhibition of motility, and agglutination of the *Arthrobacter*. The material is not believed to be related to bacteriocins from *Pseudomonas*.

MICROBACTERIUM

The position of *Microbacterium* as a separate genus among the coryneforms is, as stated earlier, somewhat precarious. Evidence has shown that at least some species of *Microbacterium* should be transferred to the genus *Corynebacterium* based on cell wall composition. *M. flavum* and *M. thermosphactum* are similar in cell wall structure to *C. diphtheriae*. However, it would appear that some doubt must be cast on lumping an organism whose claim to fame is sausage flavor deterioration in the same genus as that of a human pathogen causing an infection of the throat with production of a "powerful exotoxin." Cell wall composition as a basis for classification of *Microbacterium* will be examined further.

Cell Wall Composition

M. lacticum and *M. liquefaciens* have essentially the same cell wall composition with some minor differences in components. *M. lacticum* contains amino acids alanine, glutamic acid, glycine, aspartic acid, and lysine. Composition of the cell wall of *M. liquefaciens* includes the same amino acids with exception of ornithine instead of lysine. The carbohydrate of *M. liquefaciens* is essentially rhamnose, whereas this carbohydrate is found along with galactose and sometimes mannose in the cell wall of *M. lacticum*. *M. flavum* differs from the other two species in that it has the amino acid DL-diaminopimelic acid as well as alanine and glutamic acid, and the carbohydrates galactose, mannose, and arabinose. These components are of the same pattern as that of the human and animal pathogenic *Corynebacterium* species. *M. flavum* also has enzyme patterns in starch gel similar to those of the pathogens, and does not differ greatly in heat resistance from *C. diphtheriae*. All these features tend toward placing *M. flavum* in with *Corynebacterium*, while *M. lacticum* and *M. liquefaciens* are retained as *Microbacterium* species (Robinson 1966B). Although they believed that their comparisons were not complete, and hence possibly not entirely correct, da Silva and Holt (1965) found that *M. lacticum* formed a cluster with plant pathogenic corynebacteria at a similarity level of 82%. They also pointed out that the metabolic pathways for glucose oxidation differ between phytopathogenic *Corynebacterium* species and *M. lacticum*. *M. lacticum* has a cell wall composition that is similar to that of *Arthrobacter*; Davis and Newton (1969) suggested that it be included in their *Arthrobacter* group.

While several investigators have commented on the nebulous status of *M. thermosphactum* since its description was given by McLean and Sulzbacher (1953), the organism continues to be a possible isolate from meats or other foods regardless of its name. As indicated by Barlow and Kitchell (1966), *M. thermosphactum* shares some characteristics of coryneforms while resembling homofermentative lactobacilli in other respects, thus occupying a doubtful position in classification.

M. thermosphactum was proposed as a new species by McLean and Sulzbacher (1953), who studied 46 cultures isolated from sausage and pork trimmings and suggested that these be classified in the *Lactobacteriaceae*, and in the genus *Microbacterium* because of production of catalase. That the organism carries on a pleomorphic growth cycle resembling *Arthrobacter* has also been established (Davidson *et al.* 1968). Large masses of growth are observed in old cultures on fresh transfer.

Physiological Properties

Heat Resistance.—Probably one of the most distinguishing features of *Microbacterium* species is the heat resistance of *M. lacticum* and *M. flavum* and lack of heat resistance of *M. thermosphactum*. *M. thermosphactum* produces catalase and ferments carbohydrates to produce lactic acid, in common with other *Microbacterium* species, but has low heat resistance (the most heat resistant strain survives heating at $63°C$ for 3 min, but not for 5 min). *M. lacticum* withstands heating for 2.5 min at $85°C$, and *M. flavum* resists heating for the same time period at $71.6°C$. Decimal reduction times are as follows: *M. thermosphactum*, D_{50} is 2.5 min; *M. lacticum*, D_{70} is 4 min; *M. flavum*, D_{65} is 2 min (Davidson *et al.* 1968). When *Microbacterium* cultures were heated for 10 min, eighteen strains of *M. lacticum* were able to survive as high a temperature as $75°C$ and five survived up to $85°C$; *M. liquefaciens* survived $65°C$ but not $70°C$; and *M. flavum* was resistant to $60°C$, but not $65°C$ (Robinson 1966A). Attempts to increase heat resistance of *M. lacticum* were unsuccessful (Anagnostopoulos *et al.* 1966).

Temperature Limits.—As given by McLean and Sulzbacher (1953), optimum growth temperatures for *M. thermosphactum* were between 20 and $22°C$; three strains of *Microbacterium* tested by Brownlie (1966) had an optimum near $25°C$ (aerobic conditions on APT medium). Doetsch and Pelczar (1948) gave an optimum temperature of 30°C for *Microbacterium*. *M. thermosphactum* grows on veal infusion agar after 7 days of incubation at 0–1°C, but the other *Microbacterium* species fail to grow at 4°C on Heart Infusion Agar. *M. thermosphactum* does not grow at 37°C even if moved to a lower temperature later, but *M. lacticum* and *M. flavum* do grow at 37°C. Psychrophilic strains of *Microbacterium* have been isolated from sausage and other meats and generally do not grow at temperatures above 35°C (Brownlie 1966), but have an optimum growth rate at 25°C.

Oxygen Requirements.—It is of interest in food spoilage with regard to *M. thermosphactum* that reduced oxygen supply, and possibly increases in carbon dioxide concentration of the atmosphere, appear to favor development of the organism. As pointed out by Davidson *et al.* (1968) most studies in which *M. thermosphactum* has been isolated have involved packaging materials or methods and storage conditions of meat designed to limit exchange of gases. Weidemann (1965) observed spoilage of beef muscle stored in nitrogen at 0°C to be caused primarily by psychrophilic *Microbacterium* (probably *M. thermosphactum*) after 59 days. Work of Gardner *et al.* (1967) with prepackaged pork stored at 2°C and 16°C indicated an increase in proportions of *M. thermosphactum* as CO_2 levels increased in packages. Carbon dioxide is less inhibitory to *M. thermosphactum* than to more aerobic types of bacteria, such as *Pseudomonas*. However, *Microbacterium* have a faster growth rate under aerobic conditions than under anaerobic conditions. At a favorable range of water activity (0.96 to 0.99) the number of generations per hour at 25°C under aerobic conditions is at least double that in an anaerobic system (Brownlie 1966).

Fermentation Products

All *Microbacterium* from beef and known cultures of *M. thermosphactum*, *M. flavum* and *M. lacticum* tested by Davidson *et al.* (1966) were fermentative in their action on glucose, and no gas was detected, although McLean and Sulzbacher (1953) mentioned that CO_2 was formed during fermentation reactions in Eldridge fermentation tubes. The main product of glucose fermentation by *M. thermosphactum* is L(+) lactic acid (McLean and Sulzbacher 1953). Lactic acid accounted for more than 75% of the glucose utilized by *M. thermosphactum* in studies by Davidson *et al.* (1966). Other acids, amounting to about 1% of the glucose used, were acetic and propionic, with trace amounts of other fatty acids detected. Doetsch and Pelczar (1948) attempted to differentiate among species of

Microbacterium based on acid production from various carbo-hydrates; one group produced acid from glycerol and raffinose whereas *M. lacticum* and *M. flavum* did not. The first group also fermented arabinose and xylose to produce acid, similar to the later observations of McLean and Sulzbacher (1953) for *M. thermosphactum*.

Lactic acid is the predominant acid formed in milk by *M. lacticum* and *M. flavum*, which occur in pasteurized milk by virtue of their previously mentioned heat resistance, although in small numbers in comparison with other lactic organisms (Speck 1943).

Water Activity.—The most rapid rate of growth for *Microbacterium thermosphactum* and two other *Microbacterium* isolates from meat is at a high A_w of 0.99, with a limiting A_w of 0.94 (Brownlie 1966). Growth rate is less with anaerobic conditions than with aerobic conditions at all A_w levels between 0.94 and 0.99; the range of A_w for growth is also reduced. Lowering temperature from 25° to 10° to 0°C also exerts a restriction on range of A_w under anaerobic conditions (*e.g.*, at 10°C, A_w for growth, aerobic: 0.94-0.99, anaerobic: 0.96-0.99). The tolerance to A_w under aerobic conditions is similar to that for *Lactobacillus plantarum* and *Pediococcus cerevisiae*, but the limiting A_w is lower than that for *Pseudomonas fluorescens* or *Lactobacillus brevis*. Reduced A_w values were tolerated by *Microbacterium* tested by Brownlie (1966) under anaerobic conditions in contrast to other lactic acid organisms.

pH.—The ability of *Microbacterium* to grow at pH values of 5 to 9 was tested by Brownlie (1966); he found the greatest growth rate at about pH 7.0 (25°C).

All 26 strains of *Microbacterium* examined by Robinson (1966A) grew well at pH 6.8 and 7.5, but were inhibited in growth at pH 5.0. Effect of pH on growth of *Microbacterium* has no unusual implications; growth would be expected to be limited, if occurring at all, in acid foods.

Special Considerations

M. thermosphactum may give confusing reactions with the catalase and benzidine tests, both of which are influenced by medium and incubation temperature (Davidson *et al.* 1968). Strong positive reactions are obtained on APT agar at 20°C; heart infusion agar (HIA) gives a weaker catalase reaction. At 30°C, both media produce reduced rates or negative results for catalase. Young cultures may occasionally give negative results on HIA, so the failure to detect catalase may not be limited to old cultures only. Nutrient agar may produce results similar to those with HIA. The benzidine reaction, an indication of the presence of cytochromes, is also affected by growth medium. Quantitative content of cytochromes is influenced by medium and temperature, similar to effects described above for the catalase test. APT is more favorable than HI broth; 20°C incubation results in greater production of cytochromes with either medium than observed at 30°C. APT medium has added iron, and this may cause an increase in cytochrome content and catalase activity.

In heat resistance studies of two strains of *M. lacticum*, both an inhibitory and a stimulatory effect have been observed (Anagnosto-poulos *et al.* 1964). The stimulatory action of heating was considered to be due to RNA excreted as a result of heating, or carried over from the inoculum, while the inhibitory effect was thought to be caused by products of heat-induced degradation of DNA from the cell into the medium. Both effects may be superimposed, DNA degradation products may interfere with RNA synthesis while DNA degradation itself is being neutralized.

Microbacterium has been shown to be inhibited by low concentrations of undissociated nitrous acid (25 to 200 ppm $NaNO_2$). Inhibition was greater at 0°C than at 25° or 10°C, and greater retardation of growth was obtained as the pH was reduced from 7.0 to 5.0 (Brownlie 1966). No isolates of *Microbacterium* were able to grow at pH 5.5 in the presence of 200 ppm $NaNO_2$, but all grew when nitrite was lacking. In contrast, 21 of 25 lactobacilli isolated from cured meats grew in the presence of nitrite at pH 5.5.

As indicated earlier, the taxonomic position of *Microbacterium* is still open to question. *M. thermosphactum* has been suggested to be closer to the *Lactobacillaceae* than to the *Corynebacteriaceae*. However, on the basis of a positive catalase reaction, functional cytochrome, and morphological characteristics, evidence also points toward its retention in the family *Corynebacteriaceae*. Further observations of Shaw and Stead (1970) on lipid composition of *M. thermosphactum* also point to retention of the organism in the *Corynebacteriaceae*. Overall lipid composition of *M. thermosphactum* is unlike that of the lactobacilli, particularly with regard to glycolipids and unsaturated fatty acid composition.

CELLULOMONAS

The importance of *Cellulomonas* in foods remains to be seen; the genus represents soil coryneform bacteria that have as a primary characteristic the ability to attack cellulose. Some strains may occur which do not attack cellulose, but breakdown of cellulose has been considered as a criterion for "authentic" members of the genus.

Cellulomonas shows true coryneform morphology, with club-shaped cells and angular "V" formations. Aging of cultures produces shortening of the irregular rods, but coccoid cells are not formed in large proportions.

Cell Wall Composition

Cell walls of *Cellulomonas* do not contain meso-diaminopimelic acid or arabinose, or the galactose characteristic of *Arthrobacter*. The walls of six strains characterized by Keddie *et al.* (1966) contained alanine, glutamic acid and lysine as the major amino acids.

Gram Reaction

Although previous descriptions of *Cellulomonas* have been given as Gram-negative (Mulder and Antheunisse 1963) or Gram variable (Breed *et al.* 1957), the six strains examined by Keddie *et al.* (1966) were weakly Gram-positive and had cell wall amino acids similar to Gram-positive organisms. Mixtures of Gram-positive and Gram-negative rods may occur because cells are readily decolorized.

Other Characteristics

These organisms are primarily aerobic, with reduced growth under anaerobic conditions. Relationship of growth to temperature and pH is not unusual—optima appear to be at about 30°C and near neutrality. *Cellulomonas* is not resistant to pasteurization (63°C for 30 min). G and C content of DNA is 75 moles per cent (Tm).

Nutritional requirements, like cell wall composition, are quite homogeneous. Vitamin requirements are for biotin and thiamine (final concentrations in test medium, 0.001 μg/ml and 0.5 μg/ml, respectively).

Ability of *Cellulomonas* to degrade cellulose is sufficiently well correlated with other characteristics so as to serve to distinguish these organisms from other coryneforms (da Silva and Holt 1965). Non-cellulolytic strains also appear to be closely related to *C. biazotea* so as to be considered in the same genus (Bousfield 1972).

BREVIBACTERIUM

Brevibacterium is another genus in the coryneform group of bacteria that occupies a doubtful position in classification. Da Silva and Holt (1965) suggested that *B. linens* be transferred to the genus *Arthrobacter*, with a new name combination, *Arthrobacter linens*. Mulder and Antheunisse (1963) and Bousfield (1972) described *B. linens* as closely resembling *Arthrobacter globiformis*. Mulder *et al.* (1966) pointed out the relationship of *B. linens* to *Arthrobacter* found on cheese surfaces. *B. linens* has been described as an organism important in cheese ripening, and many of the coryneform bacteria isolated from cheese resemble it, particularly with regard to dependency on light for orange pigment formation. Morphological characteristics of *B. linens* are also similar to those of *Arthrobacter* although Bergey's Manual (Breed *et al.* 1957) merely describes the genus *Brevibacterium* morphologically in terms of Gram-positive, nonspore-forming, unbranched rods. *B. linens* does not have as strong a tendency to form coccoid cells as *Arthrobacter*, but does form irregular shaped rods which may show snapping division and palisade cell formations (Mulder *et al.* 1966). *B. linens* also differs from *Arthrobacter* in that it does not form glycogen as a reserve material.

Halophilic Nature

B. linens grows in salt concentrations up to 8% in yeast extract glucose medium, with some strains showing delayed growth, although only slight, at levels of NaCl up to 15% (Mulder *et al.* 1966). Enhanced enzyme activity, rather than greater permeability to nutrients, is believed responsible for the favorable effect of salt.

Proteolytic Activity

Strains of *B. linens* and orange pigmented cheese strains resembling *B. linens* hydrolyze gelatin; addition of 4% salt promotes protein hydrolysis, particularly for paracasein. Ripening of soft cheeses is believed to be due at least in part to proteolytic activity of *B. linens* and related organisms.

Nutritional Requirements

Brevibacterium and similar organisms from cheese are more exacting in amino acid requirements than is *Arthrobacter*.

CORYNEFORMS IN FOODS

Isolations and descriptions of various coryneforms in different foods are tabulated in Table 10.5.

Saprophytic coryneform bacteria have been recovered from many kinds of foods. Coryneforms have been found in sea water, on the surface of sea fish and fresh-water fish, on plants, on cheese and in milk, in poultry, eggs, and meat, and in poultry deep litter. Although all of these items are not foods, they are all at least associated with foods.

In addition to the obvious difference in pathogenicity, there are definite differences among saprophytic coryneforms and animal and plant pathogens in cell wall composition; these differences have been described earlier.

TABLE 10.5
CORYNEFORMS IN VARIOUS FOODS

Food	Organism	Reference
Dairy Products		
Milk	*Arthrobacter*	Mulder and Antheunisse (1963)
	Coryneforms	Abd-El-Malek and Gibson (1952)
	C. lacticum	Abd-El-Malek and Gibson (1952)
	Microbacterium	Abd-El-Malek and Gibson (1952)
	C. bovis	Jayne-Williams and Skerman (1966)
	C. pyogens, C. ulcerans	Skerman and Jayne-Williams (1966)
	C. lacticum	Skerman and Jayne-Williams (1966)
Milk and dairy	Coryneforms	Thomas *et al.* (1966)
equipment	Coryneforms	Thomas *et al.* (1967)
Cheese		
Menshanger Cheese	*Arthrobacter*	Mulder and Antheunisse (1963)
Different varieties	*Arthrobacter*	Mulder *et al.* (1966)
	Brevibacterium linens	Mulder *et al.* (1966)
Cheddar		Dempster (1968)
Cream		Dempster (1968)
Meat		
Beef	Probably Coryneforms	Ayres, J. C. (1960)
	Microbacterium	Wolin *et al.* (1957)
	Microbacterium	Rogers and McCleskey (1957)
	Microbacterium	Jaye *et al.* (1962)
	Microbacterium	Weidemann, J. F. (1965)
		Brownlie (1966)
		Kraft *et al.* (1966)
	M. thermosphactum	Gardner (1966)
	M. thermosphactum	Davidson *et al.* (1968)
	Coryneforms	Vanderzant and Nickelson (1969)
Fresh Pork	*M. thermosphactum*	Sulzbacher and McLean (1951)
		Gardner (1966)
		Gardner *et al.* (1967)
	Coryneforms	Vanderzant and Nickelson (1969)
Sausage	*Microbacterium*	Sulzbacher and McLean (1951)
	M. thermosphactum	Sulzbacher and McLean (1953)
	M. thermosphactum	Gardner (1966)
	M. thermosphactum	Dowdell and Board (1968)
Frankfurters	*Microbacterium*	Kraft (1951)
	M. thermosphactum	Drake *et al.* (1958)
Bacon	*Corynebacterium*	Gardner (1968)
Hams, Italian raw	Coryneforms	Giolitti *et al.* (1971)
	Arthrobacter	
	Corynebacterium	
Lamb	*M. thermosphactum*	Barlow and Ketchell (1966)
	Coryneforms	Vanderzant and Nickelson (1969)
Poultry		
Chicken	*Corynebacterium*	Barnes (1960)
Turkey giblets	*Corynebacterium*	Salzer *et al.* (1967)
	Arthrobacter	Kraft *et al.* (1966)
	Brevibacterium	

TABLE 10.5 (Continued)

Food	Organism	Reference
Turkey	Coryneforms M. thermosphactum	Barnes and Shrimpton (1968)
Eggs Liquid egg	Arthrobacter Corynebacterium	Kraft et al. (1966)
Shell eggs	Arthrobacter	Board et al. (1964)
Fish Sea fish	Arthrobacter Arthrobacter	Mulder et al. (1966) Sieburth (1967)
Haddock	Corynebacterium	Liston and Shewan (1958)
Cod	Coryneforms	Shaw and Shewan (1968)
Curing brines	Coryneforms	Bain et al. (1958)
Shrimp	Coryneforms	Vanderzant et al. (1970)
Frozen vegetables	Coryneforms	Splittstoesser et al. (1967)

Foods

Dairy Products.—Coryneform bacteria have been recognized in milk and other dairy products for many years. In earlier years, those found in market milk received much attention when colony counts were used to determine sanitation and hygienic quality of milk, which is still being done today. The ability of many of these organisms to resist high temperatures, a character that is unusual for nonspore-forming bacteria, frequently results in their occurrence in fairly high numbers in pasteurized milk. However, they seem to play little part in spoilage. The main source of these organisms in milk has been primarily poorly cleaned milking equipment and utensils. For this reason, high counts of coryneforms in milk are considered to be an index of sanitary conditions of the equipment rather than a criterion of keeping quality. In most instances, the organism known as Corynebacterium lacticum was one of the predominant species surviving pasteurization.

In an interesting study done in England (Thomas et al. 1966, 1967) consisting of surveys of farm milk supplies done 20 years apart, some changes were brought about by improved methods of disinfection of dairy equipment:

1943-48: Corynebacteria made up 67% of 1188 thermoduric bacteria in milk when equipment was cleaned by warm detergent.

1964-66: Canned milk delivery: 309 (26%) of 1183

Bulk tank delivery: 95 (13%) of 709   when chemical disinfectants were used. In the 20 years' time period, corynebacteria were displaced by aerobic sporeformers and micrococci in the organisms surviving pasteurization.

Other work in England on milk and dairy farm equipment showed a division of coryneform bacteria into four groups, based on physiological properties (Jayne-Williams and Skerman 1966; Skerman and Jayne-Williams 1966). Group I was identified as *Corynebacterium bovis*; it was lipolytic, non-hemolytic, non-proteolytic and could grow in 9% salt. Group II was not lipolytic but was slowly proteolytic and sensitive to salt. These organisms were considered to be *C. ulcerans* (now *C. diphtheriae*). Group III were lipolytic and also liquefied serum; they were identified as *C. pyogenes* (now in with *Streptococcus*). Group IV may earlier have been considered as *Microbacterium lacticum* or *M. liquefaciens*; these survived pasteurization whereas the others were heat-labile. However, at the time, the investigators called these bacteria *C. lacticum* or *C. liquefaciens*, and where they would be placed now is uncertain. The first three groups came from the cows' udders (from "aseptically" drawn milk) whereas the last group was from market milk and dairy utensils.

There are numerous reports on other isolations of coryneforms from dairy products, including *Arthrobacter* from cheese which differed physiologically from *Arthrobacter* from soil and from dairy industry waste (pigmented strains from cheese). Some discussion of these organisms has been given earlier.

Two distinct types of coryneform bacteria have been isolated from cheese surfaces: (1) colonies gray-white or light yellow or pink, short rods transforming to cocci, and (2) colonies orange, less tendency to change from rods to cocci. This type is identical with *Brevibacterium linens*; pigment development depended on presence of light in about 50% of 29 cultures examined. Cheese coryneforms are more exacting in nutrition requirements than soil *Arthrobacter*, with orange type even more so than gray-white type. The ability of cheese coryneforms to tolerate large amounts of salt is an important factor in their occurrence in foods, and in their ability to carry on proteolytic activity, since they are not inhibited by concentrations of NaCl up to 4% or, in some cases, up to 15% (*B. linens*). They grow better in salt when the temperature is increased in the range 25 to 32°C. Soil arthrobacters, on the other hand, are inhibited by 4% salt in the medium (Mulder *et al.* 1966). Such coryneforms in cheese may play a role in the ripening of soft cheeses such as limburger, Romadour, and Pont l'Evêque. These organisms also have a tolerance to low water activity, similar to soil *Arthrobacters*.

Fish.—The salt tolerant coryneforms would be expected to be found not only in cheese, but in sea water and fish from the sea, and cured meats.

Meats.—Meats have been sources of various coryneform bacteria. Species of *Microbacterium*, which, as stated earlier, is in doubtful status as a genus, have been isolated in fairly large proportions (15%) of the flora of fresh pork sausage in a survey done in 1951. Acid production by a species of *Microbacterium* was believed to cause flavor deterioration of sausage stored at refrigeration temperatures (Sulzbacher and McLean 1951). This organism was assigned the name *Microbacterium thermosphactum*, a non-heat resistant type differing from other *Microbacterium* in that respect. *M. lacticum* and *M. flavum* can resist heating at 85°C for 2.5 min and 71.5°C for 2.5 min respectively; the most heat resistant strain of *M. thermosphactum* resisted heating at 63°C for no more than 3 min. It probably occurs as a contaminant from pork trimmings.

In a survey of fresh British pork sausage, *M. thermosphactum* was predominant in most samples: the higher the total count, the greater the percentage of *M. thermosphactum* (Dowdell and Board 1968). The incidence varied from 22% to 100% of the sausage from different manufacturers—at least one sample may have been frozen previously and thus lowered the incidence. Overnight storage of freshly made sausage held at 5°C appears to increase the incidence (70% to 85% in one case; 20% to 45% for another brand), so these organisms may actually make up an increasing proportion of the flora with refrigerated storage, while Gram-negative rods tend to decrease. *Microbacterium* predominated to a greater degree (61/167) on frankfurters from retail stores with storage at 2°C than at 10°C (41/186).

Microbacterium, probably *M. thermosphactum*, has been isolated from frankfurters in retail stores in the U.S. Workers at the American Meat Institute showed that this species also has some degree of radiation resistance greater than *Pseudomonas*, the usual predominant meat spoilage organism. The radiation resistance of *Microbacterium* was in the order of 70,000-90,000 rads for an LD99 dose (99% destruction in tryptone-yeast extract broth), whereas for *Pseudomonas*, the LD99 was approximately 6500 rads (Drake *et al.* 1958). On pork, the incidence of *M. thermosphactum* also increased with increasing CO_2 concentrations in the range 3-16% of the atmosphere (Gardner *et al.* 1967).

Microbacterium thermosphactum has also been isolated from lamb stored at 5°C, particularly when the lamb was packaged in Cryovac in which the oxygen tension is greatly reduced. These organisms were facultative anaerobes, nonmotile rods, able to grow at 1°C and in

6.5% NaCl. None survived heating at 63.5°C for 2.5 min. The *Microbacterium* isolates were reported to make up 100% of the flora on lamb after 3 days at 5°C when 21 colonies were examined. On beef, a different picture was obtained; only about 20% were *Microbacterium* and the remainder were Gram-negative rods. *Microbacterium* was believed to cause spoilage of lamb within 3 days, but not beef. These studies were done in England several years ago (Barlow and Kitchell 1966). In one study done in 1958, of 189 cultures of psychrophiles, 4 of 7 Gram-positive isolates were *Corynebacterium*, while in other work, about 9% of all bacteria isolated from packaged ground beef were *Microbacterium* (Jaye *et al.* 1962). In recent work at Texas A & M University (Vanderzant and Nickelson 1969), coryneforms made up a large percent of the isolates from lamb and beef immediately after slaughter, but then decreased in numbers during holding in refrigerated storage for 3 days. On beef after slaughter, 43% (87) were coryneforms, from lamb 28% (9) were isolated; after refrigerated storage these percentages dropped to 32% (40) and 25%, which, however, was one of only four isolates examined.

In other work, *Microbacterium*, probably the species *thermosphactum*, was isolated from fresh beef (Ayres 1960). Originally this organism was tentatively classified as a lactobacillus—it was a Gram-positive to Gram variable non-motile rod, but turned out to be catalase-positive and varied in length. It was not heat resistant. The storage of fresh beef under nitrogen at 0°C may increase the proportion of *M. thermosphactum* (Wiedemann 1965), similar to the effect of increasing CO_2 level.

In raw Italian hams, lipolytic coryneforms develop readily in the fat, more so than in the lean, so that breakdown of the fat depends at least partly on microbial lipases. *Arthrobacter* has been implicated in breakdown of sulfur-containing amino acids in raw ham, and these organisms were found in Italian sour ham (Gioletti *et al.* 1971).

In reviewing the effect of packaging conditions on *Microbacterium* in meats, it appears that conditions favoring reduced oxygen pressure, such as vacuum packaging, or atmospheres of nitrogen or carbon dioxide, seem to favor growth of *Microbacterium* over other typical spoilage forms.

Poultry.—Coryneforms of various types, often not specified, have been isolated from poultry. Often, such organisms are called "diphtheroids." Such bacteria were reported on the surface of freshly killed chickens, and include genera tentatively identified as *Corynebacterium* and *Microbacterium*. In a survey conducted in a large poultry processing plant in England, Barnes (1960) recovered high numbers of psychrophilic *Cornyebacterium* spp. from the feathers and feet of birds entering the plant. Gram-positive rods isolated from chicken meat in another investigation in England were also similar to *Corynebacterium*. Organisms isolated from turkey giblets from turkeys processed in Iowa were classified into 15 genera. *Corynebacterium* was the most commonly isolated (57/278), consisting of about 20% of the isolates, with *Pseudomonas* (50/278), 17%; *Sarcina* (34/278), 12%; *Micrococcus* (26/278), 9% and *Brevibacterium* (24/278), 8.5%, following in that order. The majority of coryneforms isolated from turkey giblets were unable to grow at 5°C or 45°C, but almost one-third of them could grow at 5°C (Kraft *et al.* 1966; Salzer *et al.* 1967).

Eggs.—Eggs and egg packing materials may be contaminated with *Arthrobacter*; it was found that this genus made up about 13% of the total number of isolates (600) and followed Gram-positive cocci and Gram-negative rods in order of dominance (Board *et al.* 1964). Dust, soil, and fecal material are common sources of contamination of eggs and packing materials.

Liquid egg from different egg breaking plants yielded *Pseudomonas* (46% of isolates) and *Arthrobacter* (27/148, or 18%) as the most prevalent organisms (Kraft *et al.* 1966). *Corynebacterium* were isolated in 3 of 148 cultures examined. The 27 cultures of *Arthrobacter* showed the striking or unique morphological changes associated with coryneforms. During the first 24 hours of incubation of cultures, the rods became coccoid cells which became swollen when inoculated into fresh media and initiated cell division by lengthening, continuing to lengthen until a pair of rods was formed by "snapping" division. Irregular swelling of the rods to form club shaped cells was common. Cell length ranged from about 1 μ to 25 μ as swelling progressed. Toward the end of 5 hours of incubation, the long rods or filaments became fragmented to form short rods. However, it was not certain if the cocci that appeared were a result of fragmentation or just a reversion from longer to shorter, more circular cells after 24-48 hours. Other characteristics of the *Arthrobacter* included a weak staining reaction in general, and Gram-positive and Gram-negative cells of the same morphology. Only a small amount of acid was produced from glucose. None of the *Corynebacterium* or *Arthrobacter* grew at 44°C, but all grew at 15°C within 14 days. The high incidence of coryneforms in liquid egg is emphasized by the observation that these bacteria were among the most common contaminants in 9 of 17 samples examined. In the study on shell eggs and egg packing materials mentioned above, *Arthrobacter* did not cause spoilage of the shell eggs. *Arthrobacter* in liquid egg therefore may not affect keeping quality of the product, but may indicate the possibility of contamination by spoilage organisms present in soil or fecal material. The organisms recovered from liquid egg may have been loosened from the shell at the time the egg was broken, or entered the egg with shell particles.

Vegetables.—Work done on coryneforms isolated from frozen vegetables at Cornell University involved examination of 100 isolates from peas, beans, and corn (Splittstoesser *et al.* 1967). These isolates were compared with known *Corynebacterium, Microbacterium*, and *Arthrobacter* by the use of numerical taxonomy. Six groups, representing 75% of the isolates, resembled members of the family *Corynebacteriaceae*. Gram-positive, catalase-positive rods made up 21% of the total count of peas, 14% on snap beans, and 8% on corn. The six groups contained four resembling *Corynebacterium* and *Microbacterium* similar to those isolated from meats, poultry and dairy products, while the other two came closer to soil *Arthrobacter*. Airborne contamination of the surfaces of processing equipment was believed to be the source of these nonspore-forming bacteria, since they do not survive the heating during the blanching operation. Once introduced they become part of the processing line flora, since counts of thousands per gram on the vegetables indicate that growth must have occurred on the equipment surfaces, the food, or both. Inadequate sanitation is responsible for growth of such organisms. Many of the cultures were not closely related to the described species. The six groups joined at a similarity level of 66%, representing 75% of the vegetable isolates, and this is not an exceptionally high similarity level. The authors stated that they concluded that most of the organisms listed in Bergey's 7th could not have come from frozen vegetables. To quote them directly, "we also have developed a cynicism in that we have become suspicious of publications which report that isolates from natural sources were identified as to species, or even variety, with little or no difficulty" (a lesson for all budding food bacteriologists).

Similar relationships between Bergey's Manual (Breed, *et al.* 1957) and bacterial isolates from foods have been reported by several investigators. A recent study of numerical taxonomy of coryneform bacteria isolated from shrimp grown in ponds revealed much the same conclusion (Vanderzant *et al* 1972). Gram-positive, catalase-positive, nonspore-forming, pleomorphic rods isolated from the shrimp and pond water showed little similarity to type cultures of *Corynebacterium, Arthrobacter, Microbacterium, Propionibacterium*, or *Brevibacterium*. About 30% of the 66 isolates were able to grow at refrigeration temperatures, but their significance in spoilage of

332 FOOD MICROBIOLOGY

shrimp or other seafood is not known, just as it is not known in poultry meat. The investigators concluded that shrimp coryneforms were distinct groups, not studied previously. The same conclusion could be made repeatedly about coryneform bacteria in many kinds of foods.

From the work that has been reported, several genera of coryneforms, particularly *Corynebacterium*, *Microbacterium*, and *Arthrobacter*, and to a lesser extent, *Brevibacterium*, are present in many kinds of food. However, their significance, if any, still remains to be seen in many cases. They may play a possible role in spoilage or ripening of cheese and sausage products, they may serve as indicators of poor hygienic condition of dairy products or equipment, but in other instances they seem to be present merely as contaminants with no special function. In certain cases, as with *Microbacterium* in cured, salted, or packaged meats, their presence depends on the processing operations such as addition of salt or lowering of oxygen pressure, or possible reduction of water activity. This last factor again may be related to salt content of the food. As mentioned in other instances, they represent an indication of high levels of bacterial contamination of product.

BIBLIOGRAPHY

ABD-EL-MALEK, Y., and GIBSON, T. 1952. Studies on the bacteriology of milk. III. The corynebacteria of milk. J. Dairy Res. *19*, 153–159.
ALEXANDER, M. 1967. *In* The Ecology of Soil Bacteria, T. R. G. Gray, and D. Parkinson (Editors). Liverpool Univ. Press, England.
ANAGNOSTOPOULOS, G. C., SEAMAN, A., and WOODBINE, M. 1964. Observations on heat-induced inhibitory and stimulatory factors on *Microbacterium lacticum*. Microbial Inhibitors in Food. 4th Intern. Symp. Food Microbiol. Goteborg, Sweden. Almquist & Wiksell, Uppsala, pp. 353–368.
ANAGNOSTOPOULOS, G. C., SEAMAN, A., and WOODBINE, M. 1966. An attempt to increase the heat resistance of *Microbacterium lacticum*. J. Appl. Bacteriol. *29*, 207–212.
BAIN, N., HODGKISS, W., and SHEWAN, J. M. 1958. The bacteriology of brines used in smoke curing of fish. pp. 103–116. Proc. 2nd Intern. Symp. Food Microbiol. (1957). Dept. Scientific Industrial Res., Food Investigation. H. M. Stationery Office, London.
BARKSDALE, L. 1970. *Corynebacterium diphtheriae* and its relatives. Bacteriol. Rev. *34*, 378–342.
BARLOW, J., and KITCHELL, A. G. 1966. A note on the spoilage of prepacked lamb chops by *Microbacterium thermosphactum*. J. Appl. Bacteriol. *29*, 185–188.
BARNES, E. M. 1960. Bacteriological problems in broiler preparation and storage. Royal Soc. Health J. *80*, 145–148.
BARNES, E. M., and SHRIMPTON, D. H. 1968. The effect of processing and marketing procedures on the bacteriological condition and shelf life of eviscerated turkeys. British Poultry Sci. *9*, 243–251.
BOARD, R. G., AYRES, J. C., KRAFT, A. A., and FORSYTHE, R. H. 1964.

The microbiological contamination of egg shells and egg packing materials. Poultry Sci. *43*, 584–595.

BOUSFIELD, I. J. 1972. A taxonomic study of some coryneform bacteria. J. Gen. Microbiol. *71*, 441–455.

BREED, R. S., MURRAY, E. G. D., and SMITH, N. R. 1957. Bergey's Manual of Determinative Bacteriology, 7th Edition. The Williams & Wilkins Co., Baltimore, Md.

BRIGHTON, W. D. 1966. The dissolved oxygen and respiration rates in pellicle and submerged cultures of *Corynbacterium diphtheriae*. J. Appl. Bacteriol. *29*, 197–206.

BROWNLIE, L. E. 1966. Effect of some environmental factors on psychrophilic microbacteria. J. Appl. Bacteriol. *29*, 447–454.

CLARK, F. E. 1952. The generic classification of the soil corynebacteria. Intern. Bull. Bacteriol. Nomenclature Taxonomy *2*, 45–56.

CONN, H. J., and DIMMICK, I. 1947. Soil bacteria similar in morphology to *Mycobacterium* and *Corynebacterium*. J. Bacteriol. *54*, 291–303.

CUMMINS, C. S., and HARRIS, H. 1956. The chemical composition of the cell wall in some Gram-positive bacteria and its possible value as a taxonomic character. J. Gen. Microbiol. *14*, 583–600.

CUMMINS, C. S., and HARRIS, H. 1959. Taxonomic position of *Arthrobacter*. Nature *184*, 831–832.

DA SILVA, G. A. N. 1964. Taxonomic relationships among certain coryneform bacteria with special reference to the plant pathogens. M. S. Thesis. Iowa State Univ., Ames.

DA SILVA, G. A. N. and HOLT, J. G. 1965. Numerical taxonomy of certain coryneform bacteria. J. Bacteriol. *90*, 921–927.

DAVIDSON, C. M., MOBBS, P., and STUBBS, M. 1968. Some morphological and physiological properties of *Microbacterium thermosphactum*. J. Appl. Bacteriol. *31*, 551–559.

DAVIS, G. H. G., and NEWTON, K. G. 1969. Numerical taxonomy of some named coryneform bacteria. J. General Microbiology *56*, 195–214.

DEMPSTER, J. F. 1968. Distribution of psychrophilic microorganisms in different dairy environments. J. Appl. Bacteriol. *31*, 290–301.

DOETSCH, R. N., and PELCZAR, M. J., JR. 1948. The microbacteria. I. Morphological and physiological characteristics. J. Bacteriol. *56*, 37.

DOETSCH, R. N., and RAKOSKY, J. 1950. Is there a *Microbacterium liquefaciens*? Bacteriol. Proc. *G16*, 38.

DOWDELL, M. J., and BOARD, R. G. 1968. A microbiological survey of British fresh sausage. J. Appl. Bacteriol. *31*, 378–396.

DRAKE, S. D., EVANS, J. B., and NIVEN, C. F., JR. 1958. Microbial flora of packaged frankfurters and their radiation resistance. Food Res. *23*, 291–296.

FARRELL, J., and ROSE, A. H. 1967. Temperature effects on microorganisms. *In* Thermobiology, A. H. Rose (Editor). Academic Press, New York.

FIEDLER, F., SCHLEIFER, K., and KANDLER, O. 1973. Amino acid sequence of the threonine-containing mureins of coryneform bacteria. J. Bacteriol. *113*, 8–17.

GARDNER, G. A. 1966. A selective medium for the enumeration of *Microbacterium thermosphactum* in meat and meat products. J. Appl. Bacteriol. *29*, 455–460.

GARDNER, G. A. 1968. Effects of pasteurization or added sulphite on the microbiology of stored vacuum packed baconburgers. J. Appl. Bacteriol. *31*, 462–478.

GARDNER, G. A. 1969. Physiological and morphological characteristics of *Kurthia zopfii* isolated from meat products. J. Appl. Bacteriol. *32*, 371–380.

GARDNER, G. A., and CARSON, A. W. 1967. Relationship between carbon dioxide production and growth of pure strains of bacteria on porcine muscle. J. Appl. Bacteriol. *30*, 500–510.

GARDNER, G. A., CARSON, A. W., and PATTON, J. 1967. Bacteriology of prepacked pork with reference to the gas composition within the pack. J. Appl. Bacteriol. *30*, 321-333.

GIOLITTI, G., CANTOTI, C. A., BIANCHI, M. A., and RENON, P. 1971. Microbiology and chemical changes in raw hams of Italian type. J. Appl. Bacteriol. *34*, 51-61.

GORDON, R. E. 1966. Some strains in search of a genus—*Corynebacterium, Mycobacterium, Nocardia* or what? J. Gen. Microbiol. *43*, 329-343.

HARRIGAN, W. F. 1966. The nutritional requirements and biochemical reactions of *Corynebacterium bovis*. J. Appl. Bacteriol. *29*, 380-394.

HARRINGTON, B. J. 1966. A numerical taxonomical study of some corynebacteria and related organisms. J. Gen. Microbiol. *45*, 31-40.

HERMAN, G. J., and WEAVER, R. E. 1970. Corynebacterium. *In* Manual of Clinical Microbiology. J. E. Blair, E. H. Lennette, and J. P. Truant (Editors). The Williams & Wilkins Co., Baltimore, Md.

HILL, L. R. 1966. An index to deoxyribonucleic acid base compositions of bacterial species. J. Gen. Microbiol. *44*, 419-437.

JAYE, M., KITTAKA, R. S., and ORDAL, Z. J. 1962. The effect of temperature and packaging material on the storage life and bacterial flora of ground beef. Food Technol. *16*, 95-98.

JAYNE-WILLIAMS, D. J., and SKERMAN, T. M. 1966. Comparative studies on coryneform bacteria from milk and dairy sources. J. Appl. Bacteriol. *29*, 72-92.

JENSEN, H. L. 1952. The coryneform bacteria. Ann. Rev. Microbiol. *6*, 77-90.

JENSEN, H. G. 1966. Some introductory remarks on the coryneform bacteria. J. Appl. Bacteriol. *29*, 13-16.

KEDDIE, R. M., LEASK, B. G. S., and GRAINGER, J. M. 1966. A comparison of coryneform bacteria from soil and herbage; cell wall composition and nutrition. J. Appl. Bacteriol. *29*, 17-43.

KRAFT, A. A. 1951. Private communication quoted by Ayres, J. C. 1951. Some bacteriological aspects of spoilage of self-service meats. Proc. 3rd Res. Conf. Am. Meat. Inst. Found., Chicago, pp. 39-53.

KRAFT, A. A., et al. 1966. Coryneform bacteria in poultry, eggs and meat. J. Appl. Bacteriol. *29*, 161-166.

LELLIOTT, R. A. 1966. The plant pathogenic coryneform bacteria. J. Appl. Bacteriol. *29*, 114-118.

LISTON, J., and SHEWAN, J. M. 1958. Bacteria brought into brines on fish. Proc. 2nd Intern. Symp. Food Microbiol. 1957. The Microbiology of Fish & Meat Curing Brines. Dept. of Sci. Ind. Res., Food Invest. H. M. Stationery Office, London.

MCBRIDE, M. E., MONTES, L. F., and KNOX, J. M. 1970. The characterization of fluorescent skin diphtheroids. Can. J. Microbiol. *16*, 941-946.

MCLEAN, R. A., and SULZBACHER, W. L. 1953. *Microbacterium thermosphactum*, spec. nov; a non-heat resistant bacterium from fresh pork sausage. J. Bacteriol. *65*, 428-433.

MORRIS, J. G. 1960. Studies on the metabolism of *Arthrobacter globiformis*. J. Gen. Microbiol. *22*, 564-582.

MOSSEL, D. A. A. 1971. Physiological and metabolic attributes of microbial groups associated with foods. J. Appl. Bacteriol. *34*, 95-118.

MULDER, E. G. et al. 1966. The relationship between *Brevibacterium linens* and bacteria of the genus *Arthrobacter*. J. Appl. Bacteriol. *29*, 44-71.

MULDER, E. G., and ANTHEUNISSE, J. 1963. Morphology, physiology and ecology of *Arthrobacter*. Ann. Inst. Pasteur. Paris *105*, 46-74. (French)

MULDER, E. G., ANTHEUNISSE, J., and CROMBACH, W. H. J. 1971. Microbial aspects of pollution in the food and dairy industries. *In* Microbial Aspects of Pollution, G. Sykes, and F. A. Skinner (Editors). Academic Press, New York.

MULDER, E. G., and ZEVENHUIZEN, L. P. T. 1967. Coryneform bacteria of the *Arthrobacter* type and their reserve material. Arch. Mikrobiol. *59*, 345-354.

OWENS, J. D., and KEDDIE, R. M. 1968. A note on the vitamin requirements of some coryneform bacteria from soil and herbage. J. Appl. Bacteriol. *31*, 344-348.

OWENS, J. D., and KEDDIE, R. M. 1969. The nitrogen nutrition of soil and herbage coryneform bacteria. J. Appl. Bacteriol. *32*, 338-347.

PIERSON, M. D., COLLINS-THOMPSON, D. L., and ORDAL, Z. J. 1970. Microbiological, sensory, and pigment changes of aerobically packaged beef. Food Technol. *24*, 129-133.

ROBINSON, K. 1966A. Some observations on the taxonomy of the genus *Microbacterium*. I. Cultural and physiological reactions and heat resistance. J. Appl. Bacteriol. *29*, 607-615.

ROBINSON, K. 1966B. Some observations on the taxonomy of the genus *Microbacterium*. II. Cell wall analysis, gel electrophoresis, and serology. J. Appl. Bacteriol. *29*, 616-624.

ROBINSON, K. 1966C. An examination of *Corynebacterium* spp. by gel electrophoresis. J. Appl. Bacteriol. *29*, 179-184.

ROBINSON, K. 1968. The use of cell wall analysis and gel electrophoresis for the identification of coryneform bacteria. *In* Identification Methods for Microbiologists, Part B. G. M. Gibbs, and D. A. Shapton (Editors). Academic Press, New York.

ROGERS, R. E., and McCLESKEY, C. S. 1957. Bacteriological quality of ground beef in retail markets. Food Technol. *11*, 318-320.

ROTH, N. G., and WHEATON, R. B. 1962. Continuity of psychrophilic and mesophilic growth characteristics in the genus *Arthrobacter*. J. Bacteriol. *83*, 551-555.

SALZER, R. H., KRAFT, A. A., and AYRES, J. C. 1967. Microorganisms isolated from turkey giblets. Poultry Sci. *46*, 611-615.

SCHEFFERLE, H. 1966. Coryneform bacteria in deep poultry litter. J. Appl. Bacteriol. *29*, 147-160.

SCHLEIFER, K. H. 1970. The murein types in the genus *Microbacterium*. Mikrobiol. *71*, 271-282. (German)

SCHLEIFER, K. H., and KANDLER, O. 1972. Peptidoglycan types of bacterial cell walls and their taxonomic implications. Bacteriol. Rev. *36*, 407-477.

SHAW, B. G., and SHEWAN, J. M. 1968. Psychrophilic spoilage bacteria of fish. J. Appl. Bacteriol. *31*, 89-96.

SHAW, N., and STEAD, D. 1970. A study of the lipid composition of *Microbacterium thermosphactum* as a guide to its taxonomy. J. Appl. Bacteriol. *33*, 470-473.

SIEBURTH, J. MCN. 1965. Hiemal development of a psychrophilic bacterial flora in a temperate estuary. Bacteriol. Proc. *G14*, 16, Am. Society Microbiology, Washington, D.C.

SIEBURTH, J. MCN. 1966. "Aggluticidin," an *Arthrobacter*-inhibitor produced by marine pseudomonads. Bacteriol. Proc. *G32*, 21, Am. Society Microbiology, Washington, D.C.

SIEBURTH, J. MCN. 1967. Inhibition and agglutination of *Arthrobacters* by Pseudomonads. J. Bacteriol. *93*, 1911-1916.

SKERMAN, T. M., and JAYNE-WILLIAMS, D. J. 1966. Nutrition of coryneform bacteria from milk and dairy sources. J. Appl. Bacteriol. *29*, 167-178.

SMITH, J. E. 1966. *Corynebacterium* species as animal pathogens. J. Appl. Bacteriol. *29*, 119-130.

SNEATH, P. H., and COWAN, S. T. 1958. An electro-taxonomic survey of bacteria. J. Gen. Microbiol. *19*, 551-565.

SPECK, M. L. 1943. A study of the genus *Microbacterium*. J. Dairy Sci. *26*, 533-543.

SPLITTSTOESSER, D. F. 1970. Predominant organisms on raw plant foods. J. Milk Food Technol. *33*, 500-505.

SPLITTSTOESSER, D. F., WEXLER, M., WHITE, J., and COLWELL, R. R. 1967. Numerical taxonomy of gram-positive and catalase positive rods isolated from frozen vegetables. Appl. Microbiol. *15*, 158-162.

STEVENSON, I. L. 1961. Growth studies on *Arthrobacter globiformis*. Can. J. Microbiol. *7*, 569-575.

STEVENSON, I. L. 1963. Some observations on the so-called "cystites" of the genus *Arthrobacter*. Can. J. Microbiol. *9*, 467-472.

STEVENSON, I. L. 1967. Utilization of aromatic hydrocarbons by *Arthrobacter* spp. Can. J. Microbiol. *13*, 205-211.

SULZBACHER, W. L., and MCLEAN, R. A. 1951. The bacterial flora of fresh pork sausage. Food Technol. *5*, 7-8.

SUNDMAN, V. 1958. Morphological comparison of some *Arthrobacter* species. Can. J. Microbiol. *4*, 221-224.

THOMAS, S. B., DRUCE, R. G., and KING, K. P. 1966. The microflora of poorly cleansed farm dairy equipment. J. Appl. Bacteriol. *29*, 409-422.

THOMAS, S. B., DRUCE, R. G., PETERS, G. J., and GRIFFITHS, D. G. 1967. Incidence and significance of thermoduric bacteria in farm milk supplies: a reappraisal and review. J. Appl. Bacteriol. *30*, 265-298.

TOMLINSON, A. J. H. 1966. Human pathogenic coryneform bacteria: their differentiation and significance in public health today. J. Appl. Bacteriol. *29*, 131-137.

VANDERZANT, C., JUDKINS, P. W., NICKELSON, R., and FITZHUGH, H. A., JR. 1972. Numerical taxonomy of coryneform bacteria isolated from pond-reared shrimp (*Penaeus aztecus*) and pond water. Appl. Microbiol. *23*, 38-45.

VANDERZANT, C., MATTHYS, A. W., and COBB, B. F., III. 1973. Microbiological, chemical and organoleptic characteristics of frozen breaded shrimp. J. Milk Food Technol. *36*, 253-261.

VANDERZANT, C., and NICKELSON, R. 1969. A microbiological examination of muscle tissue of beef, pork, and lamb carcasses. J. Milk Food Technol. *32*, 357-361.

VELDKAMP, H. 1970. Saprophytic coryneform bacteria. Ann. Rev. Microbiol. *24*, 209-240.

VELDKAMP, H. 1970. Enrichment cultures of prokaryotic organisms. *In* Methods in Microbiology, Vol. 3a, J. R. Norris, and D. W. Ribbons (Editors). Academic Press, New York.

VELDKAMP, H., VENEMA, P. A. A., HARDER, W., and KONINGS, W. N. 1966. Production of riboflavin by *Arthrobacter globiformis*. J. Appl. Bacteriol. *29*, 107-113.

WEIDEMANN, J. R. 1965. A note on the microflora of beef muscle stored in nitrogen at $0°$. J. Appl. Bacteriol. *28*, 365-367.

WERNER, H. 1966. The gram-positive nonsporing anaerobic bacteria of the human intestine with particular reference to Corynebacteria and Bifidobacteria. J. Appl. Bacteriol. *29*, 138-146.

WOLIN, E. F., EVANS, J. B., and NIVEN, C. F. 1957. The microbiology of fresh and irradiated beef. Food Res. *22*, 682-686.

ZAGALLO, A. C., and WANG, C. H. 1967. Comparative carbohydrate catabolism in *Corynebacteria*. J. Gen. Microbiol. *47*, 347-357.

CHAPTER 11

D. F. Splittstoesser | Gram-negative Nonspore-forming Rods

In this chapter we will be considering the influence of the organisms on food preservation. Of the 100 or so presently-recognized genera of Gram-negative rods, only a relatively few are significant food spoilage organisms. The most troublesome group (at least in societies which rely heavily on refrigeration) is the aerobes that grow relatively rapidly at low temperatures. Because of their importance, much of this chapter has been devoted to them—the psychrophiles. Less attention has been given to the second large group, the *Enterobacteriaceae*, because they may present less of a problem with respect to food spoilage, and because certain members of the family have been dealt with in other chapters.

PSYCHROPHILES

Certain of the Gram-negative rods have the ability to grow at low temperatures and have long been recognized as the principal spoilage agents of refrigerated foods such as red meats, poultry, fish, eggs, and dairy products. The organisms are common contaminants of fresh and sea water and once were commonly referred to as water bacteria. They are widely distributed, however, with soil, plants, and animals also serving as natural habitats.

Classification

The separation of these organisms in early editions of Bergey's Manual of Determinative Bacteriology was based primarily on color. Members of the genus *Pseudomonas* produced a water-soluble green pigment, *Flavobacterium* formed yellow colonies, and species of *Achromobacter* were nonpigmented. In later schemes (Breed *et al.* 1948), flagellation took precedence over pigmentation. The genus *Pseudomonas* then contained both pigmented and nonpigmented organisms that possessed polar flagella (a few species were nonmotile), while the achromobacters had peritrichous flagella or were nonmotile.

More recent classifications are as follows.

Pseudomonas.—According to current taxonomy, members of the genus are Gram-negative rods that possess monotrichous or multitrichous, polar flagella. Their energy-yielding metabolism is respiratory; thus they are obligate aerobes. Most species possess cytochrome c, a property indicated by a positive oxidase test.

In the 8th Edition of Bergey's Manual (Buchanan and Gibbons 1974), separation of species is based largely on the taxonomic study of Stanier *et al.* (1966). Major characters used for species differentiation are the requirement for growth factors, the accumulation of poly-β-hydroxybutyrate as an intracellular carbon reserve, and the ability to use DL-arginine and betaine as sole sources of carbon. Further breakdown is based partly on pigmentation, the use of various carbon sources for growth, growth temperature, and the arginine dihydrolase reaction. The latter is the ability to break down arginine to ornithine, carbon dioxide, and ammonia under anaerobic conditions. The psychrophiles described in this edition of the Manual are biotypes of *P. fluorescens* and closely related species. The organisms produce fluorescent pigments and give a positive arginine dihydrolase reaction. They do not accumulate poly-β-hydroxybutyrate. Certain well known psychrophilic pseudomonads, *e.g.*, *P. fragi*, are not described in the 8th Edition because it was concluded that information about them was not sufficiently complete.

Numerous microbiologists have used the system of Shewan *et al.* (1960) for the classification of psychrophilic, Gram-negative rods isolated from a variety of foods. This scheme, originally developed for marine organisms, defined *Pseudomonas* as motile rods possessing polar flagella. They were oxidase positive and were resistant to 2.5 IU of penicillin. The genus was subdivided into four groups on the bases of the oxidation-fermentation test of Hugh and Leifson (1953) and pigmentation. Groups I and II were oxidative according to the Hugh and Leifson test, Group III gave an alkaline reaction, and Group IV produced no detectable change. Group I contained the organisms that produced a diffusible green fluorescent pigment; the members of the other groups were non-pigmented.

Corlett *et al.* (1965) followed the classification of Shewan but used the computer analysis of a large battery of tests (obtained by replica plating) to identify the psychrophilic organisms present in ground beef and Dover sole. Identification was based on resistance to seven antibiotics, growth on selective media such as SS and Staphylococcus 110 agars, and on colonial appearance and cell morphology. It was concluded that this procedure permitted the rapid identification of psychrophiles as well as other microbial contaminants of foods.

Achromobacter.—Although no longer recognized as a genus in the 8th Edition of Bergey's Manual, the term *Achromobacter* is retained here because this is the nomenclature used in much of the literature on food spoilage.

As stated before, the genus was originally reserved for the non-pigmented rods and in the 7th Edition of Bergey's Manual (Breed *et al.* 1957) achromobacters were described as Gram-negative organisms that were nonmotile or motile by means of peritrichous flagella. A property of some species was that they did not produce acids from carbohydrates.

Many of the isolates designated as achromobacters from fish (Shewan *et al.* 1960), poultry (Barnes and Shrimpton 1958), meats (Brown and Weidemann 1958), and other refrigerated foods have been nonmotile, short, almost coccoidal rods. Additional features used to distinguish them from nonpigmented pseudomonads have been their sensitivity to 2.5 IU of penicillin (Shewan *et al.* 1960) and their failure to metabolize arginine under anaerobic conditions (Thornley 1960). The group has given a variable oxidase test.

The achromobacters from foods were not closely related to the type species in Bergey's Manual, *A. liquefaciens*, in that this organism was described as being motile with peritrichous flagella (Breed *et al.* 1957). Questions, therefore, were raised about the classification of this genus and after an extensive taxonomic study on various groups (Thornley 1967), it was proposed that many species of *Achromobacter* should be placed in the *Acinetobacter*, a genus proposed by Brisou and Prévot (1954).

Acinetobacter-Moraxella.—These genera are composed of nonmotile, oxidative, Gram-negative rods that are almost spherical in shape. *Acinetobacter* and *Moraxella* are differentiated in that the former gives a positive oxidase test and is more resistant to penicillin (Buchanan and Gibbons 1974). It is likely that many of the organisms that contaminate foods are quite different from those described in the 8th Edition of Bergey's Manual; for example, while it is reported that strains of *Acinetobacter calcoaceticus*, the single described species, are resistant to penicillin, the nonmotile, *Acinetobacter*-like bacteria from fish and other chilled foods are usually very sensitive, being inhibited by as little as 2.5 IU (Shewan *et al.* 1954). Furthermore, although four of the five described species of *Moraxella* are pathogenic for warm-blooded animals, it is likely that the psychrophiles from foods that are oxidase-negative, and therefore could be classified as species of *Moraxella*, are strictly saprophytic.

Alcaligenes.—Members of the genus are short, motile rods that possess peritrichous flagella. *Alcaligenes* produces little or no acid from carbohydrates and causes many media, for example litmus milk, to turn alkaline; hence the name. The organisms resemble many of the other psychrophiles in that they are obligate aerobes, give a positive oxidase test, and generally are nonpigmented. It has been proposed that some of the motile organisms formerly classified as Achromobacter species should be included in this genus.

Flavobacterium.—The Gram-negative, psychrophilic rods that form yellow-pigmented colonies have generally been designated as species of *Flavobacterium* by the food microbiologist. The genus is not homogeneous, however, and it is believed that some species may be more closely related to the genus *Cytophaga* (Weeks 1969).

The flavobacteria are aerobic organisms that are either nonmotile or possess peritrichous flagella. It is the nonmotile species that may be confused with *Cytophaga* since the creeping motility of the latter may be difficult to detect. The cytophagas, which also are found in soil, water, and marine environments, have the ability to decompose a variety of complex polysaccharides such as agar, chitin and cellulose.

Another aerobic, Gram-negative rod that produces a water-insoluble, yellow pigment is *Xanthomonas*. The organisms are plant pathogens and thus could be encountered when vegetables and other edible plants are cultured. *Xanthomonas* might be confused with *Flavobacterium* if identification were based merely on the Gram stain and pigmentation of colonies. The genus is closely related to *Pseudomonas* in that the rods possess a polar flagellum, a property that permits their differentiation from *Flavobacterium*.

Food Spoilage

The manifestations of spoilage resulting from the growth of Gram-negative, psychrophilic rods vary with the food type.

Fish.—Much of the work on spoilage at low temperatures has been with fish and other seafoods, perhaps because they are some of the more vulnerable foods. Fish undergo a number of visual changes as a result of microbial growth: various discolorations may develop, the amount of slime increases, the flesh softens, and juice may be exuded (Reay and Shewan 1969). A sweetish, fruity odor often is the first sign of spoilage; this changes to a "stale fish" and finally to a putrid odor as microbial growth progresses.

At the time of obvious spoilage, bacterial counts (predominantly nonpigmented pseudomonads) are in the range of 10^6 to 10^7 per gram (Reay and Shewan 1969; Thomson *et al.* 1974). The microorganisms responsible for the most objectionable changes, however, make up only 10 to 20% of this population. This percentage of spoilage organisms changes very little during storage. For example, Shaw and Shewan (1968) found that 16% of the organisms from fresh cod had the ability to produce strong off-odors when inoculated into sterile fish press juice or raw muscle; after the cod had been stored 12 days on ice, the proportion was about the same, 13%.

Studies in which pure cultures have been inoculated into sterile fish muscle have shown that individual isolates differ considerably in the odors that they generate (Herbert et al. 1971). Some of the volatiles were fruity while others consisted primarily of ammonia or reduced sulfur compounds. Many of the psychrophilic spoilage organisms could not produce the odor of stale fish by the reduction of trimethylamine oxide to trimethylamine. This observation is of practical interest because the testing for trimethylamine has been advocated as a means of evaluating the wholesomeness of fish.

The compounds responsible for the various off-odors of fish have been identified by using gas-liquid chromatography and other analytical procedures. The volatile substances produced by *Pseudomonas fragi* when grown on sterile fish muscle included dimethyl sulfide, acetaldehyde, ethyl acetate, ethanol, dimethyl disulfide, methyl mercaptan, ethyl butyrate, ethyl hexanoate, and butanone (Miller et al. 1973A). The fruity odor that is evident during the early stages of fish spoilage was attributed to the ethyl acetate, ethyl butyrate and ethyl hexanoate.

Other species of Gram-negative rods generate greater quantities of sulfur-containing compounds and thus are believed to be responsible for the putrid odors that develop during the more advanced stages of spoilage. Volatiles produced by *P. putrefaciens*, *P. fluorescens*, and an isolate designated as *Achromobacter* included methyl mercaptan, dimethyl disulfide, dimethyl trisulfide, 3-methyl-1-butanol, and trimethylamine (Miller et al. 1973B).

Red Meats and Poultry.—The initial signs of spoilage of refrigerated poultry and red meats are off-odors followed by the production of slime, rancidity, and various discolorations. Abnormal odors may become evident when the bacterial population reaches a level of 10^7 per cm^2; slime usually develops before the count reaches 10^8 per cm^2 (Ayres 1960; Barnes and Thornley 1966).

The Gram-negative spoilage organisms are quite similar to those found on fish and numerous workers have used the scheme of Shewan et al. (1960) for their classification. Most studies have indicated that pseudomonads and the nonmotile rods presently designated as *Acinetobacter* are the predominant spoilage organisms.

Although these foods differ significantly from fish as substrates, similar changes have been observed during spoilage. When sterile poultry breast muscle was inoculated with different cultures of *Pseudomonas*, for example, the same fruity and sulfide-like odors that develop on fish were produced (McMeekin 1975). Unlike the findings with fish, however, the percentage of off-odor-producing bacteria increased during the period the poultry was stored; from 16% at 0 days to 80% after 16 days at 2°C.

Eggs.—Psychrophilic, Gram-negative rods are the chief spoilage bacteria of eggs stored at refrigeration temperatures. The types of rot often are characterized according to color: black, red, yellow, green, and colorless rots are described in the literature. Studies on 228 cultures from 81 abnormal eggs revealed that the most common spoilage organisms were species of *Alcaligenes*, *Pseudomonas*, *Flavobacterium*, and *Achromobacterium* (Florian and Trussell 1957). Experiments with these isolates showed that many were incapable of penetrating the inner shell membrane when the membrane was intact. Two species that were able to penetrate the membrane and cause an infection were *Alcaligenes bookeri*, a cause of yellow rot, and *Pseudomonas fluorescens* which is responsible for a fluorescent green rot. The importance of these organisms was indicated by the fact that 50% of the infected eggs yielded *A. bookeri* while 25% were spoiled by *P. fluorescens*.

Dairy Products.—Pseudomonads and other Gram-negative rods have been responsible for a number of defects of milk and milk products (Foster *et al.* 1957; Frazier 1967; and Witter 1961).

The organisms have been a problem in both pasteurized and raw milk. Although psychrophiles possess little heat resistance, they often are present in low numbers in pasteurized milk, probably because of re-contamination following the heat treatment. The most common defect is a bitter flavor due to proteolysis. This may occur when pasteurized milk has been held for an extended period or when bulk raw milk has been collected or stored under unsanitary conditions. Less common types of milk spoilage are brown, blue or yellow discolorations due to the growth of different species of *Pseudomonas*, and ropiness resulting from slime produced by *Alcaligenes viscolactis* (a species described in the 7th but not the 8th Edition of Bergey's Manual).

Many of the Gram-negative rods possess lipases that can cause rancidity of cream and butter. Other off-flavors that may be produced in milkfat have been described as fishy, putrid, and skunk-like. Various species of *Pseudomonas* are often the causal organisms. Regular butter is less susceptible to spoilage than is cream or unsalted butter because the psychrophiles generally are inhibited by the concentration of sodium chloride present in the aqueous phase, usually 15% or higher. When growth of the bacteria has occurred in cream, the taints will be transmitted to the butter. The use of pasteurized, sweet cream in butter making has reduced this problem.

Cottage cheese is the one cheese susceptible to spoilage by the psychrophilic rods. Problems occur when the acid content of the cheese is low, usually because of a poor starter culture. Spoilage is generally manifest by a gelatinous-slimy growth on the curd. Species that have been isolated from spoiled cottage cheese include *Achromobacter butyri*, *A. eurydice*, *Alcaligenes metalcaligenes*, *Pseudomonas desmolyticum*, *P. fragi*, *P. fluorescens*, and *P. tralucida* (Bonner and Harmon 1957).

Frozen Foods.—Under certain conditions relatively low populations of psychrophilic pseudomonads can adversely affect the quality of frozen foods. It has been observed that chicken pies held at defrost temperatures of 35, 40 and 50° F became organoleptically unacceptable even though the aerobic plate counts were under 10^4 per gram (Peterson 1961). The liberation of extracellular enzymes, particularly proteinases and amylases, was found to be responsible. Studies on a strain of *Pseudomonas fluorescens* isolated from a frozen chicken pie showed that cell washings and cell-free extracts prepared from this organism could induce some of the same changes. Thus these preparations were able to reduce the viscosity of starch gels, impart a curdled appearance to gravy, cause a separation of fat, and discolor the meat and vegetables. It was concluded that "once the enzymes of psychrophilic bacteria are produced and liberated in the product, lowering the temperature will decrease the rate at which damage occurs to the product, but will not stop it and definitely will not remedy damage already done."

Most frozen foods undoubtedly are contaminated with psychrophilic rods and some may yield relatively high populations. For example, about 18% of the isolates from samples of frozen peas having a median aerobic plate count of 11×10^4 per gram were identified as species of *Pseudomonas*, *Achromobacter*, and *Flavobacterium* (Splittstoesser and Gadjo 1966). The effect of their presence on the quality of frozen vegetables, as well as on the quality of most other frozen foods, has not been established.

Miscellaneous Foods.—Various other foods are subject to spoilage by these Gram-negative rods. Certain halophilic strains of *Pseudomonas* have been implicated in the spoilage of salt fish; others have been isolated from defective hams and dried beef. Plant pathogenic pseudomonads cause spoilage of vegetables. Pectolytic achromobacters were one of the groups found responsible for the softening of ripe olives (Vaughn *et al.* 1969). Growth of psychrophilic strains of *Pseudomonas*, *Flavobacterium*, and

Achromobacter in maple sap results in cloudy syrup and lower sap yields; the latter because of plugged tree tap holes (Sheneman and Costilow 1959).

Enumeration

A variety of media have been found to be satisfactory for the enumeration of these organisms. Tryptone soy, heart infusion, and tryptone glucose yeast extract agars are some of the types that often are used (Barnes and Shrimpton 1958; Heather and Vanderzant 1957). When seafoods are cultured, sodium chloride may be added to the medium (Corlett *et al.* 1965) or sea water may be used (Thomson *et al.* 1974) to encourage growth of marine psychrophiles. Barnes and Thornley (1966) have recommended the use of surface plating techniques rather than pour plates because some species are destroyed by temperatures below that of melted agar.

Numerous incubation temperatures and times have been used. An incubation of 30°C for 48 hr will permit the recovery of most psychrophiles (Ingraham and Stokes 1959) as well as many of the mesophilic organisms. Incubation conditions that select for psychrophilic species are 14 days at 1°C (Barnes and Shrimpton 1958) and 10 days at 5°C (Heather and Vanderzant 1957).

Physiology

The discussion of the group's physiology will be restricted to some of the properties of importance to food preservation; in particular, some of the factors affecting growth and death.

Growth.—Some factors affecting growth are the following.

Temperature.—Most psychrophilic, Gram-negative rods can produce appreciable growth within 14 days at 0°C (Ingraham and Stokes 1959). The minimum temperature permitting growth is about -10°C. Many of the organisms cannot grow at this temperature, however. In studies on pseudomonads, *Moraxella*-like organisms, and flavobacteria, all grew at 5.25°C but only certain pseudomonads and moraxellas produced detectable growth within 20 days at 1.1° and -1.3°C (Shaw and Shewan 1968). *Pseudomonas* species belonging to Groups III and IV (see page 338) grew most rapidly at the lower temperature.

The optimum growth temperature when based on that which affords fastest growth is usually between 25 and 30°C. If the criterion is maximum cell yields, the optimum temperature range would be somewhat lower.

The maximum growth temperature for these organisms is often below 37°C and many strains fail to grow at 32° or 34°C. The maximum as well as the minimum temperature may be influenced by the physiological state of the cell. To cite an example, cultures of *Pseudomonas fluorescens*, *P. fragi*, and *P. putrefaciens* subjected to a sub-lethal heat treatment gave significantly lower viable counts at incubation temperatures of 5 and 32°C as compared to 25°C than did nonstressed, control cultures (Heather and Vanderzant 1957).

Available Water.—While most Gram-negative psychrophiles require relatively high levels of available water—an a_w of 0.96 to 0.97—some strains apparently can grow under conditions of higher osmotic tension. Proof of this is that pseudomonads have been responsible for the spoilage of salt fish, and various marine *Flavobacterium* species can be cultured in broth containing 6% sodium chloride.

Wodzinski and Frazier (1960, 1961A, 1961B) observed that certain conditions must be favorable before an organism can grow at its lowest potential a_w. A strain of *Pseudomonas fluorescens*, having an optimum a_w of 0.9989, could only grow at a_w 0.9650 when the pH of the medium and the incubation temperature were most favorable— pH 7.0 and 30°C. Changing the pH and temperature when the organism was cultured on low a_w media resulted both in an extended lag phase and a longer generation time. Their research also showed an interaction with the gaseous environment: increasing the partial pressure of carbon dioxide to 5% or reducing that of oxygen to 10% enhanced growth in the low a_w media.

Oxygen.—The pseudomonads and other Gram-negative psychrophilic rods are obligate aerobes; oxygen is required for their growth. An exception to this is that when other electron acceptors such as nitrate are present some of the organisms can grow anaerobically.

Growth of the aerobic spoilage organisms can be controlled by vacuum packaging or the displacement of air with an inert gas. As an example, pseudomonads which predominated on poultry stored in unsealed bags under refrigeration were replaced by facultative *Enterobacter* species when the poultry was held in air-evacuated, sealed pouches (Arafa and Chen 1975).

pH.—Many of the low-temperature organisms prefer a pH near neutrality and numerous strains are completely inhibited by a reaction that is only moderately acid. For example, four species of *Pseudomonas*, two *Achromobacters*, and one *Alcaligenes* isolated from spoiled cottage cheese failed to grow in pH 5.2 media; only *P. desmolyticum* and *P. tralucida* grew in media adjusted to pH 5.4 (Bonner and Harmon 1957). Heather and Vanderzant (1957) found that heat-stressed pseudomonads were more sensitive to an adverse pH, both acid and alkaline, than were nonheated controls.

The studies of Barnes and Impey (1968) illustrate a practical effect of pH. Sterile chicken breast and leg muscle both supported the growth of *Pseudomonas*, while *Acinetobacter* strains grew only on leg muscle. The explanation was that the leg muscle had a pH of 6.4 to 6.7 while the breast muscle was considerably more acid, having a pH of 5.7 to 5.9.

Death.—The following are some factors affecting death.

Heat Resistance.—These organisms possess little thermal resistance; therefore, their presence in foods that have been pasteurized, or subjected to some other heat treatment such as a blanch, generally reflects contamination following this process.

Bonner and Harmon (1957) studied the resistance of two achromobacters, one alcaligenes, and four pseudomonads isolated from spoiled cottage cheese. In one trial in which 24 hr skim milk cultures were heated at 61.7°C for 5, 10, and 20 min, all were destroyed in 20 min and only *Pseudomonas tralucida* survived 10 min at this temperature. The most sensitive species were *P. fluorescens* and *P. desmolyticum* which failed to survive five min at this temperature. In other studies, the ability of the organisms to survive 48.9°C for 15 min, a process used in the manufacture of cottage cheese, was determined. The heating medium was found to be very important in that the seven species survived when heated in skimmilk of pH 6.7 while all were destroyed in whey of pH 4.5.

The report of Barnes and Thornley (1966) that psychrophilic species may be destroyed by the temperature of melted agar is in agreement with the observation of Hayes (1963) that many marine *Flavobacteria* isolates failed to survive 45°C for 15 min.

Desiccation.—One would expect some Gram-negative rods to survive the different drying processes, even though they are relatively sensitive to dehydration (Silverman and Goldblith 1965). This is borne out by studies on various dehydrated foods. Samples of onion and garlic dried from the non-frozen state gave counts ranging from 20 to over 200×10^3 psychrophiles per gram (Vaughn 1970). Presumably many of these bacteria were Gram-negative rods. Freeze-dried shrimp yielded few Gram-negative, oxidase-positive rods immediately after rehydration; following 14 days at 4°C, however, these organisms made up 87% of the aerobic flora (Pablo *et al.* 1967). This was evidence that a low number of the psychrophiles had survived the freeze drying process.

Radiation.—Numerous studies have been concerned with extending the shelf-life of poultry, fish and other refrigerated foods by exposing them to ionizing radiations. In general the results have indicated that psychrophilic rods possess little radiation resistance.

For example the treatment of poultry with 0.5 megarads (Mrad) reduced the viable microbial population about four $\log_{10}$ cycles, which would indicate an average D-value for the mixed microflora of about 0.125 Mrad (Thornley *et al.* 1960). Some data indicate the nonmotile, nonpigmented rods (formerly classified as achromobacters) to be more resistant to gamma rays than the pseudomonads. Thus, before irradiation the microflora of Dover sole consisted of 60% pseudomonads, 16% flavobacteria, and 7.4% achromobacters; after exposure to 0.1 Mrad, the achromobacters made up 48% of the population, the flavobacteria 1.3% and the pseudomonads under 1% (Corlett *et al.* 1965A). Low dose gamma irradiation can result in unconventional spoilage of refrigerated foods because of the elimination of the pseudomonads.

Psychrophilic rods also have a low resistance to ultraviolet rays, a property that has been capitalized upon by the maple syrup industry. By using the sun's rays, bacterial build-up in the sap can be suppressed. In one study, maple syrup inoculated with a mixture of two *Pseudomonas* and two *Flavobacterium* cultures showed a reduction in viable microorganisms from 30,000 to under 10 per ml when exposed to sunlight for 3 to 6 hours. Exposure was accomplished by using containers and tubing made of polyethylene or other plastics (Frank and Willits 1960).

Antibiotics.—The work of Shewan *et al.* (1954) using antibiotic discs indicated that the pseudomonads were less sensitive to penicillin, streptomycin, and chloramphenicol than were the nonmotile organisms, classified at that time as achromobacters. Their data also showed the pigmented pseudomonads to be more resistant to oxytetracycline than the nonpigmented species.

Corlett *et al.* (1965) tested reference cultures against eight antibiotics: penicillin, 3 IU/ml; tylosin, 10 µg/ml; vancomycin, 10 µg/ml; streptomycin, 10 µg/ml; neomycin, 75 µg/ml; colistin, 100 µg/ml; oxytetracycline, 5 µg/ml; and chloramphenicol, 10 µg/ml. The pseudomonads were resistant to penicillin and vancomycin and were sensitive to neomycin and colistin; the other antibiotics gave variable responses. The achromobacters were resistant to tylosin, vancomycin, oxytetracycline and penicillin (partially), and were sensitive to the other four compounds. The single *Flavobacterium* tested, *F. capsulatum*, was resistant to all but neomycin and oxytetracycline. The two cytophagas were sensitive to all but vancomycin which gave variable results.

The treatment of refrigerated poultry and fish with tetracycline antibiotics alters the spoilage flora. With poultry, the normally more-sensitive, nonpigmented pseudomonads were replaced by pigmented strains as well as by nonmotile achromobacters (Thornley *et al.* 1960). Similar observations have been made with fish. Lee and Sinnhuber (1967) found the pseudomonads that contaminate ocean perch to be more readily inhibited by chlortetracycline than the flavobacteria and achromobacters. After storage of antibiotic-treated fish at 7°C, however, organisms developed that could tolerate 500 ppm chlortetracycline.

Sanitizers.—Psychrophiles possess little resistance to chlorine and, as a result, hypochlorites and other forms have been widely used for their control. Several studies on pseudomonads, achromobacters, and alcaligenes species, isolated from cottage cheese, illustrate the effect of chlorine on these organisms. Bonner and Harmon (1957) obtained complete destruction of seven test organisms in one minute or less when one ml of culture was added to 100 ml of 50 ppm hypochlorite solution. When suspended in 10 ppm hypochlorite, only *Achromobacter eurydice* was destroyed in 10 min while in 25 ppm, *A. butyri* was the single species to survive a 10 min exposure. The data of Davis and Babel (1954) also show marked differences between species in that the time required to destroy approximately 10^6 cells/ml in a solution of 50 ppm chlorine ranged from under 5 sec for an achromobacter to over 60 sec for a pseudomonad.

The concentration of hypochlorite required to eliminate psychrophiles from water supplies has been reported to range from 2 to 10 ppm (Witter 1961).

Gram-negative organisms, including the psychrophilic rods, are relatively resistant to quarternary ammonium compounds. All of the isolates of Bonner and Harmon (1957) survived an exposure of 10 min to 50 ppm, but were destroyed by this concentration of chlorine. The germicidal activity of quaternary ammonium compounds is influenced by pH. *Pseudomonas aeruginosa* was most susceptible at an acid pH; other species responded differently (Soike *et al.* 1952).

ENTEROBACTERIACEAE

Although perhaps best known as indicators of fecal contamination and as food-transmitted pathogens, members of this family also rate about second among the groups of Gram-negative rods as important food spoilage organisms.

Characteristics

This discussion will be limited to properties of some of the more troublesome members of the family. Additional information is presented in Chapter 9 that discusses the indicator bacteria, and in Chapter 3 that describes the genera *Salmonella* and *Shigella*.

A most important test for distinguishing *Enterobacteriaceae* from the pseudomonads and other previously discussed Gram-negative genera is the oxidation-fermentation test of Hugh and Leifson (1953). The *Enterobacteriaceae* are fermentative while members of the psychrophilic group are oxidative. In addition to acid, many species of *Enterobacteriaceae* produce carbon dioxide and hydrogen gas when grown on glucose.

Other useful determinations are the oxidase test of Kovacs (1956) and the arginine dihydrolase reaction. The *Enterobacteriaceae* are oxidase negative and most do not metabolize arginine under anaerobic conditions (Thornley 1960). Motile species can be differentiated from *Pseudomonas* in that they possess peritrichous flagella.

The separation of genera within the family is based on a large number of reactions which include the ability to produce acid from various sugars, the products of glucose metabolism, the utilization of citrate and other organic acids as a sole source of carbon, the production of H_2S and other protein reactions, and tolerance to KCN (Buchanan and Gibbons 1974).

Coliforms, the *Enterobacteriaceae* that produce both acid and gas from lactose, are the organisms perhaps most commonly incriminated in food spoilage; members of the genera *Enterobacter* and *Klebsiella* present the greatest problems. Prior to the 8th Edition of Bergey's Manual, these organisms were in the genus *Aerobacter* (no longer recognized) and this is the name that is used in much of the food literature. Typical *Enterobacter* and *Klebsiella* species give negative indol and methyl red reactions, provide a positive Voges-Proskauer test, and are able to utilize citrate as a sole source of carbon. Klebsiella species are nonmotile whereas *Enterobacter* have peritrichous flagella.

Other members of the family that may cause food spoilage belong to the genera *Serratia*, *Proteus*, and *Erwinia*. Typical strains of *Serratia* produce a red to pink pigment, prodigiosin, that facilitates their identification. Characteristics of *Proteus* species are swarming on the surface of solid media and the fact that most give a positive test for the enzyme urease. Many *Erwinia* are plant pathogens and some species have the ability to degrade pectic substances.

Food Spoilage

Coliforms and other *Enterobacteriaceae* are common contaminants of refrigerated foods such as poultry (Arafa and Chen 1975), fish (McMeekin 1975), and red meats (Tompkin 1973). Although the organisms have the ability to grow at 5°C or even lower, their growth rates at these temperatures are relatively slow. As a result, the group usually makes up only a small portion of the total spoilage microflora. This does not mean, however, that coliforms are of only minor importance as spoilage organisms of refrigerated foods: many species produce very pronounced off-flavors, often described as barny or unclean, and therefore even a small amount of metabolic activity may make the food unacceptable.

Under certain conditions coliforms may outgrow the more psychrophilic Gram-negative rods on refrigerated foods. Arafa and Chen (1975) found that because coliforms are facultative anaerobes, they predominated when oxygen was absent. Over 95% of their isolates from poultry were *Enterobacter* species when the food was stored under a vacuum in polyethylene pouches; pseudomonads predominated, as usual, when similar samples were stored under aerobic conditions.

Enterobacter aerogenes and other copious gas producers have been responsible for a variety of food defects. Gassiness in raw milk has been a problem, while in cheese, early gas formation, often when still in the press, results in objectionable eye formation or even a blowing apart of the curd. In the pickle fermentation, the growth of coliforms in cucumbers has been implicated as a cause of bloaters. There also are reports of a gassy spoilage of olives due to these organisms.

Various foods are subject to spoilage by other genera of *Enterobacteriaceae*. *Proteus*, a strongly putrefactive group, has been a spoilage agent of eggs, seafoods, and meats. The organisms are responsible, for example, for one type of black rot of eggs in which the signs are gas, a black hard yolk, and a brownish liquified white. *Serratia* has caused the surface discoloration of many foods due to the red pigment produced by many strains. During the Middle Ages, the appearance of *Serratia* on Eucharistic wafers created considerable panic since the slimy, red growth was mistaken for blood. *Erwinia*, a pathogen for a variety of plants, is a cause of soft rot of market fruits and vegetables.

ACETIC ACID BACTERIA

This group of Gram-negative rods has the ability to produce acetic acid from ethanol, hence its name. A unique property of the organisms is their tolerance to high concentrations of ethanol and their ability to carry out its oxidation in an acid medium, one of pH 4.5 or lower. This latter characteristic excludes from the group certain pseudomonads that can oxidize ethanol under more neutral conditions.

Characteristics

The acetic acid bacteria include two genera, *Gluconobacter* and *Acetobacter*. The genera are not closely related as evidenced by their placement in Bergey's Manual: *Gluconobacter* is a member of the *Pseudomonodaceae* while *Acetobacter* has not been assigned to any family (Buchanan and Gibbons 1974).

Distinguishing features of *Gluconobacter* are that motile strains possess polar flagella and that the genus oxidizes ethanol only to acetic acid. Because of their inability to oxidize acetate, gluconobacters have been referred to as "underoxidizers." Some strains form dark brown colonies and produce water-soluble pigments. In general they have a relatively low optimum growth temperature, 20 to 25°C, and many cannot grow at temperatures above 37°C.

Acetobacter is the genus of "overoxidizers" in that the organisms have the ability to oxidize ethanol all the way to carbon dioxide and water. Lactate as well as acetate is oxidized completely. A second major difference from the gluconobacters is that motile acetobacters have peritrichous flagella. Most strains are nonpigmented and their optimum growth temperature is within the range of 30 to 35°C.

Species within both genera have the ability to synthesize various polysaccharides from glucose and other sugars. *Acetobacter aceti* subsp. *xylinium* and a subspecies of *A. pasteurianus* form extracellular layers of cellulose, and strains of *Gluconobacter oxydans* produce levans or dextrans when grown in media containing sucrose or raffinose.

Another property shared by *Gluconobacter* and *Acetobacter* is their ability to partially oxidize a variety of compounds to products possessing the same carbon structure; for example, glucose to gluconic acid, gluconic acid to 5-ketogluconic acid, mannitol to fructose, glycerol to dihydroxyacetone, and propanol to propionic acid. Although the cell derives energy from these oxidations, the compounds do not serve as a source of carbon for synthesis of cellular materials.

Food Spoilage

Foods most susceptible to spoilage by acetic acid bacteria are wine, beer, and fruit products.

In wines, the principal result of their activity is an increase in the level of acetic acid, often expressed as volatile acid. The ester ethyl acetate is readily detected as a vinegary odor (Amerine and Joslyn 1970) and thus only small amounts of acetic acid need be produced to harm wine quality. In the U.S., the maximum levels permitted by

Federal laws for white and red wines are, respectively, 0.12 and 0.14 g of acetic acid per 100 cc.

Under certain conditions, acetic acid bacteria may also increase the fixed (nonvolatile) acid content of wine by producing gluconic acid from glucose. This is most apt to occur when acetification takes place early in the yeast fermentation, before much alcohol has been produced. If yeast metabolism is inhibited by the acetic acid, after a time ethanol, the preferred substrate of the bacteria, will be depleted. When this occurs, the acetic acid bacteria will oxidize residual sugars (Vaughn 1955).

In addition to increasing the amount of volatile acid in wine, growth of the organisms may produce an off-flavor that has been described as "mousey."

Defects in beer resulting from the growth of acetic acid bacteria are sourness, ropiness, and turbidity. The acetic acid produced from ethanol is responsible for the sourness, while ropiness is due to the formation of polysaccharides. Growth of the organisms during the alcohol fermentation or other earlier stages can present problems in clarification because the bacteria are difficult to remove by the usual brewery filtration procedures.

The spoilage of unfermented fruit products is not a serious problem. When growth does occur, the effects may be sourness and other off-flavors, along with sliminess and even the formation of a thick pellicle.

Growth and Controls

Oxygen.—Because the acetic acid bacteria are obligate aerobes, the universal method for preventing their growth is to eliminate oxygen from the food. In wine making this is accomplished by the yeasts during the period of vigorous fermentation; not only do they consume oxygen in their metabolism but the gas is sparged from the juice by the carbon dioxide that is generated. One potential danger site during fermentation is the pomace cap that forms when grapes are fermented on their skins; when exposed to air it often permits rapid acetification. To maintain anaerobic conditions, the cap may be submerged by the continuous pumping over of wine or by weighting it down with a wooden lattice cover.

Following fermentation the wine is stored in sealed tanks that are completely filled. When a full tank cannot be maintained, the headspace atmosphere is often replaced with CO_2.

Ethanol.—The organisms possess considerable tolerance to alcohol; the concentration has to be above 12 to 13% before growth is completely inhibited. This undoubtedly explains why alcohol-fortified Spanish sherries are not converted to vinegar during the long period they are exposed to oxygen when stored in partially filled casks.

pH.—The ability to oxidize ethanol in an acid medium is a property unique to the group. Because many strains can grow at a pH under 3.5, the reaction of most beer, wine, and other fruit products is not sufficiently low to prevent spoilage.

SO_2.—The acetic acid bacteria are relatively sensitive to sulfur dioxide in that most are inhibited by 50 to 75 ppm. Advantage has been taken of this sensitivity along with the organism's need for oxygen by maintaining a higher level of sulfur dioxide in the uppermost layer when wine is being stored in tanks.

BIBLIOGRAPHY

AMERINE, M. A., and JOSLYN, M. A. 1970. Table Wines; The Technology of Their Production, 2nd Edition. Univ. Calif. Press, Berkeley.

ARAFA, A. S., and CHEN, T. C. 1975. Effect of vacuum packaging on microorganisms on cut-up chickens and in chicken products. J. Food Sci. 40, 50–52.

AYRES, J. C. 1960. The relationship of organisms of the genus Pseudomonas to the spoilage of meat, poultry and eggs. J. Appl. Bacteriol. 23, 471–486.

BARNES, E. M., and IMPEY, C. S. 1968. Psychrophilic spoilage bacteria of poultry. J. Appl. Bacteriol. 31, 97–107.

BARNES, E. M., and SHRIMPTON, D. H. 1958. The effect of the tetracycline compounds on the storage life and microbiology of chilled eviscerated poultry. J. Appl. Bacteriol. 21, 313–329.

BARNES, E. M., and THORNLEY, M. J. 1966. The spoilage flora of eviscerated chickens stored at different temperatures. J. Food Technol. 1, 113–119.

BONNER, M. D., and HARMON, L. G. 1957. Characteristics of organisms contributing to spoilage in cottage cheese. J. Dairy Sci. 40, 1599–1611.

BREED, R. S., MURRAY, E. G. D., and HITCHENS, A. P. 1948. Bergey's Manual of Determinative Bacteriology, 6th Edition. The Williams & Wilkins Co., Baltimore, Md.

BREED, R. S., MURRAY, E. G. D., and SMITH, N. R. 1957. Bergey's Manual of Determinative Bacteriology, 7th Edition. The Williams & Wilkins Co., Baltimore, Md.

BRISOU, J., and PREVOT, A. R. 1954. Studies in bacterial systematics. X. Revision of the species grouped in the genus Achromobacter. Ann. Inst. Pasteur (Paris) 86, 722–728. (French)

BROWN, A. D., and WEIDEMANN, J. F. 1958. The taxonomy of the psychrophilic meat-spoilage bacteria: a reassessment. J. Appl. Bacteriol. 21, 11–17.

BUCHANAN, R. E., and Gibbons, N. E. 1974. Bergey's Manual of Determinative Bacteriology, 8th Edition. The Williams & Wilkins Co., Baltimore, Md.

CORLETT, D. A., Jr., LEE, J. S., and SINNHUBER, R. O. 1965A. Application of replica plating and computer analysis for rapid identification of bacteria in some foods. I. Identification scheme. Appl. Microbiol. 13, 808–817.

CORLETT, D. A., Jr., LEE, J. S., and SINNHUBER, R. O. 1965B. Application of replica plating and computer analysis for rapid identification of bacteria in some foods. II. Analysis of microbial flora in irradiated Dover sole (Microstomus pacificus). Appl. Microbiol. 13, 818–822.

DAVIS, P. A., and BABEL, F. J. 1954. Slime formation on cottage cheese. J. Dairy Sci. *37*, 176–184.

ELLIOTT, R. P., and MICHENER, H. D. 1965. Psychrophilic microorganisms in foods. A review. U.S. Dept. Agr. Tech. Bull. *1320*.

FLORIAN, M. L. E., and TRUSSELL, P. L. 1957. Bacterial spoilage of shell eggs. IV. Identification of spoilage organisms. Food Technol. *11*, 56–60.

FOSTER, E. M. *et al.* 1957. Dairy Microbiology. Prentice-Hall, New Jersey.

FRANK, H. A., and WILLITS, C. O. 1960. Maple syrup. XIII. Sterilizing effect of sunlight on maple sap in transparent tubes. Appl. Microbiol. *8*, 141–145.

FRAZIER, W. C. 1967. Food Microbiology, 2nd Edition. McGraw-Hill Book Co., New York.

HAYES, P. R. 1963. Studies on marine flavobacteria. J. Gen. Microbiol. *30*, 1–19.

HEATHER, C. D., and VANDERZANT, C. 1957. Effect of temperature and time of incubating and pH of plating medium on enumerating heat-treated psychrophilic bacteria. J. Dairy Sci. *40*, 1079–1086.

HERBERT, R. A., HENDRIE, M. S., GIBSON, D. M., and SHEWAN, J. M. 1971. Bacteria active in the spoilage of certain seafoods. J. Appl. Bacteriol. *34*, 41–50.

HUGH, R., and LEIFSON, E. 1953. The taxonomic significance of fermentative versus oxidative metabolism of carbohydrates by various Gram-negative bacteria. J. Bacteriol. *66*, 24–26.

INGRAHAM, J. L., and STOKES, J. L. 1959. Psychrophilic bacteria. Bacteriol. Rev. *23*, 97–108.

KOVACS, N. 1956. Identification of *Pseudomonas pyocyanea* by the oxidase reaction. Nature *178*, 703.

LEE, J. S., and SINNHUBER, R. O. 1967. Selection of microbial populations in chlortetracycline-treated ocean perch (*Sebastodes alutus*). Appl. Microbiol. *15*, 543–546.

McMEEKIN, T. A. 1975. Spoilage association of chicken breast muscle. Appl. Microbiol. *29*, 44–47.

MILLER, A., III, SCANLAN, R. A., LEE, J. S., and LIBBEY, L. M. 1973A. Identification of the volatile compounds produced in sterile fish muscle (*Sebastes melanops*) by *Pseudomonas fragi*. Appl. Microbiol. *25*, 952–955.

MILLER, A., SCANLAN, R. A., LEE, J. S., and LIBBEY, L. M. 1973B. Volatile compounds produced in sterile fish muscle (*Sebastes melanops*) by *Pseudomonas putrefaciens*, *Pseudomonas fluorescens*, and an *Achromobacter* species. Appl. Microbiol. *26*, 18–21.

PABLO, I. S., SINSKEY, A. J., and SILVERMAN, G. J. 1967. Selection of microorganisms due to freeze-drying. Food Technol. *21*, 748–754.

PETERSON, A. C. 1961. An ecological study of frozen foods. *In* Proceedings of Low Temperature Microbiology Symposium. Campbell Soup Co., Camden, New Jersey.

REAY, G. A., and SHEWAN, J. M. 1969. The spoilage of fish and its preservation by chilling. Advan. Food Res. *2*, 343–398.

SHAW, B. G., and SHEWAN, J. M. 1968. Psychrophilic spoilage bacteria of fish. J. Appl. Bacteriol. *31*, 89–96.

SHENEMAN, J. M., and COSTILOW, R. N. 1959. Identification of microorganisms from maple tree tap holes. Food Res. *24*, 146–151.

SHEWAN, J. M., HOBBS, G., and HODGKISS, W. 1960. A determinative scheme for the identification of certain genera of gram-negative bacteria, with special reference to the Pseudomonadaceae. J. Appl. Bacteriol. *23*, 379–390.

SHEWAN, J. M., HODGKISS, W., and LISTON, J. 1954. A method for the rapid identification of certain nonpathogenic asporogenous bacilli. Nature *173*, 208–209.

SILVERMAN, G. J., and GOLDBLITH, S. A. 1965. The microbiology of freeze-dried foods. *In* Advances in Applied Microbiology, Vol. 7, W. W. Umbreit (Editor). Academic Press, New York.

SOIKE, K. F., MILLER, D. D., and ELLIKER, P. R. 1952. Effect of pH of solution on germicidal activity of quaternary ammonium compounds. J. Dairy Sci. *35*, 764–771.

SPLITTSTOESSER, D. F., and GADJO, I. 1966. The groups of microorganisms composing the "total" count population in frozen vegetables. J. Food Sci. *31*, 234–239.

STANIER, R. Y., PALLERONI, N. J., and DOUDOROFF, M. 1966. The aerobic pseudomonads: a taxonomic study. J. Gen. Microbiol. *43*, 159–271.

THOMSON, A. B., DAVIS, H. K., EARLY, J. C, and BURT, J. R. 1974. Spoilage and spoilage indicators in queen scallops. J. Food Technol. *9*, 381–390.

THORNLEY, M. J. 1960. The differentiation of *Pseudomonas* from other gram-negative bacteria on the basis of arginine metabolism. J. Appl. Bacteriol. *23*, 37–52.

THORNLEY, M. J. 1967. A taxonomic study of *Acinetobacter* and related genera. J. Gen. Microbiol. *49*, 211–257.

THORNLEY, M. J., INGRAM, M., and BARNES, E. M. 1960. The effects of antibiotics and irradiation on the *Pseudomonas-Achromobacter* flora of chilled poultry. J. Appl. Bacteriol. *23*, 487–498.

TOMPKIN, R. B. 1973. Refrigeration temperature. As an environmental factor influencing the microbial quality of food—a review. Food Technol. *27*, No. 12, 54–58.

VAUGHN, R. H. 1955. Bacterial spoilage of wines with special reference to California conditions. Advan. Food Res. *6*, 67–108.

VAUGHN, R. H. *et al*. 1969. Gram-negative bacteria associated with sloughing, a softening of California ripe olives. J. Food Sci. *34*, 224–227.

VAUGHN, R. H. 1970. Incidence of various bacteria in dehydrated onions and garlic. Food Technol. *24*, 189–191.

WEEKS, O. B. 1969. Problems concerning the relationship of cytophagas and flavobacteria. J. Appl. Bacteriol. *32*, 13–18.

WITTER, L. D. 1961. Psychrophilic bacteria—a review. J. Dairy Sci. *44*, 983–1015.

WODZINSKI, R. J., and FRAZIER, W. C. 1960. Moisture requirements of bacteria. I. Influence of temperature and pH on requirements of *Pseudomonas fluorescens*. J. Bacteriol. *79*, 572–578.

WODZINSKI, R. J., and FRAZIER, W. C. 1961A. Moisture requirements of bacteria. IV. Influence of temperature and increased partial pressure of carbon dioxide on requirements of three species of bacteria. J. Bacteriol. *81*, 401–408.

WODZINSKI, R. J., and FRAZIER, W. C. 1961B. Moisture requirements of bacteria. V. Influence of temperature and decreased partial pressure of oxygen on requirements of three species of bacteria. J. Bacteriol. *81*, 409–415.

CHAPTER 12

Homer W. Walker | Aerobic and Anaerobic Spore-forming
Bacteria and Food Spoilage

Endospores occur in several genera of bacteria. The spore-forming
bacteria of concern to the food microbiologist, however, are found in
two genera: *Clostridium* and *Bacillus*. Those Gram-positive rods that
produce endospores and whose terminal electron acceptor is molecular oxygen are commonly referred to as aerobes and have been
classified as *Bacillus* species. Those Gram-positive rods that produce
endospores and that cannot utilize molecular oxygen as a terminal
electron acceptor are referred to as anaerobes and have been
classified as *Clostridium* species. Endospores have been observed in
several other species of bacteria, but these species are not of
particular concern in food microbiology.

CHARACTERISTICS OF THE SPORE

The bacterial endospore is a resting or dormant stage which is the
end result of many changes in cellular structure during the
transformation of a vegetative cell into a spore. One vegetative cell
gives rise to one spore during this growth cycle, and the spore
contains the components needed for the production of a vegetative
cell when conditions are favorable.

The endospore is distinguished from other bacterial cells by
showing refractility when examined by phase-contrast microscopy,
by not staining with basic dyes, and by being resistant to the action
of many enzymes. The bacterial spore is also unique in that it
contains large quantities of dipicolinic acid (DPA); this compound
has been found only rarely in other organisms and has been related
to the heat resistance of the spore. The bacterial spore is notorious
for its resistance to heat and also for its ability to survive to a much
greater extent than vegetative cells exposure to cold, drying,
chemicals, ultraviolet light, and other destructive agents. The ability
to survive adverse environmental conditions has made these organisms of particular importance in the processing and preservation
of foods in which they may cause spoilage or may be a public health
hazard.

The endospore is not an essential stage in growth of spore-forming
bacteria. Cultures can be maintained indefinitely in the vegetative
state. Spore formation usually occurs after the logarithmic growth
phase and is favored, generally, by conditions that encourage the

growth of vegetative cells, although the optimum range of conditions usually is more restrictive for spore formation than for cell growth. The observation has been made that spores are formed by healthy cells facing starvation. A desirable addendum to this observation is that specific sporulation factors or environmental conditions also may be required for sporulation to occur.

The dormant state of the spore is an inactive phase characterized by minimal metabolic activity. The spore can remain in this inactive but viable condition for years and then under suitable conditions can germinate and produce vegetative cells. Some confusion exists in the terminology used to describe the resting state of the spore. Terms frequently used are *dormancy* and *cryptobiosis*. Generally, the term cryptobiosis is reserved to describe the condition of no or arrested metabolism of the spore. Dormancy is differentiated from cryptobiosis in that it describes the failure of viable spores to germinate under apparently favorable conditions.

ACTIVATION OF SPORES

The changing of a dormant spore into a vegetative cell can be described as occurring in three successive stages: activation, germination, and outgrowth. Activation of the bacterial spore results from some treatment which does not itself initiate germination; but, after the treatment, the spore population will germinate more rapidly, more completely, or both. In other words, activation is needed when dormant spores fail to germinate when placed under conditions favorable for development and growth of vegetative cells. Exposure to heat is a common means of activating dormant spores. Heat treatment is not required by all bacterial spores; but in those instances where heat activation is effective, the time for germination decreases and frequently the minimum temperature at which germination occurs decreases. The main influence of heat activation is a reduction in the lag period preceding germination. Heat activation is a reversible process; the ability to germinate rapidly, and any new metabolic activity, may be lost when activated spores are stored for several days.

Activation of spores also may be achieved by exposure to low pH; for example, reversible activation occurs in spores of *Bacillus stearothermophilus* when they are exposed to pH 1.5. Compounds which reduce disulfide bonds may activate spores after prolonged incubation. Other treatments which can activate some spores are exposure to a mixture of calcium and dipicolinic acid, treatment with polar solvents such as dimethylformamide and dimethylsulfoxide, and aging of spores; aged spores sometimes show similarities to spores that have been activated by various other means. The activated spore retains most of the spore properties such as resistance to adverse conditions, refractility and lack of staining with basic dyes. Apparently activation does not require metabolism in the cell, and most probably consists of reversible changes in the configuration of macromolecules necessary for maintaining the integrity of the spore.

GERMINATION AND OUTGROWTH OF SPORES

The next step in the breaking of dormancy is germination. Activated spores will go through this irreversible process when exposed to the proper environment. No macromolecular synthesis occurs during germination, but degradative changes apparently occur during this stage as evidenced by the excretion of fragments of spore components into the medium. Spores may excrete material which can amount to as much as 30% of their dry weight. The exudate consists mainly of calcium, dipicolinic acid, fragments of de-polymerized murein and small amounts of amino acids, peptides, proteins and carbohydrate material.

In addition to the loss of cellular components during germination, a number of other changes are taking place: swelling of the spore, loss of refractility, loss of resistance to heat and other environmental stresses, an increase in permeability and stainability by basic dyes, and an increase in metabolic activities. The germinated spore, however, is cytologically distinct from the vegetative cell and lacks many of the typical macromolecules and enzymatic activities found in the mature vegetative cell.

Some of the changes that occur during the germinating process have been used to quantitatively measure the number of germinated spores in a suspension. The change in optical density of a spore suspension is commonly used to measure germination. The change in optical density is attributed to the decrease in refractility of the germinated spores and a concomitant decrease in the amount of light scattered by the germinated spore. Loss of heat resistance is also used successfully to determine the extent of germination. Germinated spores will not survive heat treatment and spores surviving the heat treatment may be enumerated by the plate count. Direct microscopic counts of spores that have lost refractility or that take up a stain are theoretically feasible but usually are not as satisfactory as the previous techniques.

The final step in the development of a vegetative cell from a dormant spore is outgrowth. Outgrowth involves the synthesis of new cell structures which result in the emergence of a vegetative cell.

THE GENUS *BACILLUS*

Organisms in the genus *Bacillus* are rod-shaped, Gram-positive cells which may measure from 0.3 to 2.2 μm by 1.2-7.0 μm in length. The majority of these organisms are motile with the flagella being typically lateral. An outstanding characteristic of this genus is the formation of heat-resistant endospores with not more than one spore formed in a sporangial cell. Sporulation occurs in the presence of oxygen.

Differentiation of species within the genus *Bacillus* has been somewhat easier since the publication of a monograph by Smith *et al.* (1952) in which they describe their studies on aerobic spore-forming bacteria; they placed the species of the genus *Bacillus* in three groups on the basis of morphology and physiology. The classification presented in Bergey's Manual of Determinative Bacteriology (Breed *et al.* 1957) closely follows that employed by these workers. Wolf and Barker (1968) have followed the classification of Smith *et al.* (1952) but have made some rearrangements based on difficulties they encountered because of variability in characteristics of some species. Table 12.1 shows the divisions made in the works cited above. Wolf and Barker (1968) have modified the scheme of Smith *et al.* (1952) by transferring *B. pantothenticus* from

TABLE 12.1
GROUPS OF *BACILLUS* SPECIES ON BASIS OF MORPHOLOGICAL CHARACTERISTICS
(SMITH *ET AL*. 1952; WOLF AND BARKER 1968).

	Group I	Group II	Group III
Spores:	Oval or cylindrical	Oval or cylindrical	Spherical
Sporangia:	Not definitely swollen	Distinctly swollen	Swollen (usually)
Spore wall:	Thin, not easily stained	Thick, easily stained	

Species:	A. Diameter of vegetative rods 0.9 μ or more; Protoplasm of young cells vacuolated
	   B. megaterium
	   B. cereus and variants
	
	B. Diameter of vegetative rods less than 0.9 μ; protoplasm of young cells not vacuolated
	   B. coagulans
	   B. licheniformis
	   B. subtilis and variants
	   B. pumilus
	   B. lentus
	   B. firmers
		B. stearothermophilus
		B. coagulans
		B. laterosporus
		B. polymyxa
		B. macerans
		B. brevis
		B. circulans complex
		B. alvei
		B. pantothenticus
			B. sphaericus
			B. pasteurii

group III to group II. Group III is characterized by its tendency to biochemical inertness, which they felt was not descriptive for *B. pantothenticus*. They also observed two sporangial types of *B. coagulans* which they suggest might represent distinct species. They have resolved this question for the time being by including *B. coagulans* in both groups I and II.

SPECIES OF *BACILLUS* IN FOODS

Bacillus cereus and Its Variants

B. cereus was one of the first species of this genus to be described; however, much confusion exists in the earlier literature because it has been isolated and described under a number of different names and because it was confused with *B. subtilis* for some time. For these reasons, the early literature on these organisms is difficult to evaluate in view of present day classification.

Both Smith *et al.* (1952) and Wolf and Barker (1968) have included several variants in their classification. They include *B. cereus* var. *mycoides* which typically forms rhizoidal colonies; *B. cereus* var. *thuringiensis* which forms crystals within the sporangium and is an insect pathogen; *B. cereus* var. *anthracis* which is an animal pathogen. It is generally agreed that a close relationship exists between these three variants and *B. cereus*; but some workers believe there is justification for maintaining *B. anthracis* and *B. thuringiensis* as independent species.

B. cereus is described as aerobic or facultatively anaerobic and capable of growing in glucose broth under anaerobic conditions. Gas is usually produced from nitrates under anaerobic conditions; no gas is produced if ammonium ion is used as the nitrogen source. As is true of many spore-forming bacteria, *B. cereus* is widely distributed in nature and has been found in air, soil, water, milk, dust, cocoa powder, meat brines, semi-preserved meats, spices and so on.

Bacillus cereus and its variants under proper circumstances can produce a condition known as "broken cream" or "bitty" cream in both raw and pasteurized milk. The cream breaks into particles of varying size which cannot be re-emulsified like the cream of normal milk. In hot tea or coffee, particles of the cream float on the surface and present an objectionable appearance. Milk in the initial stages is normal in flavor and acidity; later, a firm non-acid clot is formed. *B. cereus* usually does not attack lactose but produces a rennet-like enzyme that coagulates milk at relatively low acidity. This is referred to as sweet-curdling of milk; other organisms are capable of doing this also. *B. cereus* will slowly hydrolyze casein. Broken cream is associated with the production of lecithinase by *B. cereus*. Microscopic examination of cream in which *B. cereus* has grown shows rupture of the fat globule membrane. This rupture is attributed to hydrolysis of lecithin in the membrane. Non-lecithinase producing strains do not rupture the membrane nor do they produce broken cream.

Most bacilli do not readily form spores in milk but diluted milk in thin films provides excellent conditions for sporulation of *B. cereus* at room temperature (22°C). Immediate washing and sterilization of equipment and utensils helps control contamination. Proper refrigeration will prevent growth of this organism; *B. cereus* is unlikely to develop if the milk is held at 15°C or below. The optimum growth temperature for *B. cereus* is 30°C and the range for growth is 15 to 50°C.

Bacillus subtilis and Variants

Considerable confusion existed concerning the identity of *B. subtilis* until studies by Conn (1930), Gibson (1937, 1944) and Smith *et al.* (1952) clarified the situation. This confusion existed because of the great variations in cultural characteristics shown by various strains, the application of different names to these variants, confusion with *B. cereus*, and inadequate descriptions of isolates. Some names which are now considered as synonymous with *B. subtilis* and its variants are: *B. vulgatus*, *B. mesentericus*, *B. aterrimus*, *B. niger*, *B. panis*, and *B. natto*. Strains that produce a red pigment have been referred to as *B. globigii*. *B. aterrimus* or *B. subtilis* var. *aterrimus* produces a blue-black to black pigment in media containing a readily utilizable carbohydrate. *B. niger* or *B. subtilis* var. *niger* produces a black pigment only if tyrosine is present. Pigment formation is variable and may be lost. Cultures of *B. subtilis* and related bacilli may have such features as tough pellicles and wrinkled growths which may be pigmented on solid media, but these characteristics are of limited use for classification.

In the past, a type of spoilage called "ropiness" was a serious problem in both homemade and commercially produced bread; the problem was particularly prevalent during warm, humid weather. The condition was traced to the survival and outgrowth of spores of *B. subtilis* in the product. The internal temperature of bread may reach 100°C for a brief time during baking, which is not sufficient to eliminate the spores.

Ropiness in bread first becomes evident as brownish spots or patches accompanied by an unpleasant odor in the interior of the loaf. Sticky disintegration of the crumb follows caused by the hydrolysis of bread proteins or carbohydrates or both; the organism also can form a gum as a result of the metabolism of the sugar

362 FOOD MICROBIOLOGY

released by the decomposition of the starch by amylase. The bread may become markedly alkaline, markedly acid, or remain neutral depending on the strain involved—unaffected bread has a slightly acid reaction. In the alkaline condition, the characteristic stringiness, from which the name is derived, develops rapidly. The odor of ropy bread in the advanced stages of spoilage has been described as resembling that of rotting fruit, bad cheese or decomposing overripe melons.

Even though spores are present in the product, ropiness will not develop unless the environment is warm and moist and the acidity of the bread is low. Control of the condition can be achieved in several ways. Reduction of the number of spores in the product by good sanitation and quality control for ingredients such as flour, yeast cultures and malt products is of prime importance. Storage of the final product under good ventilation at a temperature of 18°C or below prevents growth. Rope-causing bacilli grow poorly, if at all, between pH 4.6 and 5.5 and dough with a pH between 4.5 and 5.0 appears to be safe from spoilage. Various acids have been used and tested for adjusting the pH of dough; various phosphate salts are used as acidulants in flours. In addition, propionic acid and its salts are used in baked goods to inhibit the development of ropiness in bread. The level of propionate required is a function of acidity; smaller amounts are required as the pH of the dough is decreased.

Development of blue-black spots throughout a loaf of bread has been attributed to *B. subtilis* var. *aterrimus* in several instances. Bread inoculated with this organism developed spots within 24 to 48 hours at 37°C and 48 to 72 hours at 30°C. Control can be achieved by methods used to prevent development of ropy bread.

B. subtilis has been isolated a number of times from canned cured meats and semi-preserved meats, but is seldom a cause of spoilage in canned foods because it grows poorly, if at all, under anaerobic conditions. One instance of swelling of canned ham has been attributed to *B. subtilis*. The isolate resembled *B. subtilis* except that it grew at thermophilic temperatures and produced gas in carbohydrate broth containing nitrate. In the classification developed since that time, this organism might logically be considered *B. licheniformis*. Strains of *B. subtilis* isolated from canned hams are capable of growing well in 7.5% sodium chloride and certain strains have shown some growth in 12.5% sodium chloride.

Extreme sliminess of the surface and an offensive odor developed on carcasses of squab duckling when they were shipped in a moist condition at 10°C. A nonmotile variant of *B. subtilis* that produced a large capsule was implicated. Spoilage of this type was controlled by refrigeration at 4°C.

Milk heated at temperatures above 100°C for various times has been produced and referred to as commercially sterilized milk. Bacterial spores may survive this treatment but cause little spoilage in marketed products indicating that the product is consumed before spoilage occurs or conditions such as storage temperature are unsuitable for growth. At ambient temperature, 5 to 10% of bottled sterilized milk may spoil within 7 to 14 days. In commercially sterilized milk, B. subtilis is usually the predominant spore-forming organism. The most important and most common spoilage attributed to B. subtilis is sweet curdling; and in some instances it is responsible for the development of off-flavor such as "bad-egg" and digestion of the protein.

Sweet curdling in evaporated milk has been attributed also to strains of B. subtilis. The pH of the milk remains normal, but the contents of the can gel into a smooth, custard-like consistency. Thinning and development of a bitter taste resembling quinine in canned cream has also been attributed to B. subtilis. Again the question arises if these organisms might not be placed elsewhere in the newer classification because of their growth under anaerobic conditions. A number of early workers report the occurrence and growth of B. subtilis in canned foods; the conditions described would lead to the assumption that today many of these organisms would not be classified as B. subtilis.

Brines containing soft cucumbers sometimes contain B. subtilis (B. mesentericus fuscus). These organisms grow readily in 9% salt, 0.2% acetic acid, and 0.3% lactic acid; cultural filtrates cause spoilage of firm, desalted cucumbers in 24 hours. This organism probably plays a very minor role in softening under normal conditions. The pectic enzymes of B. subtilis are inhibited by 2% salt but salt tolerance of the enzyme can be increased to 7% by adaptation; softening occurs at much higher levels of salt. In addition, the pectolytic enzyme of B. subtilis is most active at pH 8.6 with a range of 6.0 to 11.0; it loses activity rapidly under acid conditions. Softening of cucumbers would not be likely to occur under commercial salt-stock solution unless the brine remained at pH 5.5 or above and the desirable lactic fermentation was retarded and prevented lowering of the pH. In addition, commercial cucumber fermentations occur between pH 3 and 4 and the softening enzymes in brines are most active in the pH range of 4 to 5. This organism possibly could be involved in the softening of olives being processed for California ripe or green-ripe olives or under certain conditions in Spanish-type olives because of the alkaline conditions created by the use of sodium hydroxide.

Natto, a fermented food used in some areas of the orient, is prepared by fermenting soybeans with B. subtilis (B. natto). Cooked beans become covered with a stringy viscous polysaccharide and develop a strong odor of ammonia. The final product has a strong persistent flavor and is used as a side dish with rice.

Bacillus licheniformis

B. *licheniformis* does not differ greatly from B. *subtilis*; the differences between the two organisms reported by Smith *et al.* (1952) were the abilities of B. *licheniformis* to grow well anaerobically, to form high acidity in glucose broth, and to form gas in alkaline nitrate broth. These activities are not typical for B. *subtilis*. Also some evidence is available that the antigens in spores of the two organisms differ.

The relative heat resistance of the spores of these two organisms is difficult to assess but apparently the spores of many strains of B. *licheniformis* are less resistant than spores of B. *subtilis*. For example, in studies on commercially sterilized milk, B. *licheniformis* was the organism most frequently recovered from raw bulk milk, but B. *subtilis* predominated in the heat treated product. Some exceptions exist, however, as evidenced by the survival and recovery of B. *licheniformis* from other heat processed foods. Under certain circumstances, spores of this organism are rapidly destroyed at 100°C; on the other hand, one strain has a D value at 100°C of 4.10 minutes in skim milk. D value is the number of minutes required to destroy 90% of the population at a given temperature.

B. *licheniformis* belongs to a group of spore-forming bacteria which produces a soft coagulum with rapid digestion in commercially sterilized milk when the product is incubated at 30°C. Inoculation of canned milk products with spores of this organism has resulted in slightly "blown" cans.

Although spores of B. *licheniformis* frequently survive in semi-preserved and canned meats, they are not a common cause of spoilage of such products. Nevertheless, this organism can produce gas in certain canned meats because of its ability to grow anaerobically with the production of acid and gas in the presence of nitrates and sugar. In glucose broth under anaerobic conditions, limited gas and enough acid may be produced to reduce the pH to as low as 5.0. Growth does not occur in canned foods with a pH of less than 4.6. Most strains isolated from canned cured meats grow readily in 7.5% sodium chloride and many can tolerate 15% sodium chloride.

B. *licheniformis* has caused soft swells in cans of banana puree incubated at 55°C. Strains isolated from these cans differed from others in their ability to grow at 60°C but not at 65°C. The maximum temperature range is usually cited as being between 50°C and 56°C. Two strains isolated from a can of roasted veal 113 years old grew well at 55°C and strains isolated from sugar processing plants have been cultivated at 60°C. Obviously, strains of this organism can cause spoilage under conditions favoring thermophilic growth and are sometimes a nuisance when trying to detect true thermophiles.

This organism as well as strains of B. subtilis has been implicated in the production of viscous substances in bread and in sugar cane and sugar beet juices. The condition in bread is caused by the production of levan and is not to be confused with ropiness in bread which is caused by strains of B. subtilis. B. licheniformis is capable of synthesizing levan from sucrose. Certain gummy masses in cane juices described as resembling frog's spawn and the concomitant formation of gas is favored by the high concentration of sucrose, the presence of nitrate in the juice, and the favorable pH created by the addition of calcium carbonate to the system. Sucrose solutions may become so viscous as to interfere with factory operations.

B. licheniformis can be added to food products along with such items as sugar, starch, spices, and cocoa powder. In cocoa powder, for example, slightly less than half of the total aerobic spore-forming bacilli may be of this species. Cocoa powder could contaminate such products as chocolate flavored milk; but the numbers added from this source are generally small. In addition, adequate refrigeration of such a product after processing prevents growth and subsequent spoilage.

Bacillus pumilus

The properties of B. pumilus approach those of certain variants of B. subtilis. In fact, B. subtilis, B. licheniformis and B. pumilus are closely related. Gibson (1944) with some reservations separated B. pumilus from B. subtilis on the basis of no starch hydrolysis and no nitrate reduction by B. pumilus. It does not produce gas from nitrate under anaerobic conditions and experiences very scant, if any, growth under anaerobic conditions. This species grows well in the range of 28°C to 40°C; the maximum temperature for growth is in the range of 45°C to 50°C.

B. pumilus has been recovered from various foods and food ingredients but has not been characterized as causing spoilage. It has been recovered from starch, sugar, spices, raw and pasteurized milk, cucumber and olive fermentations, and canned hams. Poly-galacturonase-like enzymes produced by this organism are similar to those found in B. subtilis; but, as with B. subtilis, it is probably not involved with the softening of cucumbers undergoing a normal acid fermentation. They might contribute to softening certain types of olives in which sodium hydroxide is used and alkaline conditions prevail.

Bacillus megaterium

Several useful characteristics for the identification of *B. megaterium* are the ability to grow on ammonium basal medium, requirement of no growth factors, and lack of any lecithinase activity. It cannot grow in glucose under anaerobic conditions nor does it produce gas from nitrates under anaerobic conditions. Most of the cultures produce some pigment varying from cream to yellow, pink, brown, or black on some medium or other. Pigment formation, however, is too unstable to be used for classification.

Some of the synonyms for this species are: *B. tumescens, B. oxalaticus, B. ruminatus, B. graveolans, B. simplex,* and *B. cohaerans.* For many years *B. megatherium* was used; but the International Committee on Bacteriological Nomenclature in 1951 ruled that *B. megaterium* is correct.

This organism has been isolated at various times from foods but is not a significant cause of spoilage. It has been isolated from such items as commercially sterilized milk, evaporated and raw milk, sauerkraut, cheese and beef. Coagulation of evaporated milk accompanied by gas and a cheesy odor, and spoilage and swelling of cans of spiced ham and luncheon meat have been attributed to *B. megaterium*; however, gas production from carbohydrate is not a typical reaction for this organism. Souring of fresh beef described as a stinking acid fermentation has been credited to this species; this spoilage occurs over a wide temperature range even as low as 0°C to 4.5°C. Souring of hams and picnic shoulders by this organism occurs when the meat is not properly refrigerated before curing. The maximum growth temperature ranges from 40°C to 45°C; and optimal growth occurs between 28°C and 37°C.

Bacillus coagulans

B. coagulans was first isolated from coagulated evaporated milk by Hammer (1915). Berry (1933) described *B. thermoacidurans* as the cause of off-flavor in commercially canned tomato juice. Later these two organisms were shown to be identical; the name "*B. coagulans*" was maintained because of prior usage. Other names which are synonymous include *B. dextrolacticus, B. thermoacidificans,* and *Lactobacillus cereale.* The terminology of *B. coagulans* var. *thermoacidurans* is invalid since *B. coagulans* and *B. thermoacidurans* are synonymous.

Variations in sporangial types and in spore morphology create some difficulty in placing this organism in the appropriate grouping (Table 12.1). Smith *et al.* (1952) placed it in Group I based mostly on physiological reactions. Wolf and Barker (1968) placed it in both Groups I and II; they found two distinct types and at least four subsidiary types. Their work implies a need for another species and may also explain some of the variability in certain characteristics reported in the literature.

B. coagulans grows best in the temperature range of 40°C to 60°C; the optimum temperature is in the range of 45°C to 55°C. The upper limiting temperature for growth varies for different strains. Virtually all strains can grow at 60°C but a few grow poorly, if at all, at 55°C and above. No growth at 65°C is typical although growth has been reported at temperatures of 63°C to 65°C. *B. coagulans* is not considered an obligate thermophile because of growth at low temperatures. Limited growth may occur as low as 18°C; growth at room temperature is extremely slow, very slow at 25°C, and good at 30°C. The lowest temperature at which appreciable growth can occur apparently lies between 25 and 30°C.

Spores of *B. coagulans* are of low heat resistance when compared with spores of obligate thermophiles. Most strains can be destroyed in 0.7 min at 121°C; some strains isolated from vegetables which spoiled during storage tests came from cans treated at 121°C for 1.2 to 3.0 min. Thermal resistance of spores is greater in evaporated milk than in tomato juice. These differences in heat resistance in different products may be more apparent than real because the influence of pH, lipids, proteins and other constituents in the food product have not been evaluated.

B. coagulans produces acid and no gas in glucose broth under aerobic conditions when peptone is used as the nitrogen source, according to Breed *et al.* (1957). Others have reported, however, that under aerobic conditions, glucose is converted to carbon dioxide, acetic acid, and lactic acid and that under anaerobic conditions no gas is produced and lactic acid is the main product.

B. coagulans is one of the organisms that causes "flat-sour" spoilage of foods, particularly those in a pH range of 3.7 to 4.5. Flat-sour spoilage refers to the production of acid, predominantly lactic acid, in the canned food product without the production of gas; thus spoiled cans do not bulge or swell. Flat-sour spoilage of evaporated milk typically shows no bulging of the can and the contents are usually firmly curdled, but on occasion a soft, flaky curd with considerable whey is formed. The milk may have a sweetish, cheesy odor resembling that of Swiss cheese; the flavor has been described as sour and cheesy. Considerable increase in acidity occurs, mainly from the production of d-lactic acid. Similar spoilage has occurred in a sucrose-containing milk product used for addition to coffee; spoilage occurred during storage at 32°C.

B. *coagulans* is also responsible for flat-sour spoilage of canned tomato juice. The flavor of the product has been described as "acid," "like musty hay," "amyl acetate," "phenolic," "medicinal," "vinegary," or "bitter to sour flavor." The off-flavor can sometimes be detected before any changes in pH are evident, and develops most rapidly around 37°C.

The pH of tomato juice normally falls in the range of 3.8 to 4.5. When canned tomato juice was first produced, this low pH with a hot fill temperature approximating 100°C was thought to be sufficient to produce a commercially sterile product; spores of B. *coagulans*, however, can survive this heat treatment and cause spoilage. This organism ordinarily does not grow in tomato juice if the pH of the juice has been adjusted to 4.2 to 4.3 with an acid such as citric. Growth of this facultative anaerobe has been reported at a pH as low as 4.19 in tomato juice. A low concentration of spores and a pH of 4.3 in tomato juice will inhibit growth of most strains of B. *coagulans*. Acidification of tomato juice to pH 4.1 seems adequate to control growth in all cases. Presterilization by a high temperature-short time process is used to control this type of spoilage.

Bacillus stearothermophilus

The obligate thermophile B. *stearothermophilus* was first isolated from canned corn and string beans by Donk (1920). B. *kaustophilus* and B. *calidolactis* are considered synonymous with B. *stearothermophilus* by Smith et al. (1952). Grinsted and Clegg (1955) believe that B. *calidolactis* is a distinct species because it hydrolyzes gelatin and produces swollen sporangia during spore formation. B. *calidolactis* was isolated originally from skim milk held at high temperatures for extended periods of time before drying. It was described as a strict thermophile that rapidly coagulated milk at 71°C as a result of the production of d-lactic acid. Fields and Harris (1972), on the other hand, concluded from their observations of cultural and biochemical activities and the guanine plus cytosine content that B. *calidolactis* and B. *stearothermophilus* are synonymous. Studies by others suggest the need for establishing other species to accommodate certain aerobic, obligately thermophilic bacilli that cannot be classified satisfactorily as B. *stearothermophilus*.

B. *stearothermophilus* has the distinctive ability of growing at 65°C; some strains are capable of growing at 70°C or above. Growth at 65°C is considered a fairly reliable and stable characteristic of this species; good growth occurs from 50 to 65°C. In attempting to establish lower and upper limiting temperatures, as with other organisms, a certain amount of variation has been observed. For various strains, minimum temperatures of 45°C, 40-41°C, 37°C and 33°C have been reported. No growth occurs at 28°C. After storage for four years at 28°C, many strains are capable of growing at lower temperatures but never as low as 28°C. For some strains, maximum growth temperatures of 70°C, 72°C, and 75°C have been reported. No growth of this organism was observed in cans of vegetables incubated for months at 37°C; it was suggested that at this temperature heated spores tend to die or to return to dormancy. This effect is not the same as autosterilization in which accumulated metabolic products cause the rapid dying off of an organism after growth has occurred in the can and caused spoilage.

Spores of organisms with high maximum growth temperatures generally have a high heat resistance, and spores of *B. stearothermophilus* are among the most resistant. The degree of resistance varies depending on the environmental conditions during sporulation and during heating of the spores. A treatment of 121°C for 20 min under most conditions is considered to be satisfactory for total destruction of this organism; survival beyond this treatment has been observed, however. This resistance favors the survival of *B. stearothermophilus* over that of *B. coagulans* in the processing of nonacid foods; *B. coagulans* may survive in acid foods such as tomato juice which is given a less rigorous heat treatment. *B. stearothermophilus*, however, will not grow at the pH found in tomato juice.

B. stearothermophilus does not grow in culture media of pH 5.0, which differentiates it from *B. coagulans*, and it usually does not grow in canned foods of pH 5.3 or lower. However, the final pH of glucose broth in which various strains have grown may be as low as 4.5. Cans of asparagus spoiled by this organism have shown a pH of 4.4 to 4.6; cans of spoiled corn a pH of 4.8. Autosterilization may occur since vegetative cells of this organism will die off in excessive acidity.

Flat-sour spoilage ordinarily is not detected until the can is opened and the contents examined. Cans of asparagus spoiled by this organism contain turbid brine and have a sour odor. Starchy products such as corn may become sloppy; this change is attributable to hydrolysis of the starch.

Bacillus polymyxa and Bacillus macerans

These two organisms are unique in the genus *Bacillus* in that they produce both acid and gas during the aerobic metabolism of carbohydrates; they are sometimes grouped together and referred to as the "aerobacilli." *B. polymyxa* has been referred to at various times as *B. asterosporous, B. aerosporus,* and *Aerobacillus asterosporus. B. betanigrificans, B. schuylkiliensis, B. soli* and *B. vagans* are synonyms for *B. macerans.*

The sporangia of these organisms are characteristically bulging, and the spores are oval, located terminally to subterminally in the sporangium. Biotin is an essential growth factor for both with thiamin being an additional requirement for *B. macerans.* Gas production may be more variable and erratic with *B. macerans* than with *B. polymyxa.* Variants intermediate between *B. macerans, B. circulans,* and *B. alvei* occur. Additional differences between these two organisms are that *B. polymyxa* yields good growth on proteose-peptone acid agar slants, produces acetyl-methylcarbinol and does not produce crystalline dextrins; *B. macerans* does not grow on proteose-peptone acid agar or produce acetyl-methlycarbinol but does produce crystalline dextrins. Also *B. macerans* normally has a higher maximum growth temperature than *B. polymyxa.*

B. macerans and *B. polymyxa* are not considered major causes of canned food spoilage even though some outbreaks of spoilage have been attributed to them. The main characteristic of such outbreaks is the production of gas consisting chiefly of carbon dioxide and some hydrogen. Sometimes disintegration of the tissue of the canned fruit or vegetable occurs as well as the formation of a gummy mass of polysaccharide. Spoilage by these organisms has been observed in commercially canned and home canned fruits and vegetables such as peas, cling peaches, mixed diced fruit, apricots, tomatoes, asparagus, spinach, and canned banana puree. The spores of these organisms ordinarily do not survive commercial heat processes, and their presence usually results from leakage into the can during cooling because of defective seals or leaks in the container.

Cultures of *B. polymyxa* grow well in the presence of 2% NaCl; many are inhibited by 4% NaCl and none will grow in 6% NaCl. Growth occurs in 35% sucrose but not in 45%. The minimum pH for growth is in the range of 3.8 to 4.0. Good growth occurs at temperatures of 28°C to 35°C; the maximum growth temperature for most strains is 40°C with no growth at 45°C.

B. polymyxa has been isolated from rotted potatoes in which the interior was a gummy mass. Rotting with the evolution of much gas occurs when this organism is inoculated onto slices of fresh potatoes, carrots, onions, cucumbers, and iris stems and if the slices are incubated between room temperature and 37°C. Most bacterial potato rots are caused by nonspore-forming bacteria, and spoilage by *B. polymyxa* is not common. This species is capable of splitting pectic substances and elaborates enzymes that utilize polygalacturonic acid as a substrate. These enzymes are most active at pH values of 8.3 and above and temperatures in the range of 45 to 50°C. Spoilage of foods by these enzymes would occur under unusual circumstances.

Bacillus macerans has been recovered from spoiled cans of marine products, meat products and, in conjunction with *B. polymyxa*, in canned fruits. Acid and gas are normally produced. Certain strains of *B. macerans* (originally named *B. betanigrificans*) can cause blackening of canned beets in the presence of iron. These strains upon laboratory storage tend to lose the ability to produce black pigment and are indistinguishable from other strains of *B. macerans*.

Maximum temperature for growth of this organism is usually 45 to 50°C; some strains can grow up to 54°C. Growth does not occur below a pH range of 3.8 to 4.0. *B. macerans* can grow in 25% sucrose solutions, but not above, and in 2% but not 4% sodium chloride.

Bacillus circulans

B. circulans is so named because of the rotary motion of the cells within colonies; this rotary motion may give rise to motile colonies. Some cultures of this organism produce crystalline dextrins which are typical of *B. macerans*; others resemble *B. polymyxa* in that they produce large amounts of slime. *B. circulans* is also described as being Gram negative. Organisms included in the *B. circulans* group are heterogeneous and the present classification is inadequate.

Organisms from this group are not frequent causes of spoilage, but they have been isolated from canned foods, sugar, starch, spices, commercially sterilized milk and canned meat and fish products. Cardboard or oxidized flavors in commercially sterilized milk have been attributed to *B. circulans*. Certain strains can produce slight acidity without coagulation or digestion, although others may cause digestion. A halophile tolerating 15% sodium chloride and producing blown cans of bacon was placed in this group. It grew in glucose broth if nitrate or nitrite were present; these compounds were reduced to nitrogen gas. It grew well at 37°C and 45°C, producing gas within two days; no growth occurred at 50°C within 5 days. Sporulation was poor and the spores were easily destroyed by heat.

Other species of the genus *Bacillus* have been recovered from foods. They may cause spoilage under favorable conditions, but they occur mainly as incidental contaminants.

SPECIES OF CLOSTRIDIUM IN FOODS

Bacteria classified in the genus *Clostridium* are described as anaerobic; some, however, can tolerate limited levels of oxygen. They usually stain Gram-positive or variable; the age of the culture is important since cultures tend to stain Gram-negative as they age. Sporulation is difficult to observe in some species, particularly when they are cultivated on laboratory media. The production of spores ordinarily causes swelling of the sporangium with a few exceptions. Most of the species are motile by means of peritrichous flagella; a few species are non-motile. For convenience, the various *Clostridium* species are sometimes grouped on the basis of their ability to decompose proteins or ferment carbohydrates. On this basis, four groups become apparent: 1) proteolytic but not saccharolytic; 2) saccharolytic but not proteolytic; 3) saccharolytic and proteolytic; and 4) neither. *C histolyticum* is strongly proteolytic and shows little saccharolytic activity, while *C. butyricum* would fall into the strongly saccharolytic group. *C. sporogenes*, *C. bifermentans* and certain strains of *C. botulinum* exhibit both properties. *C. perfringens* manifests both saccharolytic and proteolytic properties, but the saccharolytic activity predominates.

Saccharolytic types ferment sugars to produce a variety of organic acids and alcohols such as acetic, butyric, and propionic acids and butyl, ethyl, and isopropyl alcohols. Carbohydrates are usually fermented by clostridia with the production of gas. Proteolytic species attack proteins and amino acids producing such malodorous compounds as hydrogen sulfide, mercaptans, and skatole. Proteolytic clostridia frequently completely peptonize casein in milk cultures and cause blackening of cooked meat cultures if hydrogen sulfide is produced. The putrefactive or proteolytic clostridia can be grouped on the basis of utilization of amino acids; this characteristic may be useful for the identification of isolates.

Species of clostridia are widely distributed in nature and have been found in soil, water, sewage, and the intestinal tracts of animals. Only a few of the species are common inhabitants of foods; these few, however, are of extreme importance in maintaining safety and quality of foods.

Clostridium sporogenes and Other Putrefactive Anaerobes

C. sporogenes and related species are putrefactive anaerobes that frequently cause spoilage of medium acid and low acid canned foods. Various strains of putrefactive anaerobes have been isolated from a variety of canned fish, fish pastes, and canned meat products, including meat pastes and pasteurized canned hams. These organisms can be recovered from meat during slaughter and processing; the numbers are generally low, ranging from less than one spore per gram to two or three spores per gram of meat. *C. sporogenes* is one of the most commonly occurring putrefactive anaerobes in pasteurized meats.

Despite the relatively low numbers of these organisms present in fresh meats, many instances can be cited in which putrefactive anaerobes have caused spoilage of canned meats and canned sea foods. *C. sporogenes* along with *C. bifermentans, C. mucosum, C. parabifermentans, C. septicum, C. paraputrificum*, and *C. putrefaciens* are associated with taint or putrefaction of country style hams and other types of meats. Spoilage is characterized by putrefaction and excessive gas formation. Meat pastes containing less than 2% salt are readily attacked and slow growth occurs in the presence of 3% salt. At the normal pH of canned luncheon meat (approximately pH 6.0), nitrite appears to be the chief preservative agent against spoilage by putrefactive anaerobes.

C. sporogenes grows vigorously in cooked meat medium with evolution of gas and digestion of meat particles. The digested meat becomes blackened and a foul odor develops. In milk cultures, the casein may be entirely digested. Proteolytic strains of *C. botulinum* and *C. sporogenes* resemble each other in many ways except *C. sporogenes* does not produce toxin.

A putrefactive anaerobe that produces highly heat-resistant spores was isolated from spoiled corn by Cameron in 1927; its identity was in some doubt and was referred to as Putrefactive Anaerobe (P.A.) 3679. This organism is accepted now as a strain of *C. sporogenes* and is used as a test organism for determination of heat processes for canned foods. *C. sporogenes* produces both smooth and rough variants; *C. parasporogenes* has been used for the smooth variant.

C. sporogenes, C. perfringens, and *C. butyricum* are the three most commonly occurring anaerobic spore-formers in milk. These clostridia usually do not multiply in raw or pasteurized milk; the E_h or oxidation-reduction potential of pasteurized milk usually is in the range of +230 mV to 290 mV, which does not favor the growth of most clostridia. Other organisms, however, may grow in the milk and lower the E_h sufficiently to permit growth of clostridia.

C. sporogenes is probably the most common cause of spoilage in processed cheese, although other clostridia have been implicated in sporadic outbreaks of spoilage. Spore-forming bacteria such as this organism can survive heat treatments given processed cheese mixtures. The spoiled cheese is usually badly swollen or puffed with numerous gas holes in the center of the cheese; a very obnoxious, penetrating odor is evident. In regions of concentrated growth, the cheese consists of a soft, white material. Typical spoilage in experimental batches of cheese heavily inoculated with the organism occurs only when skim milk powder or casein digest is added. Skim milk powder may be a prime source of contamination and it also furnishes nutrients for the growth of the organism in the cheese.

C. sporogenes causes spoilage of other types of cheese also. For example, it can produce blowing of Grana cheese. A defect in Swiss cheese referred to as "white stinker" has been traced to this organism. The defect occurs in cheese which hasn't developed sufficient acidity, or when eye formation in the cheese is slow and the cheese is kept for a long time in a warm cellar. *C. sporogenes* requires a warmer storage temperature and a higher water activity than most clostridia to cause spoilage. The development of gas in cheese a few weeks after manufacture, referred to as "late gas" defect, is brought about by several species of *Clostridium* including *C. sporogenes, C. lentoputrescens, C. pasteurianum* and *C. butyricum.*

These clostridia do not grow readily in cheese because of their susceptibility to acid and salt and their inability to compete with starter organisms. The lower pH limit for growth of *C. sporogenes* lies between 5.3 and 5.4. In processed cheese containing added skim milk powder, growth is limited at pH 5.6. Differences as to the amount of sodium chloride needed to inhibit growth are attributable to the test media used and other conditions of cultivation. Salt concentrations of 6.5% sodium chloride in glucose agar, 12% in pork infusion medium, and 7.6% in processed cheese inhibit growth.

"Zapatera" is a spoilage that occurs during the fermentation of Spanish olives. Saccharolytic-proteolytic types of clostridia have been isolated from these outbreaks; *C. bifermentans* and *C. sporogenes* are the predominating types. No one organism but several organisms of the saccharolytic-proteolytic type are probably responsible for zapatera spoilage. Additional confusion may exist as to the cause of zapatera because of the frequent inclusion of butyric acid spoilage under this category.

Zapatera is characterized by the development of a "cheesey" or "sagey" odor in the early stages; this odor eventually develops into a penetrating, fecal stench. Deterioration develops when the desired lactic acid fermentation ceases before the pH of the brine reaches 4.5. Normal brines contain only acetic and lactic acids; brines from zapatera spoilage contain such acids as formic, propionic, butyric, and succinic.

C. sporogenes along with other closely related anaerobes may be instrumental in causing explosion of chocolate candies; yeasts may also cause this type of spoilage. Explosion may occur ten days to two weeks after manufacture. The chocolate coating is cracked and the center of the candy is exposed. The flavor is impaired by the development of rancidity and other off flavors. Possible sources of the contamination are sugar and egg albumin. The defect can be controlled by use of a filling that will not permit growth of these gas-forming anaerobes.

Clostridium putrefaciens

C. putrefaciens has been associated with spoilage of hams and has been isolated from hogs during slaughter. It produces a typical sweet-sour odor in tainted hams rather than putrefaction. In laboratory cultures, meat particles may not be altered for some time; old meat cultures do not show liquefaction, but a marked softening of the meat without reduction in bulk becomes evident. Thirty to 40% of the total nitrogen in meats is converted to amino acids and ammonia. Ham spoiled by growth of *C. putrefaciens* produced no ill effects when eaten after cooking.

C. putrefaciens is one of the few known clostridia capable of growing at psychrotrophic temperatures. The optimum temperature for growth lies between 20 to 25°C; growth develops poorly, if at all, at 37°C. The minimum growth temperature lies somewhere between 0 to 4°C. Some isolates grow well at 0°C, with rapid growth at 8 to 10°C and extensive growth in hams held at 1 to 2°C; other strains apparently will not grow at 4°C.

This organism hydrolyzes gelatin, casein, and egg albumen; it produces hydrogen sulfide and is one of the non-motile species of *Clostridium*. In meat broth, usually no noticeable change occurs within 24 hours, but after 48 hours slight turbidity and a few gas bubbles among the meat particles may appear. Gas consisting of hydrogen and carbon dioxide is formed from glucose; maximal production of gas occurs at neutral or slightly alkaline reaction.

Spores develop in about one week at 20 to 25°C and in about 2 weeks at 8 to 10°C. The spores do not demonstrate great resistance; they will survive 80°C for 20 min but not 100°C for 10 min. Sodium chloride and potassium nitrate at levels of 3% in growth medium will inhibit growth at 20 to 25°C.

Clostridium butyricum and Other Butyric Acid Producing Clostridia

Clostridium species that belong to the butyric group have been implicated in the spoilage of various foods including cheeses, syrups, canned fruits, tomato juice, olives, and grape juice. *C. butyricum* is isolated frequently from pasteurized canned hams. During their growth, they ferment carbohydrates to produce a number of compounds such as fatty acids, aldehydes, and alcohols; the principal compounds are butyric acid and butyl alcohol. The butyric fermentation is accompanied by the production of hydrogen and carbon dioxide. Fermentation characteristics may change with alterations in cultural conditions and with laboratory storage.

Species which are included in this group are *C. butyricum*, *C. butylicum*, *C. acetobutylicum*, *C. beijerinckii*, *C. saccharobutyricum*, *C. multifermentans*, *C. pasteurianum* and *C. tyrobutyricum*. The last two species differ from the others in that they ferment only a limited number of carbohydrates (glucose, sucrose, and mannitol) and do not ferment lactose. The others ferment a greater number of carbohydrates, including lactose. *C. saccharobutyricum* is similar to *C. butyricum* and in some instances has been classified as such.

Butyric acid anaerobes can multiply in some cheeses producing gas and an unpleasant taste due to the production of butyric acid. Late gassing in such cheeses is frequently caused by lactate-fermenting clostridia; both *C. butyricum* and *C. tyrobutyricum* ferment lactate with the production of gas. Some disagreement exists as to the differentiation of these organisms but enough differences have been observed to accept them as distinct species. Both of these species ferment glucose and lactate; *C. butyricum* ferments lactose while the ability of *C. tryobutyricum* to ferment lactose is variable. *C. tyrobutyricum* has a pH range of 4.4 to 7.5 for germination of spores with an optimal range of 5.8 to 6.0 as compared to a pH range of 5.0 to 8.0 and an optimum of 6.5 for *C. butyricum*. *C. butyricum* is more sensitive to sodium chloride than is *C. tyrobutyricum*. Growth of *C. tyrobutyricum* at a lower pH and higher salt concentrations than *C. butyricum* may explain why *C. tyrobutyricum* is recovered more frequently than *C. butyricum* from cheeses with butyric acid defects.

Development of butyric clostridia is observed mainly in those cheeses characterized by a curd with a high solids content and a ripening of long duration. Examples of such cheeses are Gruyére, Edam, Cheddar and Swiss types. Growth of these clostridia in cheese is manifested by the appearance of a strong rancid flavor and the development of holes in higher numbers and greater size than in normal cheese. Swelling of the cheese and bursting may occur. Swelling usually appears eight to ten days after making the cheese.

The presence of these clostridia in cheese has been related to the feeding of the dairy cow, particularly the feeding of silage containing spores of this group. The number of these spores in milk normally is low and rarely exceeds 20 per ml; only 5 cells/ml of milk, however, are sufficient to cause spoilage.

A butyric fermentation of olives by clostridia can occur under certain conditions and has been observed in green olives undergoing the Spanish or Sicilian type of pickling process. This type of spoilage usually occurs in the early stages of fermentation when appreciable quantities of glucose and mannitol are present in the brine. The desirable lactic acid bacteria are low in numbers and the pH is high. Under these conditions, butyric anaerobes grow and cause spoilage. In early stages of this type of spoilage, the odor is that of butyric acid or rancid butter; subsequently the odor becomes more pronounced, developing into an ill-smelling stench. There has been a practice in the olive industry to designate all malodorous fermentations as "zapatera." The butyric fermentation is not true "zapatera," which is associated with saccharolytic-proteolytic clostridia.

Saccharolytic, butyric anaerobes have been isolated from butyric spoilage of olives; none of them grew at pH 4.3. Species associated with this spoilage and production of gas pockets in olives are *C. beijerinckii, C. multifermentans, C. fallax, C. butyricum,* and *C. acetobutylicum. C. beijerinckii,* the predominant organism, is the least tolerant of low pH; most strains do not grow below pH 4.5. Of this group, *C. multifermentans* is most tolerant to pH. *C. acetobutylicum* grows in the presence of 6% sodium chloride, a higher concentration than the others tolerate.

C. pasteurianum causes spoilage of acid or medium acid canned foods that is characterized by the presence of butyric acid, a cheesy odor, and swelling of the can as the result of production of carbon dioxide and hydrogen. Butyric fermentation of canned tomatoes by *C. pasteurianum,* and occasionally by *C. butyricum,* is associated with high pH values in the final product. In addition to tomatoes, such spoilage has been observed in canned pears, figs, nectarines, apricots and pineapples as well as juices made from some of these items.

Spores of *C. pasteurianum* and *C. butyricum* are not classified as highly heat-resistant and spoilage can be controlled by processing the canned product to the necessary center temperature which is dependent upon the pH of the product. With butyric spoilage, pH 4.5 seems to be a critical point below which spoilage does not occur, depending on the numbers of organisms present. The pH of the product can be adjusted by addition of citric acid. Delayed appearance of swells due to these organisms in commercial packs has been attributed to high pH accompanied by relatively small numbers of organisms per can.

The optimum temperature for growth of *C. pasteurianum* is 30°C; growth at 37°C occurs slowly. It will grow readily in laboratory media with a pH above 4.0 and slowly in media with a pH range of 3.6 to 4.0. Growth has occurred in pear juice with a pH of 3.55. It is more tolerant of sucrose and sodium chloride than are closely related organisms.

Thermophilic Anaerobes

Thermophilic spore-forming bacteria usually produce spores that are more heat resistant than those of mesophilic spore-formers;

thermophiles, therefore, may be more prevalent in underprocessed foods than mesophiles. The most frequently encountered thermophilic anaerobic spore-formers are non-proteolytic, saccharolytic and do not produce hydrogen sulfide. These organisms are referred to as "thermophilic anaerobes" or "T.A.'s." They are responsible for spoilage of non-acid canned foods called "hard swell." *C. thermosaccharolyticum* is typical of the T.A.'s; it grows rapidly, producing acid and abundant gas consisting mainly of hydrogen and carbon dioxide. This species is capable of growth over a range of 30 to 65°C with some strains unable to grow below 37°C. Optimum growth occurs at 55°C. *C. thermosaccharolyticum* coagulates milk and production of gas causes a channeling of the clot. It produces acetic acid and butyric acid, which impart an unpleasant odor to the spoiled food. In addition to being associated with spoiled foods, it has been isolated with some frequency from crystallized sugar. Strains of this organism may be used to evaluate processes for canned foods.

C. thermoaceticum is another thermophile that has been isolated from canned vegetables. It has a minimum growth temperature of 49°C and a maximum in the range of 63 to 69°C. This organism does not produce gas and is only slightly saccharolytic. It grows slowly and may require as much as 10 days at 55°C to manifest growth in a food item; in laboratory cultures, 48 hours may be required before growth becomes evident. The odor of the food in which it grows is not changed greatly. In glucose medium it produces acetic acid and a pH of 3.0. The lowest pH observed in baby food from which it was isolated was 4.2 and the highest was 5.7. It can produce hydrogen sulfide. The usual canning procedures are not sufficient to destroy spores, but its importance as a spoilage organism is restricted because of its high minimum growth temperature.

The thermophilic spore-forming anaerobe known for many years as *C. nigrificans* has been reclassified as *Desulfotomaculum nigrificans*. The organism was isolated by Werkman and Weaver (1927) from spoiled canned corn. They classified it as a *Clostridium* but more recent studies show it to be Gram-negative, whereas most *Clostridium* species are Gram-positive, to differ in DNA base composition from other *Clostridium* species, and to contain cytochromes which are absent in *Clostridium* species.

This organism is found in soils, compost heaps, thermal spring water, and sugar and it produces "sulfur stinker" spoilage of canned vegetables. Spoilage of canned foods by this organism is characterized by blackening of the contents of the can and production of hydrogen sulfide which imparts an unpleasant odor to the product. No external signs of spoilage are evident. Acid canned foods are not susceptible to spoilage; spoilage of this type has been reported in canned corn, peas, baby clams and mushrooms. Cans of corn spoiled in this manner contain a bluish gray liquid with numerous darkened kernels floating about. Blackening of the germ occurs due to the presence of iron and the formation of the sulfide. After several months the liquid may assume a normal color and the germs become gray. Blackening is more pronounced in canned peas than in corn.

D. nigrificans is a non-saccharolytic organism which is only feebly proteolytic. Cysteine is attacked resulting in the formation of hydrogen sulfide. Deep agar colonies of this organism are black particularly if ferrous salts are present. The maximum growth temperature ranges from 65 to 70°C, depending upon the strain, with a minimum of 30 to 31°C and an optimum of 55°C. The pH for optimum growth is 7.2-7.4 with no growth evident at pH values of 5.8 or 7.6. Spores of this organism have survived 100°C for as long as eight hours at pH 7.0. Processing of corn at 118°C for 70 min in No. 2 cans has been reported as ineffective in controlling this type of spoilage.

BIBLIOGRAPHY

ALDERTON, G., THOMPSON, P. A., and SNELL, N. 1964. Heat adaptation and ion exchange in *Bacillus megaterium* spores. Science *143*, 141-143.

ALLEN, P. W. 1919. "Rope" producing organisms in the manufacture of bread. Abstr. Bacteriol. *3*, 4.

ASCHEHOUG, V., and JANSEN, E. 1950. Studies on putrefactive anaerobes as spoilage agents in canned foods. Food Res. *15*, 62-67.

AYRES, J. C., and ADAMS, A. T. 1953. Occurrence and nature of bacteria in canned beef. Food Technol. 7, 318-323.

BABAD, J., and BOROS, D. L. 1961. Bacterial spoilage in canned process cheese. Israel J. Agr. Res. *11*, 57-63.

BALTZER, J., and WILSON, D. C. 1965. The occurrence of clostridia on bacon slaughter lines. J. Appl. Bacteriol. *28*, 119-124.

BECKER, M. E., and PEDERSON, C. S. 1950. The physiological characters of *Bacillus coagulans* (*Bacillus thermoacidurans*). J. Bacteriol. *59*, 717-725.

BEERENS, H., CASTEL, M. M., and PUT, H. M. C. 1962. Characteristics for the identification of some *Clostridium* of the butyricum group. Ann. Inst. Pasteur, Lille *103*, 117. (French)

BEERENS, H., and DES ROSIERS, A. 1968. Study on the thermophilic *Clostridium*. Ann. Inst. Pasteur, Lille *19*, 64-121. (French)

BERGERE, J. L., GOUET, P., HERMIER, J., and MOCQUOT, G. 1968. The butyric group of *Clostridium* in milk products. Ann. Inst. Pasteur, Lille *19*, 42-54. (French)

BERGERE, J. L., and HERMIER, J. 1970. Spore properties of clostridia occurring in cheese. J. Appl. Bacteriol. *33*, 167-179.

BERRY, R. N. 1933. Some new heat resistant, acid tolerant organisms causing spoilage in tomato juice. J. Bacteriol. *25*, 72-73.

BILLING, E., and CUTHBERT, W. A. 1958. "Bitty" cream: the occurrence and significance of *Bacillus cereus* spores in raw milk supplies. J. Appl. Bacteriol. *21*, 65-78.

BOHTER, C. W. 1963. Microbial spoilage of canned foods. *In* Microbiological Quality of Foods, L. W. Slanetz, C. O. Chichester, A. R. Gaufin, and Z. J. Ordal (Editors). Academic Press, New York.

BOLTJES, F. Y. K. 1955. *Bacillus* in semi-preserved meats. Ann. Inst. Pasteur, Lille *7*, 89-94.

BOWEN, J. F., STRACHAN, C. C., and MOYLS, A. W. 1954. Further studies of butyric fermentation in canned tomatoes. Food Technol. *8*, 471-473.

BOYER, E. A. 1923. A study of the spoilage of hams and other pork products. Am. Food J. *18*, 197-200.

BOYER, E. A. 1926. A contribution to the bacteriological study of ham souring. J. Agr. Res. *33*, 761-768.

BREED, R. S., MURRAY, E. G. D., SMITH, N. R. 1957. Bergey's Manual of Determinative Bacteriology, 7th Edition. The Williams & Wilkins Co., Baltimore, Md.

BUNYEA, H. 1921. A souring of beef caused by *Bacillus megatherium*. J. Agr. Res. *21*, 689-698.

BUNZELL, H. H., and FORBES, M. 1930. A method for testing for ropiness of bread. Cereal Chem. *7*, 465-472.

BURKE, M. V., STEINKRAUS, K. H., and AYRES, J. C. 1950. Methods of determining the incidence of putrefactive anaerobic spores in meat product. Food Technol. *4*, 21-25.

BUTTIAUX, R., and BEERENS, H. 1955. Gas-producing mesophilic clostridia in canned meats, with improved techniques for their identification. J. Appl. Bacteriol. *18*, 581-590.

BUTTIAUX, R., and FLAMENT, J. 1951. Bacteriological analysis of semi-preserved meat. Ann. Inst. Pasteur, Lille *4*, 145-180. (French)

CAMERON, E. J., and BIGELOW, W. D. 1931. Elimination of thermophilic bacteria from sugar. Ind. Eng. Chem. *23*, 1330-1333.

CAMERON, E. J., and ESTY, J. R. 1926. The examination of spoiled canned foods. II. Classification of flat sour organisms from nonacid foods. J. Infect. Diseases *39*, 89-105.

CAMERON, E. J., ESTY, J. R., and WILLIAMS, C. C. 1936. The cause of "black beets": an example of oligodynamic action as a contributory cause of spoilage. Food Res. *1*, 73-81.

CAMPBELL, L. L., and POSTGATE, J. R. 1964. Classification of the spore-forming sulfate-reducing bacteria. Bacteriol. Rev. *29*, 359-363.

CANDY, M. R., and NICHOLS, A. A. 1956. Some bacteriological aspects of commercially sterilized milk. II. Type of spore-forming bacteria isolated. J. Dairy Res. *23*, 329-335.

CLARK, F. M., and DEHR, A. 1947. A study of butyric acid-producing anaerobes isolated from spoiled canned tomatoes. Food Res. *12*, 122-128.

CONN, H. J. 1930. The identity of *Bacillus subtilis*. J. Infect. Diseases *46*, 341-350.

CROSSLEY, E. L. 1938. The bacteriology of meat and fish pastes, including a new method of detection of certain anaerobic bacteria. J. Hyg. *38*, 205.

DELMOUZOS, J. G., STADTMAN, F. H., and VAUGHN, R. H. 1953. Malodorous fermentation, acidic constituents of zapatera olives. J. Agr. Food Chem. *1*, 333-334.

DEMAIN, A. L., and PHAFF, H. J. 1957. Current knowledge concerning the causes of softening of cucumbers during curing. J. Agr. Food Chem. *5*, 60-64.

DONK, P. J. 1919. Some organisms causing spoilage in canned foods with special reference to flat sours. Abstr. Bacteriol. *3*, 4-5.

DONK, P. J. 1920. A highly resistant thermophilic organism. J. Bacteriol. *5*, 373-374.

DONOVAN, K. O. 1959. The occurrence of *Bacillus cereus* in milk and on dairy equipment. J. Appl. Bacteriol. 22, 131–137.
DORNER, W. 1942. Cheese defects and cheese diseases. J. Bacteriol. 43, 46–47.
DOWSON, W. J. 1943. Spore-forming bacteria in potatoes. Nature 152, 331.
DRAKE, S. D., EVANS, J. B., and NIVEN, JR., C. F. 1960. The effect of heat and irradiation on the microflora of canned hams. Food Res. 25, 270–278.
DUNCAN, C. L., and FOSTER, E. M. 1968. Role of curing agents in the preservation of shelf-stable canned meat products. Appl. Microbiol. 16, 401–405.
EDDY, B. P., and INGRAM, M. 1956. A salt-tolerant denitrifying *Bacillus* strain which 'blows' canned bacon. J. Appl. Bacteriol. 19, 62–70.
FABIAN, F. W., and JOHNSON, E. A. 1938. Experimental work on cucumber fermentation. Mich. Agr. Expt. Sta. Tech. Bull. 157.
FARMILOE, F. J., CORNFORD, S. J., COPPOCK, J. B., and INGRAM, M. 1954. The survival of *Bacillus subtilis* spores in the baking of bread. J. Sci. Food Agr. 5, 292–304.
FIELDS, M. L., and HARRIS, O. 1972. Identification of named cultures of *Bacillus calidolactis* as *Bacillus stearothermophilus*. J. Bacteriol. 110, 772–774.
FISHER, E. A., and HALTON, P. 1928. A study of "rope" in bread. Cereal Chem. 5, 192–208.
FOSTER, E. M. et al. 1957. Dairy Microbiology. Prentice-Hall, New Jersey.
FRANK, H. A. 1955. The influence of cationic environments on the thermal resistance of *Bacillus coagulans*. Food Res. 20, 315–321.
FRANKLIN, J. G. 1970. Spores in milk: problems associated with UHT processing. J. Appl Bacteriol. 33, 180–191.
FRANKLIN, J. G., WILLIAMS, D. J., and CLEGG, L. F. L. 1956. A survey of the number and types of aerobic mesophilic spores in milk before and after commercial sterilization. J. Appl. Bacteriol. 19, 46–53.
FURIA, T. E. 1968. Handbook of Food Additives. The Chemical Rubber Co., Cleveland.
GABIS, D. A., LANGLOIS, B. E., and RUDNICK, A. W. 1970. Microbiological examination of cocoa powder. Appl. Microbiol. 20, 644–645.
GALESLOOT, T. E. 1961. Concerning the anaerobic bacilli which cause spoilage in processed cheese. Ned. Melk-en Zuiveltijdschr. 15, 263–281.
GALLOWAY, L. D., and RYMER, T. E. 1967. Note on blue-black spots in bread. J. Food Technol. 2, 95–96.
GARVIE, E. I., and STONE, M. J. 1953. The associative growth of *Bacillus cereus* and *Streptococcus lactis* in milk. J. Dairy Res. 20, 29–35.
GIBSON, T. 1937. The identity of *Bacillus subtilis* and its differentiation from other spore-forming bacteria. Soc. Agr. Bacteriol. Proc. Abstr.
GIBSON, T. 1943. The *Bacillus subtilis* group in relation to industrial products. Proc. Soc. Agr. Bacteriol., 13–15.
GIBSON, T. 1944. A study of *Bacillus subtilis* and related organisms. J. Dairy Res. 13, 248–260.
GIBSON, T., and TOPPING, L. E. 1938. Further studies of the aerobic spore-forming bacilli. Soc. Agr. Bacteriol. Proc. Abstr., 43–44.
GILLESPY, T. G., and THORPE, R. H. 1968. Occurrence and significance of thermophiles in canned foods. J. Appl. Bacteriol. 31, 59–65.
GILLILAND, J. R., and VAUGHN, R. H. 1943. Characteristics of butyric acid bacteria from olives. J. Bacteriol. 46, 315–322.
GORDON, R. E., and SMITH, N. R. 1949. Aerobic sporeforming bacteria capable of growing at high temperatures. J. Bacteriol. 58, 327–341.
GOUDKOV, A. V., and SHARPE, M. E. 1965. Clostridia in dairying. J. Appl. Bacteriol. 28, 63–73.
GOULD, G. W., and HURST, A. 1969. The Bacterial Spore. Academic Press, New York.

GRANVILLE, A., and FIÉVEZ, L. 1958. Survival of *Bacillus* in meat brines. *In* The Microbiology of Fish and Meat Curing Brines, B. P. Eddy (Editor). Stationery Office, London. (French)

GREENBERG, R. A. *et al.* 1966. Incidence of mesophilic spores in raw pork, beef, and chicken in processing plants in the United States and Canada. Appl. Microbiol. *14*, 789–793.

GRINSTED, E., and CLEGG, L. F. L. 1955. Spore-forming organisms in commercial sterilized milk. J. Dairy Res. *22*, 178–190.

GROSS, C. E., VINTON, C., and STUMBO, C. R. 1946. Bacteriological studies relating to thermal processing of canned meats. V. Characteristics of putrefactive anaerobes used in thermal resistance studies. Food Res. *11*, 405–410.

HACHISUKA, Y. N. *et al.* 1955. Studies on spore germination. Effect of nitrogen sources on spore germination. J. Bacteriol. *69*, 399–406.

HAMILTON, I. R., and JOHNSTON, R. A. 1961. Studies on cucumber softening under commercial salt-stock conditions in Ontario. II. Pectolytic microorganisms isolated. Appl. Microbiol. *9*, 128–134.

HAMMER, B. W. 1915. Bacteriological studies on the coagulation of evaporated milk. Iowa Agr. Expt. Sta. Res. Bull. *19*, 119–131.

HEIMPEL, A. M. 1967. A critical review of *B. thuringiensis* and other crystalliferous bacteria. Ann. Rev. Entomol. *12*, 287.

HIRSCH, A., McCLINTOCK, M., and MOCQUOT, G. 1952. Observations on the influence of inhibitory substances produced by the lactobacilli of Gruyère cheese on the development of anaerobic sporeformers. J. Dairy Res. *19*, 179–186.

HOFFMAN, C., SCHWEITZER, T. R., and DALBY, G. 1937. Control of rope in bread. Ind. Eng. Chem. *29*, 464–467.

HOOD, E. G., and BOWEN, J. F. 1950. A new type of bacterial spoilage in Canadian process cheese. Sci. Agr. *30*, 38–42.

HOOD, E. G., and SMITH, K. N. 1951. Bacterial spoilage in process cheese. Sci. Agr. *31*, 530–540.

HUSSONG, R. V., and HAMMER, B. W. 1928. A thermophile coagulating milk under practical conditions. J. Bacteriol. *15*, 179–188.

HUSSONG, R. V., and HAMMER, B. W. 1931. Observations of *Bacillus calidolactis.* Iowa State Coll. J. Sci. *6*, 89–92.

INGRAM, M. 1952. Internal bacterial taints ('bone taint' or 'souring') of cured pork legs. J. Hyg. *50*, 165.

INGRAM, M., and DAINTY, R. H. 1971. Changes caused by microbes in spoilage of meats. J. Appl. Bacteriol. *34*, 21–39.

INGRAM, M., and HOBBS, B. C. 1954. The bacteriology of "pasteurized" canned hams. J. Roy. Sanit. Inst. *74*, 1151–1163.

INTERN. ASSOC. MICROBIOLOGISTS. 1951. Bacteriological nomenclature and taxonomy. Opinion I. The correct spelling of the specific epithet in the species name *Bacillus megaterium* De Bary 1884. Intern. Bull. Bacteriol. Nomenclature Taxonomy *1*, 35–36.

JANSEN, E., and ASCHEHOUG, V. 1951. *Bacillus* as spoilage organisms in canned foods. Food Res. *16*, 457–461.

JAYNES, J. A., PFLUG, I. F., and HARMON, L. G. 1961. Effect of pH and brine concentration on gas production by a putrefactive anaerobe (PA) 3679 in a processed cheese spread. J. Dairy Sci. *44*, 1265–1271.

JENSEN, L. B., and HESS, W. R. 1941. A study of ham souring. Food Res. *6*, 273–326.

JENSEN, L. B., WOOD, I. H., and JANSEN, C. E. 1934. Swelling in canned chopped hams. Ind. Eng. Chem. *26*, 1118–1120.

JONES, A. H., and FERGUSON, W. E. 1961. Factors affecting the development of *"Bacillus coagulans"* in fresh tomatoes and in canned tomato juice. Food Technol. *15*, 107–111.

SPORE-FORMING BACTERIA AND FOOD SPOILAGE **383**

JONES, O., and PEARCE, E. 1954. Bacteria in wholesome canned foods. J. Appl. Bacteriol. *17*, 272–277.
KAWATAMORI, T., and VAUGHN, R. H. 1956. Species of *Clostridium* associated with zapatera spoilage of olives. Food Res. *21*, 481.
KELLY, C. D. 1926. Bacteria causing spoilage of evaporated milk. Proc. Trans. Roy. Soc. Canada (3rd Ser.) *20* (Sec. V), 387–394.
KNAYSI, G. 1948. The endospore of bacteria. Bacteriol. Rev. *12*, 19–77.
KNOCK, G. G., LAMBRECHTS, M. S. J., HUNTER, R. C., and RILEY, F. R. 1959. Souring of South African tomato juice by *Bacillus coagulans*. J. Sci. Agr. *10*, 337–342.
LAMANNA, C. 1942. Relation of maximum growth temperature to resistance to heat. J. Bacteriol. *44*, 29–35.
LAMANNA, C. 1942. The status of *Bacillus subtilis*, including a note on the separation of precipitinogens from bacterial spores. J. Bacteriol. *44*, 611–617.
LIN, C. C., and LIN, K. C. 1970. Spoilage bacteria in canned foods. II. Sulfide spoilage bacteria in canned mushrooms and a versatile medium for the enumeration of *Clostridium nigrificans*. Appl. Microbiol. *19*, 283–286.
LIN, C. C., WU, B. K., and LIN, D. K. 1968. Spoilage bacteria in canned foods. I. Flat sour spoilage bacteria in canned asparagus and the thermal death time. Appl. Microbiol. *16*, 45–47.
LLOYD, D. J., CLARK, A. B., and McCREA, E. D. 1920-1921. On rope (and sourness) in bread. Together with a method of estimating heat resistant spores in flour. J. Hyg. *19*, 380–393.
MALLMAN, W. L. 1932. The spoilage of dressed ducks by sliminess. J. Agr. Res. *44*, 913–918.
MARGALITH, P., and SHOENFELD, R. 1962. Thermophilic aerobacillus from banana purée. Appl. Microbiol. *10*, 309–310.
MATTEUZZI, D., and ANNIBALDI, S. 1970. Blowing in Grana Cheese caused by *Clostridium tyrobutyricum*. XVIII. Intern. Dairy Congr. *IE*, 143.
McBRYDE, C. N. 1911. A bacteriological survey of ham souring. U.S. Bur. Animal Ind. Bull. *132*.
McCLUNG, L. S. 1935. Studies on anaerobic bacteria. IV. Taxonomy of cultures of a thermophilic species causing "swells" of canned foods. J. Bacteriol. *29*, 189–204.
McKENZIE, D. A. 1945. Bacteriological aspects of sterilized milk. Dairy Ind. *10*, 334–339.
MEAD, G. C. 1971. The amino acid-fermenting clostridia. J. Gen. Microbiol. *67*, 47–56.
MIKOLAJCIK, E. M. 1970. Thermodestruction of *Bacillus* spores in milk. J. Milk Food Technol. *33*, 61–63.
MOL, J. H. H., and TIMMERS, C. A. 1970. Assessment of the stability of pasteurized comminuted meat products. J. Appl. Bacteriol. *33*, 233–247.
MORGAN, G. F. V. 1943. Some problems in the sterilized milk industry. Proc. Soc. Agr. Bacteriol., 31–33.
MOSSEL, D. A. A., and MOL, J. H. H. 1956. A typical case of delayed spoilage in a dairy product incubation test. Appl. Microbiol. *4*, 69–70.
MUNDT, J. O., and KITCHEN, H. M. 1951. Taint in Southern country-style hams. Food Res. *16*, 233–238.
NAGEL, C. W., and VAUGHN, R. H. 1961. The characteristics of a polygalacturonase produced by *Bacillus polymyxa*. Arch. Biochem. Biophys. *93*, 344–352.
NAGEL, C. W., and WILSON, T. M. 1970. Pectic acid lyases of *Bacillus polymyxa*. Appl. Microbiol. *20*, 374–383.
NEILSON, N. E., and CHRISTIE, H. W. 1957. *Bacillus stearothermophilus* in herring stickwater. Can. J. Microbiol. *3*, 937–938.
NICHOLS, A. A. 1939. Bacteriological studies of canned milk products. J. Dairy Res. *10*, 231–249.

NICHOLS, A. A., HOWAT, G. R., and JACKSON, C. J. 1937. Bitterness and thinning in canned cream. J. Dairy Res. 8, 331–346.
NORTJE, B. K., and VAUGHN, R. H. 1953. The pectolytic activity of species of the genus Bacillus: Qualitative studies with Bacillus subtilis and Bacillus pumilus in relation to the softening of olives and pickles. Food Res. 18, 57–69.
PARSON, L. B., and STURGES, W. S. 1927. Quantitative aspects of the metabolism of anaerobes. I. Proteolysis by Clostridium putrefaciens compared with that of other anaerobes. J. Bacteriol. 14, 181–192.
PEDERSON, C. S. 1971. Microbiology of Food Fermentations. Avi Publishing Co., Westport, Conn.
PERIGO, J. A., WHITING, E., and BASHFORD, T. E. 1967. Observations on the inhibition of vegetative cells of Clostridium sporogenes by nitrite which has been autoclaved in a laboratory medium, discussed in the context of sub-lethally processed cured meats. J. Food Technol. 2, 377–397.
PORTER, R., McCLESKEY, C. S., and LEVINE, M. 1937. The facultative sporulating bacteria producing gas from lactose. J. Bacteriol. 33, 163–183.
PUT, H. M. C., and WYBINGA, S. G. 1963. The occurrence of Bacillus coagulans with high heat resistance. J. Appl. Bacteriol. 26, 428–434.
REED, G. B. 1945. Clostridium parasporogenes, an invalid species. J. Bacteriol. 49, 503–505.
RICE, A. C., and PEDERSON, C. S. 1954. Factors influencing growth of Bacillus coagulans in canned juice. I. Size of inoculum and oxygen concentration. Food Res. 19, 115–123.
RICHMOND, B., and FIELDS, M. L. 1966. Distribution of thermophilic aerobic sporeforming bacteria in food ingredients. Appl. Microbiol. 14, 623–626.
RIDGWAY, J. D. 1955. Some recent observations on the bacteriology of sterilized milk. J. Appl. Bacteriol. 18, 374–387.
RIDGWAY, J. D. 1958. The incidence and thermal resistance of mesophilic spores found in milk and related environments. J. Appl. Bacteriol. 21, 118–127.
RIEMANN, H. 1963. Safe heat processing of canned cured meats with regard to bacterial spores. Food Technol. 17, 39–49.
ROBERTS, T. A., and HOBBS, G. 1968. Low temperature growth characteristics of clostridia. J. Appl. Bacteriol. 31, 75–88.
ROSS, H. E. 1965. Clostridium putrefaciens: a neglected anaerobe. J. Appl. Bacteriol. 28, 49–51.
RUYLE, E. H., and SOGNEFEST, P. 1951. Thermal resistance of a facultative aerobic spore-forming bacterium in evaporated milk. J. Milk Food Technol. 14, 173–175.
SARLES, W. B., and HAMMER, B. W. 1932. Observations on Bacillus coagulans. J. Bacteriol. 23, 301–314.
SCARR, M. P. 1968. Thermophiles in sugar. J. Appl. Bacteriol. 31, 66–74.
SMITH, L. D. S. 1970. Clostridia. In Manual of Clinical Microbiology, J. E. Blair, E. H. Lennette, and J. P. Truant (Editors). American Society of Microbiology, Bethesda, Md.
SMITH, N. R., and GORDON, R. E. 1959. Illegitimate combinations, Bacillus cereus var. terminalis and Bacillus coagulans var. thermoacidurans. J. Bacteriol. 77, 810.
SMITH, N. R., GORDON, R. E., and CLARKE, F. E. 1952. Aerobic sporeforming bacteria. U.S. Dept. Agr. Misc. Publ. 559.
SMYTH, H. F. 1927. A bacteriologic study of the Spanish green olives. J. Bacteriol. 13, 56–57.
SOGNEFEST, P., and JACKSON, J. M. 1947. Pre-sterilization of canned tomato juice. Food Technol. 1, 78–84.
SPIEGELBERG, C. H. 1940A. Clostridium pasteurianum associated with spoilage of an acid canned fruit. Food Res. 5, 115–130.
SPIEGELBERG, C. H. 1940B. Some factors in the spoilage of an acid canned fruit. Food Res. 5, 439–455.

SPIEGELBERG, C. H. 1944. Sugar and salt tolerance of *Clostridium pasteur-ianum* and some related anaerobes. J. Bacteriol. *48*, 13-30.

STARR, M. P., and MORAN, F. 1962. Eliminative split of pectic substances by phytopathogenic soft-rot bacteria. Science *135*, 920-921.

STEINKRAUS, K. H., and AYRES, J. C. 1964. Incidence of putrefactive anaerobic spores in meat. J. Food Sci. *29*, 87-93.

STONE, M. J. 1952A. The action of the lecithinase of *Bacillus cereus* on the globule membrane of milk fat. J. Dairy Res. *19*, 311-315.

STONE, M. J. 1952B. The effect of temperature on the development of broken cream. J. Dairy Res. *19*, 302-310.

STONE, M. J., and ROWLANDS, A. 1952. 'Broken' or 'bitty' cream in raw and pasteurized milk. J. Dairy Res. *19*, 51-62.

STUMBO, C. R., GROSS, C. E., and VINTON, C. 1945. Bacteriological studies relating to thermal processing of canned meats. Food Res. *10*, 260-272.

STURGES, W. S., and DRAKE, E. T. 1927. A complete description of *Clostridium putrefaciens* (McBryde). J. Bacteriol. *14*, 175-179.

SUSSMAN, A. S., and HALVORSON, H. O. 1966. Spores, Their Dormancy and Germination. Harper and Row, New York.

TANIKAWA, E., MOTOHIRO, T., and AKIBA, M. 1967. Causes of can swelling and blackening of canned baby clams. II. Bacterial action involved in can swelling and blackening of baby clams. J. Food Sci. *32*, 231-234.

TANNER, F. W., ECHELBERGER, E. E., and CLARKE, F. M. 1932. Auto-sterilization as a problem in the bacteriological examination of foods. Proc. Soc. Exptl. Biol. Med. *29*, 1000-1001.

TANNER, F. W., and EVANS, F. E. 1933. Effect of meat curing solutions on anaerobic bacteria. I. Sodium chloride. Zentralblatt Bakt Parasitenk. Infektionskrank-heiten (Zweite Abt.) *88*, 44-54.

THEOPHILUS, D. R., and HAMMER, B. W. 1938. Influence of growth temperature on the thermal resistance of some bacteria from evaporated milk. Iowa State Agr. Expt. Sta. Res. Bull. *224*.

TOWNSEND, C. T. 1938. Spore-forming anaerobes causing spoilage in canned acid foods. J. Bacteriol. *36*, 315-316.

TOWNSEND, C. T. 1939. Spore-forming anaerobes causing spoilage in acid canned foods. Food Res. *4*, 231-237.

TOWNSEND, C. T., ESTY, J. R., and BASELT, F. C. 1938. Heat resistance studies on spores of putrefactive anaerobes in relation to determination of safe processes for canned foods. Food Res. *3*, 323-346.

TRAMER, J. 1964. The inhibitory action of nisin on *Bacillus stearo-thermophilus*. *In* Microbial Inhibitors in Food, N. Molin, and A. Erichsen (Editors). Almquist and Wiksell, Stockholm.

VAUGHN, R. H., KREULEVITCH, I. H., and MERCER, W. A. 1952. Spoilage of canned foods caused by the *Bacillus macerans—polymyxa* group of bacteria. Food Res. *17*, 560-570.

VAUGHN, R. H., LEVINSON, J. H., NAGEL, C. W., and KRUMPERMAN, P. H. 1954. Sources and types of aerobic microorganisms associated with the softening of fermenting cucumbers. Food Res. *19*, 494-502.

VAUGHN, R. H., and STADTMAN, T. C. 1946. A note on pH tolerance of *Aerobacter aerogenes* as related to natural ecology and decomposition of acid food products. J. Bacteriol. *51*, 263.

VERHOEVEN, W. 1950. On a sporeforming bacterium causing the swelling of cans containing cured ham. Antonie van Leeuwenhoek J. Microbiol. Serol. *16*, 269-281.

WEINZIRL, J. 1922. The cause of explosion in chocolate candies. J. Bacteriol. 7 599-604.

WEINZIRL, J. 1927. Sugar as a source of the anaerobes causing explosion of chocolate candies. J. Bacteriol. *13*, 203-207.

WERKMAN, C. H. 1929. Bacteriological studies on sulfid spoilage of canned vegetables. Iowa State Coll. Agr. Mech. Arts Res. Bull. *117*, 163-180.

WERKMAN, C. H., and WEAVER, H. J. 1927. Studies in the bacteriology of sulphur stinker spoilage of canned sweet corn. J. Sci. Iowa State Coll. *2*, 57–67.

WHEATON, E. J., BURROUGHS, J. D., and HAYS, G. L. 1957. Flat sour spoilage of tomato juice and its control with subtilin. Food Technol. *11*, 286–289.

WILLIS, A. T. 1964. Anaerobic Bacteriology in Clinical Medicine, 2nd Edition. Butterworths Inc., Washington.

WOESE, C. R., and MOROWITZ, H. J. 1958. Kinetics of release of dipicolinic acid from spores of *Bacillus subtilis*. J. Bacteriol. *76*, 81–83.

WOLF, J., and BARKER, A. N. 1968. The genus *Bacillus*: aids to the identification of its species. *In* Identification Methods for Microbiologists, Part B, B. M. Gibbs, and D. A. Shapton (Editors). Academic Press, London.

XEZONES, H., SEGMILLER, J. L., and HUTCHINGS, I. J. 1965. Processing requirements for a heat-tolerant anaerobe. Food Technol. *19*, 1001.

CHAPTER 13

H. B. Naylor | Micrococci

TAXONOMICAL CONSIDERATIONS

Cohn, a German botanist who developed the first usable classification scheme for bacteria, introduced the genus name *Micrococcus* in 1872. Since that time, the genus *Micrococcus* has been retained in most of the classification schemes that have been proposed for bacteria (Cowan 1962). Descriptions of the organisms included in the genus, however, have varied widely, and in most instances inadequate criteria have been used in the designation of species within the genus. As a result, the taxonomy of these organisms has been in a confused state for a long time, and continues to be controversial up to now. Renewed interest in the problem in recent years has resulted in several studies which have contributed greatly to a better understanding of the natural relationships among these organisms, and should serve as the basis for a much improved classification system (Baird-Parker 1962, 1963, 1965; Kocur *et al.* 1971; Schliefer and Kandler 1972; Lachica *et al.* 1971).

Before considering the role of micrococci in foods, it is essential that some guide lines be drawn as to the general characteristics of the organisms under consideration. In the eighth edition of *Bergey's Manual of Determinative Bacteriology* (1974) some rather drastic changes were made in the *Micrococcaceae* family when compared with the seventh edition of the Manual (1957). Most of the changes are based on sound evidence generated within the past 15 years. All obligately anaerobic cocci, including certain *Sarcina* species, were removed from the *Micrococcaceae* family and placed with more closely related groups of organisms. The aerobic *Sarcina* species were transferred to the genus *Micrococcus* since there is good evidence that the only major difference between these organisms is the ability of *Sarcina* species to produce cubical packets of cells, a characteristic which is often difficult to demonstrate with cultures maintained under laboratory conditions. *S. urea*, a sporeformer, was given the genus name *Sporosarcina* and classified along with the other endospore-producing bacteria in the family *Bacillaceae*. The genus *Gaffkya* was eliminated completely. One of the two previously recognized species in this genus, *G. homari*, which causes a disease of lobsters, has basic characteristics typical of organisms included in the genus *Aerococcus* of the *Streptococcaceae* family. The other species, *G. tetragena*, is basically indistinguishable from organisms included in the genus *Staphylococcus* and was classified as such. With these changes, only two of the genera previously included in the family *Micrococcaceae* were retained; namely, *Staphylococcus* and *Micrococcus*. One new genus, *Planococcus*, was added to accommodate a group of motile cocci isolated from marine environments which usually produce a yellow-brown pigment and have other characteristics which differ from those of either of the other two genera.

As a result of these changes, the family *Micrococcaceae* now contains a more homogeneous group of organisms, the major distinguishing characteristics of which can be stated as follows: Gram-positive, asporogenous, catalase and/or other heme-protein positive, aerobic or facultatively anaerobic cocci, which may or may not be motile. Other characteristics which are useful for associating an unknown culture with this family include pigment production and cell arrangement. Production of water-insoluble carotenoid pigments which are various shades of yellow, yellow-brown, orange, various shades of red, or violet is rather common among these organisms. The characteristic growth pattern of most of the organisms results in the production of cell clumps, either irregular masses or regular packets, although some types may grow predominantly as pairs or tetrads.

There are certain precautions that should be taken when attempting to identify an unknown organism with the family *Micrococcaceae* (Baird-Parker 1965). Sufficient microscopic observations should be made of cells at different stages of the growth cycle and under different growth conditions to provide sound evidence that the organism really is a coccus. For example, some organisms may appear to be typical cocci when observed during the stationary phase of growth, but assume the rod form during the period of cell division. Precautions should also be taken when determining the Gram staining reaction. This should be done on cultures during the logarithmic phase of growth since there is a tendency for the cells of some Gram-positive cultures to lose this characteristic very soon after reaching the stationary phase. A third precaution involves the catalase test. Although detection of catalase activity is relied upon heavily to differentiate the heme-protein synthesizing cocci of the family *Micrococcaceae* from the cocci of the family *Streptococcaceae*, the latter of which are incapable of synthesizing such compounds, this test is not infallible. Some organisms in the latter group have nonheme catalase activity, and there are reported cases of organisms in the family *Micrococcaceae* which fail to decompose hydrogen peroxide, either due to inability to synthesize catalase or to suppressed catalase activity attributed to growth conditions. Such organisms contain other heme-protein components such as cytochromes, however, which can be detected by other procedures. One such procedure is the benzidine test proposed by Deibel and Evans (1960). This test is no more difficult to use than the simple test for catalase activity, and is superior to it because the presence or absence of heme-proteins is detected rather than catalase activity.

The organisms now included in the family *Micrococcaceae* which are of greatest recognized importance in food microbiology are classified either in the genus *Staphylococcus* or the genus *Micrococcus*. These organisms have several characteristics in common, and are often found growing together under certain environmental conditions. One natural habitat is the skin and the mucous membranes of the upper respiratory tract of warm-blooded animals, including humans. The initial contamination of raw animal food materials with these organisms is often from such sources.

Although staphylococci and micrococci have been classified together in a single genus called either *Staphylococcus* or *Micrococcus* by various taxonomists in the past, there now exists sufficient evidence of basic differences between these organisms to justify dividing them into two genera. Unfortunately, certain of the basic differences can only be detected by use of special laboratory techniques that are too complicated for routine use in identifying unknown cultures. For example, the combined results obtained in several laboratories have established that the guanine (G) plus cytosine (C) content of the deoxyribonucleic acid (DNA) of staphylococci falls within the range of 30 to 37%, whereas this varies from 55 to 75% for the micrococci. This significant difference in DNA composition is probably the best single criterion to use to separate these organisms into the two genera, but the methods for determining it are too time consuming for routine application. Another basic difference between these organisms has been elucidated from studies of the structure of their cell walls (Schleifer and Kandler 1972). The interpeptide bridges in the peptidoglycan polymer of the staphylococcal walls are different from those of micrococcal walls. Whereas variations in the structure of these bridges occur among the micrococci, they are of one basic type for all of the staphylococci that have been studied so far, and this type has not been found in any of the micrococci. Again, the procedures that must be employed to determine such differences do not lend themselves to routine use.

Of the tests that are practical for routine use in microbiological

laboratories, Baird-Parker (1962, 1963, 1965) found in his comprehensive study of *Micrococcaceae* organisms that a test similar to that
proposed earlier by Evans *et al.* (1955) for production of acid from
glucose under anaerobic conditions was the most reliable for dividing
the cultures into two broad groups. He grouped the facultatively
anaerobic organisms, capable of producing acid from glucose under
anaerobic conditions, in the genus *Staphylococcus,* and the obligately
aerobic organisms, incapable of producing acid from glucose under
anaerobic conditions, in the genus *Micrococcus.* This is a simple,
easily applied test, but certain precautions must be taken when using
it. Since some micrococci are able to produce acid by an oxidative
process, strict anaerobic conditions must be maintained from the
beginning to the completion of the test. The recommendations of the
International Subcommittee on Staphylococci and Micrococci (Subcommittee 1965) as to medium composition and procedures to be
used when performing the test should be followed explicitly. When
carefully performed, this test gives reliable results with a high
percentage of *Micrococcaceae* isolates, although it has been clearly
shown by several workers that the test fails with certain isolates if
the G+C content in the DNA is used as the primary criterion for
separating micrococci from staphylococci. The problem arises with
some cultures which ferment glucose very slowly under anaerobic
conditions, but have DNA with a G+C content in the 30 to 37%
range, which is typical for staphylococci. A recent test proposed by
Evans and Kloos (1972) in which a semisolid thioglycolate medium
is used to detect growth may prove to be a partial answer to the
problem. It is claimed that facultatively anaerobic *Micrococcaceae*
cultures produce visible growth in this medium whether or not they
are capable of fermenting glucose rapidly.

Another easily applied test which is useful for differentiating
staphylococci and micrococci is the lysostaphin test. Lysostaphin is
an enzyme discovered by Schindler and Schuhardt (1964) which
causes lysis of staphylococci but not of micrococci. Presumably,
lysostaphin acts specifically on the type of interpeptide bridges
found in the peptidoglycan polymer of staphylococcal, but not
micrococcal, cell walls. Data from several sources, some of which are
summarized by Lachica *et al.* (1971), indicates that this test is at
least as dependable as the test for anaerobic fermentation of glucose,
and perhaps more so.

Other laboratory tests suitable for routine use have been proposed
and undoubtedly new tests will be developed for separating
staphylococci from micrococci. It is not likely, however, that any
single test will ever be found to be perfect for this purpose.

Consequently, to minimize errors in identification a combination of two or more tests, each of which has a high degree of reliability, should be used.

Since organisms of the genus *Staphylococcus* are considered in another chapter of this book, the following discussion will be confined to the organisms presently classified in the genus *Micrococcus*. To summarize their major characteristics, these are the organisms of the family *Micrococcaceae* that are aerobic and unable to grow and ferment glucose under strictly anaerobic conditions, are not lysed by lysostaphin, are nonmotile, and have DNA with 55 to 75% G+C. The micrococci are mainly saprophytic, nonpathogenic, and nontoxigenic organisms which are widely disseminated in nature. Division of the genus into easily recognized species has been, and continues to be, a difficult problem. Many different species names have been used to designate the various organisms, but in most cases the criteria used for separating species have proven to be inadequate for one reason or another. As a result, it is virtually imposssible to assign newly isolated cultures to specific species on the basis of descriptions given in any of the taxonomical schemes proposed up to and including the 7th edition of *Bergey's Manual for Determinative Bacteriology* (1957).

In recent years there has been renewed interest in the taxonomy of these organisms, and considerable progress has resulted from the research work of many different scientists. The comprehensive studies of large numbers of cultures by Baird-Parker (1962, 1963, 1965) resulted in a new classification scheme which is superior to any proposed previously. He was able to divide the organisms having the characteristics typical of micrococci into eight subgroups based on morphological, physiological, and biochemical characteristics. When he applied this system to a large number of named cultures received from various sources throughout the world, Baird-Parker (1965) was able to classify all but a small percentage of them. It was also found that cultures with very similar characteristics had been given several different species names by the workers who had originally isolated them.

Baird-Parker chose not to attach species names to his subgroups of micrococci because of the uncertainty as to what constitutes a bacterial species. Some of the major characteristics of the various subgroups will be considered here, but for a detailed account the reader should consult the original publication of Baird-Parker (1965). Organisms placed in Baird-Parker's (BP) subgroups 7 and 8 produce little if any acid from glucose under aerobic conditions which sets them apart from other micrococci. They have DNA with G+C content in the approximate range of 66 to 75%. These are the subgroups into which the aerobic *Sarcina* species were placed, along with other cocci with the same characteristics except for that of producing cubical packets of cells. BP subgroup 8 contains the pink to red-pigmented nonmotile and nonhalophilic organisms, whereas BP subgroup 7 contains both white and yellow-pigmented cultures. There is also a distinct difference in the peptidoglycan structure of the cell walls between the organisms in the two subgroups (Schleifer and Kandler 1972). The organisms placed in BP subgroups 5 and 6 produce acid but no acetoin from glucose under aerobic conditions. They are also able to utilize a wider range of carbohydrates than most of the other micrococci. Separation of the two subgroups is based primarily on the production of phosphatase by BP subgroup 6 organisms.

In addition to acid, the micrococci included in BP subgroups 1 through 4 produce acetoin as one endproduct of aerobic glucose metabolism. Further division of these organisms into subgroups is based on several additional characteristics. In general, the characteristics of micrococci placed in these subgroups are very similar to those of coagulase-negative staphylococci. Consequently, when the standard test for anaerobic production of acid from glucose is used as the only criterion for separating micrococci from staphylococci, some cultures may be classified as micrococci in BP subgroups 1 through 4 and then subsequently may be found to have G+C content in the DNA, peptidoglycan structure, and lysostaphin sensitivity typical of staphylococci. These cultures differ from typical staphylococci in that they ferment glucose very slowly under anaerobic conditions. The problems involved in the classification of these organisms have not been completely resolved.

The designation of named species of micrococci remains a difficult problem. Ideally, this should be done on the basis of genetic relatedness using the modern tools of microbial genetics, such as transformation and transduction. A start has been made in this direction (Kloos 1969; Schleifer *et al.* 1972), which has resulted in genetic confirmation of the validity of the species, *M. luteus*, as described by Kocur *et al.* (1972). Yellow-pigmented cultures previously known by several different species names and included in BP subgroup 7 were shown to be closely related genetically and to have nearly identical physiological and morphological characteristics. Two other species which are now widely accepted but which have not been subjected to critical genetic analysis are *M. roseus* and *M. varians. Micrococcus roseus* (Kocur and Pácová 1970) corresponds to the BP subgroup 8 organisms while *M. varians* (Kocur and Martinec

1972) corresponds to the yellow-pigmented organisms included in BP subgroup 5. There is still some question as to the validity of any other named species of *Micrococcus* at this time.

This discussion was not intended as a critical review of the literature on the taxonomy of micrococci since many of the contributions to our knowledge of this subject have not been referred to. Rather, the discussion was intended only to make the reader aware of some of the problems confronting the food microbiologist who attempts to identify micrococci. As yet, no completely satisfactory scheme has been devised for this purpose although significant progress has been made toward this ultimate goal during the past twenty years.

Despite the recognized shortcomings of Baird-Parker's (1965) system for grouping the micrococci, it offers advantages for microbiologists who are not primarily interested in taxonomy, but still must categorize the organisms with which they are working. This can be done on the basis of results obtained with relatively unsophisticated tests which can be performed in a routine manner. The system also avoids the use of certain species names where the validity of the species has not been adequately established as yet, and it removes the temptation to introduce new species names prematurely.

GENERAL CHARACTERISTICS

Micrococci are saprophytic organisms which are very widely distributed in nature. They are found in both fresh and salt water, in air, and soil, in domestic sewage and animal faeces, and on plant surfaces. They are normal inhabitants of the skin of humans and other warm-blooded animals, and are usually present in the slime layer on fish. Thus it is not surprising that micrococci are present on or in essentially all fresh raw food materials destined for human consumption. They are rarely the primary cause of spoilage of raw foods, however, since they are unable to compete with more rapidly growing organisms present in the mixed flora. On the other hand, when considered in the broadest sense, micrococci are well equipped to survive and grow under certain conditions which will inhibit or markedly restrict the growth of many competing organisms. Thus, micrococci assume importance in foods under conditions which give them a selective advantage such as occurs with certain methods used in food preservation. Some of the characteristics of micrococci which bear on this point will be discussed briefly before considering specific food products in which these organisms play a role, either desirable or undesirable.

Many of the micrococci have a rather high tolerance for sodium chloride. They are capable of growing in nutrient solutions containing concentrations of this salt in the general range of about 10 to 25% which is inhibitory to most other microorganisms. Micrococci will often be found to be a significant part of the microbial flora of those foods in which salt concentrations in this 10 to 25% range are depended upon as the major means of preservation.

The ability to utilize the nitrate ion as an electron acceptor and to reduce it to the nitrite level or beyond is characteristic of a high percentage of strains of micrococci classified in BP subgroups 1 through 6. Such organisms can grow in the absence of gaseous oxygen when nitrate is available, and the production of nitrite ions affords some selective advantage since many organisms are inhibited by nitrite. In the preparation of a variety of cured meat products nitrate may be added along with sodium chloride in the pickling solutions. Certain micrococci can grow in such solutions if the temperature is above $0°C$, and they assist in the production of nitrite, which is essential for fixation of meat pigment. Nitrite also has an inhibitory effect on certain undesirable bacteria such as clostridia.

Some of the micrococci are quite resistant to heat when compared with other asporogenous bacteria. If present in a low-acid product such as milk, they survive the normal pasteurization process of $62.8°C$ for 30 min or $71.7°C$ for 15 sec. Thermoduric micrococci usually make up a significant percentage of the total viable bacterial population in freshly pasteurized milk, but under normal conditions of milk production and processing the actual number of these organisms present in the product is small and of no practical significance. If milk handling equipment is improperly cleaned and sanitized either on the producing farm or in the processing plant, a rapid build-up of thermoduric micrococci may occur. They are well adapted to grow at room temperature on wet food-handling surfaces even if the concentration of nutrients is quite low. The resultant contamination of the raw milk may cause the standard plate count on the pasteurized product to exceed the legal limit of 2×10^4 /ml. Thermoduric micrococci are of no recognized public health significance, but they may contribute to spoilage of pasteurized milk or cream if the storage temperature exceeds 7 or $8°C$.

Micrococci are more resistant to dehydration than are most other asporogenous bacteria. They are the most common kind of airborne bacteria, and this is one source of these organisms in foods. Thermoduric micrococci are among the most prevalent types of viable asporogenous bacteria found in dehydrated foods. When the preheat treatment applied to a food is insufficient to kill the micrococci and appreciable numbers of these organisms are present at the time of drying, then products dried by methods other than the atmospheric drum process will contain large numbers of viable micrococci.

Work on the application of ionizing radiations to the preservation of foods has established that at least some micrococci are remarkably resistant to the process. Anderson *et al.* (1956) were the first to isolate such an organism which was subsequently named *Micrococcus radiodurans*. Studies on this and several other strains isolated from various sources have clearly shown that *M. radiodurans* is more resistant to ionizing radiations than are the spores of *Clostridium botulinum* type A, which were formerly believed to be the most resistant. Much of the information on the general characteristics of this organism and on the factors that influence its resistance to ionizing radiations has been summarized by Jay (1970) and will not be repeated here.

While *M. radiodurans* is of considerable interest to scientists studying the basic factors that determine resistance to radiations, it is of little practical significance in connection with radiation sterilization of foods because it is readily destroyed by heat. From their study of thermal death characteristics of this organism, Duggan *et al.* (1963) concluded that heating a food product to a uniform temperature of 65.5°C for 1 to 2 min prior to irradiation will eliminate *M. radiodurans* for all practical purposes.

Other micrococci, although not possessing the extreme resistance exhibited by *M. radiodurans*, are more resistant to ionizing radiations than are most other asporogenous bacteria. They have been found to predominate in the surviving flora on fresh fish fillets exposed to low dosages of ionizing radiation (0.3 to 0.6 M rads) for the purpose of prolonging storage life under refrigeration (Abrahamsson *et al.* 1965; Kazanas, 1968; Pelroy *et al.* 1967; Corlett *et al.* 1965). It has also been found that the surviving micrococci cannot grow at the usual storage temperatures (0.5C to 6°C) employed, so they are not involved in the spoilage of this kind of product under normal conditions.

Most micrococci are also quite sensitive to acid conditions. With few exceptions they are inhibited at about pH 5.0. Many of them are able to grow at pH values up to about 8.5 and some will grow at even higher pH values. Under conditions favorable for their growth, micrococci can reduce the acidity of foods containing organic acids such as lactic or acetic which they are capable of oxidizing. On the other hand, the micrococci corresponding to BP subgroups 1 through

6 produce some acid from metabolizable sugars and higher alcohols under conditions of restricted oxygen supply.

With regard to the effect of temperature on growth, the micrococci are mesophilic organisms that grow optimally in the general range of 25°C to 35°C. No thermophilic micrococci have been reported, and for most strains the upper limit for growth is in the 40°C to 45°C range. Some micrococci can grow at temperatures as low as about 2°C, and the majority of the cultures listed by Baird-Parker (1965) grew at 10°C. Although psychrotrophic micrococci have the potential for growth under aerobic conditions on low-acid foods stored at temperatures within this range, they are seldom found in significant numbers on foods that have spoiled under refrigeration. The reason for this is that many of the fresh foods stored under refrigeration contain a mixed flora of microorganisms including psychrotrophic Gram-negative bacteria, such as pseudomonads, which out-grow the micrococci. If these organisms are inhibited in some way, a large population of micrococci may develop. As an example, in the production of certain types of brine-treated meat products, the brine contains sufficient sodium chloride to inhibit the pseudomonads and most other psychrotrophic Gram-negative bacteria. Under these conditions, salt-tolerant micrococci may grow extensively on the surface of the meat both before and after it is removed from the brine.

As a group, the micrococci are not noted for their ability to cause rapid and extensive chemical changes in foods. They are relatively inert when compared with bacteria such as the coliforms, the pseudomonads or the lactic acid bacteria. Micrococci do have the ability to cause certain chemical changes which are of some practical importance in foods. As mentioned previously, all micrococci, because of their aerobic nature, can oxidize a variety of organic substrates including certain acids, sugars, and higher alcohols. Many of the micrococci can reduce nitrate to nitrite, and they are of practical importance in the production of various cured and fermented meat products. Some micrococci produce proteolytic enzymes which will hydrolyze gelatin, or casein, or both. Presumably, other food proteins may also be attacked by these enzymes, but very little information is available on this point. Finally, some micrococci produce lipolytic enzymes which will hydrolyze food lipids such as milk-fat and lard. Such organisms can cause hydrolytic rancidity under conditions favoring their growth on fatty foods.

ROLE IN SPECIFIC FOODS

Micrococci are present in nearly all raw foods, and they can be recovered from many kinds of processed food products. Despite their wide-spread occurrence, micrococci are of limited importance in most foods, but there are some foods in which they do play an active role.

Brick and Other Surface-Ripened Cheeses

Brick cheese is representative of a class of cheeses ripened under conditions that favor extensive growth of microorganisms on the surface. Among the cheeses in this category are port du salut, liederkranz, limburger, tilsiter, and several other varieties. Although these cheeses have distinctly different characteristics at the completion of the ripening process, the general sequence of organisms growing on the surface during the early stage of ripening is quite similar for all. A major factor involved is the general method used in salting the freshly formed cheese. This is done either by floating the cheese in a strong brine (20 to 25% sodium chloride solution) or by rubbing dry salt on the surfaces. With either method, there is a high concentration of salt at the surface of the cheese during the early stages of ripening which limits growth of microorganisms to those with a high salt tolerance.

The salted cheese at about pH 5 is placed in curing rooms at a temperature of 10 to 15°C with a relative humidity of about 90%. Under these conditions, salt-tolerant yeasts appear first. As the lactic acid is oxidized by the yeasts, the pH at the surface of the cheese rises sufficiently to allow salt-tolerant micrococci to compete, and they become one of the predominant types of organisms in the mixed flora. Micrococci also oxidize lactic acid and contribute to the continued rise in pH. As the pH approaches 5.8, salt tolerant bacteria of the *Brevibacterium* genus, either *B. linens* or *B. erythrogenes* or both, find conditions favorable for growth and soon become the predominant type of organism in the surface smear. These strongly proteolytic organisms are believed to be largely responsible for the characteristic limburger-like flavor of most types of cheese ripened by this general procedure.

What contribution, if any, the micrococci make to the flavor is not known. In the case of the mild flavored brick cheese, development of the surface flora is often interrupted before the pH has risen sufficiently to allow growth of the *Brevibacterium* species. At this stage, micrococci are predominant in the smear. It was concluded by Lubert and Frazier (1955), who inoculated cheese made under carefully controlled conditions with pure cultures of yeast and various micrococci isolated from brick cheese, that the micrococci are responsible for the characteristic flavor. They found that cheese inoculated only with a mycoderma yeast did not develop typical flavor, and that the micrococci, which were identified at that time as strains of *M. varians*, *M. caseolyticus*, and *M. freudenreichii*, could not grow on cheese which had not been inoculated with the yeast. When fresh cheese was inoculated simultaneously with yeast and micrococci, the yeast grew first and the micrococci grew later. Under these conditions, the typical brick cheese flavor developed. From these and other experiments, it was concluded that the yeast stimulated growth of the micrococci by lowering the acidity of the cheese and also by producing growth factors required by the micrococci. The nature of the flavor components contributed to the cheese by the micrococci is not known.

Cheddar Cheese

It is generally conceded that cheddar cheese made from pasteurized milk does not ripen as quickly or develop as much desired flavor as does top quality cheese made from raw milk. Among the many possible reasons for this is the destruction of heat-sensitive organisms present in the raw milk, some of which might be of importance in flavor development. Both heat-sensitive and thermoduric strains of micrococci are generally found in raw milk, and at times they are present in rather large numbers in cheese made from raw milk. Alford and Frazier (1950A) reported that two-day old cheddar cheese made from raw milk had counts of nonlactic organisms ranging from 6×10^5 to 1×10^7 per gram. They estimated that about 80% of these organisms were micrococci with white-pigmented types predominating. By contrast, cheese made from pasteurized milk consistently had $<5 \times 10^4$ nonlactic organisms per gram, and the predominant micrococci produced a yellow pigment. Based on the classification system in use at that time, the predominant organisms isolated from the raw milk cheese were identified as varieties of *M. freudenreichii* and *M. caseolyticus*.

Alford and Frazier (1950B) then studied the effect of selected strains of the *M. freudenreichii* type on the development of flavor in cheese made from milk which was inoculated with the test organisms after pasteurization. The strains selected were lipolytic, capable of producing acid coagulation of milk, and able to grow in competition with lactic starter organisms. They grew actively during the cheese-making operation, increasing in numbers from 1×10^6 /ml in the milk to 2×10^8 /gram of cheese at pressing. It was found that some test strains, but not all, greatly accelerated the development of flavor in the cheese, although it was not clearly stated whether or not the flavor was typical for cheddar cheese. As an aside, it is interesting to speculate as to how the organisms used in this study would be classified by the system now in use. The ability of the test strains to grow extensively in competition with the lactic starter organisms in the cheese strongly suggests that they were facultative anaerobes. If so, they would now be classified in the genus *Staphylococcus* (varieties of *S. epidermidis*?), a genus which was not recognized in *Bergey's Manual* at the time this work was done.

Robertson and Perry (1961) found that a selected micrococcus strain, when added to pasteurized milk just before making cheddar cheese, improved the flavor of the ripened cheese in 28 out of 33 experimental lots made at different times during the cheese-making season in New Zealand. With the other five lots there was no detectable difference in cheddar flavor between experimental and control cheeses. In this work, the organism under study did not multiply during the cheese-making operation, and there was a rapid decrease in numbers of viable cells during the ripening period. No species name was suggested for this organism, but judging from the rather complete description of biochemical and physiological characteristics given, it corresponded quite closely to micrococci included in BP subgroup 2. It was salt tolerant, lipolytic, coagulase negative, nonpigmented, acetoin positive, and capable of producing considerable acid from lactose under aerobic conditions. The latter characteristic, coupled with its inability to grow during the cheese-making operation, indicates that the organism was an obligate aerobe, and hence would also be classified as a micrococcus under the present classification system. Of particular interest was the finding that the selected micrococcus enhanced the cheese flavor even when added to the milk at the low level of approximately 1×10^5 viable cells per ml. It has been a rather common assumption in the past that only those organisms that grow extensively during cheese making and/or ripening play a significant role in flavor development.

Salt Cured Meat Products

A time-honored method of preserving beef and pork involves the addition of sodium chloride to the meat in sufficient quantity to inhibit the growth of most microorganisms, particularly those capable of causing putrefactive spoilage. Although sodium chloride is the major preservative, other factors also contribute to the keeping quality of many of the cured products. Nitrite, nitrate, or a combination of the two is usually added along with the salt. Whether added as such or produced from nitrate by bacterial action, nitrite is necessary for the fixation of the red color of the meat. In addition, it is inhibitory to many microorganisms, including *Clostridium botulinum*, and also may contribute to the flavor of the finished products. Sugar is generally added to the curing mixture largely because of its desirable effect on flavor and texture of the products, but it serves also as a readily available energy source for growth of salt-tolerant bacteria. Other additives are used for various purposes other than preservation. Other factors that aid in limiting microbial growth are refrigeration during the curing process, and the smoking treatment that is applied to many of the salt-cured products. In addition to the germicidal effects of certain smoke constituents, the moderate heat treatment and resultant drying of the meat surfaces also aid in preservation.

There are many variations of the salt-curing method used by different processors in the preparation of products of this general type, but the technological aspects are beyond the scope of this discussion. For details, the reader is referred to the recent book by Kramlich *et al.* (1973). There are two basic methods for curing primal cuts of meat such as the bellies, hams, shoulders, loins, and jowls of hogs, and the briskets and certain other cuts of beef. These are the fast or short-cure method, and the slow or long-cure procedure.

With the newer short-cure method, the curing solution is pumped into the meat either through a tube inserted into the main artery of cuts such as hams, or through hypodermic-like needles that penetrate directly into the tissues. Both methods of pumping may be applied to hams in order to assure rapid and uniform distribution of the solution. Sufficient nitrite is included in the solution to assure color fixation, thus making it unnecessary to depend on microbial reduction of nitrate during curing. With this procedure, which is now widely used in the U.S., there is essentially no opportunity for micrococci or any other bacteria to grow during the curing operation. Some products may be smoked immediately after pumping while others may be pumped and then covered with curing solution for a short time before smoking.

With the older slow-cure methods, the dry mixture of curing agents is placed directly on the meat, in which case it dissolves in the liquid drawn from the tissues, or it is first dissolved in water to make a concentrated brine in which the meat is immersed. In either case, depending on the thickness of the cuts, from several days to a few weeks are required for uniform distribution of the curing agents throughout the meat. The temperature should be maintained at $4°C$ or lower during the curing period. Under these conditions, psychrotrophic micrococci are generally found to be the predominant type of bacteria on the meat surfaces and in the brine. They continue to grow on the meat after removal from the brine, and are present in large numbers in the superficial slime that forms on some types of cured products. A high percentage of the micrococci growing under these conditions have been shown to be lipolytic and capable of reducing nitrate (Kitchell 1962). Since reduction of nitrate is relied upon as the major source of nitrite in the slow-cure process, the micrococci play an important role in color fixation.

Micrococci also have been implicated in the spoilage of certain types of cured meat products. An example is Wiltshire bacon which is produced mainly in Great Britain and some European countries. Under conditions that allow excessive growth of micrococci to occur, a rancid, cheesy flavor is produced which may be due to extensive hydrolysis of fat by these organisms. Much of the work on the microbiology of Wiltshire bacon has been summarized by Kitchell (1962) and by Ingram and Dainty (1971).

Dry, Fermented Sausage

Fermented dry sausages of several types are produced primarily in European countries. Although there are many variations in the manufacturing process, basically it involves mixing the curing ingredients (salt, nitrate, and glucose) and spices of various kinds with coarsely ground meat. Sausages prepared with the mixture are allowed to undergo fermentation at approximately 20°C until sufficient acid has been produced to inhibit spoilage organisms. The sausages are slowly air-dried under controlled conditions until the water activity is lowered to a point where bacterial growth is completely inhibited. The entire process takes several weeks for completion. Even though this type of sausage is not cooked, and it may or may not be smoked, it has very good keeping quality.

The general quality of dry fermented sausage is largely determined during the early stage of the fermentation. Ideally, rapid reduction of nitrate and rapid production of acid should take place. With the natural fermentation where the ingredients are depended upon to supply suitable organisms, unsatisfactory results are quite often obtained. Niinivaara (1955) found that the fermentation could be satisfactorily controlled by inoculating the sausage mixture with a strain of micrococcus selected from a successful batch of naturally fermented sausage. The culture used as a starter was selected for its ability to grow rapidly at high salt concentrations, to reduce nitrate quickly and extensively, and to produce acid from glucose (Niinivaara et al. 1964). Such starter cultures have been produced commercially and sold under the trade name "Baktofermente."

BIBLIOGRAPHY

ABRAHAMSSON, K., DESILVA, N. N., and MOLIN, N. 1965. Toxin production by Clostridium botulinum, type E, in vacuum-packed, irradiated fresh fish in relation to changes of the associated microflora. Can. J. Microbiol. *11*, 523-529.

ALFORD, J. A., and FRAZIER, W. C. 1950A. Occurrence of micrococci in cheddar cheese made from raw and from pasteurized milk. J. Dairy Sci. *33*, 107-113.

ALFORD, J. A., and FRAZIER, W. C. 1950B. Effect of micrococci on the development of flavor when added to cheddar cheese made from pasteurized milk. J. Dairy Sci. *33*, 115-120.

ANDERSON, A. W. *et al.* 1956. Studies on a radio-resistant micrococcus. I. Isolation, morphology, cultural characteristics, and resistance to gamma radiation. Food Technol. *10*, 575-578.

BAIRD-PARKER, A. C. 1962. The occurrence and enumeration, according to a new classification, of micrococci and staphylococci in bacon and on human and pig skin. J. Appl. Bacteriol. *25*, 352-361.

BAIRD-PARKER, A. C. 1963. A classification of micrococci and staphylococci based on physiological and biochemical tests. J. Gen. Microbiol. *30*, 409-427.

BAIRD-PARKER, A. C. 1965. The classification of staphylococci and micrococci from world-wide sources. J. Gen. Microbiol. *38*, 363-387.

BREED, R. S, MURRAY, E. G. D., and SMITH, N. R. 1957. Bergey's Manual of Determinative Bacteriology, 7th Edition. The Williams & Wilkins Co., Baltimore, Md.

BUCHANAN, R. E., and GIBBONS, N. E. 1974. Bergey's Manual of Determinative Bacteriology, 8th Edition. The Williams & Wilkins Co., Baltimore, Md.

CORLETT, D. A., JR., LEE, J. S., and SINNHUBER, R. O. 1965. Application of replica plating and computer analysis for rapid identification of bacteria in some foods. II. Analysis of microbial flora in irradiated Dover sole *(Microstomus pacificus)*. Appl. Microbiol. *13*, 818-822.

COWAN, S. T. 1962. An introduction to chaos, or the classification of micrococci and staphylococci. J. Appl. Bacteriol. *25*, 324-340.

DEIBEL, R. H., and EVANS, J. B. 1960. Modified benzidine test for the detection of cytochrome-containing respiratory systems in microorganisms. J. Bacteriol. *79*, 356-360.

DUGGAN, D. E., ANDERSON, A. W., and ELLIKER, P. R. 1963. Inactivation-rate studies on a radiation-resistant spoilage microorganism. III. Thermal inactivation rates in beef. J. Food Sci. *28*, 130-134.

EVANS, J. B., BRADFORD, W. L., and NIVEN, C. F., JR. 1955. Comments concerning the taxonomy of the genera *Micrococcus* and *Staphylococcus*. Intern. Bull. Bacteriol. Nomenclature Taxonomy *5*, 61-66.

EVANS, J. B., and KLOOS, W. E. 1972. Use of shake cultures in a semisolid thioglycolate medium for differentiating staphylococci from micrococci. Appl. Microbiol. *23*, 326-331.

FRAZIER, W. C. 1967. Food Microbiology, 2nd Edition. McGraw-Hill Book Co., New York.

INGRAM, M., and DAINTY, R. H. 1971. Changes caused by microbes in spoilage of meats. J. Appl. Bacteriol. *34*, 21-39.

JAY, J. M. 1970. Modern Food Microbiology. Van Nostrand Reinhold, New York.

KAZANAS, N. 1968. Proteolytic activity of microorganisms isolated from freshwater fish. Appl. Microbiol. *16*, 128-132.

KITCHELL, A. G. 1962. Micrococci and coagulase negative staphylococci in cured meats and meat products. J. Appl. Bacteriol. *25*, 416-431.

KLOOS, W. E. 1969. Transformation of *Micrococcus lysodeikticus* by various members of the family *Micrococcaceae*. J. Gen. Microbiol. *59*, 247-255.

KOCUR, M., BERGAN, T., and MORTENSEN, N. 1971. DNA base composition of gram-positive cocci. J. Gen. Microbiol. *69*, 167-183.

KOCUR, M., and MARTINEC, T. 1972. Taxonomic status of *Micrococcus varians* Migula 1900 and designation of the neotype strain. Intern. J. Syst. Bacteriol. *22*, 228-232.

KOCUR, M., and PÁCOVÁ, Z. 1970. The taxonomic status of *Micrococcus roseus* Flügge, 1886. Intern. J. Syst. Bacteriol. *20*, 233-240.
KOCUR, M., PACOVA, Z., and MARTINEC, T. 1972. Taxonomic status of *Micrococcus luteus* (Schroeter 1872) Cohn 1872, and designation of the neotype strain. Intern. J. Syst. Bacteriol. *22*, 218-223.
KRAMLICH, W. E., PEARSON, A. M., and TAUBER, F. W. 1973. Processed Meats. Avi Publishing Co., Westport, Conn.
LACHICA, R. Y. F., HOEPRICH, P. D., and GENIGEORGIS, C., 1971. Nuclease production and lysostaphin susceptibility of *Staphylococcus aureus* and other catalase-positive cocci. Appl. Microbiol. *21*, 823-826.
LUBERT, D. J., and FRAZIER, W. C. 1955. Microbiology of the surface ripening of brick cheese. J. Dairy Sci. *38*, 981-990.
NIINIVAARA, F. P. 1955. The influence of pure bacterial cultures on the maturing and red color development of raw sausage. Acta Agral. Fennica *84*, 95-101. (German)
NIINIVAARA, F. P., POHJA, M. S., and KOMUTAINEN, S. E. 1964. Some aspects about using bacterial pure cultures in the manufacture of fermented sausages. Food Technol. *18*, 25-31.
PELROY, G. A., SEMAN, J. P., JR., and EKLUND, M. W. 1967. Changes in the microflora of irradiated petrole sole (*Eopsetta jordani*) fillets stored aerobically at $0.5°C$. Appl. Microbiol. *15*, 92-96.
ROBERTSON, P. S., and PERRY, K. D. 1961. Enhancement of the flavour of cheddar cheese by adding a strain of micrococcus to the milk. J. Dairy Res. *28*, 245-253.
SCHINDLER, C. A., and SCHUHARDT, V. T. 1964. Lysostaphin: a new bacteriolytic agent for the staphylococcus. Proc. Natl. Acad. Sci. *51*, 414-421.
SCHLEIFER, K. H., and KANDLER, O. 1972. Peptidoglycan types of bacterial cell walls and their taxonomic implications. Bacteriol. Rev. *36*, 407-477.
SCHLEIFER, K. H., KLOOS, W. E., and MOORE, A. 1972. Taxonomic status of *Micrococcus luteus* (Schroeter 1872) Cohn 1872: Correlation between peptidoglycan type and genetic compatibility. Intern. J. Syst. Bacteriol. *22*, 224-227.
SUBCOMMITTEE. 1965. Minutes of first meeting of subcommittee on taxonomy of staphylococci and micrococci. Intern. Bull. Bacteriol. Nomenclature Taxonomy *15*, 107-108.
———1974. Bergey's Manual of Determinative Bacteriology. 8th ed. The Williams and Wilkins Co., Baltimore.

CHAPTER 14

J. R. Stamer | Lactic Acid Bacteria

The members of the genera *Lactobacillus*, *Leuconostoc*, *Pediococcus*, and *Streptococcus* are collectively and commonly referred to as the "lactic acid bacteria."

From an ecological viewpoint, the lactic acid bacteria occupy no singular natural habitat. Their isolation from spoiled food products, contributions to desirable food fermentations of plant and animal origin, and their limited associations with pathogenicity attest to their ubiquitous distribution in our environment.

The distribution patterns of the lactics, as described by Mundt and Hammer (1968) and a discussion pertaining to the ecology of the lactic streptococci by Sandine *et al.* (1972), suggest that the pathways of dissemination occur via a cyclic route involving the distribution of intestinal wastes to plant materials with a subsequent return to the host by the ingestion or handling of the food product. Their wide ranges of distribution, coupled with their ability to grow on a myriad of organic substrates under acidic, basic, aerobic, and anaerobic conditions make the lactic acid bacteria formidable competitors in all areas of food processing. Some examples of foods which differ markedly in physical and chemical composition, but yet are susceptible to attack by the lactic acid bacteria are shown in Table 14.1.

TAXONOMY

From a taxonomic viewpoint this somewhat loosely defined group of microorganisms has presented to microbiologists severe challenges in establishing rigid parameters so essential for definitive classification purposes. Therefore, the classification schemes formulated as a result of the continued advances in microbial physiology, anatomy, genetics, and biochemistry have periodically been subjected to extensive revisions. In retrospect a comparison of the 6th Edition of Bergey's Manual of Determinative Bacteriology (Breed *et al.* 1948) to the 7th Edition (1957), makes the extent of such revisions strikingly apparent. In the earlier edition the lactic group, comprised of seven genera, was incorporated under the single family *Lactobacteriaceae*. In the latter edition not only was the family name changed to *Lactobacillaceae*, but the family was expanded to include ten genera. In the 8th Edition of Bergey's Manual the lactic group is divided into

TABLE 14.1

EXAMPLES OF UNDESIRABLE EFFECTS PRODUCED BY LACTIC ACID BACTERIA UPON SOME FOOD COMMODITIES

Product	Undesirable Trait	Organisms Implicated	References
Dairy	Malty flavor of cream	*Streptococcus lactis* var. *maltigenes*	Frazier 1958
	Slit-open defect in cheddar cheese	*Leuconostoc citrovorum*	Holmes *et al.* 1968
Meats	Green discoloration Surface slime	*Lactobacillus viridescens* *Streptococcus* spp, *Lactobacillus* spp.	Niven *et al.* 1949
Fish marinades	Gas production; Ropiness	*Lactobacillus buchneri* *Lactobacillus brevis* *Lactobacillus casei*	Meyer 1964 Priebe 1970
Citrus juices	Gas, off-flavors	*Lactobacillus plantarum* *L. brevis, Leuconostoc dextranicum*	Hays and Riester 1952
Mayonnaise	Gas, off-flavors	*Lactobacillus plantarum* *L. fructovorans*	Kurtzman *et al.* 1971
Pickles	Bloaters	*Lactobacillus brevis*	Etchells *et al.* 1964
Sauerkraut	Slime formation	*Leuconostoc mesenteroides*	Pederson 1960
Tomato	Gas, swollen cans	*Lactobacillus brevis*	Pederson 1929
Sugar	Slime formation Acidification	*Leuconostoc mesenteroides* *Lactobacillus confusus*	Tilbury 1968; Sharpe *et al.* 1972
Beer, wine, and ciders	Haze formation, reduction or increase in organic acids, Production of diacetyl & acetylmethylcarbinol	*Lactobacillus fermenti, L. pastorianus, L. trichodes, L. plantarum, Leuconostoc oenos, Pediococcus cerevisiae*	Haas 1960; Rankine and Bridson 1971; Carr and Davies 1970

two families, *Streptococcaeceae* and *Lactobacillaceae*. The former family consists of the genera *Streptococcus, Leuconostoc, Pediococcus, Aerococcus*, and *Gemella*, whereas the latter family is comprised of the genus *Lactobacillus*. Although the taxonomic characteristics which define the boundaries of the newly revised classification scheme are both subtle and marked, all species within the family groups share a common denominator, namely the ability to produce lactic acid from a fermentable carbohydrate.

MORPHOLOGY

From the standpoint of morphology, the members of the family are Gram-positive, nonsporulating rods or cocci occurring singly, in pairs or chains, and occasionally as tetrads. Although partial differentiation of the genera may appear to be a simple exercise in staining and microscopy, morphological confirmation may be difficult to achieve at times because of the lack of sharp visual demarcation between cocci and very short rods. This inability or reticence to establish exact morphological identification has produced the descriptive term, "cocco-bacillus" (Perry and Sharpe 1960). This capacity of the cell to assume such indeterminate bacillary or coccoid morphologies may be directed in part by environmental conditions. For example, cultures freshly isolated from vegetables sources and of questionable morphological definition can be induced to elongate several-fold, by presumably imposing upon the cells conditions which lead to excessive biological stress. These aberrant morphologies may be observed to occur as a result of the incorporation of organic acids into a complex growth medium, increasing the concentration of hydrogen ions, vigorous aeration, the use of gaseous atmospheres such as carbon dioxide or nitrogen, and elevated temperatures for incubation. Although these external stimuli may exert profound effects upon morphology, it does not imply that cocci may be transformed into rods (or vice versa) for if this were so, then the division between the genera within this family would no longer exist. This is to emphasize that the results of the initial direct microscopic examination of foods should be viewed with reservation, and that the diverse chemical nature and conditions of storage of a particular product may distort the true morphological characteristics of the bacteria in question, thereby producing erroneous interpretations.

NUTRITIONAL REQUIREMENTS

In order to establish and maintain their pronounced fermentative capacities, the lactic acid bacteria require a vast spectrum of organic and inorganic compounds for growth activities. Their fastidious nutritional demands, as evidenced by their absolute requirements for preformed organic precursors such as amino acids, purines, pyrimidines, and vitamins, have made them particularly useful agents for analytical bio-assay purposes (Tittsler *et al.* 1952). Needless to say, the reverse application—that is, the supplementation of a defined, deficient medium with known essential micronutrients (vitamins etc.)—can likewise assist the investigator in establishing species identification. For example, in the case of the two thiamin-requiring species *Lactobacillus buchneri* and *L. brevis*, the former species requires riboflavin for growth, whereas the latter has no such requirement. Or, in the case of *L. leichmanni* folic acid is a required adjunct; however, the closely related species *L. plantarum* is devoid of this requirement.

Since the nutritional requirements may be species specific and unknown factors essential for growth remain to be elucidated, there is currently no one synthetic medium universally acceptable for the growth of all lactic acid bacteria. However, examples of media quite frequently employed for their selection and maintenance are those developed by DeMan *et al.* (1960), Garvie (1960), Whittenbury (1966), and Sabbaj *et al.* (1971).

When grown on agar plates the lactic acid bacteria often produce diminutive colonies, ranging from pinpoint to a few millimeters in diameter. Therefore, the incorporation of acid-base indicators, such as bromocresol-green dye, or the formation of cleared zones in calcium carbonate agar media facilitate enumeration of the aciduric species.

Many species display poor surface growth on agar media and maximum growth yields are best achieved under microaerophilic to anaerobic conditions of incubation. Some species are indifferent in their responses to atmospheric oxygen while others may utilize oxygen as a terminal respiratory acceptor (Anders *et al.* 1970). Although they are generally considered to be catalase-negative, exceptions have been noted (Whittenbury 1964; Johnston and Delwiche 1965). These latter "atypical" catalase-positive responses are manifest under special conditions such as when the cultures are grown in low levels of carbohydrate (Dacre and Sharpe 1956). To circumvent peroxide toxicity, peroxide-destroying enzymes, commonly called peroxidases become a vital constituent of the respiratory network. In the case of the lactic acid bacteria, these latter enzymes catalyze the oxidation of organic substrates by hydrogen peroxide, thus eliminating the necessity for a complete catalase system. Although peroxidase activity is often difficult to demonstrate, an improved rapid technique for its detection reportedly has been devised by Bordeleau and Bartha (1969).

TYPES OF LACTIC FERMENTATIONS

In addition to serving as criteria for general classification purposes, the quantities and stereochemical configurations of the lactic acid produced serve as supplemental guidelines for establishing the identity of both genera and species.

On the basis of physiological responses the members of the lactic group may be subdivided into homofermentative and heterofermentative classes, depending upon the manner in which the bacteria degrade a utilizable hexose.

According to Doelle (1969), the homofermentative strains produce a 90% conversion of glucose to lactic acid, whereas the heterofermenters produce lesser amounts of lactic acid with corresponding increments in acetic acid, carbon dioxide and neutral components.

To differentiate between homofermentative and heterofermentative pathways of fermentation, the test described by Hansen (1968) has been successfully employed. In brief, a complex growth medium inoculated with the respective culture is covered with sterile vaseline and incubated at 37°C for two weeks. The production of carbon dioxide, as evidenced by the upward shift of the seal from the surface of the broth, is indicative of a heterofermentative response.

Again, whether these physiological responses are a completely valid means of establishing generic or species independence is a point of conjecture because the demarcation between homofermenters and heterofermenters is at times difficult to define.

Of the genera under consideration in this chapter, the cocci *Pediococcus*, *Streptococcus* are homofermenters and *Leuconostoc* is an obligate heterofermenter, whereas the rods (members of the genus *Lactobacillus*) are partitioned into two categories as a result of either heterofermentative or homofermentative responses.

Since the types of lactic acid produced, i.e. L(+), D(−), or racemic mixtures are stable phenotypic characteristics, the configuration and rotation of the acid salts serve as supplemental tests for genera or species identification. It may be seen in Table 14.2 that the genus *Streptococcus* and *Leuconostoc* produce L(+) or D(−) lactic acid only, and that the acidic configurations produced by *Pediococcus* and *Lactobacillus* are of the D, L, or DL types. Unlike the streptococci or the leuconostocs, the development of singular or racemic acid isomers by the latter two genera is species rather than genera specific. These differences in isomeric formation are un-

TABLE 14.2
GENERAL SUMMARY OF DISTINGUISHING CHARACTERISTICS OF THE GENERIC SUBDIVISIONS OF LACTIC ACID BACTERIA

	Morphology	Salt tolerance (% NaCL)	pH range	Growth (10°C)	Growth (45°C)	Optimum growth temperature	Survival (63°C/30 min)	Ammonia from Arginine	Configuration of lactic acid produced
Physiological class:									
1. Homofermentative									
Genera:									
A. *Lactobacillus*[a]	rod								
Groups:									
1. *thermobacterium*		3-6		–	+	37-45	+	–	DL, D(–)
2. *streptobacterium*		>9	4.0–7.4	+	v	28-32	–	–	DL, L(+)
B. *Pediococcus*[b]	coccus								
Species:									
1. *cerevisiae*		10		+	+	25-33	–	+	DL
2. *damnosus*		<4	4.2-8.6	+	–		–	–	DL
3. *halophilus*		10-18		+	–		–	+	L(+)
4. *parvulus*		(requires 5) >6.5		+	–		–	–	DL
C. *Streptococcus*[c]	coccus								
Groups:									
1. enterococcus		6.5	4.6-9.6	v	+	37	+	v	L(+)
2. lactic		2-4	4.2-9.2	+	–	30	–	v	
3. pyogenes		<6.5	4.6-—	–	–	37	–	+	
4. viridans		<6.5	4.0-—	–	+	37	v	–	

TABLE 14.2 (*Continued*)

	Morphology	Salt tolerance (%NaCL)	pH range	Growth (10°C)	Growth (45°C)	Optimum growth temperature	Survival (63°C/30 min)	Ammonia from Arginine	Configuration of lactic acid produced
2. Heterofermentative									
Genera:									
A. *Lactobacillus*[a]	rod								
Group:									
1. *betabacterium*		6.8	3.2-7.2	+	v	28-40	-	+	DL
B. *Leuconostoc*[d]	coccus								
Species:									
1. *cremoris*		<3	3.2-6.5	+	-	20-25	-	-	D(-)
2. *dextranicum*		<3		+	-		-	-	
3. *lactis*		<3		+	-		-	-	
4. *mesenteroides*		6.5		+	-		-	-	
5. *oenos*		-			-			-	

Classification according to: a) Orla-Jensen (1919), Sharpe *et al.* (1966); b) Coster and White (1964); c) Sherman (1937); d) Garvie (1960).

Growth: + positive; - negative; V species variation within group doubtedly due to the inherent presence of either D or L lactic dehydrogenases only, the occurrence of both D and L dehydrogenases, or the actions of lactate racemase systems.

The determination of the configuration of lactic acid derived from a fermented medium is quite a simple procedure. The acid obtained by ether extraction of the acidified medium is crystallized as the zinc salt. Upon desiccation, racemic zinc lactate, which is optically inactive, forms a trihydrate containing 18.18% water, whereas the D or L salts form dihydrates comprised of 12.89% water. Optical rotation (dextrorotary or levorotary) as determined by polarimetry provides confirmation of their optical activities.

Although the above physical criteria have been employed successfully for many years, recent advances in enzyme technology now permit the quantitative determination of the isomeric moieties of lactic acid by specific lactic dehydrogenase assays (Doelle 1971). These enzymic assays, in addition to providing quantitative determinations of either L(+) or D(−) lactic acid, offer rapidity in measurements on a microdeterminative scale.

Since the major energy-yielding mechanism of the lactic acid fermentation is derived from the degradation of polyhydroxy carbon compounds, carbohydrates serve as selective substrates for propagation purposes. Glucose and/or fructose, the hexoses so abundantly found in many food products, are used by all lactic bacteria, whereas other hexoses, di-, tri- and polysaccharides, and pentoses are utilized less frequently.

By taking advantage of the ability or inability of the organisms to ferment specific sugars, it is possible to develop screening procedures which facilitate isolation. For example, let us assume that the microscopic examination of an acidic vegetable product reveals the presence of Gram-positive rods. Since this latter observation leads us to believe that the species present in the product are members of the family *Lactobacillus*, we would like to obtain more definite information as to the nature of the species involved. Now, by plating the sample onto selective agar media containing glucose and fructose, mannitol, arabinose, or xylose, and observing growth, we can achieve the following provisional differentiation:

(1) Glucose-fructose—total lactic population
(2) Mannitol—homofermenters: (*L. casei, plantarum, salivarus*)
(3) Arabinose—heterofermenters: (*L.   brevis,   buchneri,   cellobiosus*, and *fermenti*)
         homofermenter: *L. plantarum*
(4) Xylose—heterofermenters: *L. brevis, buchneri*

An approximation of the numbers and types of species present in the sample may be obtained by correlating the counts derived from mannitol, arabinose, and xylose substrates to that of the glucose-fructose values. Thus, by the judicious selection of substrate, the numbers and types of sugars required for preliminary screening purposes can be reduced significantly. For detailed information pertaining to the many singular sugars which support the growth of specific species of *Lactobacillus*, the summary as prepared by Hansen (1968) should be consulted.

Although the results of specific sugar fermentations are often adequate for identification purposes, additional criteria such as the production of ammonia from L-arginine by the short hetero-fermentatative lactobacilli (Garvie 1960) and more refined methods employing radioisotopes (Weiss *et al*. 1968), comparative enzyme analyses (Mizushima *et al*. 1964; Buyze *et al*. 1957), cell wall composition (Slade and Slamp 1972; Kandler 1970; Kandler *et al*. 1968), or serological typing (Lancefield 1933; Sharpe 1955) serve as useful analytical tools to establishing species confirmation.

LIPOLYTIC PROPERTIES

It is generally assumed that the lactic acid bacteria are non-lipolytic, or at best, possess feeble lipases for the cleavage of triglycerides. Since these microorganisms contain lipid-synthesizing enzymes, as evidenced by the presence of appreciable amounts of lipids within the cellular membrane, it is reasonable to assume that they may also have enzymes required for the degradation of extracellular lipid materials.

Recent evidence suggests that many strains, particularly those associated with dairy fermentation, produce both lipases and esterases (Oterholm *et al*. 1968). Furthermore, using intact cells and cell-free extracts, it was shown that the two enzymes have different locations within the cell; the lipase resembling an endoenzyme, the esterase an ectoenzyme. The role of lipase and esterase activity in fat hydrolysis and its subsequent effects upon desirable and undesirable flavor development in many food products remains to be elucidated.

PROTEOLYTIC PROPERTIES

The lactic acid bacteria, with the exception of certain species of *Streptococcus* (enterococcus group), display feeble proteolytic properties. Although this inability to hydrolyze proteinaceous substrates places an added restriction upon their ability to reproduce in foods containing high levels of protein, there is evidence to support the premise that nitrogenous bases are degraded by some lactic species.

As reported by Meyer (1961, 1965), certain species of *Betabacteria* (heterofermentative lactobacilli) are responsible for producing "protein" swells in canned marinated herring. This type of spoilage arises as a result of the decarboxylation of amino acids under acidic conditions. In addition to degrading arginine, a physiological test employed for the identification of the heterofermentative lactobacilli (Table 14.2), many of the cultures isolated from fish marinades are capable of decarboxylating glutamic acid, tyrosine, and histidine. Although the mechanism of amino acid liberation is unknown, it would appear that the nitrogenous compounds arise as a result of the endogenous proteolytic enzyme actions of raw fish muscle rather than being a unique clastic property of the lactic acid bacteria.

Evidence that lactic acid bacteria provide vital and essential peptidase activities for the production of high-quality cheddar cheese has been described by Reiter *et al.* (1969). In this case, free amino acids produced by both intra and extracellular dipeptidase activities of microbial origin, generate compounds which serve as flavor enhancing agents. Although the nature of the chemical compounds and the mechanisms of flavor enhancement remain to be defined, it now appears that lactic acid starter cultures, by providing peptidase activities, provide precursors essential to cheddar cheese flavor formation.

TEMPERATURE

The lactic acid bacteria are classified as mesophiles in that they are capable of growing throughout the 20° to 40°C temperature span. Some species, however, are able to grow at temperatures above or below those considered in the mesophilic range and this has provided a physical means for establishing subgroups within the genera *Lactobacillus* and *Streptococcus*. The use of temperature as a criterion for segregation was first suggested by Orla-Jensen (1919) for the differentiation of the sub-genera *Thermobacterium* and *Streptobacterium* within the family *Lactobacillaceae*. Later Sherman (1937) proposed similar temperature limits for members of the genus *Streptococcus*. It should be pointed out that although growth temperatures have long served as principal guidelines for establishing divisions within the genera, some investigators feel that this singular characteristic does not justify the creation of subdivisions within the classification scheme.

The *Thermobacterium* group includes those homofermentative lactobacilli able to grow at the elevated temperatures of 45°C or higher. Unlike their tolerance to these higher temperatures, the species grow less well at 20°C and indeed, many strains fail to grow at this lower temperature. However, this tolerance to higher temperatures cannot be relied upon as being a characteristic unique to the *Thermobacteria*, because some of the non-thermobacteria such as *L. plantarum* and *L. casei* (members of the sub-genus *Streptobacterium*) can also grow at 45°C. In order to circumvent this mutual response to a single elevated temperature, incubation at two temperature extremes (15° and 45°C) serves to provide differentiation because the streptobacteria, unlike the thermobacteria, grow at 15°C and even as low as 6°C.

In conjunction with their abilities to grow at higher temperatures, some species of *Thermobacteria* are capable of surviving pasteurization (63°C for 30 min). This heat resistance, as in the case of *L. thermophilus*, can be a problem in the dairy industry if the holding temperatures of pasteurized milk is poorly controlled and high enough to support the growth necessary for a lactic acid fermentation.

If it is valid to differentiate by temperature requirements, then this physical characteristic may be used to distinguish a limited number of species included in the heterofermentative sub-genus, *Betabacterium*. On the basis of growth temperature, *L. fermenti* (growth at 45°C, no growth at 15°C) more nearly resembles the thermobacteria, whereas *L. brevis*, *L. cellobiosus*, and *L. viridescens* are more closely allied to the streptobacteria which show the opposite temperature responses.

The ability of lactics to grow at low temperatures has produced problems to processors of refrigerated foods such as meats and unbaked bakery products. *Lactobacillus viridescens* is the prime causative agent for producing green taints in cured meat products. Refrigerated doughs (rolls, biscuits, and breads) are spoiled by a less well defined lactic acid flora. In less advanced stages of spoilage, sour flavors may develop while in more severe cases, gas may bulge the container or even split the seams with an extrusion of dough. Bacteria isolated from such spoiled doughs belong primarily to the genera *Lactobacillus* and *Leuconostoc*. *Streptococci*, none which were representative enterococci, have also been isolated from bakery products (Hesseltine *et al.*, 1969).

In contrast to these deleterious effects, many of these same microorganisms are undoubtedly responsible for the leavening and souring actions in the dough used in the San Francisco sourdough bread process (Kline and Sugihara 1971). Although the organisms responsible for this fermentation have not been entirely elucidated, the sourdough bacteria appear to be heterofermentative lactobacilli.

Thus, what appears to be an unwanted condition in a segment of one industry becomes a capital gain for another.

As mentioned previously, the capacity of the various species of *Streptococci* to grow at prescribed temperatures serves as a supplemental characteristic for classification purposes. The temperature criteria, *i.e.*, growth responses at 10° and 45°C, are similar to those used to differentiate *Lactobacilli* and therefore provide a means of segregating the genus into four groups; namely, enterococcus, lactic, pyogenes, and viridans.

In considering the genus as a whole, the species of the latter two groups are generally considered to play less significant roles in food microbiology than their allied counterparts. In the case of the pyogenic group, *S. pyogenes* (the pathogen responsible for rheumatic fever, scarlet fever, and nasal and respiratory infections) has been shown to be transmitted via infected food sources (Boissard and Fry 1955; Moore 1955; and Otte and Ritzerfeld 1960). However, since the pyogenic group is intolerant to either 10° or 45°C temperatures and is killed by pasteurization processes, infectious transmission can be abated by establishing high standards for personal hygiene and rigidly controlling temperatures for food storage.

In the case of the viridans group, these members are primarily animal parasites and although they can be recovered from the human mouth, throat and intestinal tract, they have little or no ability to produce diseases.

The enterococcus group have as their primary habitat the intestinal tract of man and warm-blooded animals. In addition to serving as indicators for fecal contamination these species possess remarkably great physiological endurances as evidenced by their abilities to survive pasteurization (63°C for 30 min), initiate growth at 10° and 45°C, and their tolerances to high pH (9.6) and salt concentrations in excess of 6.5%.

Since the enterococci possess such diverse properties of endurance and also because of their taxonomic proximity to the more pathogenic species, they have been suspected to be the causative agents for numerous food poisoning episodes. Although incriminated in such outbreaks, it has not been conclusively proved that they were the agents solely responsible for the observed malady (Shattock 1962). Furthermore, attempts to confirm the role of biotoxic principles by the administration of massive doses of *S. faecalis* and *S. faecium* to human volunteers were completely ineffective in inducing the usual symptoms of food poisoning (Silliker and Deibel 1960).

Although their role in food poisoning remains to be elucidated, it appears that the enterococci can play detrimental or beneficial roles in food fermentations. For example, from the standpoint of food spoilage it was shown that streptococci can be responsible for the production of sour odors and flavors in pasteurized ham (Niven 1963), but it is interesting to note that these same types of microorganisms play a vital role in inducing the desirable aromacity in sausage fermentations. Furthermore, Dahlberg and Kosikowsky (1949), while studying the effects of *S. faecalis* upon the cheddar cheese fermentation, found that some strains because of their proteolytic properties enhanced the body and flavor characteristics of the final product. Thus, the pros and cons of accepting these cultures for food fermentations with their alleged undesirable pathogenic properties remain to be resolved.

The streptococci of the lactic group are used as starter cultures for the induction of souring and enhancing the flavor of milk and cream. The species of *S. lactis* and *S. cremoris* are characterized by their low temperature range, unusual among streptococci; they grow well at 10°C but fail to develop at temperatures in excess of 40°C.

To achieve their status as good dairy fermenters, these cultures must produce lactic acid rapidly. Retardation of growth in milk due to factors such as bacterial competition, antibiotics, bacteriocides or an unfavorable temperature may yield a product which suffers from musty, cabbage, and other unclean flavors.

EFFECTS OF SALT, pH, AND ANAEROBIOSIS UPON PURE AND MIXED POPULATIONS

In vegetable fermentations the inability to control the growth of an undesirable microflora permits the production of objectionable end products, some of which might even be toxic. The lactic acid bacteria with their powerful fermentative properties and their tolerance to levels of acidity and salt (often inhibitory to other organisms) provide biological mechanisms for repressing the growth of other species.

Certain chemical and physical conditions are available for restricting the growth of undesirable organisms so that the lactic acid bacteria can predominate. These regulatory factors may include pasteurization of raw product, control of incubation temperature, the concentration of sodium chloride, pH, and the presence or absence of air.

To prepare vegetables for fermentation, crystalline salt or brines are frequently employed. The salt serves a multifold purpose; namely, to release fermentable nutrients by osmosis, maintain a crisp texture, and impart desired flavor.

Within certain limits, salt will direct the types of microorganisms which will grow in many food products. For example, many of the

Gram-negative bacteria (*Pseudomonas, Achromobacter, Flavobacterium*, etc.) are quite intolerant to concentrations in excess of 2%, whereas the lactic acid bacteria and the coliform types are less sensitive to this commonly used inorganic ingredient. In brines where the development of lactic acid is rapid, the interactions of acids and salt provide synergistic responses which serve to deter the growth of putrefactive bacteria and the more salt-tolerant coliforms.

Of the four genera under consideration, *Leuconostoc* is less salt tolerant than members of the *Lactobacillus, Pediococcus*, and *Streptococcus*. The variations in tolerance to salt is reflected in the fermentation of vegetables such as cucumbers and cabbage. In the pickle fermentation the use of 5 to 8% brines results in the infrequent isolation of leuconostocs, the species so often encountered during the early stages of other vegetable fermentations. Therefore, this fermentation occurs chiefly as the result of the more salt-tolerant species such as *Pediococcus, L. plantarum*, and *L. brevis*. In contrast to the sensitivity of *Leuconostoc* toward salt, the morphologically related cocci, *i.e., Pediococci* and the enterococcus group of the *Streptococci* show greater resistance to increased salt levels. *S. faecalis*, a member of the enterococcus group, is capable of growing in a complex broth containing 6.5% salt, whereas some species of Pediococci may tolerate upwards of 12% brine.

The inhibitory effects of the interactions of brine and acid levels upon lag and generation times of lactic acid bacteria associated with the sauerkraut fermentation have been studied by Stamer *et al.* (1971). Under conditions simulating those found during the initial phases of the kraut fermentation, it was concluded that *Streptococcus faecalis* was the species most sensitive to a combination of low pH and salt, whereas *P. cerevisisae* and *L. plantarum* were the most tolerant species. Of the heterofermentative cultures examined, *L. brevis* was less subject to growth inhibition than was *L. mesenteroides*.

The relationship of optimum temperature requirements for growth and salt concentrations is also an interesting problem. It has been shown that the more thermophilic species, such as *L. thermophilus, L. lactis, L. helviticum, L. fermenti*, and *L. delbreuckii*, exhibit lower salt tolerances (2.5 to 4.0%) than their mesophilic counterparts (Etchells *et al.* 1964).

The specific type of treatment used for reducing competitive interference depends not only on the types and numbers of microorganisms present, but also upon the nature of the raw product. For example, in the dairy and meat industries, the raw products may be first pasteurized and then inoculated with pure or mixed lactic cultures, whereas vegetable fermentations rely upon the inocula derived directly from the "normal" lactic flora of the fresh product.

Examples of products which are fermented or stored under such highly competitive and diverse environmental conditions include beets, carrots, cauliflower, pickles, sauerkraut, fruit juices, fermented ciders and wines.

During the course of vegetable fermentations, the microbial and chemical environs are in a state of constant change. The failure to reach microbial equilibria, as evidenced by the rise and fall in viability of (a) particular species within a mixed culture system, is particularly evident in bulk fermentations conducted in filled or sealed systems. Since the tank or enclosed system is a "batch operation," as opposed to a continuous fermentation process, it is precluded that within such confined environments, there is only limited time of growth for each species and therefore, a steady state is seldom attained. This resulting vacillation in microbial and chemical interactions contributes to the onset of definite sequential growth patterns. Hence, those species observed to be so dominant during the early phases of the fermentation are relegated to minor numbers or completely replaced by the more tolerant and adaptive species.

The filling and closure of the fermenting tanks represents a major control mechanism for initiating the onset of a highly desirable anaerobic condition. Following the filling operation, the amount of available oxygen becomes limited due to the vigorous metabolic activities of the indigenous microflora and the endogenous respiration of the vegetable substrate. The resulting reduction in oxidation-reduction potential tends to exclude the aerobic microorganisms, thereby encouraging the development of the more facultative species. Although the aerobic species may initially outnumber the facultative and anaerobic species, the removal of oxygen grossly impairs their growth activities. Since some lactic acid bacteria grow at redox potentials between 0.1 and 0.4 volts (Kunkee 1967), and obligate anaerobes usually require potentials less than 0.1V, the lactic acid bacteria serve as ideal intermediaries for the induction of the fermentation under anaerobic conditions.

In addition to being able to grow under anaerobic conditions, many species (either singularly or as mixed cultures) are able to initiate growth throughout a broad pH spectrum, 9.6- 3.0. The coccoid species, *Streptococcus* and *Pediococcus* have greater tolerances to an alkaline pH than their corresponding counterparts, the rods (Table 14.2). Therefore, many foods differing markedly in hydrogen ion concentrations (vegetables, meats, and milk, pH 5- 7; fruits pH 3-4) are susceptible to lactic fermentations.

The capability of producing lactic acid with a concomitant reduction in pH has in recent years become a valuable tool for reducing the pathogenic flora of many foods. By such associative actions, these food-bioprocessing bacteria provide antagonistic mechanisms for the suppression of growth of *Staphylococcus aureus* in complex broths and milk (Kao and Frazier 1966; Gilliland and Speck 1972). *Salmonella* in sausage (Goepfert and Chung 1970) and psychrotrophic spoilage in ground beef (Reddy *et al*. 1970).

Additional evidence as to the effect of pH and substrate composition upon total bacterial populations, even to the point of selective exclusion of species within the lactic acid group, has been shown to occur in fruit juices and ciders. Carr (1959) and Carr and Davies (1970) upon examining apple ciders (pH 3.0–4.6) found the flora to consist of heterofermentative rods related to *Lactobacillus brevis*; heterofermentative cocci, *Leuconostoc*; and homofermentative bacilli, *Lactobacillus plantarum*; whereas organisms belonging to the genera *Streptococcus* and *Pediococcus* were never isolated. Further studies on the acid-tolerant isolates showed that the heterofermenters were able to grow at pH 3.2 while the homofermenters were inhibited at pH 4.0.

In addition to influencing growth, pH may alter the normal degradative pathways. For example, it was shown in the above reports that *L. brevis* when grown at pH 3.6 readily decarboxylated malic acid to yield lactic acid and carbon dioxide; however, at pH 4.8 succinic acid became the major end-product. Similar effects of pH upon the physiological regulatory mechanisms have been reported by Gunsalus and Niven (1942), who showed that *Streptococcus liquifaciens* formed greater amounts of lactic acid at pH 5 than at higher pH values. Therefore, the creation of acidic environments not only serve as growth regulators for unwanted microbial species, but influence and regulate the physiological responses of the lactic acid bacteria themselves. It should be noted, however, that the control of microbial competitors is not due to pH only because organic acids (volatile and non-volatile), when tested under conditions independent of pH, have been shown to have pronounced inhibitory effects upon growth.

Further evidence as to the adaptability of the lactic acid bacteria to more severe and hostile environments, such as combinations of low pH, low carbohydrate concentrations, and high levels of ethanol, is exemplified in their ability to grow in alcoholic beverages. For example, table wines which are from 10 to 14% ethanol, less than 0.2% residual carbohydrate, and high in acid (pH 3.0–3.8) are capable of supporting the growth of some species. This ability to adapt to an unfavorable environment is not restricted to a single genus, since isolates, characterized as members of the genera *Lactobacillus*, *Leuconostoc*, and *Pediococcus* have been recovered from many fully fermented table wines. Even greater stresses, such as the high alcohol concentrations found in fortified wines (19-20%), place no absolute degree of inhibition upon some lactic acid bacteria. Exhaustive studies on such a heterofermentative, alcohol-tolerant species, *Lactobacillus trichodes*, not only indicate its ability to grow in fortified wines with the production of excessive amounts of undesirable sediments, but show that the organism requires high levels of ethanol (15-20%) for maximum growth activities (Fornachon *et al.* 1949; Kunkee 1970).

Although it is generally assumed that leuconostocs will not grow in media with an initial pH value of 4.2 or less (Garvie 1967) and are also quite intolerant to ethanol concentration in excess of 7% (Mundt and Hammer 1966), exceptions to this rule are now noted (Rice and Mattick 1970). Recent in-depth characterizations of these alcohol-tolerant heterofermentative cocci have resulted in a proposed expansion in the numbers of species within the family *Leuconostoc* to include *L. oenos* (Garvie 1967). Because of its tolerance to wine, the latter species is used for promoting the malo-lactic fermentation in commercial wine production. The resulting secondary fermentation, that is the conversion of the dicarboxylic acid, malic, to lactic acid and carbon dioxide, produces a reduction in total acidity and a concomitant increase in pH by a microbiological mechanism. The resulting reduction in acidity, enhancement of flavor, and the assurance of a microbiologically-stable bottled product may be valuable assets to high acid wines produced in Eastern U.S. and Europe. However, in less acidic wines such as those found in Australia and California, the malo-lactic fermentation accompanied with its characteristic increase in pH, provides a more favorable substrate for the potential development of undesirable spoilage organisms. Thus, the same substrate (wine) when produced in different geographical areas is vulnerable to attack by the lactic acid bacteria, and what represents spoilage in one geographical area may be a beneficial attribute in another sector.

There are additional examples where lactic acid bacteria, either as pure or mixed cultures, play beneficial or undesirable roles dependent upon the utilitarian acceptance of the end products formed. For many years *Leuconostoc mesenteroides* and *L. dextranicum* have been recognized as the spoilage organisms in sugar and vegetable fermentation industries.

In the sugar industry *L. mesenteroides* has produced severe economic losses by the production of copious masses of slime material from the raw sugar. The mucilaginous dextran, formed by the condensation of monomeric glucose from disaccharide degradation, prevents the optimum recovery of crystalline solids. Although the extra-cellular-polysaccharides are highly objectionable in the sugar processing industry, the innocuous nature of the polymers has made them a favorable component for use as a plasma extender, a thickening agent for food products, and a chemical base for the synthesis of absorption and ion exchange resins.

Slime formation is likewise frequently encountered in certain vegetable fermentations, particularly cauliflower and sauerkraut. Since both fermentations are carried out by an uncontrolled mixed lactic acid flora, the leuconostocs often appear as the predominant species during the early phases of fermentation. During this formative stage of fermentation, the whole or shredded vegetable may become enveloped within a pronounced viscous or slimey dextran mass. Since the dextran is extremely soluble in water, it can be removed readily from the whole vegetable by high pressure rinsing prior to processing. In the case of sauerkraut, however, the rinsing step is omitted because it removes or dilutes the flavor components associated with the product. Since other lactic acid bacteria, primarily members of the genus *Lactobacillus*, are able to utilize the dextrans for growth, the objectionable slimey state is overcome by merely allowing the sauerkraut to age in the vat—another example where the end products produced by one species become a utilizable substrate for succeeding lactic species.

CHEMICAL GROWTH INHIBITORS

There are many organic and inorganic substances which invoke either lethal or biostatic effects upon microbial populations; however, because of stringent Federal regulations governing the use of such additives, only a limited number of preservatives can be added to foods used for human consumption.

Although antibiotics such as chloro- and oxytetracyclines, penicillin, bacitracin, etc., are effective chemotheraputic agents, their acceptance and use as permissable food additives remain a controversial issue. Therefore, the more traditional preservatives such as sulfur dioxide, metabisulfite, benzoic and sorbic acids still find favor as microbial inhibitors in acid foods. Of these compounds, benzoate and sorbate are primarily antimycotic agents and are without inhibitory effects on the lactic acid bacteria, whereas sulfur dioxide or metabisulfite is a potent growth inhibitor of bacteria, yeasts, and molds.

Since the efficacy of sulfur dioxide and metabisulfite depend upon the formation and availability of sulfurous ions ("free" SO_2), their inhibitory properties are dramatically enhanced in products having low pH values. Among such products which are frequently treated with sulfur dioxide are dried fruits (raisins, apricots), vegetables (pickled cauliflower and sauerkraut packaged in plastic bags), fruit juices, and wines.

In wines it has been found that the inhibitory effects of sulfur dioxide are influenced by a multitude of factors including pH, concentration of ethanol, and the levels and types of active carbonyl compounds present. For example, Fornachon et al. (1949) found that *Lactobacillus trichodes* can grow in non-sulfited wines containing 18 to 20% ethanol; however, by adding 75 ppm sulfur dioxide growth is prevented. More recently deMenzes et al. (1972) showed that *Leuconostoc oenos*, a species frequently associated with the malo-lactic fermentation, was markedly inhibited from growing in Eastern wines containing only 20 ppm sulfur dioxide. Thus, it appears that the effectiveness of sulfur dioxide depends not only upon substrate composition but also upon the tolerance of the prevailing species.

Although the above agents of known chemical composition may effectively retard microbial growth, there are present in natural products inherent inhibitors which can adversely affect food fermentations. Among these inhibitors are such compounds as acrolein, allyl isothiocyanate, butyl thiocyanate, mercaptans, and the oils of spices. However, because of their low levels of concentration within plant tissues, coupled with their high degree of volatility, these compounds often serve only as temporary inhibitors. Although of short duration, this inhibition can be disastrous because it may permit the growth of an undesirable, non-lactic flora, resulting in a product which is not only putrid but potentially unsafe for consumption.

The effect of processing and its role in potentially generating antimicrobial agents should also be considered. In recent studies by Fleming et al. (1973) it was shown that oleuropein, the bitter glucoside of green olives, gives rise to an aglycone, elenolic acid, and β-3, 4-dihyroxyphenylethyl alcohol as a result of acid hydrolysis. Although intact oleuropein possesses only feeble antibacterial properties, the former two compounds significantly inhibited the growth of several species of lactic acid bacteria associated with the fermentation of olives. Therefore, the often observed slow or delayed fermentation of green olives can be attributed in part, to the generation of inhibitors by processing procedures or the actions of degradative enzymes inherent to the plant tissues.

SUMMARY

One of the most important characteristics of the lactic acid bacteria is their ability to produce and tolerate highly acidic conditions. Therefore, foods which are produced or stored under a diversity of conditions are subject to contamination and degradation by these microorganisms.

Although some pathogenic species, of which there are few, have been incriminated as agents of food poisoning, there is insufficient evidence available to indict them as being major toxicants of foods. Therefore, to the food industry from a public health aspect the lactic acid bacteria are probably best described as "nuisance" bacteria, because they are more frequently associated with spoilage than with pathogenicity. Although spoilage is a serious, deleterious result, it is comforting to know that many of the same species responsible for these maladies are likewise the essential agents for inducing food fermentation—one of the oldest mechanisms known for preserving food.

BIBLIOGRAPHY

ANDERS, R. F., HOGG, D. M., and JAGO, G. R. 1970. Formation of hydrogen peroxide by Group N *Streptococci* and its effect on their growth and metabolism. Appl. Microbiol. *19*, 608-612.

BOISSARD, J. M., and FRY, R. M. 1955. A food-borne outbreak of infection due to *Streptococcus pyogenes*. J. Appl. Bacteriol. *18*, 478-483.

BORDELEAU, L. M., and BARTHA, R. 1969. Rapid technique for enumeration and isolation of peroxidase producing microorganisms. Appl. Microbiol. *18*, 274-275.

BREED, R. S., MURRAY, E. G. D., and SMITH, N. R. 1957. Bergey's Manual of Determinative Bacteriology, 7th Edition. The Williams & Wilkens Co., Baltimore, Md.

BUYZE, G., VAN DEN HAMER, J. A., and DEHAAN, P. G. 1957. Correlation between hexose monophosphate-shunt, glycolytic system and fermentation-type in *Lactobacilli*, Antonie van Leeuwenhoek. J. Microbiol. Serol. *23*, 245 -250.

CARR, J. G. 1959. Some special characteristics of the cider *Lactobacilli*. J. Appl. Bacteriol. *22*, 377-383.

CARR, J. G., and DAVIES, P. A. 1970. Homofermentative lactobacilli of ciders including *Lactobacillus mali nov.* spec. J. Appl. Bacteriol. *33*, 768-774.

COSTER, E., and WHITE, H. R. 1964. Further studies of the genus *Pediococcus*. J. Gen. Microbiol. *37*, 15-31.

DACRE, J. C., and SHARPE, M. E. 1956. Catalase production by *Lactobacilli*. Nature *178*, 700.

DAHLBERG, A. C., and KOSIKOWSKY, F. J. 1949. The bacterial count, gramine content and quality score of commercial American cheddar and stirred curd cheese made with *Streptococcus faecalis*. J. Dairy Sci. *32*, 630-636.

DEMAN, J. C., ROGOSA, M., and SHARPE, M. E. 1960. A medium for the cultivation of lactobacilli. J. Appl. Bacteriol. *23*, 130-135.

DEMENZES, T. J. B., SPLITTSTOESSER, D. F., and STAMER, J. R. 1972. Induced malo-lactic fermentation of New York State Wines. N.Y. Food Life Sci. *5*, 24-26.

424 FOOD MICROBIOLOGY

DOELLE, H. W. 1969. Bacterial Metabolism. Academic Press, New York.
DOELLE, H. W. 1971. Nicotinamide adenine dinucleotide-dependent and nicotinamide adenine dinucleotide-independent lactate dehydrogenases in homofermentative and heterofermentative lactic acid bacteria. J. Bacteriol. 108, 1284–1289.
ETCHELLS, J. L., COSTILOW, R. N., ANDERSON, T. E., and BELL, T. A. 1964. Pure culture fermentation of brined cucumbers. Appl. Microbiol. 12, 523–535.
FLEMING, H. P., WALTER, W. M., JR., and ETCHELLS, J. L. 1973. Antimicrobial properties of oleuropein and products of its hydrolysis from green olives. Appl. Microbiol. 26, 777–782.
FORNACHON, J. C. M., DOUGLAS, H. C., and VAUGHN, R. H. 1949. Lactobacillus trichodes nov. spec., a bacterium causing spoilage in appetizer and dessert wines. Hilgardia 19, 129–132.
FRAZIER, W. C. 1958. Food Microbiology, 2nd Edition. McGraw-Hill Book Co., New York.
GARVIE, E. I. 1960. The genus Leuconostoc and its nomenclature. J. Dairy Res. 27, 283–293.
GARVIE, E. I. 1967. Leuconostoc oenos sp. nov. J. Gen. Microbiol. 48, 431–438.
GILLILAND, S. E., and SPECK, M. L. 1972. Interactions of food starter cultures and food-borne pathogens: Lactic streptococci versus Staphylococci and Salmonellae. J. Milk Food Technol. 35, 307–310.
GOEPFERT, J. M., and CHUNG, K. C. 1970. Behavior of Salmonella during the manufacture and storage of a fermented sausage product. J. Milk Food Technol. 33, 185–191.
GUNSALUS, I. C., and NIVEN, C. F., JR. 1942. The effect of pH on the lactic acid fermentation. J. Biol. Chem. 145, 131–136.
HAAS, G. J. 1960. Microbial control methods in the brewery. Advan. Appl. Microbiol. 2, 113–162.
HANSEN, P. A. 1968. Type strains of Lactobacillus species. In A Report by the Taxomic Subcommittee on Lactobacilli and Closely Related Organisms, P. A. Hansen (Editor). American Type Culture Collection, Rockville, Md.
HAYS, G. L., and RIESTER, D. W. 1952. The control of "Off-odor" spoilage of mayonnaise and salad dressings. Appl. Microbiol. 21, 870–874.
HESSELTINE, C. W., GRAVES, R. R., ROGERS, R., and BURMEISTER, H. R. 1969. Aerobic and facultative microflora of fresh and spoiled refrigerated dough products. Appl. Microbiol. 18, 848–853.
HOLMES, B., SANDINE, W. E., and ELLIKER, P. R. 1968. Some factors influencing carbon dioxide production by Leuconostoc citrovorum. Appl. Microbiol. 16, 56–61.
JOHNSTON, M. A., and DELWICHE, E. E. 1965. Distribution and characteristics of the catalases of Lactobacillaceae. J. Bacteriol. 90, 347–351.
KANDLER, O. 1970. Amino acid sequence of the murein and taxonomy of the genera Lactobacillus, Bifidobacterium, Leuconostoc and Pediococcus. Intern. J. Syst. Bacteriol. 20, 491–507.
KANDLER, O., SCHLEIFER, K. H., and DANDL, R. 1968. Differentiation of Streptococcus faecalis Andrewes and Horder and Streptococcus faecium Orla-Jensen based on the amino acid composition of their murein. J. Bacteriol. 96, 1935–1939.
KAO, C. T., and FRAZIER, W. C. 1966. Effect of lactic acid bacteria on growth of Staphylococcus aureus. Appl. Microbiol. 14, 251–255.
KLINE, L., and SUGIHARA, T. F. 1971. Microorganisms of San Francisco sour dough bread process. II. Isolation and characterization of undescribed bacterial species responsible for the souring activity. Appl. Microbiol. 21, 459–465.
KUNKEE, R. E. 1967. Malo-lactic fermentation. Advan. Appl. Microbiol. 9, 235–279.

KUNKEE, R. E. 1970. Stimulatory effect of ethyl alcohol on growth of *Lactobacillus trichodes*. Bacteriol. Proc., p. 2, Am. Society Microbiology, Washington, D.C.

KURTZMAN, C. P., ROGERS, R., and HESSELTINE, C. W. 1971. Microbial spoilage of mayonnaise and salad dressings. Appl. Microbiol. *21*, 870–874.

LANCEFIELD, R. C. 1933. A serological differentiation of human and other groups of haemolytic streptococci. J. Exptl. Med. *57*, 571–595.

MEYER, V. 1961. Spoilage problems of canned fish. VI. The occurrence and identification of amino acid decarboxylations in the production of marinades. Veröffentl. Inst. Meeresforsch, Bremerhaven, 7, 264–276. (German)

MEYER, V. 1964. The biochemical and bacteriological causes of swelling of marinades and aspects of its prevention. Intern. Symp. Food Microbiol. 4th, Goteborg, Sweden, C.A. *63*, 12230.

MEYER, V. 1965. Marinades. *In* Fish as Food, Vol. III, Part 1, G. Borgstrom (Editor). Academic Press, New York.

MIZUSHIMA, S., HIYAMA, T., and KITAHARA, K. 1964. Quantitative studies on glycolytic enzymes in *Lactobacillus plantarum*. J. Gen. Appl. Microbiol. (Tokyo) *10*, 33–44.

MOORE, B. 1955. Streptococci and food poisoning. J. Appl. Bacteriol. *18*, 606–618.

MUNDT, J. O., and HAMMER, J. C. 1966. Suppression of *Leuconostoc mesenteroides* during isolation of *lactobacilli*. Appl. Microbiol. *14*, 1044.

MUNDT, J. O., and HAMMER, J. C. 1968. Lactobacilli on plants. J. Appl. Microbiol. *16*, 1326–1330.

NIVEN, C. F., JR. 1963. Microbial indexes of food quality. Faecal Streptococci. *In* Microbiological Quality of Foods, L. W. Slanetz *et al.* (Editors). Academic Press, New York.

NIVEN, C. F., JR., CASTELLANI, A. G., and ALLANSON, V. 1949. A study of the lactic acid bacteria that causes surface discoloration of sausage. J. Bacteriol. *58*, 633–641.

ORLA-JENSEN, S. 1919. The lactic acid bacteria. Mem. Acad. Roy. Sci. Lettres, Danemark, Sect. Sci. Ser. 8, *5*, 81–196.

OTERHOLM, A., ORDAL, Z. J., and WITTER, L. D. 1968. Glycerol ester hydrolase activity of lactic acid bacteria. Appl. Microbiol. *16*, 524–527.

OTTE, H. J., and RITZERFELD, W. 1960. Massive epidemics of tonsillitis as a result of *Streptococci* in food. Deut. Medizin. Wisch. *85*, 1625–1628. (German)

PEDERSON, C. S. 1929. The types of organisms found in spoiled tomato products. N.Y. State Agr. Expt. Sta. Tech. Bull. *150*.

PEDERSON, C. S. 1960. Sauerkraut. *In* Advances in Food Research, Vol. 10, C. O. Chichester, E. M. Mrak, and G. F. Stewart (Editors). Academic Press, New York.

PERRY, K. D., and SHARPE, M. E. 1960. Lactobacilli in raw milk and in cheddar cheese. J. Dairy Res. *27*, 267–277.

PRIEBE, K. 1970. Investigations as to the cause of ropiness in marinated baked herring. Arch. Lebensmittelhyg. *21*, 13–23. (German)

RANKINE, B. C., and BRIDSON, D. A. 1971. Bacterial spoilage in dry red wine and its relationship to malo-lactic fermentation. Australian Wine, Brew., Spirit Rev. *3*, 44–51.

REDDY, S. G., HENRICKSON, R. L., and OLSON, H. C. 1970. The influence of lactic cultures on ground beef quality. J. Food Sci. *35*, 787–791.

REITER, B., SOROKIN, Y., PICKERING, A., and HALL, A. J. 1969. Hydrolysis of fat and protein in small cheeses made under aseptic conditions. J. Dairy Res. *36*, 65–76.

RICE, A. C., and MATTICK, L. R. 1970. Natural malo-lactic fermentation in New York State Wines. Am. J. Enol. Viticult. *21*, 145–152.

SABBAJ, J., SUTTER, L., and FINEGOLD, S. M. 1971. Selective media for Group D Streptococci. Appl. Microbiol. *22*, 1008–1011.

SANDINE, W. E., RADICH, P. C., and ELLIKER, P. R. 1972. Ecology of the lactic streptococci. J. Milk Food Technol. *35*, 176–184.

SHARPE, M. E. 1955. A serological classification of lactobacilli. J. Gen. Microbiol. *12*, 107–122.

SHARPE, M. E., FRYER, T. F., and SMITH, D. D. 1966. *In* Identification Methods for Microbiologists. B. M. Gibbs, and F. A. Skinner (Editors). Academic Press, New York.

SHARPE, M. E., GARVIE, E. I., and TILBURY, R. H. 1972. Some slime-forming heterofermentative species of the genus *Lactobacillus*. Appl. Microbiol. *23*, 389–397.

SHATTOCK, P. M. F. 1962. Enterococci. *In* Chemical and Biological Hazards in Food. J. C. Ayres *et al.* (Editors). Iowa State Univ. Press, Ames, Iowa.

SHERMAN, J. M. 1937. The Streptococci. Bacteriol. Rev. *1*, 3–97.

SILLIKER, J. H., and DEIBEL, R. H. 1960. On the association of enterococci with food poisoning. Bacteriol. Proc. p. 48, Am. Soc. Microbiology, Washington, D.C.

SLADE, H. D., and SLAMP, W. C. 1972. Peptidoglycan composition and taxonomy of Group D, E, and H. *Streptococci* and *Streptococcus mutans*. J. Bacteriol. *109*, 691–695.

STAMER, J. R., STOYLA, B. O., and DUNCKEL, B. A. 1971. Growth rates and fermentation patterns of lactic acid bacteria associated with the sauerkraut fermentation. J. Milk Food Technol. *34*, 521–525.

TILBURY, R. H. 1968. Biodeterioration of harvested cane sugar. *In* Biodeterioration of Materials, A. H. Walters, and J. J. Elphick (Editors). Elsevier Publishing Co., Amsterdam.

TITTSLER, R. P. *et al.* 1952. Symposium on the lactic acid bacteria. Bacteriol. Rev. *16*, 237–260.

WEISS, N., BUSSE, M., and KANDLER, O. 1968. The origin of the end products of the lactic acid fermentation of *Lactobacillus acidophilus*. Arch Mikrobiol. *62*, 85–93. (German)

WHITTENBURY, R. 1964. Hydrogen peroxide formation and catalase activity in lactic acid bacteria. J. Gen. Microbiol. *35*, 13–26.

WHITTENBURY, R. 1966. A study of the genus *Leuconostoc*. Archiv. Mikrobiol. *53*, 317–327.

CHAPTER 15

Henry J. Peppler | Yeasts

In its most familiar role as a fermenter of sugars to carbon dioxide and ethyl alcohol, yeast usage is widespread, and, as such, yeasts are universally recognized as benefactors of mankind. As one of our busiest microbial groups, yeasts impart flavor to foods and feeds, enrich human and animal diets, and enhance our leisure and social graces. When given the opportunity, some yeasts bring about undesirable changes during fermentation processes and cause defects in raw and processed foods. And there are a few pathogenic species of yeast; however, no yeast has been incriminated in a clinical case of food intoxication.

Only 349 species of yeasts are recognized taxonomically (Lodder 1970), while more than 100,000 different microorganisms, fungi and algae have been classified so far. Though few in number and less ubiquitous than bacteria, yeasts are found in a variety of natural habitats, generally living a saprophytic life in environments favorable to their growth. Yeasts are most frequently isolated from the surfaces of leaves (Preece and Dickinson 1971), fruits and cereal grains, from honey, vineyard and orchard soils, exudates of trees, insects, aquatic environments and fabricated products. While largely a harmless group of microorganisms, a few yeasts are pathogenic to plants, animals and insects, and some species live parasitically in them (Gentles and La Touche 1969). Distribution and dispersion of yeasts is affected by animals, insects, waters and winds (Do Carmo-Sousa 1969).

Those species of yeast most frequently encountered may be described simply as oval, nonmotile, nucleated, single-celled, free-living, chemosynthetic microorganisms about $6 \times 10 \ \mu m$ in size, and reproducing mainly vegetatively and by budding. Many yeast species, however, fall outside of this definition. For example, species used in breadmaking and brewing, among others, also form sexual spores (ascospores). Bud formation, its fine structure and mechanism, has been richly illustrated micrographically by Talens *et al.* (1973).

A wide range of foods support yeast growth. In a mixed flora with bacteria and molds, which occurs often in contamination and spoilage problems, the outgrowth and metabolic activity of yeast is likely to depend on (a) the characteristics of the contaminating species; (b) the initial cell concentration; (c) the available moisture or water activity (a_w); (d) the concentration and availability of sugars

427

and other nutrients in the food or processing system; (e) low pH values; (f) the oxygen tension (low for fermentative, high for oxidative yeasts); and, (g) the storage temperature and time.

The undesirable changes and defects which yeasts produce during fermentations and in fresh and processed foods manifest themselves in many ways. Liquid products, for example, may become turbid, form a scum, film or pellicle, evolve gas, change color, emit atypical aromas, produce off-flavors, and be so altered in composition that bacterial growth flourishes. Surfaces of some foods may develop powdery coatings, pasty slimes or discolorations. Contaminating yeasts in harvested bakers' and brewers' yeasts may diminish, by dilution of the cell mass, the performance of the primary cultivated species.

Contamination of the food processing and handling chain may occur at almost every station. Foremost of the routes and sources of yeast contamination are the raw materials and process ingredients brought into the plant, insects, air-borne microtrash, careless workers and nonsterile equipment. In the latter instances, residual contamination may be due to faulty equipment design, unclean surfaces and inadequate or poorly monitored treatment procedures.

For the detection, enumeration, isolation and identification of yeasts appearing in foods, a variety of selective, differentiating and general-purpose media have been devised. Some serve better than others. To favor yeast growth while suppressing the faster-growing bacteria, culture media are acidified as a general rule. These may be made more selective with the addition of a suitable antibiotic, such as chloramphenicol, and restrictive sources and concentrations of energy and nitrogen compounds.

It is the intent of this chapter to provide the food technologist with a synoptic introduction to yeasts and their involvement in food technology. Excellent reviews focusing on yeasts in food microbiology have been published. Mrak and Phaff (1948) treated food spoilage yeasts as a special aspect of ecology and described species involved beneficially in unique fermentations. Pederson (1971) discussed the role of yeasts in food fermentations. Reed and Peppler (1973) reviewed the industrial applications of yeasts. Ingram (1958) accented the biological and environmental factors conducive to food spoilage by yeasts. Walker and Ayres (1970) compiled the comprehensive yeast literature reported for specific foods. Guthrie (1972) wrote a guide to food sanitation for industry personnel and food science students. When all has been read and said, one could capture the essence in a paragraph, such as this apt summary by Phaff, Miller and Mrak (1966):

Almost any kind of food will permit yeast to grow, provided it has not been adequately heat-treated. High concentrations of sugar, salt, organic acids, the exclusion of air, refrigeration, and application of other storage conditions will not safeguard a food from the action of yeasts, provided storage is sufficiently long. The heavier the initial contamination, the sooner spoilage symptoms become apparent. For this reason, strict observance of sanitation in food-processing plants offers the best protection against losses due to microbial spoilage.

TYPES OF YEASTS

The orderly arrangement of the diverse types of yeast follows the rules prescribed by the Botanical Code (Lanjouw 1966), one of three international codes established for naming and classifying plants, animals and bacteria. Scientific names used in this chapter will adhere, insofar as possible, to the nomenclature and groupings adopted by Lodder (1970) who edited the contributions of an international team of 13 microbiologists and taxonomists. As a natural classification, which excludes mutants and hybrids, it is founded on comparative evaluations of the classical criteria: the appearance of vegetative growth of pure cultures in standardized media; the utilization of carbon and nitrogen compounds; the biochemical requirements for fermentative and/or oxidative metabolism; and the sexual characteristics (Kreger-Van Rij 1969; van der Walt 1970). Accordingly yeasts are segregated into four major groups or families comprising 39 genera and 349 species, a classification which increases the previous one (Lodder and Kreger-Van Rij 1952) by ten new genera and 185 accepted species. In Table 15.1 are listed the more abundant genera of the ascospore-forming yeasts, also designated true, perfect or sexual yeasts. Included are notations concerning metabolic and food-associated characteristics and habitat (besides soils). In nearly all of the 13 genera listed are found yeast species involved with foods in some manner. Those species encountered in food deterioration will be discussed in the next section. Omitted from Table 15.1 are nine genera, each with a single species, none of which have been reported in spoilage of foods. They are identified below, however, for reference purposes and to complete the tally of ascosporogenous genera.

Citeromyces (*Cit.*) *matritensis* (synonym *Hansenula matri-tensis*)—from fruits.

Coccidiascus (*Co.*) *legeri*—intestinal parasite in fruit flies (Drosophila).

Lodderomyces (*Lod.*) *elongisporus* (synonym *Sacch. elongi-*

TABLE 15.1
GENERA OF ASCOSPORE-FORMING YEASTS

Genera (a) (b)	Properties, Habitat and Food Associations	
Saccharomyces (Sacch.) - 41	F. - Ox. (c);	bakers', brewers', wine, food, distillery and sugar-tolerant yeasts.
Pichia (P.) - 35	Ox.;	form films and esters.
Hansenula (H.) - 25	F. - Ox.;	form esters, extracellular mannans; nitrate utilized.
Debaryomyces (Deb.) - 18	F. - Ox.;	cheeses, meats, brines; salt tolerant.
Kluyveromyces (K.) - 18	F. - Ox.;	milk products; lactose fermented.
Endomycopsis (E.) - 10	Ox.;	cereal grains; in wood-boring insects; some pellicle formers.
Metchnikowia (M.) - 5	F. - Ox.;	in brine shrimp; film formation variable.
Schwanniomyces (Schw.) - 4	F. - Ox.;	in soils.
Schizosaccharomyces (Schiz.) - 4	F.;	maple sap, sirups, wineries; sugar tolerant.
Hanseniaspora (H'spora) - 3	F. - Ox.;	in wineries, fruit flies; alcohol sensitive.
Dekkera (D.) - 2	Ox.;	produce acetic acid, pellicles; use nitrate; in breweries.
Lipomyces (L.) - 2	Ox.;	human skin and soil inhabitant.
Nadsonia (N.) - 2	F. - Ox.;	film formers; exudates of trees.
(Nine genera omitted.) (d)		

(a) From Lodder (1970).
(b) Genus name followed by its abbreviation and number of species in it.
(c) F. = fermentation; Ox. = oxidative metabolism.
(d) Nine additional genera, each with a single species, are identified in the text.

sporus)—originally isolated from soil and orange concentrate.

Nematospora (Nem.) coryli—causes yeast spot on pepper pods, legumes and citrus fruit.

Pachysolen (Pa.) tannophilus—isolated from tanning liquors and certain broad-leafed trees.

Saccharomycodes (S'codes) ludwigii (synonym Sacch. ludwigii)—found on deciduous trees of Europe.

Saccharomycopsis (S.) guttulata—common in rabbit alimentary tract.

Wickerhamia fluorescens (synonym Kl. fluorescens)—from wild squirrels (Japan).

Wingea robertsii—in pollen pablum for carpenter bee; related to Pichia.

The principal non-spore-forming genera, also designated false, imperfect or asexual yeasts, are arranged in Table 15.2 with brief comment on food incidence and habitat, other than soils. Species occurring in foods represent only 6 of the 12 genera of asporogenous

TABLE 15.2
GENERA OF NON-SPORE-FORMING YEASTS

Genera [a] [b]	Properties, Habitat and Food Association	
Candida (C.) - 81	F. - Ox. [c] ;	fruit products, food yeasts; film formers and nitrate utilizers.
Torulopsis (T.) - 36	F. - Ox.;	dairy, fruit and vegetable products; osmotolerant; variable on nitrate.
Cryptococcus (Cr.) - 17	Ox.;	animal parasite; variable on nitrate.
Rhodotorula (Rh.) - 9	Ox.;	red or yellow pigments; some nitrate utilization; in milk, fruit and vegetable products.
Trichosporon (Tr.) - 8	Ox.;	tree sap, beer and meat contaminants.
Brettanomyces (Br.) - 7	F. - Ox.;	beer, wine spoilage; acetic acid formed; cycloheximide resistant; imperfect stage of Dekkera.
Kloeckera (Kl.) - 4	F. - Ox.;	fruit products; in fruit flies.
Pityrosporum (Pit.) - 3	Ox.;	on scalp and skin of mammals.
Sterigmatomyces (St.) - 2	Ox.;	in sea water.
(Three genera omitted.) [d]		

(a) From Lodder (1970).
(b) Genus name followed by its abbreviation, and number of species in it.
(c) F. = fermentation; Ox. = oxidative metabolism.
(d) Four additional genera, each with only one species are mentioned in the text.

yeasts. Omitted from Table 15.2 are three genera, each with a solitary species, all nonfermentative, and only one associated, but rarely, with a food:

Oosporidium (O.) margaritiferum (synonym Tr. margaritiferum)—from tree slime fluxes.

Schizoblastosporion (Schizobl.) starkeyi-henricii—from soils; related to Pityrosporum.

Trigonopsis (Trig.) variabilis—isolated from beer; triangular cells with budding at the angles; related to Torulopsis.

Of this assortment of yeasts, just a few species belonging to three genera are regularly employed industrially. Foremost in utility and beneficence is Saccharomyces cerevisiae. Its numerous and versatile culture strains provide the seed for the propagation of bakers' yeast (Reed and Peppler 1973) and dried food and fodder yeasts (Peppler 1970). They may be used in the brewing of ale, saké (Kodama 1970) and some beer (Rainbow 1970), in the production of potable distillery products (Harrison and Graham 1970), and in some winemaking (Amerine et al. 1972; Kunkee and Amerine 1970). Yeast protoplasm itself yields miscellaneous constituents such as amino acids, coenzymes, enzymes, flavor precursors, nucleotides, and the vitamin B complex (Reed and Peppler 1973). Classed with Sacch.

cerevisiae are the wine and saké fermenting species formerly designated *Sacch. ellipsoideus, Sacch. saké, Sacch. vini, Sacch. cerevisiae* var. *ellipsoideus* and *Candida robusta.* The lager beer yeast long known as *Sacch. carlsbergensis* is now a synonym of *Sacch. uvarum.* Some unwanted yeasts encountered in brewing may be strains of *Sacch. cerevisiae* and *Sacch. uvarum* (Richards 1972; Kleyn and Hough 1971). These and other *Saccharomyces* which are found in spoiled foods are identified below.

Lodder and associates also united some of the other prominent yeasts with like physiology and minor morphological differences. Certain wine and sherry yeasts known as *Sacch. beticus, Sacch. oviformis* and *Sacch. pastorianus* are all classed as *Sacch. bayanus* (Amerine *et al.* 1972; Santa Maria and Vidal, 1973). The lactose fermenting yeasts, *Sacch. fragilis, Sacch. lactis* and *Sacch. kefyr,* are now designated *Kluyveromyces fragilis, K. lactis* and *Candida kefyr,* respectively. They are associated with whey-grown food yeasts, fermented milk products and certain cheeses.

Yeasts capable of growing in honey and like sugar concentrates were formerly placed in the genus *Zygosaccharomyces;* however, they are currently classified as *Saccharomyces.* The principal species are *Sacch. bailii, Sacch. bisporus* (and its variety *mellis*), and *Sacch. rouxii.* With selected strains of these osmotolerant yeasts Steinkraus and Morse (1966) developed a process for making mead, the honey wine (Morse and Steinkraus 1971).

Most of the yeast tonnage produced as protein and vitamin enrichments for animal feeds involves species of *Candida.* With *Candida utilis* and *C. tropicalis* fodder yeast is obtained from paper mill wastes, wood hydrolyzates, molasses, distillery wastes and other process effluents. Starch is converted to yeast protein during the symbiotic growth of *C. utilis* and *Endomycopsis fibuligera,* a starch-digesting yeast (Jarl 1969). *E. fibuligera* is essential in the preparation of tapé, a fermented food of Indonesia (Djien 1972).

When grown in sauerkraut waste, *C. utilis* cells develop high concentrations of invertase (Hang *et al.* 1973). Normal alkanes in gas oils are removed by *C. intermedia, C. lipolytica,* and *C. guilliermondii* (Miller and Johnson 1966; Ahearn and Roth 1966; Gradova *et al.* 1969). These, as well as species of *Rhodotorula,* aid in the decomposition of marine oil slicks (Ahearn *et al.* 1971), and the deterioration of aircraft fuels (Engel and Swatek 1966). Some species of *Candida* stimulate the growth of lactic acid bacteria in cheeses (Lashkari 1971; Devoyod *et al.* 1971) and in cheese starters (Soulides 1955; Peppler and Frazier 1942). In a similar effect *Trichosporon* species precede the smear bacteria on the surface of limburger types of cheese (Ades and Cone 1969); and, according to Hosono and Tokita (1970, A B) two yeasts they isolated, *C. mycoderma* and *D. kloeckeri*, may contribute to limburger flavor. The psychrophilic yeast, *Candida scottii*, originally isolated from chilled beef, predominates in the cold, dry soils of Antarctica (Di Menna 1960).

Candida humicola accumulates L-serine (Yamada 1971). It is also capable of converting pesticidal arsenic compounds to trimethylarsine, a volatile toxin (Cox and Alexander 1973).

Unusually large amounts of extracellular mannan polymers (yeast gum) are synthesized by several species of *Hansenula* (*H. holstii, H. capsulata, H. minuta*) as well as *Pichia pinus* and *Torulopsis pinus* (Slodki *et al.* 1972, 1973). The formation of mannose-containing polysaccharides is characteristic of most yeast species (Spencer and Gorin 1973). Other types of yeasts can be useful in waste treatment and protein synthesis. *Trichosporon cutaneum* grows well in cheese whey (Atkin *et al.* 1967); it also degrades benzoic acid and phenol (Paynter and Bungay 1971; Rao and Bhat 1971). A methanol-utilizing yeast, *Torulopsis glabrata*, is a potential food and fodder yeast (Asthana *et al.* 1971). Twenty cultures of methanol-utilizing yeasts, including two new species of *Candida* and *Torulopsis*, were isolated from azalea blossoms and spoiled tomatoes by Oki *et al.* (1972).

Two species of yeast are capable of infecting humans. *Candida albicans*, the most common pathogenic yeast, causes thrush (candidiasis, moniliasis), which is usually a mild inflammation of oral mucous membranes. It may also be the primary agent in bronchitis, pulmonary infections and dermatitis (Gentles and La Touche, 1969). Normal persons may harbor this yeast in the mouth, intestinal tract and vagina.

The most virulent yeast is *Cryptococcus neoformans*, the causative agent of cryptococcosis, a primary pulmonary infection resembling tuberculosis. Infections originating in the lungs may spread throughout the body, and ultimately the yeast invades the meninges and the brain.

When the need for identification of species arises, methods for characterization of pure yeast cultures discussed by van der Walt (1970) will be helpful. Schemes and keys for identifying yeasts based on physiological tests were developed by Beech *et al.* (1968), Barnett and Pankhurst (1974). Authentic species of yeasts are available from worldwide culture collections. In the U.S. the largest depository of yeasts and other microorganisms has been collected by the Agricultural Research Service at the U.S. Department of Agriculture's Northern Regional Research Laboratory in Peoria, Ill. Known as the

ARS Culture Collection, it has preserved more than 9500 species and strains of yeasts (Hesseltine, *et al.* 1970). Authenticated cultures of yeasts are maintained in numerous general and special collections established in The Netherlands, England, Japan, Canada, Spain and elsewhere. The identity of 329 collections in 52 countries, and a listing of microbial species they hold, has been published by the World Federation of Culture Collections (Martin and Skerman 1972).

YEASTS IN FOOD SPOILAGE

While many foods harbor yeasts, for one reason or another, major incidents of undesirable changes by them are infrequent today. This is attributed mainly to the steady advances of food technology and marketing which focus on the elimination of microorganisms, especially the more competitive and dangerous bacteria. Effective reduction and repression of microbial loads of materials into and around the process have been achieved by placing, where appropriate, stricter standards and specifications on raw materials. Furthermore, greater control of materials in movement as well as processing, packaging and storage of food products have supplemented sound programs for plant cleanup and disinfection. Refresher seminars in health and sanitation for production employees have increased in frequency and effectiveness, especially in small companies. And added to these trends and effects is the motivating influence of intensified surveillance for health hazards and deceptive practices conducted by multiple regulatory agencies and consumer action groups.

Thus the efforts exerted towards attrition and destruction of bacterial masses also diminish and inactivate the concomitant yeasts. What few serious problems they present often result from errors of formulation and processing or from post-process contamination because of package failure, damage or storage accidents. Contamination of foods at the use level, always a possibility, should not be overlooked. Yeasts are found in abundance in sandwiches and salads at retail outlets (Christiansen and King 1971), in fresh and frozen vegetables (Winter *et al.* 1971), in fresh seafoods, packaged meats and delicatessen salads (Koburger 1971, 1972A; Fowler and Clark 1975).

The current literature in food science and technology reflects the concern for minimal, and in certain situations, zero microbial populations in foods. In recognition of the stressed and late log state of yeasts which occur sparsely in foods, conditions and methods for complete recovery and enumeration have been reexamined. Koburger (1971, 1972A, B) obtained higher counts of yeasts and molds when refrigerated food samples were plated on neutral, antibiotic-containing media than on the recommended acidified agar (American Public Health Association 1967). Nelson (1972) found that more yeast survived a sublethal heat treatment when the recovery medium was adjusted to pH 8; pH 10 was optimum for *Candida utilis*. Highest counts of bakers' yeast stressed by desiccation are obtained when the sample is rehydrated in water at 38-46°C (Peppler and Rudert 1953; Herrera *et al.* 1956).

A synopsis of recent studies concerning yeasts in food spoilage is presented in Table 15.3. It lists examples of foods containing yeast,

TABLE 15.3
YEASTS IN FOODS AND FOOD SPOILAGE

FOODS — ROLE — YEAST — REFERENCE

Catsup — gaseous fermentation — *Sacch. bailii* (syn. *Sacch. acidifaciens*).[a]

Mayonnaise, Blue cheese dressing — gaseous fermentation — *Sacch. bailii.* [b]

Cured meats — acidification — *Debaryomyces, Candida, Torulopsis.*[c]

Fresh meats — slow growth, no defect — *Debaryomyces, Candida, Torulopsis, Rhodotorula.* [d, e]

Sausages — slime — *Debaryomyces.*[e]

Ground beef — slow growth, no defect — *Candida, Torulopsis, Rhodotorula.* [f]

Poultry — slow growth, no defect — *Candida, Rhodotorula, Saccharomyces.* [g]

Smoked salmon — post process contamination — unidentified. [h]

Crabmeat — harmless growth — *Candida, Cryptococcus, Rhodotorula, Torulopsis, Trichosporon.* [i]

Pickles in brine — acid loss, softening, discoloration — *Debaryomyces.* [j]

Olives in brine — gas pocket formation; softening — *Hansenula, Saccharomyces.* [k]

Tomatoes — fermentation — *Hanseniaspora uvarum, Kl. apiculata, P. kluyveri.* [l]

Flour — harmless inhabitants — *Sacch. microellipsoides, Candida sp., H. anomala, P. burtonii.* [m]

Margarine — rancidity — *Candida lipolytica.* [n]

Lard — increased alkanals and methyl ketones — *C. lipolytica.* [o]

Beer — turbidity, off-flavors — strains of *Sacch. cerevisiae* and *Sacch. bayanus; Dekkera intermedia.* [p]

Wine — off-flavor, turbidity — *Brettanomyces, Candida, Pichia, Hansenula, Kloeckera.* [q]

Carbonated soft drinks — gaseous fermentation — osmophilic yeasts. [r]

Grana cheese — harmless microflora — *T. candida, C. pseudotropicalis.* [s]

Roquefort cheese — symbiont — *Saccharomyces, Hansenula* species. [t]

Cottage cheese — fruity flavor — unidentified. [u]

Yogurt — fruity flavor, fermentation — *Kluyveromyces fragilis.* [v]

TABLE 15.3 (*Continued*)

FOODS — ROLE — YEAST — REFERENCE

Fruit juice concentrates, honey — ester formation, fermentation — *Hansenula Sacch. rouxii, Sacch. bisporus* var. *mellis.*[w]

Syrups — maple sap fermentation — *Trichosporon* sp., *Cr. albidus.* [x]

Figs, strawberries — fermentation, softening — *Kloeckera apiculata, Hanseniaspora* sp.[y]

Dates — souring, fermentation — *Hanseniaspora* sp., *Candida guilliermondii, C. krusei, C. tropicalis.*[z]

FOOTNOTES:

(a) Mori *et al.* (1971); (b) Kurtzman *et al.* (1971), Rankine and Pilone (1973); (c) Leistner and Bem (1970); (d) Wickerham (1957), Lechowich (1971); (e) Drake *et al.* (1959), Jensen (1954), Koburger (1972); (f) Jay (1972), Koburger (1971); (g) Walker and Ayres (1959); (h) Lee and Pfeifer (1973); (i) Eklund *et al.* (1965); (j) Pederson (1971), Bell and Etchells (1956); (k) Vaughn *et al.* (1972), Mrak *et al.* (1956); (l) DeCamargo and Phaff (1957); (m) Kurtzman *et al.* (1970); (n) Bours and Mossel (1969); (o) Smith and Alford (1969); (p) Kleyn and Hough (1971), Day and Helbert (1971), Richards (1968, 1972); (q) Amerine, *et al.* (1972), Kunkee and Amerine (1970); (r) Sand (1971, 1973), Witter *et al.* (1958); (s) Bottazzi (1968); (t) Galzin and Galzy (1972), Devoyod and Sponem (1970); (u) Roth *et al.* (1971); (v) Arnott *et al.* (1974), Davis *et al.* (1971), Soulides (1956); (w) Lüthi (1959), Ingram (1958), Murdock (1964), Mrak and McClung (1940), Beech (1958); (x) Frank *et al.* (1959), Frank and Willits (1961), Sheneman and Costilow (1959), Naghski (1957); (y) Miller and Phaff (1962), Lowings (1956); (z) Zein *et al.* (1956), Mrak *et al.* (1942).

the undesirable changes produced, and the type of yeast involved. A few reports of earlier incidents are included to illustrate the variety of activities caused by different yeasts. No attempt is made here to catalog the comprehensive literature available. This was ably done by Mrak and Phaff (1948), Ingram (1958), Walker and Ayres (1970).

As innocuous inhabitants on meats, marine products, vegetables, flour, honey and cheeses, yeasts rarely cause spoilage. But upon the addition of a cheese to a salad dressing, for example, gaseous fermentation may ensue (Kurtzman *et al.* 1971). Heterofermentative lactobacilli are responsible, however, in outbreaks of gassiness where only a few hundred yeast cells are found (Peppler *et al.* 1972; Sinell and Siems 1973). And yet, English (1953) has shown that visible fermentation by osmotolerant yeasts occurs with fewer than 10,000 cells per ml at the appropriate available moisture (a_w). Bahlsen (1972) employed osmotolerant yeasts to leaven low moisture (12–18%) pastry doughs.

In Grana, a type of Italian cheese including Parmesan, yeasts found in the young cheese were regarded as part of the normal microflora (Bottazzi 1968). On the other hand, the proteolytic properties of *K. fragilis* and *C. lipolytica* have proven useful in cheese ripening (Chang *et al.* 1972; Lashkari 1971). In the manufacture and shelf-life of yogurt, however, yeasts are the biggest hazard (Davis *et al.* 1971; Tilbury *et al.* 1974).

Properly ripened honey is susceptible to spoilage by osmotolerant yeasts (*Sacch. rouxii*), particularly after granulation occurs which increases the available moisture (White *et al.* 1962). While flash pasteurization of commercial honey stabilizes it, enzymatic changes during extended storage alter its carbohydrate composition.

Softening of brined cucumbers and olives was caused by pecto-lytic, salt tolerant yeasts (Vaughn *et al.* 1972). The strains of *Sacch. cerevisiae*, *K. fragilis* and *C. pseudotropicalis* involved produced pectin esterase and polygalacturonase (Luh and Phaff 1951; Bell and Etchells 1956). Deterioration of brined sauerkraut, cucumbers and other vegetables has been hastened by the action of the film-forming, acid-utilizing yeast strains in *Debaryomyces* and *Endomycopsis* (Pederson 1971; Bell and Etchells 1956). Efforts to preserve ginger in brines undergoing natural yeast fermentations resulted in less satisfactory color and texture of the spice (Brown and Lloyd 1972).

Off-flavors in beer and orange juice may result from excessive development of diacetyl by strains of *Saccharomyces* (Haukeli and Lie 1972; Suomalainen and Ronkainen 1968; Murdock 1964). In both situations the yeast which produced diacetyl could also remove it after the main fermentation.

Wine flavor may be improved by means of a secondary bacterial fermentation in which malic acid is decarboxylated to lactic acid, usually by *Leuconostoc mesenteroides* or *L. oenos* (Amerine and Kunkee 1968; Rankine 1972). The action of deacidifying bacteria is influenced by the yeasts in the wine. According to Fornachan (1968), malo-lactic fermentation is favored by some strains of *Sacch. cerevisiae*, while certain strains of *Sacch. uvarum* and *Sacch. bayanus* are antagonistic. Yang (1973) found that grape musts can be deacidified by *Schizosaccharomyces pombe*, which metabolizes malic acid to ethanol and carbon dioxide.

Yeast contamination of medicinal products is regarded as adulteration (Berreth 1972, 1975). Occasionally yeasts cause spoilage in soft drinks (Berreth 1973, 1974; Put and Sand 1974; Sand 1974; Török and Deak 1974).

CONTROL OF YEASTS IN FOODS

Control measures for yeast nullification usually involve combina-tions of physical and chemical agents, allowing, of course, for product tolerance of the procedures applied. Beginning with plant and apparatus cleanliness, effective control is augmented by selective temperatures (for concentration, pasteurization, sterilization, freezing, desiccation), antimycotics, filtration, safeguards against recontamination, judicious storage conditions, and unimpaired product distribution. In putting it all together, the control of yeasts is frequently less troublesome than is the control of bacteria, except when a few viable survivors remain in the product. Temperature and available moisture influence profoundly the growth, metabolism and survival of yeasts. Both factors, in combination with acidity adjustments, desiccation and antimycotics, prolong the shelf life of foods. Such synergic effects have been advocated especially for products containing 20 to 30% moisture (Labuza *et al.* 1972; Brockmann 1970). Pasteurization of foods with substerilizing doses of gamma radiation is limited to a few foods in several countries; however, its application continues to receive worldwide attention (Diehl 1972).

Temperature

Yeasts commonly found in foods multiply over a wide temperature span. Slow growth is possible at $0°C$, or just above, and the maximum growth temperatures for the common yeasts are near $40°C$ (Stokes 1971). The optimum temperature for reproduction lies between 25° and 35°C. No thermophilic yeasts, *i.e.*, budding above 50°C, have been reported. Several species of *Candida*, however, are psychrophilic (Elliott and Michener 1965). They are capable of multiplying between -5 and 20°C, usually fail to grow above 20°C, and die rapidly at 30°C (Michener and Elliott 1964; Baxter and Gibbons 1962; Ingraham and Stokes 1959; Lawrence *et al.* 1959).

Death of vegetative yeast cells occurs in less than 30 min at 50 to 60°C, showing about the same lethality as vegetative bacteria cells. Yeast spores are somewhat more resistant than the parent vegetative cells. According to the summary by Stokes (1971), death rates for yeasts vary only slightly in different liquid media. In contrast 18 of 19 yeasts freshly isolated from a raw milk supply to a plant experiencing gassiness in Cheddar cheese were destroyed in 5 min at 62.5°C. The remaining one survived for 40 min at the same temperature (Atherton *et al.* 1969).

In general, yeasts survive freezing with little loss of metabolic activity. Compressed bakers' yeast can be stored essentially unimpaired for many months (Reed and Peppler 1973). Dehydrated bakers' yeast, at 8% moisture and in vacuum pack, could be subjected to several freezing and thawing cycles without affecting yeast baking activity (Felsher *et al.* 1955).

Water Activity

Microbial growth is controlled by water activity, the unbound free water available for metabolism. Water activity, designated a_w, is the

ratio of water vapor pressure above the food to vapor pressure of
pure water at the same temperature, which is assigned a value of 1.
Numerically a_w is equal to the corresponding relative humidity
(R.H.) expressed as a fraction, R.H./100 (Scott 1957; Mossel and
Ingram 1955).

Yeasts are more demanding in their moisture requirements for
growth than are molds. This is illustrated in Figure 15.1 which relates

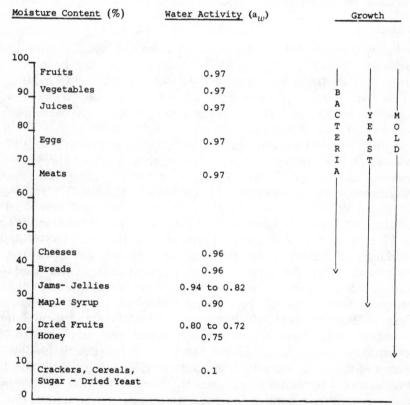

FIG. 15.1. EFFECT OF MOISTURE CONTENT AND WATER ACTIVITY OF COM-
MON FOODS ON GROWTH OF MICROORGANISMS[a]

[a] In part from Kaplow (1970)

moisture content of common foods to water activity, and the latter
to microbial growth. For yeasts, the data reported by Mossel and
Ingram (1955) is cited most often. They found a_w = 0.88 to be the
approximate water activity below which yeast growth ceases. In
studies with *Candida utilis*, Lubuza *et al.* (1972) obtained growth for
two months down to a_w 0.84. For 207 yeasts isolated from cured
meats Bem and Leistner (1970) found the lowest a_w value to be

0.87; it was characteristic of the predominant species, an osmotolerant yeast. Numerous applied studies of the activity water concept have been made; for pet foods (Kaplow 1970), for military use (Brockmann 1970), for new consumer products (Bone 1973). Intermediate moisture foods are consumable without rehydration and are stable without thermal treatment or refrigeration. The general relationship of water in foods in microbial spoilage has been discussed by Matz (1965).

Inhibitors

From the long list of chemicals inhibitory to yeast growth and metabolic activity, only a very few compounds may be used to protect foods from deterioration by yeasts (Beech and Carr 1955). Those which have been permitted in foods include sorbic acid, sorbates, sulfur dioxide, metabisulfite, diethyl pyrocarbonate, sodium benzoate, methyl paraben and propyl paraben.

Sorbic acid and its salts may be added to some foods, up to 0.3%. It is effective in cheese, orange juice, margarine, fruit, syrups, yogurt, pickles, and other foods (Deak *et al.* 1970; Balatsouras and Polymenacos 1963; Etchells *et al.* 1961; Bell *et al.* 1959; Geminder 1959; Costilow *et al.* 1955). In its action on bakers' yeast, sorbic acid is fungistatic, and not fungicidal, at the point of complete inhibition at 17 mM, or 0.19% sorbic acid (Harada and Higuchi 1968). Some producers of bakers' yeast add low levels of sorbate to the fresh household variety to suppress mold contaminants. Sorbic acid is assimilated by certain yeasts (Deak and Novak 1972). In a comprehensive review of three volumes, Lück (1970) has covered, in German, the chemistry, biochemistry, microbiology and technology of sorbic acid. There is some use of sorbic acid and sorbates in winemaking (Amerine 1972) and bakery products (Brachfeld 1969). High resistance to the action of sorbic acid is exhibited by *Sacch. acidifaciens*, a contaminant of wines (Eschenbruch 1973), and *Sacch. bailii* isolated from spoiled beverages and liquid food products (Pitt, 1974).

The antiseptic and microbicidal value of sulfur dioxide effectively aids the preservation of fruits (Bolin and Boyle 1972) and is used in winemaking (Amerine 1972). Domestic limit for wine is 350 mg/l as total sulfur dioxide, of which no more than 70 mg/l is free sulfur dioxide. Most wines, according to Amerine (1972), are below 200 mg and 50 mg, respectively, for total and free sulfur dioxide. Sulfite also controls unwanted yeasts in the raw juice. The desired wine yeasts, whether naturally present or added as cultured starters, are adaptable to as much as 150 mg SO_2/l. For commercial and home use, wine yeasts may be acclimatized to ferment in the presence of 50 to 100 mg SO_2/l (Reed and Peppler 1973).

The benzoates are effective inhibitors of yeasts and molds occurring in acidic foods below pH 4.5 (Brachfeld 1969). Sodium benzoate usage in 28 countries was tabulated by Botma (1973). In the U.S., usage is limited to 0.1% by weight, and is most effective in fruit pie fillings, jams, jellies and the preservation of high-moisture dried prunes (Schade, et al. 1973). Methylparaben (methyl-p-hydroxybenzoate) and propylparaben (propyl-p-hydroxybenzoate) are regulated food preservatives. Their use as food ingredients, either singly or in combination, is limited to 0.1% (Anon. 1973C). Other applications of the parabens were reviewed by Frank and Willits (1961) and Aalto et al. (1953). Heptyl paraben (n-heptyl-p-hydroxybenzoate) at a concentration of 12 ppm prevents yeast growth in stored sugar solutions (Badgley 1970).

Diethyl pyrocarbonate (DEP) has been in worldwide use for the preservation of beer, orange juice and white wine. In wine at normal pH it inhibits yeast at levels of 50 to 150 mg/l (Amerine 1972). Löfroth and Gejvall (1971) emphasized the importance of limiting DEP to beverages with a pH below 4.5 and low in protein content in order to minimize the formation of ethyl carbamate, a carcinogen. Use of DEP as a food preservative has been banned in the U.S. by the Food and Drug Administration (Anon. 1972A).

Experimental studies by Loncin et al. (1970) have demonstrated inhibition of most unwanted yeasts encountered in a brewery by n-octyl gallate at a concentration of 15 mg/l.

Filtration

Biological stabilization of wine, beer, carbonated beverages and bottled water may be achieved by pasteurization of the final product, as in packaged beer, or by aseptic filling after either bulk pasteurization of micro-filtration through membrane filters (Anon. 1972B, 1973A; Berger 1972). A unique method of controlling viable yeast in beer was reported by Kitamura and Yamamoto (1972). They lysed viable brewers' yeast with zymolyase, a non-glucanase type of enzyme derived from *Arthrobacter luteus*.

Irradiation

Gamma radiation treatment of foods has been under investigation for many years. Recent advances have led to approval of substerilization dosage, in the 1000 rad (Krad) to 1,000,000 rad (Mrad) range (Hlavacek 1973). Presently two foods may be irradiated in the U.S.: wheat and wheat products for insect disinfestation; and white potatoes for sprout inhibition. In The Netherlands mushrooms and potatoes may be irradiated, and limited production of irradiated onions, asparagus, strawberries, spices, shrimp and cacao beans is approved. Other developments include feeding trials with radio-sterilized products (Diehl 1973), the isolation of yeasts from irradiated crab meat (Eklund *et al.* 1966), radiopasteurization of molasses fermentation media (Iizuka *et al.* 1968), preservation of fish fillets (Kawabata *et al.* 1968), apple juice (Zehnder *et al.* 1971), grape juice (Febri *et al.* 1972) and spices (Diehl 1972; Tjaberg *et al.* 1972; Gonzalez *et al.* 1972; Vajdi and Pereira 1973).

YEAST DETECTION AND DETERMINATION

Methods for handling, culturing and examining yeasts are chiefly extensions of bacteriological techniques which were adapted by mycologists and taxonomists. In these continuing and painstaking efforts to isolate, characterize and classify yeasts, advanced micro-scopic techniques and equipment have been supplemented with an abundance of concoctions. These may be grouped into three classes: general purpose, differential and selective media. Several works which have reviewed media and methods are recommended: for general orientation and reference, Beech and Davenport (1971, 1969), Collins and Lyne (1970), Shapton and Gould (1969), Laboratory Manual for Food Canners and Processors (1968), Phaff *et al.* (1966), Recommended Methods for the Microbiological Examination of Foods (1958), and Manual for Microbiological Methods (1957); for media and methods of isolation and identification, Dunnigan (1972), van der Walt (1970), Beech *et al.* (1968) and Wickerham (1951); for isolation and differentiation of brewing and wild yeasts, Richards (1969, 1972) and Fowell (1967).

General Purpose Media

The overall nutritional requirements of yeasts are satisfied by numerous all-purpose growth media. Most of these are natural media such as malt extract, yeast extract, beer wort, molasses wort, milk, fruit juices, and others. They may be used as liquids or solidified with agar. For maintaining yeast cultures, a combination of extracts was popularized by Wickerham (1951). It contains 0.3% each of dried malt extract, dried yeast extract, 0.5% peptone, 1% glucose and 2% agar. Sabouraud Dextrose Agar (Anon. 1971) is in general use.

Differential Media

Differential media for yeasts have generally been based on yeast tolerance for acidity, antibiotics, sugars and certain analine dyes.

Most widely used for detection of yeasts in foods, and their enumeration, are potato dextrose agar (PDA) and Mycophil Agar (MA) (Difco 1971; BBL 1972). PDA is generally acidified to pH 3.5 to 4.0, preferably with 10% tartaric acid. Since some yeasts are sensitive to certain acids, the kind of acidulant used can affect the recovery of yeasts from food samples (Koburger 1971, 1972; Nelson 1972). Chudyk (1973) obtained the highest yeast counts on Yeast-Malt Extract Agar (pH 6.2). It was superior to 20 other media tested in the isolation of airborne microflora in a winery. For osmotolerant yeasts, such as occur in concentrated fruit juices and brines, Ingram (1959) recommended a medium containing 50% glucose, 1% citric acid and 1% tryptone.

Inclusion of antibiotics in PDA or MA favors yeast (and mold) recovery from the mixed populations often encountered in food and its environs. Of the numerous variations, the formulations found effective include (a) PDA plus chloramphenicol (Anon. 1971), (b) malt/glucose agar and yeast extract/glucose/malt agar containing tetracycline HCl (Kurtzman *et al.* 1971), (c) peptone/glucose/ chlorotetracycline/rose bengal agar (Overcast and Weakley 1969; Jarvis 1973), (d) yeast extract/glucose agar plus oxytetracycline (Mossel *et al.* 1970) which is recommended by the International Commission on Microbiological Specifications for Foods (Thatcher and Clark 1968, 1974), (e) MA with dihydrostreptomycin sulfate and penicillin G at pH 7.0 (Rhodes and Hrubant 1972), (f) combinations of cycloheximide and sorbic acid for the isolation of *Brettanomyces* species (van der Walt 1970). Some of these media are available in preformulated and dehydrated form (BBL 1972; Difco 1971; Wallerstein 1971).

Differential media based on yeast sensitivity or tolerance to certain dyes have been developed and applied most successfully in breweries for detection and identification of unwanted (wild) yeasts. The fuchsin-sulfite agar of Brenner *et al.* (1970), the recommended medium, and crystal violet agar (Kato 1967) detect *Sacch. diastaticus*, a common wild yeast which does not grow in lysine agar (Morris and Eddy 1957). Lysine agar is in general use, however, for detection of other contaminating yeasts except *Sacch. bayanus* (syn. *Sacch. pastorianus*) and *Sacch. cerevisiae* var. *ellipsoideus*. These two species also develop on the media of Brenner *et al.*, and Kato.

Selective Media

By choosing the appropriate energy source, such as melibiose in Yeast Nitrogen Base (YNB) for *Sacch. uvarum*, the growth of known species can be favored; in this case *Sacch. cerevisiae* is excluded. In a similar manner a medium of Yeast Carbon Base (YCB) supplemented with nitrate as the sole source of nitrogen selects *Candida utilis* in samples containing *Saccharomyces* species, which do not utilize nitrate. YNB and YCB are synthetic media developed by Wickerham (1951), and they are available in dehydrated form (Difco 1971). The inclusion of ferric ammonium citrate in Wickerham's medium causes colonies of *Candida pulcherrima* to assume shades of maroon with metallic iridescence (Wickerham 1951). Numerous similar selective media have been reviewed by Beech and Davenport (1971).

Culture Techniques

By combining membrane filtration with short incubation on a nutrient surface and staining with safranin, Richards (1970) detected microscopically as few as 1 to 3 cells in a half-pint of beer contaminated with different brewers' and wild yeasts. Wild yeasts were recognized by their microcolony morphology. The incubation step could be omitted when cell levels reached 700 to 5000 cells per filter, depending upon the type of yeast present.

Membrane filtration is preferentially employed for microbial analysis of liquids (Anon. 1973A, 1972B, 1969). Beer, juices, beverages, wine, water, among others, can be passed through membranes of small pore size (0.45 μm for beer). The trapped yeast is incubated on a culture medium and examined microscopically or subcultured for identification. A differential membrane technique for rapid detection and enumeration of living and dead yeasts in bottled wines has been developed (Neradt and Kunkee 1973; Kunkee and Neradt 1974).

Serological techniques, when combined with fluorescent microscopy, provide a rapid means of distinguishing between brewers' yeast strains and the non-brewing contamination in pitching yeast (Richards 1969, 1972; Thompson and Cameron 1971). The development and application of immunofluorescent methods to brewery problems were reviewed by Rainbow (1970).

Other suggestions and improvements that have been garnered from the laboratory study of yeasts are worthy of notation here. Instead of water as a diluent in preparing foods for microbial analysis, a 0.1% peptone solution has increased the yield of microorganisms in studies reported by Kurtzman *et al.* (1971), Straka and Stokes (1957). Several investigators have cautioned against the use of 30°C incubators for yeast cultures because the growth of many free living yeasts is retarded. Phaff *et al.* (1966) suggest incubation at 20 to 25°C and Kurtzman *et al.* (1971) incubate at 28°C; brewery laboratories prefer the lower temperature range (Berger 1972).

Three new laboratory aids may be helpful: (1) a convenient sampling and shipping system comprised of a sterile two-pocket pouch containing a swab on one side and a holding medium on the other (Gaines 1972); (2) a microtube yeast fermentation kit for identification of *Candida* species (Anon. 1973B); and, (3) a "dip-stick" type yeast and mold sampler, containing a selective dehydrated medium, for monitoring liquids (Anon. 1974).

With suitable accessories the light microscope is an adequate and indispensable tool for observing morphology, making direct counts, estimating culture viability, and assessing culture purity. This may be done directly or after treatment with appropriate dyes. For a direct count of viable cells, equal volumes of yeast suspension and methylene blue solution (0.01%), or Congo red solution (0.2%) are mixed and examined in a Petroff-Hausser and phase-counting chamber, or hemocytometer. Cells stained by either dye are nonviable cells. Campbell and Thomson (1970) successfully applied the Gram staining procedure to centrifuge sediments of wines. Glenister (1970) applied staining techniques to beer sediments.

Ultimately the choice of methods and procedures is made after a few trial runs to determine which conditions are best suited for the problem at hand. Experience is indispensable, but innovation wins a lot of ball games. In closing this synopsis of yeasts in foods, it seems appropriate to venerate the Father of Food Technology, who, in defense of his earliest experiments, said: "Fabroni has proved that heat applied to grape juice or must, destroys the fermentation of this *vegeto-animal*, which is pre-eminently *leaven*."—Francois Appert (1812).

BIBLIOGRAPHY

AALTO, T. R., FIRMAN, M. C., and RIGLER, N. E. 1953. p-Hydroxybenzoic acid esters as preservatives. J. Am. Pharm. Assoc. *42*, 449‒456.

ADES, G. L., and CONE, J. F. 1969. Proteolytic activity of *Brevibacterium linens* during ripening of Trappist-type cheese. J. Dairy Sci. *52*, 957‒961.

AHEARN, D. G., MEYERS, S. P., and STANDARD, P. G. 1971. The role of yeasts in the decomposition of oils in marine environments. Dev. Ind. Microbiol. *12*, 126‒134.

AHEARN, D. G., and ROTH, F. J. 1966. Physiology and ecology of psychrotrophic carotenogenic yeasts. Dev. Ind. Microbiol. 7, 301‒309.

AMERICAN PUBLIC HEALTH ASSOC. 1958. Recommended methods for the Microbiological Examination of Foods. 1st Edition. American Public Health Assoc., New York.

AM. PUBLIC HEALTH ASSOC. 1972. Standard Methods for the Examination of Dairy Products, 13th Edition. American Public Health Assoc., New York.

AMERINE, M. A. 1972. Quality control in the California wine industry. J. Milk Food Technol. *35*, 373‒377.

AMERINE, M. A., BERG, H. W., and CRUESS, W. V. 1972. Technology of Wine Making, 3rd Edition. Avi Publishing Co., Westport, Conn.

AMERINE, M. A., and KUNKEE, R. E. 1968. Microbiology of winemaking Ann. Rev. Microbiol. *22*, 323‒358.

ANON. 1969. Yeast detection kit. Prod. Bull. *YDK-1*, Millipore Corp., Bedford, Mass.

ANON. 1971. Microbiological culture media and reagents. Tech. Bull. *357*, Wallerstein Co., Deerfield, Ill.

ANON. 1972A. Food additives. Diethyl pyrocarbonate. Fed. Register *37*, No. 149, 15426.

ANON. 1972B. Biological stabilization of wine. Appl. Rep. *AR-72*, Millipore Corp., Bedford, Mass.

ANON. 1973A. Microbiological analysis of soft drinks. Appl. Manual *AM601*, Millipore Corp., Bedford, Mass.

ANON. 1973B. Yeast fermentation kit. *In* Microtube Systems for Bacteriology and Serology. Canalco, Inc., Rockville, Md.

ANON. 1973C. Methyl paraben and propyl paraben. Affirmation of GRAS status of direct human food ingredients. Fed. Register *38*, No. 143, 20048-20050.

ANON. 1974. Yeast and Mold Sampler. Millipore Corp., Bedford, Mass.

APPERT, F. 1812. The Art of Preserving Animal and Vegetable Substances. Black, Parry and Kingsbury, London.

ARNOTT, D. R., DUITSCHAEVER, C. L., and BULLOCK, D. H. 1974. Microbiological evaluation of yoghurt produced commercially in Ontario. J. Milk Food Technol. *37*, 11-13.

ASTHANA, H., HUMPHREY, A. E., and MORITZ, V. 1971. Growth of yeast on methanol as the sole carbon substrate. Biotechnol. Bioeng. *13*, 923-929.

ATHERTON, H. V., ADESS, M. L., and BEAULIEN, R. D. 1969. Growth and resistance characteristics of some yeast cultures isolated from raw milk. J. Dairy Sci. *52*, 896.

ATKIN, C., WITTER, L. D., and ORDAL, Z. J. 1967. Continuous propagation of *Trichosporon cutaneum* in cheese whey. Appl. Microbiol. *15*, 1339-1344.

BADGLEY, G. R. 1970. Sugar industry use for new preservative. Food Technol. *24*, 138.

BAHLSEN, W. 1972. Low moisture pastry doughs. Brit. Pat. 1,270,221.

BALATSOURAS, G. D., and POLYMENACOS, N. G. 1963. Chemical preservatives as inhibitors of yeast growth. J. Food Sci. *28*, 267-275.

BARNETT, J. A., and PANKHURST, R. J. 1974. A New Key to the Yeasts. American Elsevier Publishing Co., New York.

BAXTER, R. M., and GIBBONS, N. E. 1962. Observations on the physiology of psychrophilism in a yeast. Can. J. Microbiol. *8*, 511-517.

BBL MANUAL OF PRODUCTS AND LABORATORY PROCEDURES, 1972. 6th Edition, BioQuest, Inc., Cockeysville, Md.

BEECH, F. W. 1958. The yeast flora of apple juices and ciders. J. Appl. Bacteriol. *21*, 257-266.

BEECH, F. W., and CARR, J. G. 1955. A survey of inhibitory compounds for the separation of yeasts and bacteria in apple juices and ciders. J. Gen. Microbiol. *12*, 85-94.

BEECH, F. W., and DAVENPORT, R. R. 1969. The isolation of nonpathogenic yeasts. *In* Isolation Methods for Microbiologists, D. A. Shapton, and G. W. Gould (Editors). Academic Press, London.

BEECH, F. W., and DAVENPORT, R. R. 1971. Isolation, purification and maintenance of yeasts. *In* Methods in Microbiology, Vol. 4, C. Booth (Editor). Academic Press, London.

BEECH, F. W., DAVENPORT, R. R., GOSWELL, R. W., and BURNETT, J. K. 1968. Two simplified schemes for identifying yeast cultures. *In* Identification Methods for Microbiologists, B. M. Gibbs, and D. A. Shapton (Editors). Academic Press, London.

BELL, T. A., and ETCHELLS, J. L. 1956. Pectin hydrolysis by certain salt-tolerant yeasts. Appl. Microbiol. *4*, 196-201.

BELL, T. A., ETCHELLS, J. L., and BORG, A. F. 1959. Influence of sorbic acid on the growth of certain species of bacteria, yeasts and filamentous fungi. J. Bacteriol. *77*, 573-580.

BEM, Z., and LEISTNER, L. 1970. The water activity tolerance of yeasts present in pickled meat products. Fleischwirtschaft 50, 492-493. (German)

BERGER, D. G. 1972. Quality control in the brewing industry. J. Milk Food Technol. 35, 719-724.

BERRETH, D. A. 1972. FDA Weekly Recall Rept. WW2, 1.

BERRETH, D. A. 1973. FDA Weekly Recall Rept. FF73, 1.

BERRETH, D. A. 1974. Food and Drug Administration Weekly Report of Seizures, etc., January 30, April 17, June 26, July 10, August 28, October 9, November 27, December 4, 26. FDA Press Office, Rockville, Md.

BERRETH, D. A. 1975. Food and Drug Administration Weekly Report of Seizures, etc., January 8. FDA Press Office, Rockville, Md.

BOLIN, H. R., and BOYLE, F. P. 1972. Effect of storage and processing on sulfur dioxide in preserved fruit. Food Prod. Dev. 6, No. 7, 82, 84, 86.

BONE, D. 1973. Water activity in intermediate moisture foods. Food Technol. 27, No. 4, 71-76.

BOTMA, Y. 1973. The use of benzoic acid and its salts. Naarden News 24, No. 5, 1-12. Naarden, Inc., Owings Mills, Md.

BOTTAZZI, V. 1968. Microbiology of Grana cheese. VII. Study of yeasts during early ripening. Scienza Tecnica Lattiero-Casearia 19, 353-359. Food Sci. Technol. Abstr. 1, 4B129, 1969.

BOURS, J., and MOSSEL, D. A. A. 1969. A comparison of methods for the determination of lipolytic properties of yeasts isolated from margarine; demonstration of lipase and urease. Antonie van Leeuwenhoek 35, Suppl., I, 29-30.

BOYD, J. W., and TARR, H. L. A. 1955. Inhibition of mold and yeast development in fish products. Food Technol. 9, 411-412.

BRACHFELD, B. A. 1969. Antimicrobial food additives. Baker's Dig. 43, No. 5, 60-62.

BRENNER, M. W., KARPISCAK, M., STERN, H., and HSU, W. P. 1970. Differential medium for detection of wild yeast in the brewery. Proc. Am. Soc. Brew. Chemists, 79-88.

BROCKMANN, M. 1970. Development of intermediate moisture foods for military use. Food Technol. 24, 896-900.

BROWN, B. I., and LLOYD, A. C. 1972. Investigation of ginger storage in salt brine. J. Food Technol. 7, 309-321.

CAMPBELL, I., and THOMSON, J. W. 1970. Measurement of low levels of yeast contamination in beers and wines. J. Inst. Brew. 76, 465-469.

CHANG, J. E., YOSHINO, U., and TSUGO, T. 1972. Cheese ripened mainly with yeast. II. Proteolytic activity of Saccharomyces fragilis. Nippon Chikusan Gakkai-Ho 43, 193-197. Chem. Abstr. 77, 32934.

CHRISTIANSEN, L. N., and KING, N. S. 1971. The microbial content of some salads and sandwiches at retail outlets. J. Milk Food Technol. 34, 289-243.

CHUDYK, R. V. 1973. Media for isolating Ontario winery microflora. J. Inst. Brew. 79, 509-512.

COLLINS, C. H., and LYNE, P. M. 1970. Microbiological Methods, 3rd Edition. University Park Press, Baltimore, Md.

COSTILOW, R. N., FERGUSON, W. E., and RAY, S. 1955. Sorbic acid as a selective agent in cucumber fermentations. Appl. Microbiol. 3, 341-345.

COX, D. P., and ALEXANDER, M. 1973. Effect of phosphate and other anions on trimethylarsine production by Candida humicola. Appl. Microbiol. 25, 408-413.

DAVIS, J. G., ASHTON, T. R., and McCASKILL, M. 1971. Enumeration and viability of Lactobacillus bulgaricus and Streptococcus thermophilus in yogurts. Diary Ind. 36, 569-575.

DAY, T. A., and HELBERT, J. R. 1971. Identification of Dekkera intermedia from spoiled beer. Proc. Am. Soc. Brew. Chem., 96-104.

DEAK, T., and NOVAK, E. K. 1972. Assimilation of sorbic acid by yeasts. Acta Aliment. 1, 87-104. Chem. Abstr. 77, 43705.

DEAK, T., TUSKE, M., and NOVAK, E. K. 1970. Effect of sorbic acid on the growth of some species of yeast. Acta Microbiol. *17*, 237–256. Chem. Abstr. *74*, 95966.

DE CAMARGO, R., and PHAFF, H. J. 1957. Yeasts occurring in Drosophila flies and in fermenting tomato fruits in Northern California. Food Res. *22*, 367–372.

DEVOYOD, J. J. *et al.* 1971. Microbial combinations in Roquefort cheese. Lait *51*, 399–415.

DEVOYOD, J. J., and SPONEM, D. 1970. Yeasts in microbial flora of Roquefort cheese. Lait *50*, 524–530.

DIEHL, J. F. 1972. Sterilization of food by radiation. Z. Bakteriol. IB *156*, 157–170. (German)

DIEHL, J. F. 1973. Irradiated food. Science *180*, 214–215.

DIFCO MANUAL, 1971. 9th Edition. Difco Laboratories, Inc., Detroit, Mich.

DI MENNA, M. E. 1960. Yeasts from Antarctica. J. Gen. Microbiol. *23*, 295–300.

DJIEN, K. S. 1972. Tapé fermentation. Appl. Microbiol. *23*, 976–978.

DO CARMO-SOUSA, L. 1969. Distribution of yeasts in nature. *In* The Yeasts, Vol. 1, A. H. Rose, and J. S. Harrison (Editors). Academic Press, New York.

DRAKE, S. D., EVANS, J. B., and NIVEN, C. F. 1959. The identity of yeasts in the surface flora of packaged frankfurters. Food Res. *24*, 243–246.

DUNNIGAN, A. P. 1972. Bacteriological Analytical Manual for Foods, 3rd Edition. Food and Drug Administration, Bureau of Foods, U.S. Govt. Printing Office, Washington, D.C.

ECKLUND, M. W., SPINELLI, J., MIYAUCHI, D., and DASSOW, J. 1966. Development of yeast on irradiated Pacific crab meat. J. Food Sci. *31*, 424–431.

ECKLUND, M. W., SPINELLI, J., MIYAUCHI, D., and GRONINGER, H. 1965. Characteristics of yeasts isolated from Pacific crab meat. Appl. Microbiol. *13*, 985–990.

ELLIOTT, R. P., and MICHENER, H. D. 1965. Factors affecting the growth of psychrophilic microorganisms in foods. U.S. Dept. Agr. Tech. Bull. *1320*.

ENGEL, W. B., and SWATEK, F. E. 1966. Some ecological aspects of hydrocarbon contamination and associated corrosion in aircraft. Dev. Ind. Microbiol. *7*, 354–366.

ENGLISH, M. P. 1953. The fermentation of malt extract by an osmophilic yeast. J. Gen. Microbiol. *9*, 15–25.

ESCHENBRUCH, R. 1973. Contamination of wine by *Saccharomyces acidifaciens*, a yeast highly resistant to Baycovin, sulphur dioxide and sorbic acid. Wynboer *496*, 23–24.

ETCHELLS, J. L., BORG, A. F., and BELL, T. A. 1961. Influence of sorbic acid on populations and species of yeasts occurring in cucumber fermentation. Appl. Microbiol. *9*, 139–144.

FABRI, I., VAS, K., and STEHLIK, G. 1972. Effect of pimarcin, ascorbic acid and grape juice proteins on radiation tolerance of wine yeasts in grape juice. Acta Aliment. *1*, 17–27. Chem. Abstr. *77*, 18228.

FELSHER, A. R., KOCH, R. B., and LARSEN, R. A. 1955. The storage stability of vacuum-packed active dry yeast. Cereal Chem. *32*, 117–124.

FORNACHON, J. C. M. 1968. Influence of different yeasts on the growth of lactic acid bacteria in wine. J. Sci. Food Agr. *19*, 374–378.

FOWELL, R. R. 1967. Infection control in yeast factories and breweries. Process. Biochem. *2*, No. 12, 11–15.

FOWLER, J. L., and CLARK, W. S. 1975. Microbiology of delicatessen salads. J. Milk Food Technol. *38*, 146–149.

FRANK, H. A., NAGHSKI, J., REED, L. L., and WILLITS, C. O. 1959. Maple syrup. XII. Effect of zinc on the growth of microorganisms in maple sap. Appl. Microbiol. *7*, 152–155.

FRANK, H. A., and WILLITS, C. O. 1961. Maple Syrup. XVII. Prevention of mold and yeast growth in maple syrup by chemical inhibitors. Food Technol. *15*, 1-3.

GAINES, J. 1972. Trans-Cul^{TM} systems for culture sample collection. Food Drug Packaging *27*, No. 6, 3, 25.

GALZIN, M., and GALZY, P. 1972. Yeasts in Roquefort cheese. Ind. Aliment. Agr. *89*, 19-29.

GEMINDER, J. J. 1959. Use of potassium and sodium sorbate in extending shelf-life of smoked fish. Food Technol. *13*, 459-461.

GENTLES, J. C., and LA TOUCHE, C. J. 1969. Yeasts as human and animal pathogens. *In* The Yeasts, Vol. I, A. H. Rose, and J. S. Harrison (Editors). Academic Press, London.

GLENISTER, P. R. 1970. Useful staining techniques for the study of yeast, beer, and beer sediments. Proc. Am. Soc. Brew. Chem., 163-167.

GONZÁLEZ, O. N., DIMAVNAHAN, L. B., PILAC, L. M., and ALABASTRO, V. O. 1972. Effects of gamma radiation on peanuts, onions and ginger. Philippine J. Sci. *98*, 279-293. FSTA *5*, 1G32.

GRADOVA, N. B., *et al.* 1969. Biosynthesis of protein and fatty substances from petroleum hydrocarbons. Biotechnol. Bioeng. Symp. *1*, 99-104.

GUTHRIE, R. K. 1972. Food Sanitation. Avi Publishing Co., Westport, Conn.

HANG, Y. D., SPLITTSTOESSER, D. F., and LANDSCHOOT, R. L. 1972. Sauerkraut waste: a favorable medium for cultivating yeasts. Appl. Microbiol. *24*, 1007-1008.

HANG, Y. D., SPLITTSTOESSER, D. F., and LANDSCHOOT, R. L. 1973. Production of yeast invertase from sauerkraut waste. Appl. Microbiol. *25*, 501-502.

HARADA, K., and HIGUCHI, R. 1968. Studies on sorbic acid. IV. Inhibition of the respiration in yeast. Agr. Biol. Chem. (Tokyo) *32*, 940-946.

HARRISON, J. S., and GRAHAM, J. C. J. 1970. Yeasts in distillery practice. *In* The Yeasts, Vol. 3, A. H. Rose, and J. S. Harrison (Editors). Academic Press, London.

HAUKELI, A. D., and LIE, S. 1972. Production of diacetyl, 2-acetolactate and acetoin by yeasts during fermentation. J. Inst. Brew. *78*, 229-232.

HERRERA, T., PETERSON, W. H., COOPER, E. J., and PEPPLER, H. J. 1956. Loss of cell constituents on reconstitution of active dry yeast. Arch. Biochem. Biophys. *63*, 131-143.

HESSELTINE, C. W., HAYNES, W. C., WICKERHAM, L. J., and ELLIS, J. J. 1970. History, policy and significance of the ARS culture collection. *In* Culture Collections of Microorganisms, H. Tisuka, and T. Hasegawa (Editors). Proc. Intern. Conf. Culture Collections, 1968, Tokyo.

HLAVACEK, R. G. 1973. New advances in irradiated foods. Food Process. *34*, No. 4, F4-F10.

HOSONO, A., and TOKITA, F. 1970A. Role of yeasts in volatile flavor substances in Limburg cheese. Japan J. Zootech. Sci. *41*, 131-134. Dairy Sci. Abstr. *32*, 667.

HOSONO, A., and TOKITA, F. 1970B. Lipolytic properties of *Candida mycoderma* and *Debaryomyces kloeckeri* isolated from Limburger cheese and some properties of the lipases produced by these yeasts. Nippon Chikusan Gakkai-Ho *41*, 519-529. Chem. Abstr. *76*, 43210.

IIZUKA, H., SHIBABE, S., and ITO, H. 1968. Effect of gamma irradiation on fermentation medium consisting mainly of cane molasses. Food Irradiation *3*, 116-122. Food Sci. Technol. Abstr. *5*, 1H184, 1973.

INGRAHAM, J. L., and STOKES, J. L. 1959. Psychrophilic bacteria. Bacteriol. Rev. *23*, 97-108.

INGRAM, M. 1958. Yeasts in food spoilage. *In* The Chemistry and Biology of Yeasts, A. H. Cook (Editor). Academic Press, New York.

INGRAM, M. 1959. Comparisons of different media for counting sugar tolerant yeasts in concentrated orange juice. J. Appl. Bacteriol. *22*, 234-247.

JARL, K. 1969. The Symba yeast process. Food Technol. 23, 1009-1012.

JARVIS, B. 1973. Comparison of an improved rose bengal-chlortetracycline agar with other media for the selective isolation and enumeration of moulds and yeasts in foods. J. Appl. Bacteriol. 36, 723-727.

JAY, J. M. 1972. Mechanism and detection of microbial spoilage in meats at low temperatures; a status report. J. Milk Food Technol. 35, 467-471.

JENSEN, L. B. 1954. Microbiology of Meats, 3rd Edition. Garrard Press, Champaign, Ill.

KAPLOW, M. 1970. Commercial development of intermediate moisture foods. Food Technol. 24, 889-893.

KATO, S. 1967. Measurement of infectious wild yeasts in beer by means of crystal violet medium. Bull. Brew. Sci. (Tokyo), 13, 19-24. J. Inst. Brew. 74, 475.

KAWABATA, T., KOZIMA, T., and OKITSU, T. 1968. Effect of irradiation and preservatives on the keeping quality of fish fillets. Food Irradiation 3, 40-48. Food Sci. Technol. Abstr. 5, 1R59, 1973.

KITAMURA, K., and YAMAMOTO, Y. 1972. Purification and properties of an enzyme, zymolyase, which lyses viable yeast cells. Arch. Biochem. Biophys. 153, 403-406.

KLEYN, J., and HOUGH, J. 1971. The microbiology of brewing. Ann. Rev. Microbiol. 25, 583-608.

KOBURGER, J. A. 1971. Fungi in foods. II. Some observations on acidulants used to adjust media pH for yeasts and mold counts. J. Milk Food Technol. 34, 475-477.

KOBURGER, J. A. 1972A. Fungi in foods. III. Enumeration of lipolytic and proteolytic organisms. J. Milk Food Technol. 35, 117-118.

KOBURGER, J. A., 1972B. Fungi in foods. IV. Effect of plating medium pH on counts. J. Milk Food Technol. 35, 659-660.

KOBURGER, J. A. 1973. Fungi in foods. V. Response of natural populations to incubation temperatures between 12° and 32°C. J. Milk Food Technol. 36, 434-435.

KODAMA, K. 1970. Saké yeast. In The Yeasts, Vol. 3, A. H. Rose, and J. S. Harrison (Editors). Academic Press, London.

KREGER-VAN RIJ, N. J. W. 1969. Taxonomy and systematics of yeasts. In The Yeasts, Vol. 1, A. H. Rose, and J. S. Harrison (Editors). Academic Press, London.

KUNKEE, R. E., and AMERINE, M. A. 1970. Yeasts in winemaking. In The Yeasts, Vol. 3, A. H. Rose, and J. S. Harrison (Editors). Academic Press, London.

KUNKEE, R. E., and NERADT, F. 1974. A rapid method for detection of viable yeast in bottled wines. Wines & Vines 55, No. 12, 36, 38-39.

KURTZMAN, C. P., ROGERS, R., and HESSELTINE, C. W. 1971. Microbiological spoilage of mayonnaise and salad dressings. Appl. Microbiol. 21, 870-874.

KURTZMAN, C. P., WICKERHAM, L. J., and HESSELTINE, C. W. 1970. Yeasts from wheat and flour. Mycologia 62, 542-547.

NATL. CANNERS ASSOC. RES. LABORATORIES. 1968. Laboratory Manual for Food Canners and Processors. Avi Publishing Co., Westport, Conn.

LABUZA, T. P., CASSIL, S., and SINSKEY, A. J. 1972. Stability of intermediate moisture foods. 2. Microbiology. J. Food Sci. 37, 160-162.

LANJOUW, J. 1966. International Code of Botanical Nomenclature. Kemink en Zoon, Utrecht.

LASHKARI, B. Z. 1971. Improvements in or relating to making cheese. Brit. Pat. 1,251,654.

LAWRENCE, N. L., WILSON, D. C., and PEDERSON, C. S. 1959. The growth of yeasts in grape juice stored at low temperatures. II. The types of yeasts in grape juice stored at low temperatures. II. The types of yeasts and their growth in pure culture. Appl. Microbiol. 7, 7-11.

LECHOWICH, R. V. 1971. Microbiology of meat. *In* The Science of Meat and Meat Products, 2nd Edition, J. F. Price (Editor). W. H. Freeman & Co., San Francisco.

LEE, J. S., and PFEIFER, D. K. 1973. Aerobic microbial flora of smoked salmon. J. Milk Food Technol. *36*, 143-145.

LEISTNER, L., and BEM, Z. 1970. Presence and significance of yeasts in pickled meat products. Fleischwirtschaft *50*, 350-351. (German)

LODDER, J. 1970. The Yeasts: A Taxonomic Study, 2nd Edition. North-Holland Publishing Co., Amsterdam, Holland.

LODDER, J., and KREGER-VAN RIJ, N. J. W. 1952. The Yeasts: A Taxonomic Study. North-Holland Publishing Co., Amsterdam, Holland.

LÖFROTH, G., and GEJVALL, T. 1971. Diethyl pyrocarbonate: Formation of urethan in treated beverages. Science *174*, 1248-1250.

LONCIN, M., KOZULIS, J. A., and BAYNE, P. D. 1970. N-Octyl gallate—a new beer microbiological inhibitor. Proc. Am. Soc. Brew. Chemists, 89-101.

LOWINGS, P. H. 1956. The fungal contamination of Kentish strawberry fruits in 1955. Appl. Microbiol. *4*, 84-88.

LÜCK, E. 1970. Sorbic Acid—Chemistry, Biochemistry, Microbiology, Technology, Vol. III. B. Behr's Verlag, Hamburg. (German)

LUH, B. S., and PHAFF, H. J. 1951. Studies in polygalacturonase of certain yeasts. Arch. Biochem. Biophys. *33*, 212-227.

LÜTHI, H. 1959. Microorganisms in noncitrus juices. Advan. Food Res. *9*, 221-284.

MANUAL OF MICROBIOLOGICAL METHODS, 1957. American Society for Microbiology, McGraw-Hill, New York.

MARTIN, S. M. and SKERMAN, V. B. D. 1972. World Directory of Collections of Cultures of Microorganisms. John Wiley & Sons, New York.

MATZ, S. A. 1965. Water in foods. Avi Publishing Co., Westport, Conn.

MICHENER, H. D., and ELLIOTT, R. P. 1964. Minimum growth temperatures for food-poisoning, fecal-indicator and psychrophilic microorganisms. Advan. Food Res. *13*, 349-396.

MILLER, T. L., and JOHNSON, M. J. 1966. Utilization of alkanes by yeasts. Biotechnol. Bioeng. *8*, 549-565.

MILLER, M. W., and PHAFF, H. J. 1962. Successive microbial populations in *Calimyrna* figs. Appl. Microbiol. *10*, 394-400.

MORI, H., NASUNO, S., and IGUCHI, N. 1971. A yeast isolated from tomato ketchup. J. Ferment. Technol. *49*, 180-187.

MORRIS, E. O., and EDDY, A. A. 1957. Method of the measurement of wild yeast infection in pitching yeast. J. Inst. Brew. *63*, 34-35.

MORSE, R. A., and STEINKRAUS, K. H. 1971. Honey wine. U.S. Pat. 3,598,607.

MOSSEL, D. A. A., and INGRAM, M. 1955. The physiology of the microbial spoilage of foods. J. Appl. Bacteriol. *18*, 232-268.

MOSSEL, D. A. A. *et al.* 1970. Oxytetracycline-glucose-yeast extract agar for selective enumeration of moulds and yeasts in foods and clinical material. J. Appl. Bacteriol. *33*, 454-457.

MRAK, E. M., and McCLUNG, L. S. 1940. Yeasts occurring on grapes and in grape products in California. J. Bacteriol. *40*, 395-407.

MRAK, E. M., and PHAFF, H. J. 1948. Yeasts. Ann. Rev. Microbiol. *2*, 1-46.

MRAK, E. M., PHAFF, H. J., VAUGHN, R. H., and HANSEN, H. N. 1942. Yeasts occurring in souring figs. J. Bacteriol. *44*, 441-450.

MRAK, E. M., VAUGHN, R. H., MILLER, M. W., and PHAFF, H. J. 1956. Yeasts occurring in brines during the fermentation and storage of green olives. Food Technol. *10*, 416-419.

MURDOCK, D. I. 1964. Voges-Proskauer-positive yeasts isolated from frozen orange juice. J. Food Sci. *29*, 354-359.

NAGHSKI, J., and WILLITS, C. O. 1959. Process for enhancing flavor of maple

syrup. U.S. Pat. 2,880,094. March 31.
NELSON, F. E. 1972. Plating medium pH as a factor in apparent survival of sublethally stressed yeasts. Appl. Microbiol. 24, 236–239.
NERADT, F., and KUNKEE, R. E. 1973. Rapid method for detection of viable yeast in bottled wines. Weinberg u. Keller 20, 1–12.
OKI, T., KOUNO, K., KITAL, A., and OZAKI, A. 1972. New yeasts capable of assimilating methanol. J. Gen. Appl. Microbiol. 18, 295–305.
OVERCAST, W. W., and WEAKLEY, D. J. 1969. Aureomycin-rose bengal agar for enumeration of yeast and mold in cottage cheese. J. Milk Food Technol. 32, 442–444.
PAYNTER, M. J. B., and BUNGAY, H. R. 1971. Effect of microbial interactions on performance of waste treatment processes. Biotechnol. Bioeng. Symp. 2, 51–61.
PEDERSON, C. S. 1971. Microbiology of Food Fermentations. Avi Publishing Co., Westport, Conn.
PEPPLER, H. J. 1970. Food Yeasts. In The Yeasts, Vol. 3, A. H. Rose, and J. S. Harrison (Editors). Academic Press, London.
PEPPLER, H. J., DECKER, R. L., and MANTSCH, R. F. 1972. Unpublished experiments.
PEPPLER, H. J., and FRAZIER, W. C. 1942. Influence of a film yeast, Candida krusei, on the heat resistance of certain lactic acid bacteria grown in symbiosis with it. J. Bacteriol. 43, 181–192.
PEPPLER, H. J., and RUDERT, F. J. 1953. Comparative evaluation of some methods for estimation of the quality of active dry yeast. Cereal Chem. 30, 146–152.
PHAFF, H. J., MILLER, M. W., and MRAK, E. M. 1966. The Life of Yeasts. Harvard Univ. Press, Cambridge, Mass.
PITT, J. I. 1974. Resistance of some food spoilage yeasts to preservatives. Food Technol. Australia 26, 238–239, 241.
PREECE, T. F., and DICKINSON, C. H. 1971. Ecology of Leaf Surface Microorganisms. Academic Press, New York.
PUT, H. M. C., and SAND, F. E. M. J. 1974. A method for determination of heat resistance of yeasts causing spoilage in soft drinks. Proc. Fourth Intern. Symp. Yeasts, Vienna, Part I, B 35.
RAINBOW, C. 1970. Brewer's yeasts. In The Yeasts, Vol. 3, A. H. Rose, and J. S. Harrison (Editors). Academic Press, New York.
RANKINE, B. C. 1972. Influence of yeast strain and malo-lactic fermentation on composition and quality of table wines. Am. J. Enol. Viticult. 23, 152–158.
RANKINE, B. C., and PILONE, D. A. 1973. Saccharomyces bailii, a resistant yeast causing serious spoilage of bottled table wine. Am. J. Enol. Viticult. 24, No. 2, 55–58.
RAO, B. V, and BHAT, J. V. 1971. Characteristics of yeast isolated from phenol and catechol-adapted activated sludges. Antonie van Leeuwenhoek, J. Microbiol. Serol. 37, 303–312.
REED, G., and PEPPLER, H. J. 1973. Yeast Technology. Avi Publishing Co., Westport, Conn.
RHODES, R. A., and HRUBANT, G. R. 1972. Microbial population of feedlot waste and associated sites. Appl. Microbiol. 24, 369–377.
RICHARDS, M. 1968. The incidence and significance of wild Saccharomyces contaminants in the brewery. J. Inst. Brew. 74, 433–435.
RICHARDS, M. 1969. The rapid detection of brewery contaminants belonging to the genus Saccharomyces—examination of lager yeasts. J. Inst. Brew. 75, 476–480.
RICHARDS, M. 1970. Routine accelerated membrane filter method for examination of ultra-low levels of yeast contaminants in beer. Wallerstein Lab. Commun. 33, 97–103.

RICHARDS, M. 1972. Brewery spoilage microorganisms. Brewers Dig. *47*, No. 2, 58–59, 62–64.

ROTH, L. A., CLEGG, L. F. L., and STILES, M. E. 1971. Coliforms and shelf life of commercially produced cottage cheese. Can. Inst. Food Technol. J. *4*, 107–111.

SAND, F. E. M. J. 1971. Hygienic aspects of soft drink bottling lines. Brauwelt *111*, 1788–1800. Food Sci. Technol. Abstr. *4*, 3H440, 1972. (German)

SAND, F. E. M. J. 1973. Osmophilic yeast in the refrigerated beverage industry. Brauwelt *113*, 320–327, 414–419. Food Sci. Technol. Abstr. *5*, 7H109, 1973. (German)

SAND, F. E. M. J. 1974. An ecological survey of yeasts within a soft drinks plant. Proc. Fourth Intern. Symp. Yeasts, Vienna, Part I, F 7.

SANTA MARIA, J., and VIDAL, D. 1973. Genetic control of "flor" formation by *Saccharomyces*. J. Bacteriol. *113*, 1078–1080.

SCHADE, J. E., STAFFORD, A. E., and KING, A. D. 1973. Preservation of high-moisture dried prunes with sodium benzoate instead of potassium sorbate. J. Sci. Food Agr. *24*, 905–911.

SCOTT, W. J. 1957. Water relations of food spoilage microorganisms. Advan. Food Res. *7*, 83–127.

SHAPTON, D. A., and GOULD, G. W. 1969. Isolation Methods for Microbiologists. Academic Press, New York.

SHENEMAN, J. M., and COSTILOW, R. N. 1959. Identification of microorganisms from maple tree tapholes. Food Res. *24*, 146–159.

SINELL, H. J., and SIEMS, H. 1973. Gas formation by yeasts responsible for spoilage of mayonnaise and mayonnaise-containing preparations. Archiv. Lebensmittelhygiene *24*, 14–17. (German)

SLODKI, M. E., SMILEY, M. J., and HENSLEY, D. E. 1973. Production of mannan polymers by fermentation. U.S. Pat. 3,713,979.

SLODKI, M. E., WARD, R. M., and CADMUS, M. C. 1972. Extracellular mannans from yeasts. Dev. Ind. Microbiol. *13*, 428–435.

SMITH, J. L., and ALFORD, J. A. 1969. Action of microorganisms on the peroxides and carbonyls of fresh lard. J. Food Sci. *34*, 75–78.

SOULIDES, D. A. 1955. A synergism between yoghurt bacteria and yeasts and the effect of their association upon the viability of the bacteria. Appl. Microbiol. *3*, 129–131.

SOULIDES, D. A. 1956. Lactose-fermenting yeasts in yoghurt and their effect upon the product and the bacterial flora. Appl. Microbiol. *4*, 274–276.

SPENCER, J. F. T., and GORIN, P. A. J. 1973. Mannose-containing poly-saccharides of yeasts. Biotechnol. Bioeng. *15*, 1–12.

STEINKRAUS, K. H., and MORSE, R. A. 1966. Factors influencing the fermentation of honey in mead production. J. Apicult. Res. *5*, 17–26.

STOKES, J. L. 1971. Influence of temperature on the growth and metabolism of yeasts. *In* The Yeasts, Vol. 2, A. H. Rose, and J. S. Harrison (Editors). Academic Press, New York.

STRAKA, R. P., and STOKES, J. L. 1957. Rapid destruction of bacteria in commonly used diluents and its elimination. Appl. Microbiol. *5*, 21–25.

SUOMALAINEN, H., and RONKAINEN, P. 1968. Mechanism of diacetyl formation in yeast fermentation. Nature *220*, 792–793.

TALENS, L. T., MIRANDA, M., and MILLER, M. W. 1973. Electron micrography of bud formation in *Metschnikowia krissii*. J. Bacteriol. *114*, 413–423.

THATCHER, F. S., and CLARK, D. S. 1968. Microorganisms in Foods, Vol. I. Univ. Toronto Press, Ontario, Canada.

THATCHER, F. S., and CLARK, D. S. 1974. Microorganisms in Foods, Vol. II. Univ. Toronto Press, Ontario, Canada.

THOMPSON, C. C., and CAMERON, A. D. 1971. The differentiation of primary brewing yeasts by immunofluorescence. J. Inst. Brew. *77*, 24–27.

454   FOOD MICROBIOLOGY

TILBURY, R. H. *et al.* 1974. Taxonomy of yeasts in yoghurts and other dairy products. Proc. Fourth Intern. Symp. Yeasts, Vienna, Part I, F 8.

TJABERG, T. B., UNDERDAL, B., and LUNDE, G. 1972. The effect of ionizing radiation on the microbiological content and volatile constituents of spices. J. Appl. Bacteriol. *35*, 473-478.

TÖRÖK, T., and DEAK, T. 1974. A comparative study on yeasts from Hungarian soft drinks and on the methods of their enumeration. Proc. Fourth Intern. Symp. Yeasts, Vienna, Part I, B 38.

VAJDI, M., and PEREIRA, R. R. 1973. Comparative effects of ethylene oxide, gamma irradiation and microwave treatments on selected spices. J. Food Sci. *38*, 893-895.

VAN DER WALT, J. P., 1970. Criteria and methods used in classification. *In* The Yeasts, 2nd Edition, J. Lodder (Editor), North-Holland Publishing Co., Amsterdam.

VAUGHN, R. H., STEVENSON, K. E., DAVE, B. A., and PARK, H. C. 1972. Fermenting yeasts associated with softening and gas-pocket formation in olives. Appl. Microbiol. *23*, 316-320.

WALKER, H. W., and AYRES, J. C. 1959. Characteristics of yeasts isolated from processed poultry and the influence of tetracyclines on their growth. Appl. Microbiol. *7*, 251-255.

WALKER, H. W., and AYRES, J. C. 1970. Yeasts as spoilage organisms. *In* The Yeasts, Vol. 3, A. H. Rose, and J. S. Harrison (Editors). Academic Press, New York.

WALLERSTEIN, 1971. Microbiological culture media and reagents. Tech. Bull. *357*. Wallerstein Co., Deerfield, Ill.

WHITE, J. W., RIETHOF, M. L., SUBERS, M. H., and KUSHNIR, I. 1962. Composition of American honeys. U.S. Dept. Agr. Tech. Bull. *1261*.

WICKERHAM, L. J. 1951. Taxonomy of yeasts. U.S. Dept. Agr. Tech. Bull. *1029*.

WICKERHAM, L. J. 1957. Presence of nitrite-assimilating species of *Debaryomyces* in lunch meats. J. Bacteriol. *74*, 832-833.

WINTER, F. H., YORK, G. K., and EL-NAKHAL, H. 1971. Microbial contamination on foods and equipment. Appl. Microbiol. *22*, 89-92.

WITTER, L. D., BERRY, J. M., and FOLINAZZO, J. F. 1958. The viability of *Escherichia coli* and a spoilage yeast in carbonated beverages. Food Res. *23*, 133-142.

YAMADA, H. 1971. L-Serine. Ger. Offen. 2,108, 214. Sept. 16.

YANG, H. Y. 1973. Deacidification of grape must with *Schizosaccharomyces pombe.* Amer. J. Enol. Viticult. *24*, 1-4.

ZEHNDER, H. J., BALKAY, A. M., and CLARKE, I. D. 1971. Microbiological aspects of the preservation of apple juice by high dose rate irradiation. Lebensm.-Wissenschaft Technol. *4*, No. 3, 73-75.

ZEIN, G. N., SHEHATA, A. M. E., and SEDKY, A. 1956. Studies on Egyptian dates. I. Yeasts isolated from souring soft varieties. Food Technol. *10*, 405-407.

CHAPTER 16

D. F. Splittstoesser
and
D. B. Prest

Molds

FOOD CONSIDERATIONS

Susceptible Foods

Filamentous fungi (molds) represent a very important group of food spoilage organisms. In general, the foods most susceptible to this type of spoilage are those which for some reason or other do not support good growth of other microorganisms. Molds grow more slowly than many bacteria and yeasts, and therefore under broadly favorable conditions the latter will make the food unfit to eat before mold development is evident. For example, hamburger, an excellent growth medium for many organisms, is almost always spoiled by bacteria. Conditions that may favor mold growth over that of other microorganisms are high acidity, limited available water and low incubation temperatures. These along with other factors will be discussed later.

Some of the foods that are more susceptible to attack from molds are fresh fruits and vegetables, bakery goods, meats and certain dairy products. As illustrated in Table 16.1 a large number of different

TABLE 16.1
SOME OF THE MORE IMPORTANT GENERA
CAUSING SPOILAGE OF DIFFERENT FOODS

Bakery Products	Dairy Products	Fruits & Vegetables	Meats
Aspergillus	Alternaria	Alternaris	Alternaria
Monilia	Cladosporium	Aspergillus	Aspergillus
Mucor	Geotrichum	Botrytis	Boletus
Rhizopus	Monilia	Cladosporium	Botrytis
Penicillium	Oospora	Colletotrichum	Cladosporium
Sporotrichum	Penicillium	Diplodia	Fusarium
		Fusarium	Monilia
		Monilia	Mucor
		Mucor	Oospora
		Oospora	Penicillium
		Penicillium	Rhizopus
		Phomopsis	Sporotrichum
		Phytophthora	Rhamnidium
		Rhizopus	
		Sclerotinia	
		Trichoderma	

genera are involved and certain molds may be important spoilage organisms for a variety of food types.

Fungi are the main cause of market diseases of fresh fruits and vegetables (Vaughn 1963); thus they are largely responsible for spoilage from the time of harvest, through the various marketing steps, until the product is consumed (Smoot, *et al.* 1971). Some of the spoilage organisms are true plant pathogens in that they have the ability to invade and establish themselves on healthy, viable plant tissue. An example is *Phytophthora infestans*, the causal organism of late blight of the potato. Other fungi are saprophytes in that their development is restricted to dead or damaged plant tissue. Species of penicillia and aspergilli generally fall into this category.

The growth of fungi on fruits and vegetables usually results in severe tissue breakdown; that is, some form of rot. The names given to the different rots often reflect the appearance of the spoiled food; thus we have diseases such as blue, grey and black rots, watery soft rots, and stem end rots. Tissue disintegration is caused mainly by the decomposition of intercellular layers, the middle lamellae, although the breakdown of cell walls and the degradation of cell protoplasts may also occur (Husain and Kelman 1959). Initial tissue masceration is caused by numerous endotype fungal enzymes (transeliminases and esterases) that attack the pectic substances which are the main constituents of the lamellae (Codner 1971). Following exhaustion of the more labile cell components, cellulases degrade the plant cell walls while proteinases, amylases and other carbohydrate-degrading enzymes destroy the protoplasm.

Molds also are the principle spoilage agents of bread and other bakery products. Mold growth on these foods results in musty off-flavors as well as the objectionable appearance of colonies and more diffuse mycelial growth. Since the organisms generally are destroyed by baking, most spoilage must be attributed to contamination that occurs after this process. Mold spores may be introduced during various post-baking operations such as cooling, slicing or wrapping; also in the kitchen of the consumer after the package has been opened.

Both fresh and cured meats are vulnerable to molds. Some of the defects resulting from their growth are whiskers, stickiness, rancidity, and various discolorations originating from pigmented species (Jensen 1954). Traditionally, fresh beef and bacon have been two of the more susceptible products. Fungi developed on beef during aging in cold storage, often a period of several weeks, while bacon became moldy because it had not received sufficient heat to destroy spores. The current practice of shorter storage periods has reduced the opportunity for mold growth on fresh beef (Ayers 1955) and vacuum packaging has eliminated much of the problem with bacon (Ingram 1963).

Cheese and butter are the dairy products most often spoiled by molds (Foster, *et al*. 1957). The acidity and relatively low moisture content of many finished cheeses results in molds being the most common cause of spoilage. The defects produced on both products are various discolorations and off-flavors including rancidity.

While the above foods represent some that are plagued most often by molds, a wide variety of others are subject to occasional spoilage by this group of organisms. Frozen foods such as meat pies, for example, may become moldy when stored for an extended time in freezers that permit significant temperature fluctuation (Gunderson 1961; Hanson and Fletcher 1958). Low moisture foods such as prunes, licorice and jellies may be spoiled by *Xeromyces bisporus* and other xerophilic species (Dallyn and Everton 1969). Softening of vegetables may occur when pectinolytic species develop on the surface of brines. Surface growth can also be a problem with other bulk-stored foods such as fruit juices and sauces (Dakin and Stock 1968). Even canned foods are not completely exempt since several molds can survive the thermal process given various fruit products (Hull 1939; Williams *et al*. 1941).

Distribution

Filamentous fungi are widely distributed in the environment. Important contributing factors undoubtedly are their capacity to utilize a great variety of carbonaceous materials, their ability to develop in the presence of very low concentrations of nutrients, and their tolerance to many adverse conditions. Their ubiquity is very evident in environments of high relative humidity where it is not uncommon to find fungal growth on walls, clothes, and metal surfaces; even on "clean" glass such as the lenses of optical equipment (Smith 1969).

Fungi account for the largest part of the total microbial protoplasmic mass of well-aerated soils (Alexander 1961) and it is likely that soil may serve as a prime respository for many species. The principal means for the dispersal of molds to different areas is via asexual spores such as zoospores, sporangiospores and conidia (Warcus 1967).

The presence of spores in the air and on the raw product are two sources of mold contamination in food processing plants. Good quality raw materials may harbor fewer molds than those of poor quality. Marshall and Walkley (1951) found that sound apples contained 12 to 1700 molds per square centimeter while damaged fruit yielded counts of over 200,000 per apple. Studies on the atmosphere of a poultry processing plant revealed populations of 31 to 62 molds per cubic foot of air (Kotula and Kinner 1964). The counts in this factory decreased as the processing day progressed, suggesting that the birds were not the original source of the organisms. In a dairy, growth of molds in floor drains was partly responsible for the amount of airborne contamination that was found, an average count of 12 per cubic foot of air (Heldman *et al.* 1965).

Factors Affecting Growth

Nutrition.—Most molds have relatively simple nutritional requirements in that they can utilize inorganic sources for all nutrients except carbon (Smith 1969). Although species and even strains may differ in their carbon needs, some general statements can be made. Thus most species grow equally well on the hexose sugars, glucose, fructose and mannose, while xylose is the more widely utilized pentose (Cochrane 1958). Of the disaccharides, maltose and sucrose are excellent carbon sources for many species; fewer are able to utilize lactose. As to longer chained polysaccharides, many fungi grow well on starch and the ability to degrade cellulose is not uncommon. Additional carbohydrates that are utilized by many species are the sugar acids and alcohols, pectic substances, hemicelluloses and chitins. Other carbon sources include organic acids, amino acids, lipids, polyphenolic and other aromatic compounds.

Molds are efficient converters of substrate carbon into cellular material; in general, more efficient than bacteria (Foster 1949). Their efficiency, expressed as dry weight of mycelia synthesized per weight of carbohydrate utilized, often is in the range of 25 to 50%. To achieve this efficiency, growth conditions must be such as to minimize the diversion of substrate carbon into waste metabolic products.

Most fungi grow well in media containing an inorganic compound as the sole source of nitrogen (Cochrane 1958). Although ammonia often appears to be the preferred form, many species also readily utilize nitrate and nitrite (Foster 1949). Some organic compounds that can serve as a source of nitrogen for many molds are amino acids, amides such as asparagine, urea, purines and pyrimidines.

The other required nutrients are, in general, those that are needed by all organisms: phosphorous, potassium, sulfur and magnesium along with trace amounts of elements such as copper, iron, zinc and manganese.

Although a carbon-energy source is usually the only organic compound required for growth, some strains may lose their ability to synthesize certain amino acids and growth factors and consequently become dependent upon an external source (Cochrane 1958; Foster

1949). The most common vitamin requirement of cultures isolated from nature is for thiamin followed by biotin and pyridoxine. The requirements for pantothenic acid, niacin, riboflavin and para-aminobenzoic acid have also been encountered but apparently are relatively rare.

Hydrogen Ion Concentration.—While a slightly acid medium generally is considered to be optimum for fungi (Smith 1969), many species are able to grow almost equally well over a wide range of hydrogen ion concentrations. For example, *Aspergillus niger, A. flavus, Penicillium roqueforti, P. expansum* and *Geotrichum candidum* exhibited similar growth rates on media ranging in pH from 5.5 to 8.4 (Brancato and Golding 1953). Furthermore, many species are able to grow relatively well at pH values considerably below that which is optimum; as an extreme example, certain strains of *A. niger* produce significant growth in media having an initial pH as low as 1.8 (Foster 1949). Although the spores of fungi are probably affected more by extremes of pH than are vegetative forms, still the range over which many can germinate is relatively wide, especially in comparison to bacterial spores. Ascospores of *Byssochlamys fulva* are able to germinate in menstrua of pH 1.8 to 8.0 with the optimum being about pH 3.5 (Hull 1939). It is likely that many fungi are somewhat less aciduric as evidenced by the fact that the plating of foods on pH 3.5 agar media yielded lower mold counts than similar nonacidified media that relied upon antibiotics rather than a low pH to inhibit bacterial growth (Koburger 1971).

Water.—The water activity, a_w, of a food determines whether sufficient moisture is available to support growth of molds and other microorganisms. Although molds, in general, need less available water for growth than many other microorganisms (Scott 1957), the requirements of different species may vary considerably. Snow (1949) found that spores of *Botrytis cinerea, Rhizopus nigrificans* and two species of *Mucor* germinated at 93% but not at 90% R.H. Spores that germinated at intermediate relative humidities, 80 to 90%, were produced by *Tricothecium roseum, Cladosporium herbarum*, six species of penicillia and four aspergilli. The organisms tolerating the lowest available water, under 0.75, were members of the *Aspergillus glaucus* group: *A. chevalieri* var. *intermedius, A. restrictus, A. candidus, A. amstelodami* and *A. echinulatus*. Another mold, *Xeromyces bisporus*, has been found to grow slowly at the very low a_w of 0.62 (Scott 1957). This organism also has a low optimum a_w, 0.97, which means that it is a true xerophile as are some of the aspergilli.

Various factors may affect the ability of a mold spore to

germinate and produce vegetative growth in a low moisture environment. Development at the lowest possible a_w usually requires that the other conditions that influence growth such as pH and temperature are optimal (Schelhorn 1951; Scott 1957).

Both the length of the latent period before spore germination and the rate of mycelial development are retarded by a low a_w. At 75% R.H., germination of conidia of the *A. glaucus* group required 15 to 18 days, while an incubation period of two years was required for limited germination of *A. echinulatus* conidia held at 66% R.H. (Snow 1949). The germ tubes that developed during the long incubation of the *A. echinulatus* spores showed an abnormal morphology and it was suspected that they may not have had the capacity to produce normal vegetative growth. It must be concluded from these different studies that several years may be required before mold growth becomes evident on a low moisture food.

Temperature.—Molds grow over a wide temperature range. Certain species are true thermophiles in that they fail to develop at temperatures below 30°C and do well at 50°C (Cochrane 1958) while others can grow at very low temperatures, below the freezing point of water. Some of the more common species such as aspergilli and penicillia have optima in the range of 20 to 25°C as do many others of importance in food spoilage. Because of this, some recommended methods for the enumeration of molds in foods call for an incubation temperature of 21°C when samples are cultured on agar media (Sharf, 1966).

As a result of their ability to grow at low temperatures, molds are common spoilage organisms of refrigerated foods and occasionally even develop on frozen products. Some species such as *Cladosporium* and *Sporotrichum* can grow at temperatures as low as -6.7°C (Michener and Elliott 1964). Development at a temperature below the freezing point of water requires an ability to grow at a low a_w since solutions become relatively concentrated as pure water is frozen out. A solution in equilibrium with ice at -10°C, for example has an a_w of 0.9074 (Scott 1957). Growth is, of course, very slow at low temperatures; for example, nine months were required for turkey pies to develop visible mold colonies when stored at -6.7°C (Hanson and Fletcher 1958).

Oxygen.—Molds are aerobic organisms in that they require oxygen for growth under most conditions (Cochrane 1958). While the exclusion or reduction of air from a food can prevent or delay their development, often the presence of only a small amount of oxygen will support sufficient growth for spoilage. Thus a heat-resistant *Penicillium* produced new vegetative growth under a vacuum of 25 inches (Williams *et al.* 1941) and in an atmosphere containing only 0.05% oxygen (Ruyle *et al.* 1946). In other studies, no inhibition of three penicillia, two aspergilli, and *Oospora lactis* was detected unless the concentration of dissolved oxygen under standard conditions was below 0.08 volumes per 1000 ml (Miller and Golding 1949). The composition of the growth medium can influence oxygen require- ments: *Fusarium oxysporum* grew in the complete absence of oxygen when the medium contained electron acceptors such as nitrate, selenite, ferric ions or an unidentified constituent of yeast extract (Gunner and Alexander 1964). It would appear, therefore, that the degree of anaerobiosis attained under commercial processing conditions often is not adequate to prevent growth of molds in many foods.

Inhibition and Destruction

Heat.—Most molds possess little heat resistance. The spores of various species such as *Monilia sitophila*, *Penicillium expansum*, *Rhizopus nigrificans*, a *Fusarium* and three aspergilli failed to survive when heated in water for only five minutes at 60°C (Jensen 1954).

Some of the organisms that are more tolerant to heat owe their resistance to sclerotia, structures that are hard mycelial masses composed of thick-walled cells. Sclerotia can be very resistant as evidenced by those of a *Penicillium*, isolated from spoiled canned blueberries, that survived four and one-half hours at 85°C (Williams *et al.* 1941). The ascospores produced by this organism showed little resistance in that they were destroyed by a treatment of only ten min at 82°C.

A few molds do form ascospores that are relatively difficult to destroy. Those produced by the two species of *Byssochlamys*, *B. fulva* and *B. nivea*, are perhaps the most resistant of all since they have been responsible for numerous spoilage outbreaks of canned acid foods. Studies on their heat resistance have revealed decimal reduction times (D-values) of over 45 min at 85°C and over 15 min at 90°C when spores were heated in a fruit syrup (Gillespy 1936-73). King *et al.* (1969) found the D-values of five different strains heated in grape juice at 87.8°C to range from 8.3 to 11.3 min. In this latter study, a 6.7 centigrade degree decrease in the heating temperature resulted in a 10-fold increase in the D-value.

Byssochlamys was originally recognized as a food spoilage organism in Great Britain and at one time the problem appeared to be restricted to that country. In recent years, however, spoilage outbreaks have occurred in many parts of the world including the

U.S. The mold has been isolated from various sections of North America (Yates and Ferguson 1963; King *et al.* 1969; Splittstoesser *et al.* 1971).

Other molds whose spores are somewhat more heat resistant are *Phialophora*, *Thermoascus aurantiacum*, *Aspergillus fischeri*, and *Penicillium vermiculatum*. Some of these organisms have been recovered when fruit and other material were heated prior to being cultured on media that were selective for fungi. *Paecilomyces varioti*, the imperfect stage of *Byssochlamys*, has also been isolated from heated materials. It is believed, however, that the isolates were actually strains of *Byssochlamys* that were identified incorrectly because few or no asci were produced when the organisms were subcultured.

Electromagnetic Irradiation.—Ultraviolet light is used to destroy mold spores that are present on the surface of various foods such as bread and cakes following baking, on raw meats during aging, on bacon and other processed meats during packaging, and on cheese during storage and packing. It also has been used to prevent growth on the surface of bulk-stored products such as fruit juice. In addition to treating foods, the processor has used ultraviolet lamps to reduce contamination of equipment surfaces such as the blades of bread slicers, to prevent mold growth on walls and other damp areas, and to reduce the population of spores present in the factory atmosphere.

The resistance of mold spores to ultraviolet radiation varies with the species. In general those producing deeply pigmented spores are most difficult to kill. Research on seven species isolated from bread showed that 1.5 to 6 min were required to achieve sterility when spore-seeded plates were exposed to a 15-watt source from a distance of 25 cm (Conklin 1944). The dosage required for 100% destruction of the different species ranged from 2025 to 8100 microwatts per square centimeter of irradiated surface. *Aspergillus ruber* was most easily killed while the dark spores of *A. niger* were most resistant. The somewhat lengthy exposure needed to destroy a high percentage of spores may explain the observation that the use of ultraviolet lamps in a cheese curing room did not significantly reduce the amount of airborne mold contamination (Smith 1942).

The effect of ionizing radiations on molds has received only limited study. It appears, however, that species differ considerably in their ability to survive this type of radiation. Thus the studies of Sommer *et al.* (1964, 1964A) on the resistance of spores of twelve fruit-decay fungi indicated approximate D-values ranging from 0.02 megarads for *Trichoderma viride* to 0.23 megarads for species of

Cladosporium. The D-values for some of the other more resistant species were: *Botrytis cinerea*, 0.13; *Diplodia natalensis*, 0.14; *Alternaria citri* and *Rhizopus stolonifer*, 0.20.

Carbon Dioxide.—Both spore germination and mycelial growth may be inhibited by high concentrations of carbon dioxide. In the studies of Brown (1922) an atmosphere containing 40% CO_2 completely prevented the spores of *Aspergillus repens* and a *Mucor* from germinating during an incubation of seven days. Concentrations of 50 to 60% were needed to prevent germination of other species such as *Botrytis cinerea, Phoma roseola* and *Rhizopus nigricans* while over 80% was required for *Penicillium glaucum* and a *Fusarium*. The rate of colony growth of many of these organisms was retarded by concentrations of CO_2 as low as 10%.

Carbon dioxide has been used to extend the shelf life of certain foods. The Böhi process, in which fruit juice is permeated with CO_2 at 15°C until it contains a concentration of 1.5% by weight, has been used in Europe (Luthi 1959). The gas also will inhibit mold development on fruits that tolerate relatively high concentrations of CO_2 during controlled atmosphere storage; an atmosphere containing 50%, for example, suppressed mold development on black currants for eight weeks (Smith 1963).

Organic Acids.—It has been known for many years that certain acids possess fungistatic and fungicidal properties (Kiesel 1913). In general the longer chained saturated acids are most inhibitory with 11 or 12 carbon atoms being the optimal length. Branched acids are less effective than straight chains containing the same number of carbon atoms and the substitution of hydrogens with hydroxyl groups reduces their antifugal properties. Dicarboxylic acids are less fungicidal than those that possess only one carboxyl group. It is the undissociated acid that is the active molecule, therefore effectiveness increases with a decrease in pH. Many acids exhibit little activity at neutral reactions.

The acids and salts most commonly used as food preservatives are propionic, sorbic and benzoic (Food Protection Committee, 1965). Propionates are added to bakery products such as bread and cakes and to certain processed cheeses. Sorbates are used in nonyeast raised bakery goods, on the surface of cheeses in consumer-size packages, and in a great variety of other foods such as salads, fruit cocktails, and various beverages including wine. Benzoic acid and its sodium salt are most effective at a low pH and therefore are most widely used in highly acidic foods such as fruit products and carbonated beverages. The activity of methylparaben and propylparaben, two benzoic acid derivatives, is less pH dependent and therefore these compounds are used as preservatives of lower acid foods such as candy and baked goods.

Sulfur dioxide.—The role of sulfur dioxide in fruit processing is to prevent browning and other oxidative changes as well as to serve as an antiseptic. The compound is especially effective against molds, which undoubtedly explains why molds are not important spoilage organisms of sulfur dioxide-containing foods such as wines. A concentration of two percent in the atmosphere of refrigerated cars and storage rooms has been maintained to retard mold growth on market grapes. On dehydrated fruit, it prevents browning reactions and inhibits the growth of xerophilic molds. In winemaking sulfur dioxide is added to the must and finished wine as the liquefied gas or as the alkali salt of sulfite, bisulfite or metabisulfite. Although 350 mg per liter of total sulfur dioxide are permitted in the U.S., most wines contain 100 mg per liter or less. As with the organic acids, the fungicidal activity of sulfur dioxide increases as the pH is reduced.

Miscellaneous.—Numerous other compounds are used or have been proposed for use as mold inhibitors (Schelhorn 1951). Some, such as sodium tetraborate, sodium bicarbonate, sodium silicate, sodium o-phenylphenate and 2-aminopyridine, have been incorporated into dips for treating the exterior surface of foods. Others have been impregnated into packaging materials such as paper wrappings. Fungicides used for this purpose have included caprylic acid, iodine, diphenyl and o-phenylphenol. Some antifungal compounds that have been of recent interest are vitamin K_5 (4-amino-2-methyl-1-naphthol) and the macrolide antibiotics, nystatin, rimocidin, and pimaricin (Ayres *et al.* 1964).

ISOLATION AND IDENTIFICATION

Methods

The methods used for the isolation of fungi from foods may vary depending whether spoiled or normal foods are being investigated. When mold development is evident, spores or mycelial fragments can be transferred directly to a suitable medium using a needle or other inoculating device. Isolation from a normal food, on the other hand, generally requires the inoculation of the food or food homogenate into an agar medium, often one that is selective for fungi. With certain foods such as juices, hyphae and spores can be concentrated by centrifugation or filtration prior to culturing.

Commonly used media for fungi are malt extract, potato dextrose and corn meal agars (Booth 1971A). Bacterial growth on these media can be inhibited by acidification to a pH of about 3.5 or by the incorporation of broad spectrum antibiotics such as streptomycin or chloramphenicol (Koburger 1971). The inoculated plates are usually incubated at room temperature because the majority of fungi develop well in this range. If thermotolerant fungi are suspected, the cultures may be incubated over a range of temperatures from that of the room to 45°C. Yeast glucose and yeast starch agars have been suggested for the higher temperatures since they do not dry out as rapidly as malt extract and potato dextrose agar (Booth 1971).

It is essential that the organisms be isolated in pure culture for identification. The primary isolates are examined directly under the steroscopic microscope at magnifications of 10-100X for evidence of sporulation. As soon as spores are visible, they are touched with a sterile inoculating needle, and streaked onto the surface of a fresh agar plate and incubated. This process is repeated until a pure culture is obtained. The use of Czapek's solution agar is helpful in the isolation of molds since growth is less profuse and thus the selection of desired spore types is facilitated.

Both the colony characteristics and the development of the culture as viewed under the microscope should be studied. Commonly "spot" inoculations (Raper and Thom 1949) of agar plates are made by suspending the inoculum (masses of conidia or bits of mycelium) in soft nonnutrient agar (0.5%) and transferring a small amount on the tip of a sterilized needle to the surface of the medium. Although single colonies on a plate are often desirable, cultures having two or more colonies may be helpful; the mature growth of the organism as well as its fruiting habits appear to be enhanced and may be studied in the area where the colonies grow toward each other. The opposite side of the colony is not under the influence of the adjacent colony; thus the former may be used to study the habits of vegetative growth and of the more immature forms.

A careful examination of the cultural characteristics should be made. The type or texture of growth, the degree and kind of sporulation, the pigment produced on the surface and on the reverse side of the colony, the extent and rate of growth at the temperature of incubation—all of these factors should be noted. To study the microscopic detail of the culture, a small amount of growth may be removed aseptically, placed in a drop of 70% alcohol, gently teased apart to spread the material, a drop of lacto-phenol solution added, and the preparation then covered with a cover glass.

Sometimes it is helpful to determine how a particular structure is formed or developed. The slide technique of Riddell (1950), a method by which mounts can be made with little disturbance to the fungus, has proved useful for this purpose. A slide supported on a V-bent glass rod resting on a filter paper and a cover glass are sterilized in a glass petri dish. A small block of the agar medium to be used is placed aseptically onto the glass slide. The block is then inoculated on each of the four sides and covered with the sterile cover glass. The preparation is then incubated in a moist atmosphere by adding a small amount of sterile 20% glycerol or sterile distilled water to the petri dish to prevent drying. The cover glass may be removed carefully when there is adequate growth and sporulation of the culture. A drop of 95% alcohol is added to fix the material to the cover glass. When the preparation is almost dry, a drop of lacto-phenol solution is added and the cover glass is gently lowered to a clean glass slide. A similar preparation may be made from the slide by carefully removing the agar block which was inoculated with the fungus, fixing the material with a drop of alcohol, adding a drop of lacto-phenol solution and covering with a cover glass.

At times it may be desirable to determine the number of spores present. The spores are concentrated either by centrifugation or filtration, washed free of the extraneous material and the concentrate adjusted to a known volume. A drop of the well-mixed final suspension is placed on a counting chamber, such as the Neubauer haemocytometer, and the number of spores counted. The total number present may be readily calculated by considering the number found in the volume counted and the dilution of the concentrate. The Howard mold counting apparatus may also be used to determine the number of mold filaments present in a food. An alternate procedure is the preparation of dilution plates.

General Morphology

Most fungi are composed of a mass of filaments or threads (mycelium) called hypae (sing., hypha). Two types of mycelium are found, one of which is divided by cross walls (septate) and the other lacking cross walls (aseptate or coenocytic). Root-like structures (rhizoids) may be present, connected by "runners" (stolons). The mycelium not only anchors the mold to the medium and provides nutrients but also bears the reproductive structures. Reproduction may be by sexual or asexual methods.

In the *Zygomycetes*, the sexual or perfect stage is represented by the formation of zygospores. These are thick-walled structures produced as the result of fusion of two compatible gametangia, either from the same thallus (homothallic) or from two different thalli (heterothallic). In the *Ascomycetes*, the sexual stage is characterized by the production of sac-like cells (asci), each ascus usually containing eight ascospores. The asci may be borne in a fruiting body or ascocarp. Ascocarps may be closed or open structures. A cleistothecium is a completely closed ascocarp while a perithecium is usually subglobose to flask shaped and has a pore or an ostiole. An apothecium, on the other hand, is an open ascocarp and often is cup-shaped or saucer-like in appearance. The *Basidiomycetes*, which will not be discussed here, have a perfect stage in which four basidiospores are borne on a specialized cell called a basidium. No sexual stage is known in the *Deuteromycetes* (Fungi Imperfecti).

Asexual reproduction also takes different forms. In some, a large number of spores (sporangiospores) are produced in a sac-like structure (sporangium) borne on a supportive structure or portion of hyphae called a sporangiophore. The tip of the sporangiophore generally extends into the sporangium and this portion of the sporangiophore is referred to as the columella. Those forms in which the sporangiophore does not project into the sporangium are said to be acolumellate. Sporangioles are acolumellate sporangia containing a small number of spores. In other forms of asexual reproduction, conidia are found. These may arise from special fertile hyphae called conidiophores or from special cells (phialides or sterigmata). If the asexual spores result from the fragmentation of the hyphae into individual cells, the spores are referred to as arthrospores. Some fungi have a specialized sporogenous cell known as an annellophore. Except for the first cell which is formed terminally, the remaining spores develop as a blown-out end of the cell through the scar left from the previous spores. The scars produced by the formation of the conidia leave a series of rings at the tip of the sporogenous cell. Another type of spore is the chlamydospore. Chlamydospores are thick-walled asexual spores formed from the rounding up of a cell or cells and may be found between two cells of septate hyphae (intercalary) or terminally on the hyphae.

CLASSIFICATION

Only those filamentous fungi which are commonly considered to be food spoilage organisms will be considered. A complete classification and description is beyond the scope of this presentation. Following the schemes of classification, a brief generic description is given for these molds. A list of selected references is included at the end of this chapter which should prove to be helpful in identification and classification.

Keys to Classes and Orders

Separation of Classes.—The classes of fungi are distinguished on the basis of 1) the type of sexual spore; 2) the type of mycelium; and

3) the type of asexual spore. The letter or number on the right is the section/description that the reader should refer to next.

Partial classification of the Eumycotina

a. Mycelium when present generally aseptate; sexual reproduction by zygospores; asexual reproduction by sporangiospores, modified sporangia or conidia.

Class: Zygomycetes

a. Mycelium septate; sexual reproduction when present by spores borne in asci or on basidia.

b

b. Sexual reproduction by spores borne in asci.

Class: Ascomycetes

b. Asci not present.

c

c. Spores borne on basidia.

Class: Basidiomycetes

c. No sexual stage known.

Class: Deuteromycetes

Zygomycetes.—In the Zygomycetes the order Mucorales contains several organisms important in foods. Mucorales are characterized by the production of profuse mycelium which is usually aseptate but sometimes forms septa, particularly at the base of reproductive structures. Asexual reproduction is mainly by spores borne in sporangia but occasionally by conidia. In some genera, a special type of sporangium, a merosporangium, is present. Merosporangia are cylindrical outgrowths from a swollen tip of the sporangiophore, each merosporangium containing a limited number of spores in a chain-like series. Members of this order grow well at room temperature and plates may be examined for sporulation after a period of 3–10 days.

Keys to Genera

An abbreviated classification of those genera often associated with foods is given below. The reader is referred to the work of Hesseltine (1955) for a complete classification of the order.

1. Rhizoids and stolons produced.

2

1. Rhizoids and stolons not produced.

3

2. Sporangia spherical; sporangiophores formed mostly opposite the rhizoids; zygospores not surrounded by a hyphal net.

Rhizopus

2. Sporangia pear-shaped; sporangiophores formed not opposite the rhizoids; zygospores surrounded by a loose hyphal net.

Absidia

3. Sporangia borne on loosely coiled branches of sporangiophores of indefinite length.

Circinella

3. Sporangia not borne on loosely coiled branches of the sporangiophore.

4

4. Sporangioles produced

5

4. Sporangioles not produced.

6

5. Spherical sporangioles produced at the tips of dichotomous branches; terminal sporangium columellate.

Thamnidium

5. Cylindrical sporangioles (merosporangia) borne on the surface of an inflated sporangiophore.

Syncephalastrum

6. Columellate sporangia produced; zygospores not enclosed by hyphae.

7

6. Acolumellate sporangia produced; zygospores generally enclosed by hyphae.

Mortierella

7. Zygospores on short branches of the sporangiophores; homothallic; heterogamous.

Zygorhynchus

7. Zygospores on separate hyphae; heterothallic or homothallic; isogamous.

Mucor

The Ascomycetes.—Many of the Ascomycetes are important plant pathogens. Only three which are important forms in food microbiology have been included. They may be differentiated on the basis of their ascocarp.

1. No outside covering or wall surrounding the asci.

Byssochlamys

1. An outside covering or wall present.

2

2. Ascocarps without a pore or ostiole.

Monascus

2. Ascocarps with an ostiole.

Neurospora

The Deuteromycetes.—Those fungi which lack a known sexual stage are placed in the form class Deuteromycetes, also called the Fungi Imperfecti. The term "form" is used to denote that there may not be a true relationship between the organisms that have been grouped together. The form class Deuteromycetes is further subdivided into form orders, form families and form genera on the basis of both the asexual reproductive structures and the vegetative mycelium.

The form orders are established on the basis of the type of asexual reproductive structures. A pycnidium is a globose or flask-shaped fruiting body which is lined with conidiophores. In an acervulus, the conidiophores are borne rather compactly from the upper surface of a mat-like mass of vegetative hyphae. A sporodochium consists of a cushion-shaped or pulvinate mass of hyphae from which short conidiophores arise. If the conidiophores are fused together, the structure is called a synnema or coremium, the conidia being borne either only at the apex or at the apex and along the sides as well.

1. Conidia and conidiophores produced in pycnidia.

Sphaeropsidales

1. Conidia not produced in pycnidia.

2

2. Conidia and conidiophores produced in acervuli.

Melanconiales

2. Conidia not produced in acervuli.

3

3. Conidia produced on conidiophores or directly on other hyphae; conidiophores may be loosely arranged and free or joined together in either a synnema or sporodochium.

Moniliales

3. No conidia produced.

Mycelia Sterilia

The form order Sphaeropsidales contains the genera *Diplodia* and *Phomopsis* which are commonly isolated from fruits and vegetables. These two genera may be readily distinguished from each other on the basis of spore types. In *Diplodia*, the pycnidia are black, ostiolate, usually found singly, globose, immersed in the substrate and bursting through the surface of the substrate, and contain simple, slender conidiophores which bear two-celled dark brown, ellipsoid or ovate conidia. The pycnidia in *Phomopsis* are also dark in color, ostiolate and are found immersed in and bursting through the surface of the substrate but they contain two types of hyaline spores. One type of spore is elongated and is often curved or bent while the other is short and ellipsoidal or ovoid in shape.

A representative of the form order Melanconiales often present in vegetables and fruits is *Colletotrichum*. This organism is characterized by the production of acervuli which bear simple, elongate conidiophores with one-celled hyaline, ovoid or oblong conidia and which are surrounded by stiff colorless to dark colored spines or bristles (setae).

The form order Moniliales contains most of the common species of molds belonging to the Deuteromycetes. In the classification given below for those organisms of this order important in food spoilage, emphasis is placed on the sporulating structure, spore morphology and color. The organisms are grouped among three of the form families of the order: the Moniliaceae, the Dematiaceae and the Tuberculiaceae. The first two are separated on the basis of pigmentation and the last by the production of sporodochia.

Form order Moniliaceae: The conidia and the conidiophores are hyaline or brightly colored; conidiophores may be lacking or well developed, but, if present, remain separate.

1. Conidiophores absent; conidia formed by fragmentation of the mycelium.
Geotrichum

1. Conidiophores present; conidia not formed by mycelial fragmentation.
2

2. Conidiophores not differing greatly from the vegetative mycelium.
3

2. Conidiophores morphologically distinct from the vegetative mycelium.
4

3. Conidiophores simple; conidia hyaline, borne in succession, the oldest at the tip of the chain.
Oidium

3. Conidiophores simple or branched; conidia pink, gray or tan, borne in succession, the youngest at the tip of the chain.
Monilia

4. Conidia borne from phialides.
5

4. Conidia not borne from phialides.
8

5. Conidiophores simple or branched; conidia in chains.
6

5. Conidiophores branched at the tips; conidia not in chains but collected in green slime balls.
Trichoderma

6. Conidiophores arise from foot cells, generally aseptate, terminating in a bulbous tip (vesicle) which bears the phialides.

Aspergillus

6. Conidiophores not arising from foot cells, generally septate, not terminating in vesicles.

7

7. Phialides irregularly produced, some in verticils, divergent tips bending away from the main axis; colonies never green.

Paecilomyces

7. Phialides in verticils, tips not divergent; colonies variously colored.

Penicillium

8. Conidia borne in chains from sporogenous cells ringed at tip (annellophores).

Scopulariopsis

8. Conidia not present in chains, not borne from annellophores.

9

9. Conidia borne on pegs, frequently in clusters, on the ends of conidiophores; hyaline or colored.

Botrytis

9. Conidia not borne on pegs.

10

10. Conidia one-celled, borne laterally or terminally from sporo-genous cells of the hyphae.

Sporotrichum

10. Conidia two-celled, borne apically on slender conidiophores, not end to end.

Trichothecium

Form family Dematiaceae: Conidiophores and/or conidia containing dark pigment.

1. Conidia borne from phialides which often flare at the tips.

Phialophora

1. Conidia not borne from phialides.

2

2. Conidia 1 or 2 celled, dark, ovoid to ellipsoidal.

3

2. Conidia contain more than two cells.

4

3. Conidia borne in succession, the youngest at the tip, often in branching chains, colonies dark green to grey green.

Cladosporium

3. Conidia borne on the sides of the hyphae (blastospores); colonies at first yeast-like but becoming black and leathery.

Aureobasidium

4. Conidia with transverse septations only.

Helminthosporium

4. Conidia with both transverse and longitudinal septations.

5

5. Conidia frequently in chains, variable in shape, but generally with an elongated apex (beak).

Alternaria

5. Conidia not in chains, variable in shape, but generally ellipsoidal to oval.

Stemphyllium

Form family Tuberculiaceae: Conidiophores are united into sporodochia. Species of the genus *Fusarium* are the most important of this family as far as food-spoiling organisms are concerned. This genus is characterized by the production of sickle or crescent shaped, septate macroconidia on sporodochia.

Description of Important Genera

Absidia: The aseptate mycelium branches profusely, producing arched stolons with rhizoids. The sporangiophores arise internodally between the rhizoids. The sporangia are pear shaped, columellate, and with a well developed swelling just below the sporangium. The sporangiospores are small, round, or oval. In those that reproduce sexually, roughened zygospores are borne aerially by suspensor cells of about equal size. One or both of the suspensor cells produce appendages which are often loosely coiled or curved and which envelope the zygospore.

Alternaria: The conidiophores are dark, rarely branched, and borne from septate mycelium. The spores have both transverse and longitudinal septations, are dark in color, and are wider at the base than the apex or ellipsoid to ovate in shape. Many are elongated at the apex giving the appearance of a beak, particularly on natural substrates. The spores are produced in chains, the youngest at the tip.

Aspergillus: The septate, vegetative mycelium may be uncolored or brightly colored. The conidiophores are generally not branched and arise perpendicularly from foot cells in the mycelium. The apex of the conidiophore enlarges to form variously shaped vesicles from which the conidium producing cells (sterigmata) radiate in either a single or double series. The terminal sterigmata produce unbranched chains of conidia. The spore bearing heads may be of various shapes—clavate, globose, radiate, or columnar. The conidia are single celled, smooth or rough, and variously colored. Some species produce a perfect stage forming closed fruiting structures (cleistothecia) in which the asci containing eight ascospores are borne. A few species produce sclerotia which are firm masses of mycelium, often more or less rounded in shape, variously colored, and which usually do not contain spores.

Aureobasidium: The colonies are variable with the septate mycelium first being white and somewhat yeast-like in appearance but becoming brown to black and shiny to leathery in age. The conidia are ovoid, single celled, borne laterally from the mycelium and reproduce by budding.

Botrytis: The condiophores may be simple or branched but are generally dark in color. When viewed in large numbers, the conidia are commonly grey. The conidia are oval to globose in shape, and are borne on short sterigmata in clusters on the swollen terminal cells of the conidiophores. Sclerotia often develop.

Byssochlamys: This organism is characterized by the production of clusters of asci borne without a trace of a wall or peridium covering them. Each ascus contains eight ascospores. Conidial structures are of the *Paecilomyces* type.

Cladosporium: The conidiophores are variously branched. The conidia are dark, generally single celled but may be two-celled, ovoid, cylindrical, globose to irregular in shape, and form branched chains. Slide cultures should be prepared or the edge of the colony studied to observe the tree-like clusters of conidia because the spore-bearing structures tend to break up when making mounts for examination.

Circinella: The mycelium is aseptate and may be hyaline or colored. Terminal sporangia are not formed. The sporangia are columellate and are borne on branches of the sporangiophore which are loosely coiled or circinate. When zygospores are formed, they are nearly smooth walled and are produced between suspensor cells of about equal size. No appendages are found on the suspensor cells.

Fusarium: Sickle or crescent shaped macroconidia are produced which generally contain several cells. The macroconidia are borne in sporodochia. Microconidia or small, round, oval or irregularly shaped cells are also produced. The mycelium is often pigmented pink, purple or yellow. Chlamydospores and sclerotia may be found in some strains.

Geotrichum: A well developed septate mycelium is formed which breaks up into short, cylindrical cells or arthrospores. These spores have truncate or flattened ends.

Helminthosporium: The septate mycelium may be light or dark. The conidiophores are septate, simple or branched, bearing multiseptate conidia terminally or laterally along the growing tip. The conidia are dark, cylindrical to ellipsoidal in shape, but sometimes curved, with rounded ends, and typically containing more than three cells.

Monascus: Colonies are characterized by the production of cleistothecia filled with eight-spored asci. The walls of the asci disintegrate on ripening so that the mature cleistothecium appears to be full of loose spores.

Monilia: Some species of this genus are the imperfect stage of *Neurospora* and *Monilinia*. The colony may appear white, grey, or often pinkish. The septate mycelium produces long branched conidiophores which break up into oval or short cylindrical conidia produced in succession, the youngest at the tip.

Montierella: The aseptate mycelium is not typically aerial but remains close to the agar surface. The sporangiophores may be simple or branched, taper at the tip, and bear acolumellate sporangia. When zygospores are produced, they are enclosed in a hyphal mass.

Mucor: The colonies are variously colored but generally are yellowish to grey. The mycelium is abundant, aseptate and much branched. The sporangiophores are simple or somewhat branched, bearing terminal sporangia which are spherical and columellate. Most species are heterothallic. When zygospores are produced, they are borne on suspensor cells which are of about equal size and which lack appendages.

Neurospora: The conidial stage of this organism is *Monilia*. The mycelium is septate, branched and bears chains of oval conidia on conidiophores which are generally branched. Most species are heterothallic, and hence, the asexual stage is more commonly found. If compatible strains do come in contact, dark colored, beaked, pear shaped perithecia are formed. The asci contain brown or black ascospores.

Oidium: This organism represents the imperfect stage of the Erysiphaceae of the powdery mildews. The mycelium is septate and white. The conidiophores are simple and bear hyaline conidia in chains, the oldest at the apex.

Paecilomyces: The conidiophores arise from septate mycelium terminating in slender spore bearing tips which tend to bend away from the main axis. The phialides may also be irregularly distributed along the aerial hyphae. Aleuriospores or large globose to ovate macrospores are often found in the basal mycelium. The conidia may be rough or smooth, ellipsoid, ovate or cylindrical. The colonies are never green. The genus is considered to be the imperfect stage of *Byssochlamys*.

Penicillium: The septate mycelium may be colorless or brightly colored. The colony texture may be velvety, floccose (cottony), funiculose (with aerial ropes of hyphae), or the conidiophores may be united into fasicles (bundles). The conidiophores are septate, unbranched, with walls thick or thin, smooth or rough. The conidiophores terminate in brush-like fruiting structures (penicilli). The penicilli may be monoverticillate or biverticillate in which other cells, called metulae, support the sterigmata which produce the conidia. The conidia may be smooth or rough, globose, subglobose, elliptical or ovoid in shape, and are variously colored when viewed in mass. A few produce a perfect stage forming cleistothecia. Others may form sclerotia.

Phialophora: The mycelium is septate and dark in color. Flask-shaped sporogenous cells arise from branches of the mycelium or laterally from the mycelium. The sporogenous cells generally have a flared tip with a collarette present. The conidia are single celled and may be hyaline or pigmented.

Rhizopus: The aseptate mycelium grows profusely and rapidly producing rhizoids and stolons. The sporangiophores develop opposite the rhizoids and bear large, terminal, globose, columellate sporangia. Zygospores, when present, are formed between suspensor cells of about equal size and lack appendages.

Scopulariopsis: Although the colonies may be variously pigmented, they are never green. The septate mycelium bears short conidiophores which produce irregular fruiting structures, some appearing penicillate. The sporogenous cells are annellophores, producing conidia in chain, the oldest at the apex of the series. The conidia are smooth or rough, single celled, and round to lemon shaped with a truncate or flattened base.

Sporotrichum: The colonies are often white but may be yellow or gray. The septate mycelium bears conidia either laterally or terminally on sporogenous cells. The conidia are single celled, globose to ovoid in shape, and are often rough walled.

Stemphyllium: The conidiophores arise from the septate mycelium and are dark in color, usually bearing a single spore which has both longitudinal and transverse septations. The spores are variable in shape but frequently are rounded at both ends. The spore wall may be smooth or rough.

Syncephalastrum: The mycelium is abundantly branched and aseptate. Cylindrical sporangioles (merosporangia) containing 2-12 spores in a series are borne on the surface of inflated sporangiophores.

Thamnidium: The mycelium is extensive and aseptate. The sporangiophores bear terminal, columellate sporangia and, lower on the same sporangiophore, whorls of acolumellate sporangioles borne at the tips of dichotomous branches.

Trichoderma: The rapidly growing colonies are white to yellow green or green in color and have a tufted or cushion-like appearance.

The mycelium is septate and bears irregularly branched conidiophores. The phialides occur singly or in clusters. The conidia are single celled, ovoid, ellipsoid, or cylindrical in shape, and are collected in balls of slime at the tips of the phialides.

Trichothecium: The colonies are white at first but often become pink in age. Septate conidiophores arise from the mycelium bearing groups of conidia at their tips. The conidia may be borne singly or in short chains but the chains are never composed of cells end to end. The conidia are two-celled and ovoid to ellipsoid in shape.

Zygorhynchus: A highly branched, aseptate, aerial mycelium is produced. The sporangiophores bear columellate, terminal sporangia. All species are homothallic. The zygospores are produced on short side branches of the hyphae, the gametangia and the suspensor cells being of unequal size.

BIBLIOGRAPHY

AINSWORTH, G. D. 1971. Ainsworth's and Bisby's Dictionary of the Fungi, 6th Edition. Commonwealth Mycological Institute, Kew, Surrey.

ALEXANDER, M. 1961. Introduction to Soil Microbiology. John Wiley & Sons, New York.

ALEXOPOULOS, C. 1962. Introductory Mycology. John Wiley & Sons, New York.

AYRES, J. C. 1955. Microbiological implications in the handling, slaughtering, and dressing of meat animals. Advan. Food Res. *6*, 109–161.

AYRES, J. C., KRAFT, A. A., DENISEN, E. L., and PEIRCE, L. C. 1964. The use of macrolide antifungal antibiotics in delaying spoilage of fresh small fruits and tomatoes. *In* Microbial Inhibitors in Foods, N. Molin (Editor). Almqvist and Wiksell, Stockholm.

BARNETT, H. L., and HUNTER, B. B. 1972. Illustrated Genera of the Imperfect Fungi. Burgess Publishing Co., Minneapolis, Minn.

BARRON, G. L. 1968. The Genera of the Hyphomycetes from Soil. The Williams & Wilkins Co. Baltimore, Md.

BESSEY, E. A. 1950. Morphology and Taxonomy of Fungi. The Blakiston Co., Philadelphia.

BOOTH, C. 1971. Introduction to General Methods. *In* Methods in Microbiology, Vol. 4, C. Booth (Editor). Academic Press, New York.

BOOTH, C. 1971A. Fungal Culture Media. *In* Methods in Microbiology, Vol. 4, C. Booth, (Editor). Academic Press, New York.

BRANCATO, F. P., and GOLDING, N. S. 1953. The diameter of the mold colony as a reliable measure of growth. Mycologia *45*, 848–864.

BROWN, A. H. S. and SMITH, G. 1957. The genus *Paecilomyces* Bainier and its perfect stage *Byssochlamys* Westling. Trans. Brit. Mycol. Soc. *40*, 17–89.

BROWN, W. 1922. On the germination and growth of fungi at various temperatures and in various concentrations of oxygen and carbon dioxide. Ann. Botany *36*, 257–283.

CHRISTENBERRY, G. A. 1940. A taxonomic study of the Mucorales in the Southeastern United States. J. Elisha Mitchell Sci. Soc. *56*, 333–366.

COCHRANE, V. W. 1958. Physiology of Fungi. John Wiley & Sons, New York.

CODNER, R. C. 1971. Pectinolytic and cellulolytic enzymes in the microbial modification of plant tissues. J. Appl. Bacteriol. *34*, 147–160.

CONKLIN, D. B. 1944. Ultra-violet irradiation of spores of certain molds collected from bread. Proc. Iowa Acad. Sci. *51*, 185–189.

DAKIN, J. C., and STOCK, A. C. 1968. *Moniliella acetoabutans*: Some further characteristics and industrial significance. J. Food Technol. *3*, 49-53.

DALLYN, H., and EVERTON, J. R. 1969. The xerophilic mold, *Xeromyces bisporus*, as a spoilage organism. J. Food Technol. *4*, 399-403.

ELLIS, J. J., and HESSELTINE, C. W. 1965. The genus *Absidia*: glogosespored species, Mycologia *57*, 222-235.

FOOD PROTECTION COMMITTEE. 1965. Chemicals used in food processing. Publ. *1274*. Natl. Acad. Sci.—Natl. Res. Council, Washington, D.C.

FOSTER, E. M. *et al*. 1957. Dairy Microbiology. Prentice-Hall, New Jersey.

FOSTER, J. W. 1949. Chemical Activities of Fungi. Academic Press, New York.

GILLESPY, T. G. 1936-37. Studies on the mould *Byssochlamys fulva*. Ann. Rept. Univ. Bristol Fruit Veg. Preserv. Res. Sta., Campden, England.

GILMAN, J. C. 1957. A Manual of the Soil Fungi, 2nd Edition. Iowa State Univ. Press, Ames, Iowa.

GUNDERSON, M. F. 1961. Mold problem in frozen foods. *In* Proceedings Low Temperature Microbiology Symposium. Campbell Soup Co., Camden, N. J.

GUNNER, H. B., and ALEXANDER, M. 1964. Anaerobic growth of *Fusarium oxysporum*. J. Bacteriol. *87*, 1309-1311.

HANSON, H. L., and FLETCHER, L. R. 1958. Time-temperature tolerance of frozen foods. XII. Turkey dinners and turkey pies. Food Technol. *12*, 40-43.

HELDMAN, D. R., HEDRICK. T. I., and HALL, C. W. 1965. Sources of airborne microorganisms in food processing areas—drains. J. Milk Food Technol. *28*, 41-45.

HESSELTINE, C. W. 1955. Genera of the Mucorales with notes on their synonymy. Mycologia *47*, 334-363.

HESSELTINE, C. W., and ANDERSON, A. 1956. The genus *Thamnidium* and a study of the formation of its zygospores. Am. J. Bot. *43*, 696-703.

HESSELTINE, C. W., and ELLIS, J. J. 1964. The genus *Absidia: Gongronella* and cylindrical-spored species of *Absidia*. Mycologia *56*, 568-601.

HESSELTINE, C. W., and FENNELL, D. I. 1965. The genus *Circinella* Mycologia *47*, 193-212.

HULL, R. 1939. Study of *Byssochlamys fulva* and control measures in processed fruits. Anal. Appl. Biol. *26*, 800-822.

HUSAIN, A., and KELMAN, A. 1959. Tissue is disintegrated. *In* Plant Pathology, Vol. I, J. G. Horsfall, and A. E. Dimond (Editors). Academic Press, New York.

INGRAM, M. 1963. Microbiological principles in prepacking meats. J. Appl. Bacteriol. *25*, 259-281.

JENSEN, L. B. 1954. Microbiology of Meats, 3rd Edition. Garrard Press, Urbana, Ill.

KIESEL, A. 1913. Research on the action of various acids and acid salts on the development of *Aspergillus niger*. Ann. Inst. Pasteur *27*, 391-420. (French)

KING, A. D., JR., MICHENER, H. D., and ITO, K. A. 1969. Control of *Byssochlamys* and related heat-resistant fungi in grape products. Appl. Microbiol. *18*, 166-173.

KOBURGER, J. A. 1971. Fungi in foods. II. Some observations on acidulants used to adjust media pH for yeast and mold counts. J. Milk and Food Technol. *34*, 475-477.

KOTULA, A. W., and KINNER, J. A. 1964. Airborne microorganisms in broiler processing plants. Appl. Microbiol. *12*, 179-184.

LUTHI, H. 1959. Microorganisms in noncitrus juices. Advan. Food Res. *9*, 221-284.

MARSHALL, C. R., and WALKLEY, V. T. 1951. Some aspects of microbiology applicable to commercial apple juice production. I. Distribution of microorganisms on fruit. Food Res. *16*, 448-456.

MICHENER, H. D., and ELLIOTT, R. P. 1964. Minimum growth temperatures for food-poisoning, fecal-indicator, and psychrophilic microorganisms. Advan. Food Res. *13*, 349-396.

MILLER, D. D, and GOLDING, N. S. 1949. The gas requirements of molds. V. The minimum oxygen requirements for normal growth and for germination of six mold cultures. J. Dairy Sci. *32*, 101-110.

RAPER, K. B., and FENNELL, D. I. 1965. The Genus *Aspergillus*. The Williams & Wilkins Co., Baltimore, Md.

RAPER, K. B., and THOM, C. 1949. A Manual of the Penicillia. The Williams & Wilkins Company, Baltimore, Md.

RIDDELL, R. W. 1950. Permanent stained mycological preparations obtained by slide culture. Mycologia *42*, 265-270.

RUYLE, E. H., PEARCE, W. E., and HAYS, G. L. 1946. Prevention of mold in kettled blueberries in No. 10 cans. Food Res. *11*, 274-279.

SCHELHORN, M. VON. 1951. Control of microorganisms causing spoilage in fruit and vegetable products. Advan. Food Res. *3*, 429-482.

SCOTT, W. J. 1957. Water relations of food spoilage microorganisms. Advan. Food Res. 7, 83-127.

SHARF, J. M. 1966. Recommended Methods for the Microbiological Examination of Foods, 2nd Edition. Am. Public Health Assoc., New York.

SMITH, F. R. 1942. The use of ultraviolet rays in the cheese factory and storage room. J. Dairy Sci. *25*, 525-528.

SMITH, G. 1969. An Introduction to Industrial Mycology, 6th Edition. St. Martins Press, New York.

SMITH, W. H. 1963. The use of carbon dioxide in the transport and storage of fruits and vegetables. Advan. Food Res. *12,*, 95-146.

SMOOT, J. J., HOUCK, L. G., and JOHNSON, H. B. 1971. Market diseases of citrus and other subtropical fruits. U.S. Dept. Agr., Agr. Handbook *398*.

SNOW, D. 1949. The germination of mold spores at controlled humidities. Ann. Appl. Biol. *36*, 1-13.

SOMMER, N. F., MAXIE, E. C., and FORTLAGE, R. J. 1964. Quantitative dose-response of *Prunus* fruit decay fungi to gamma irradiation. Radiation Bot. *4*, 309-316.

SOMMER, N. F., MAXIE, E. C., FORTLAGE, R. J., and ECKERT, J. W. 1964A. Sensitivity of *Citrus* fruit decay fungi to gamma irradiation. Radiation Bot. *4*, 317-322.

SPLITTSTOESSER, D. F., KUSS, F. R., HARRISON, W., and PREST, D. B. 1971. Incidence of heat-resistant molds in eastern orchards and vineyards. Appl. Microbiol. *21*, 335-337.

VAUGHN, R. H. 1963. Microbiol spoilage problems of fresh and refrigerated foods. *In* Microbiological Quality of Foods, L. W. Slanetz, C. O. Chichester, A. R. Gaufin, and Z. J. Ordal (Editors). Academic Press, New York.

WARCUS, J. H. 1967. Fungi in soil. *In* Soil Biology, A. Burges and F. Raw (Editors). Academic Press, New York.

WILLIAMS, C. C., CAMERON, E. J., and WILLIAMS, O. B. 1941. A facultative anaerobic mold of unusual heat resistance. Food Res. *6*, 69-73.

YATES, A. R., and FERGUSON, W. E. 1963. Observations on *Byssochlamys nivea* isolated from cucumber brine. Can. J. Bot. *41*, 1599-1601.

Index

Trypsinization, 157
Tuberculiaceae, 471, 473
Tylosin, 347
Typhoid fever, 140

Urov disease, 211

Vacuum packaging, 345, 456
Vancomycin, 347
Veal, 185, 278, 365
Vegetables, 69-70, 159-160, 173,
185, 207, 215-216, 258,
261-262, 277, 282, 331,
340, 343, 350, 367, 369-
370, 378-379, 416-418,
431, 434, 436-437, 439,
442, 455-456
Vibrio, 314
alginolyticus, 199, 203
parahaemolyticus, 198-209
characteristics, 199-202
epidemiology, 204-206
implicated foods, 205-206
isolation and identification, 206-
207
nomenclature, 198-199
outbreak prevention, 207-209
serology, 202
Viridicatin, 219, 222
Viruses, 257-269
biological inactivation, 265
characteristics, 257-258
control, 267-268
detection, 265-267
food-borne illnesses, 258-262
heat resistance, 262-264
radiation resistance, 264
Voges-Proskauer test, 274, 349

Water activity (a_w), 54-55, 147-148,
150, 164, 177-178, 200-
201, 225, 227, 314, 321,
324, 327, 345, 438-440,
457, 459-460

Water, contamination of, 144, 153,
360, 444
Watery soft rot, 456
Wheat, mycotoxin contamination of,
229-230
Whey, 433
Wickerhamia fluorescens, 430
Wine, 351-353, 405, 418-420, 422,
430-431, 435, 437, 440,
444, 463-464
Wingea robertsii, 430

Xanthocillin, 218-219, 222
Xanthomonas, 340
Xeromyces bisporus, 457, 459

Yeasts, 397, 421, 427-454
ascospore-forming, 429-430
classification and properties, 429-
434
control of, 437-442
enumeration, 428, 434-435
false. *See* nonspore-forming
food and fodder, 431-433
food spoilage by, 428-429, 434-
437
habitat, 427-428
heat resistance, 438
identification, 433-434
irradiation resistance, 441-442
nonspore-forming, 430-433
osmophilic, 430, 432, 435-440,
443
pathogenic, 433
psychrotrophic, 438
Yersinia, 271
Yogurt, 435-436, 440

Zapatera spoilage, 374, 377
Zearalenone, 220, 243
Zygomycetes, 468
Zygorhynchus, 469, 477
Zygosaccharomyces, 432
Zygospores, 466